The British Dam Society

# Reservoirs in a changing world

Proceedings of the 12th conference of the BDS held at
TRINITY COLLEGE, DUBLIN
4 – 8 September 2002

Edited by Paul Tedd

Thomas Telford

Conference organised by the British Dam Society

**Organising Committee:** Andy Rowland (Chairman), John Gosden, Jack O'Keeffe, Chris Scott, Paul Tedd, John Whiting

Published by Thomas Telford Publishing, Thomas Telford Ltd, 1 Heron Quay, London E14 4JD. www.thomastelford.com

Distributors for Thomas Telford books are
*USA:* ASCE Press, 1801 Alexander Bell Drive, Reston, VA 20191-4400, USA
*Japan:* Maruzen Co. Ltd, Book Department, 3–10 Nihonbashi 2-chome, Chuo-ku, Tokyo 103
*Australia:* DA Books and Journals, 648 Whitehorse Road, Mitcham 3132, Victoria

First published 2002

The cover photograph shows Pollaphuca dam, Ireland, courtesy of ESB

Also available from Thomas Telford Books
*The reservoir as an asset*. British Dam Society. ISBN 0 7277 2528 9
*Reservoir safety and the environment*. British Dam Society. ISBN 0 7277 2010 4
*Water resources and reservoir engineering*. British Dam Society. ISBN 0 7277 1692 1
*The prospect for reservoirs in the 21st century*. British Dam Society. ISBN 0 7277 2704 4
*A guide to the Reservoirs Act 1975*. DETR and ICE. ISBN 0 7277 2851 2
*Land drainage and flood defence responsibilities: 3rd edition*. ICE. ISBN 0 7277 2508 4
*Dams 2000*. British Dam Society. ISBN 0 7277 2870 9
*Reservoir engineering. Guidelines for practice*. Edward Gosschalk. ISBN 0 7277 3099 1
*Practical dam analysis*. M A M Herzog. ISBN 0 7277 2725 7

A catalogue record for this book is available from the British Library

ISBN: 0 7277 3139 4



Printed and bound in Great Britain by MPG Books, Bodmin

# Preface

The 12th Conference of the British Dam Society, *Reservoirs in a Changing World* was held at Trinity College, Dublin, in September 2002.

The proceedings of the conference contain a varied collection of 47 papers. Reservoir safety is the key theme with many papers on the performance and rehabilitation of both concrete and embankment dams although a few papers describe the construction of new dams overseas. The evolution of reservoirs in Ireland and the development of safety legislation in the UK are described. Risk assessment features in a number of papers as a method of assessing the safety of reservoirs. The likely effects of climate change on embankments are reviewed. Several papers address the seismic assessment of dams and structures. The performance of upstream asphaltic membranes in the UK and Ireland is described with reference to the major remedial works undertaken at Winscar which has been repaired with a PVC geomembrane. Repair of hydraulic structures and the hydrology of spillways are discussed.

The highlight of every BDS biennial conference is the presentation of the Geoffrey Binnie Lecture. The 2002 Lecture, *The Challenge for British Dam Engineers* by Dr Geoffrey P Sims, is published in the Society's journal Dams & Reservoirs.

# Contents

## Development and construction of dams

Some aspects of early Irish dam construction
E FLEMING 3

The contribution to society of Irish hydro-electric dams
J D O'KEEFFE 15

Design, construction and performance of Fullerton Pollan dam and reservoir, Co Donegal, Ireland
R C BRIDLE, J HOLOHAN, D GILLESPIE, D A SMITH, S FAWCETT, S McINERNEY, I C CARTER and R EVANS 31

Challenging values of dam builders
C S McCULLOCH 49

Ghazi-Barotha hydropower project: social issues and engineering design
P E JONES, J C ACKERS, and A CHAUDHURY 61

RCC Construction at Tannur dam
M AIREY 73

## Seismic performance of dams

Seismic assessment of Scottish dams
K J DEMPSTER, A C MORISON, S C GALLOCHER and S BU 87

Assessing the seismic performance of UK intake/outlet towers
W E DANIELL and C A TAYLOR 100

Seismic hazard in the UK – another look
C W SCOTT and J J BOMMER 112

A methodology for seismic investigation and analysis of dams in the UK
P RIGBY, S WALTHALL and K D GARDINER 126

## Remedial works to concrete and masonry dams

Stability reassessment and remedial works at Leixlip dam
B O'MAHONY and B HAUGH 143

Rehabilitation of old masonry dams at full reservoir level – a comparison of successful rehabilitation projects
V BETTZIECHE and C HEITEFUSS 155

Underwater work as a means for the rehabilitation of large hydraulic structures under operation and unrestricted water supply
C HEITEFUSS and H J KNY 167

**Hydraulic structures – hydrology**

Sluiceway isolation for gate replacement at Kotri Barrage in Pakistan
I E PADGETT and K F MORRISON 181

Maintaining the Thames tidal defences in a century of climate change
J LEWIN and S LAVERY 193

Flood control using the automatic tops spillway gates:
A case study of the Avis Dam, Namibia
P D TOWNSHEND and K A LUND 209

The release of large diameter draw-off and control valves
R P ENSTON and D C F LATHAM 218

Remedial works at Brent Reservoir to address leaking sluice gates
R A N HUGHES and P KELLY 224

Refurbishment of outlet tunnel and associated pipework at Piethorne Reservoir
A BRISCOE, A A GEORGE, I C CARTER and P GRUNDY 236

Langsett Reservoir: A combined analytical and CFD study of a reservoir side-spillway
D R S WOOLF and J N HACKER 247

Langsett Reservoir: Numerical simulation of hydraulic structures
D R S WOOLF and I SCHOLEFIELD 262

Rehabilitation of the Upper and Lower Bohernabreena Spillways
D E MacDONALD and J D MOLYNEUX 274

**Embankment dams performance and remedial works**

Rehabilitation of irrigation dams in Albania
J L HINKS and Y DEDJA 289

River Shannon hydro-electric scheme: failure of upstream slope of Fort Henry Embankment: Analysis
M LONG, I LYDON and E CONATY 302

River Shannon hydro-electric scheme: Fort Henry embankment, upstream slope failure and remedial work
B CASEY, M LONG, and T FITZGIBBON 314

Long term behaviour of Portumna embankments
E CONATY and M LONG 324

The influence of climate and climate change on the stability of abutment and reservoir slopes
P R VAUGHAN, N KOVACEVIC and A M RIDLEY 337

The influence of climate and climate change on the stability of embankment dam slopes
P R VAUGHAN, N KOVACEVIC and A M RIDLEY 353

Settlement of old embankment dams and reservoir drawdown
P TEDD, J A CHARLES and A C ROBERTSHAW 367

Internal erosion in European embankment dams
J A CHARLES 378

The use of temperature measurements for detection of leakage in embankment dams – British Waterways experience
D P M DUTTON 394

The successful grouting of Heapey embankment, Anglezarke reservoir
C D PARKS and S WALTHALL 403

**Performance and repair of upstream membranes**

Improving the watertightness of Winscar Reservoir
I C CARTER, J R CLAYDON and M J HILL 415

Turlough Hill – upper reservoir: condition of the lining after 30 years
B HAUGH 431

Colliford and Roadford dams: performance of the asphaltic concrete membranes and the embankments
J K HOPKINS, P TEDD and C BRAY 444

Breaclaich dam – upstream face joint bandage sealant and wavewall refurbishment works
K J DEMPSTER and N LANNEN 456

## Safety and risk

Tailings dam incidents and new methods
A D M PENMAN 471

The IMPACT Project – continuing European research on dambreak processes and the failure of flood embankments
M W MORRIS 484

A historical perspective on reservoir safety legislation in the United Kingdom
J A CHARLES 494

Risk assessment and the safety case in dam safety decisions
D N D HARTFORD and R A STEWART 510

Risk assessment - its development and relevant considerations for dam safety
J McQUAID 520

Multi-attribute performance monitoring for reservoir systems
J W HALL, J W LE MASURIER, E A BAKER, J P DAVIS and C A TAYLOR 534

Reservoir risk assessments in the north of Scotland
F TARRANT, J ACKERS and N GRAHAM-SMITH 551

Lake Sarez risk mitigation project
L J S ATTEWILL and L SPASIC-GRIL 563

Where to keep your dam documents?
J STEWART 575

The characteristics of UK puddle clay cores – a review
A I B MOFFAT 581

A review of systems used to assess dam safety
A J BROWN and J D GOSDEN 602

**Author index** 619

**Subject index** 621
**Dam index** 622

# Development and construction of dams

# Some aspects of early Irish dam construction

E FLEMING, Dublin Corporation Waterworks Division

SYNOPSIS. Only a handful of reservoirs were built in Ireland before 1870. This paper describes some of these early dams and the problems encountered in their construction and design. The benefits and negative social impacts of the dams are also discussed.

## INTRODUCTION

There are approximately 130 large raised reservoirs (as defined in the UK Reservoir Act 1975) on the island of Ireland. This small number is due to the late development of industry - the Industrial Revolution bypassed all but the North East of the country - and low density of population. Indeed the number of large reservoirs in the country only moved into double figures in the 1870's as public water supplies were developed for major towns. In this paper I propose to deal with some interesting aspects of those dams built before 1870.

## THE ORNAMENTAL LAKE

Probably the oldest large raised reservoir in Ireland is a so-called "pleasure lake" at **Luttrellestown House**, Co Dublin. It is shown on Rocque's 1760 map of County Dublin and contemporary correspondence indicates a construction date circa 1740. The reservoir is 5 hectares in extent and the main dam is over 8 metres high but it is privately owned and little detail of its construction is available. The earliest pleasure lake about which we do have engineering details was constructed some forty years later at **Dungannon Park**, Co Tyrone. According to local tradition, this was designed by Daviso de Arcort (**Davis Ducart**), a Sardinian who also designed the adjacent Coalisland Canal extension. The original dam is a masonry structure of local sandstone 1.4 m thick, gently curved in plan with a maximum height of 8 m above its foundations and impounds approximately 70 Ml. In 1988, to ensure its continued safety, a mass concrete wall, keyed into bedrock, was cast against the upstream face to provide

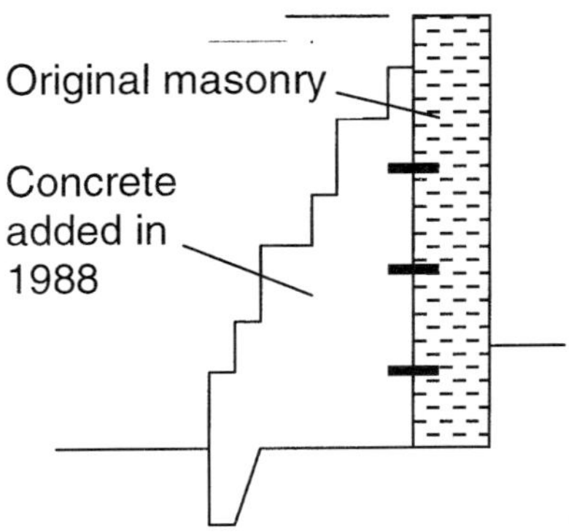


Fig 1 Section: Dungannon Park

*Reservoirs in a Changing World*, Thomas Telford, London, 2002, 3–14

additional weight. A waterproof membrane was placed between the concrete and masonry and the walls were dowelled together (Fig 1).

## CANAL EMBANKMENTS

In the hundred year period commencing in the mid-eighteenth century an extensive network of inland navigation was constructed in Ireland. The network consisted of still water canals, canalised rivers and lake navigations, most of which were interconnected. Although, unlike Britain, none of these canals required man-made storage reservoirs at their summits, their construction involved the first extensive creation of large embankments and use of puddle clay in Ireland.

A particular problem faced by the builders of the first of these canals, the **Grand Canal** linking Dublin to the river Shannon, was that of carrying the canal across the extensive Bog of Allen in the centre of Ireland. The Canal Company sought the advice of **John Smeaton** in 1773 and he advised "to avoid bogs if at all possible, but of all things going deeply into them". Smeaton's pupil, **William Jessop**, followed his master's advice and to avoid excavation, the canal was constructed by raising a huge embankment on the surface of the bog using air-dried peat, "firmly trampled and chopped", as the material. The embankment generally has a 120 m base width, a 19 m top width and is 14 m high, with the canal carried along its centre. As the locks are 30km apart, this section of the canal constitutes a large raised reservoir. These embankments have been a constant source of concern with about ten failures since 1797, the most recent of which was in 1989 when a 400 m length breached near Edenderry, Co Offaly, releasing an estimated 135 million litres of water and displacing approximately 200,000 cu.m of material (Fig 2).

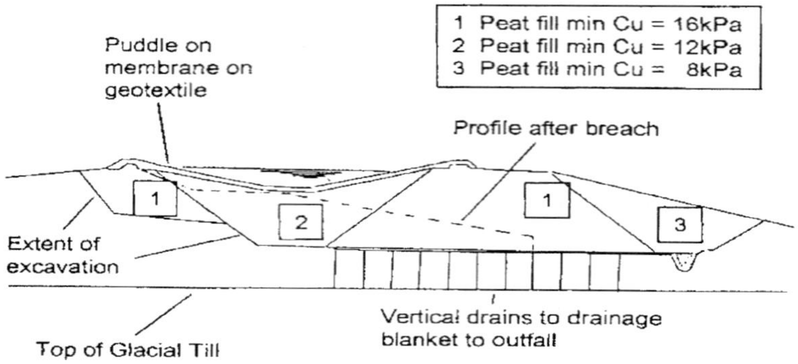


**Fig 2 Section: Grand Canal at Edenderry Breach**

The most likely mode of failure was flotation and sliding, caused by seepage of water into the bank raising the porewater pressure and softening the peat. The breach was repaired by constructing new embankments of sod peat founded on undisturbed ground. A system of vertical drains was installed to consolidate the embankment foundation, reduce porewater pressure and intercept seepage and a polyethylene membrane covered with a layer of puddle was used to seal the canal.

Jessop's method of construction involved enormous difficulties and expense and the method was abandoned for the remainder of the canal, being replaced by one of draining bogs prior to construction. A notable feature of the other canal connecting Dublin to the Shannon, the **Royal Canal**, is the massive embankment which carries it over the **Rye Water**. It is 30m high and cost over £30,000 to construct – an enormous sum at the beginning of the nineteenth century.

DAMS FOR WATERMILLS

As already mentioned, the Industrial Revolution largely bypassed Ireland in the nineteenth century. The exception to this was East Ulster where, by the early years of the century, a thriving linen industry had developed and with it the demand for power for looms and spinning mills.

In 1834 a group of mill owners on the River Bann asked **William Fairburn**, a Manchester engineer to report on means of ensuring constant water power from the river. Fairburn in turn employed **J.F. Bateman** as surveyor for his scheme and when the Bann Reservoirs Act was passed in 1836 the 26-year old Bateman was appointed as engineer to the Company. **Lough Island Reavy** in County Down was the first of the Bann Reservoirs and is of special interest because it incorporates the first large dam designed by Bateman. Construction started in 1837 and the contractor was **William Dargan**, Ireland's foremost civil engineering contractor of the mid-nineteenth century.

In this scheme an existing shallow lake was enlarged and deepened by the construction of four embankments, the largest of which has a crest length of 568 m and a maximum height of 12.5 m. It is an earthen embankment with puddle clay core and cut-off trench. Draw-off is by means of two 18-inch pipes, controlled by valves at the downstream toe and laid in an ashlar granite culvert under the embankment. The culvert has a solid masonry central stop-wall through which the pipes pass. Upstream of the stop-wall the pipes cease and the culvert is open to the reservoir (Fig. 3). It was Bateman's intention that this section of culvert be surrounded by puddle, well rammed against the ashlar granite, but the works superintendent added a rubble backing course to strengthen the side walls of the culvert. This rubble course acted as a drain and water started to leak into the culvert under a pressure of only 12-ft of water.

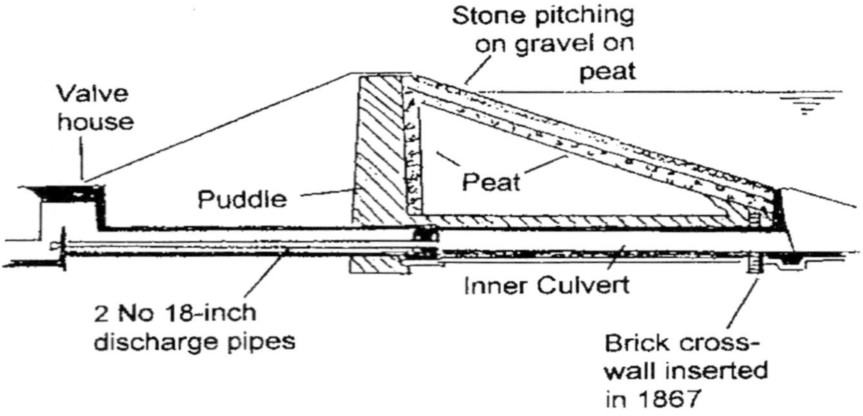


**Fig 3 Section: Lough Island Reavy Dam**

To save expense, Bateman first tried to seal the joints of the culvert wall using a hydraulic mortar made on site in accordance with directions just published by Vicat. This failed and he then resorted to picking out the joints, caulking them with oakum and pointing them with Roman cement. This was successful and although moderate leakage continued for the next twenty years, regular re-pointing always resulted in a temporary diminution of the flow. By 1867 however leakage had increased so much that the owners again called in Bateman. This time he recommended that the only long term satisfactory solution was to cut open the bank down to the culvert, remove the rubble backing and any soft material and re-build to the original specification. True to form, the mill owners baulked at the expense and in the repairs actually carried out only the upper end of the culvert was excavated, but insertion of a new cross-wall around the culvert and extensive grouting of the masonry joints was successful in curing the leak.

Lough Island Reavy also saw once again the use of peat in an Irish embankment. For this dam a layer of peat was brought up on the inner side of the puddle core and continued, below courses of gravel and stone pitching, on the slope of water face. The peat was laid dry in thin courses and trodden in, with the intention that in the event of a leak, fibrous particles of peat would be drawn into the opening where they would swell and gradually block the cavity. **John Smyth**, who supervised the 1867 repairs, also pounded small pieces of dry turf into soft spots in the puddle "thereby securing great firmness" and was an advocate of peat in dams for it's ability, by swelling, to make up for shrinkage of the puddle. Although it's use in dam construction was never popular, it is interesting that milled peat is still used today by lockkeepers in Irish canals for sealing leaks between lock walls and gates.

DAMS FOR WATER SUPPLY

During the 1850's both Dublin and Belfast grew rapidly and the need for improved water supply became imperative. Belfast was first off the mark, and **Woodburn Upper and Middle**, the first two of a series of reservoirs in the County Antrim hills built in accordance with proposals by Bateman, were completed in 1865. In the following thirty years Belfast grew so rapidly that seven further reservoirs were built, sometimes before the enabling legislation was passed! All of these are earthen embankments with puddle clay cores.

For Dublin, a Royal Commission chaired by **John Hawkshaw** recommended a supply from the river Vartry near Roundwood in County Wicklow, based on a scheme proposed by **Richard Hassard**. **Thomas Hawksley** was Dublin Corporation's consultant and his name appears on the parliamentary plans submitted in 1861. However Hawksley seems to have taken no further part in the scheme and it was Bateman and **Thomas Duncan**, Engineer of Liverpool Waterworks, who were consulted when the dam leaked during first filling.

The 500 m long, 20 m high main embankment of the **Lower Vartry Reservoir** is a typical Pennine Dam with puddle clay core (Fig. 4). As originally designed, draw-off was by means of a 33-inch pipe for water supply and a 48-inch pipe for rapid lowering of the reservoir, laid in a culvert with a brick central stop-wall through which the pipes passed. The culvert, over which the dam was constructed, was formed by building an ashlar granite top arch on a cutting through solid rock. Upstream of the stop-wall the culvert was open to the reservoir and here the scour pipe ceased whilst the supply pipe continued to a tower in the reservoir containing control valves. Both pipes were fitted with control valves at the downstream toe and there was a throttle valve just downstream of the stop-wall on the scour pipe.

First filling of the reservoir commenced in July 1866 and in November, when water depth had reached 12 m, the throttle valve on the 48-inch pipe broke and split the pipe to which it was fitted. Repairing the leak involved cutting out the pipe which, because there was no upstream control, could only be done by somehow blocking the pipe or emptying the reservoir. The first option was chosen and a month later, after several failed attempts, a large pine ball, weighed with lead to achieve neutral buoyancy, was successfully positioned by divers in the mouth of the 48-inch pipe, stopping the leak.

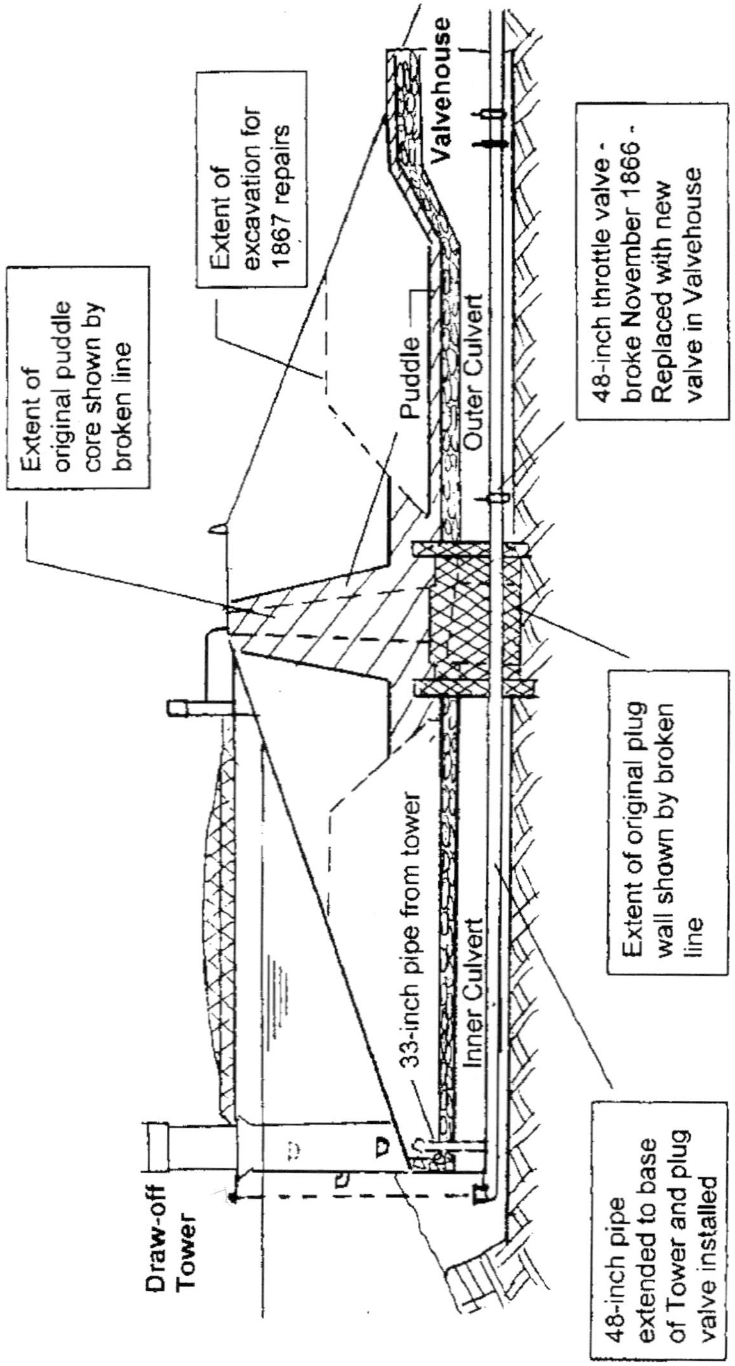


**Fig 4 Section: Lower Vartry Dam**

The work of cutting out and replacing the damaged pipe and valve now started, but before it was completed a leak appeared in the culvert. At first it was a small jet of clear water but within days the flow had become much stronger with the water deeply coloured and carrying clay and sand. In addition, a 2-m diameter subsidence appeared in the reservoir slope over the line of the tunnel. Water depth in the reservoir had now reached 16.5 m and was rising despite the 33-inch supply pipe having been opened fully. With nobody able to say why the leak had occurred, residents downstream of the dam feared the worst and evacuated their homes.

Round-the-clock efforts to reduce the water level in the reservoir commenced on 5$^{th}$ February 1867. Extra labourers were taken on, and a workforce of 850 men toiled non-stop to complete the repairs to the 48-inch pipe and to cut a gap in the overflow weir. As the hole in the slope continued to sink, it was filled with timber, straw and clay in the hope that the mixture would be drawn in and block the leakage, but to no avail. All the time the flow into the tunnel continued, changing often from joint to joint as they became choked with stones and described as "generally a 4-inch by ¾-inch jet – came with great force & noise".

One week later work on the weir was complete but the ball in the mouth of the 48-inch pipe remained stuck fast despite increasingly frantic efforts to remove it using a force pump and barge mounted winches. Crowds came from Bray and Wicklow to observe the spectacle and at least one unfortunate engineer found his hotel bed taken by a newspaper reporter. Finally on 13$^{th}$ February the ball came free and as the water level in the reservoir dropped, the leak reduced and the crisis passed.

The cause of the leak was found to be poorly compacted puddle under overhanging rock alongside the plug-wall. The bank was cut down to rock level over the central section of the culvert and rebuilt with the puddle core greatly thickened. The existing culvert plug-wall was enlarged with the construction of additional cross-walls at each end of it. These were rebated into the rock, and the gap between brick and rock filled with concrete. The 48-inch pipe was extended to the draw-off tower, controlled by a new plug valve.

Although this leak is long forgotten it was important enough to be detailed by Sir Alexander Binnie over forty years later in his book, *Rainfall Reservoirs and Water Supply*, as teaching "the possibility of leaks along the line of a culvert built under a made embankment and…the folly of placing the valves at the outer end of the pipe"

## YIELD ESTIMATION

One of the problems with which early Irish dam designers had to contend was a lack of rainfall records. It was only in the 1860's that a network of rainfall recording stations began to be established in Ireland and not until the 1880's that gauges were standardised. A small number of records were available from the late eighteenth century onwards but these tended to be of short duration and uncertain reliability. A further problem was that all of these early records were from low-level stations rather than the upland sites preferred for reservoirs.

For Lough Island Reavy Reservoir, Fairburn simply used an average annual rainfall figure of 36 inches (914 mm) for all Ireland and allowed 1/6$^{th}$ (152 mm) for evaporation and absorption losses as the basis of his calculations. An indication of Bateman's own approach to yield estimation can be seen in his installation of two rain gauges at the site despite the mill owners opposition to this expense. Although Fairburn's allowance for losses was far too low, because the assumed rainfall figure was also low (the long-term averages for this area are 1330 mm and 510 mm for rainfall and evapotranspiration respectively), the reservoir yield was better than expected.

Bateman was not as fortunate in his 1855 design for Belfast's water supply. The Woodburn scheme apparently only supplied half his estimated minimum yield during a drought in 1873. I have been unable to find details of his calculations for Belfast, but they are probably partly based on short-term rainfall records taken on-site, as this was his approach in a scheme he designed for Dublin in 1857. Bateman and Hawksley (proponents of alternative schemes) clashed on this matter at public hearings held during the 1860 Dublin Water Supply Inquiry.
By 1860 a reliable rain gauge had been in place in Dublin's Phoenix Park for twenty-three years. Hawksley put forward an empirical formula for calculating an upland catchment yield where such low-level records were available. This was based on: -

- an assumed increase of 2.5% in annual rainfall for every 100 feet gain in altitude
- three consecutive dry years with rainfall one sixth less than the long-term average
- an annual allowance of 15 inches for evaporation losses.

Bateman described this as "ridiculous", preferring to base his proportionate increase on direct comparison of rainfall readings taken at the reservoir site on a number of days with those taken simultaneously in Dublin and his assumed drought on the driest two-year period recorded. Hawksley replied "...on these points I differ very much from my friend Mr. Bateman. I have only to say that all I advance is by way of caution.........to prevent a repetition of the great mistakes which have been made in the construction of other waterworks for the supply of large cities".

I think the exceptional sequence of dry years in the 1850's strongly influenced his approach as later in his evidence he speaks of acquiring "a good deal of information we did not possess before" during the previous two or three years.

Whatever the reason, caution won out and Hawksley's formula was used in calculating the safe yield from Dublin's Vartry Reservoir, construction of which commenced in 1862. The safe yield so obtained was very conservative and only once since 1860 has any twelve-month yield dropped below the figure used in Hawksley's calculations as an annual average for a three-year dry period.

The drawback of such a conservative approach is the danger that the client, after experiencing much better than expected yields, completely disregards the original draw-off limits. This is what happened and by 1882 Parke Neville, Dublin's City Engineer, was able to report that the actual daily supply to the city was already well above the original estimate and was confidently predicting it could be increased further. Then the inevitable happened and only eleven years later, during the drought of 1893, the reservoir emptied and canal water had to be used to supply Dublin.

Whilst poor estimation of potential yield might diminish a designers reputation, the real danger of relying on short term records lay in using them for flood prediction. This is illustrated by **Robert Mallet's** 1844 proposal for the river Dodder at Dublin.

Mallet was a highly innovative engineer with a background in foundry work. His achievements ranged from investigation of metal corrosion to design and manufacture of bascule bridges and cast-iron lighthouses. In 1841 he commenced surveys of the Dodder on his own initiative. He was subsequently engaged by the Commissioners of Drainage for Ireland to investigate the feasibility of constructing a reservoir for the combined purpose of providing increased all-year waterpower for milling and at the same time controlling the violent floods on the river. He proposed a sizeable dam with a maximum height of 33 m and a crest length of 313 m. His design was generally conservative, with a freeboard of 2.4 m above the top water level - "thus ensuring the impossibility of overflow".

He carefully assessed all available rainfall records for Ireland and decided that the most reliable and appropriate was a two-year set taken at the College of Surgeons in the centre of Dublin!! Although he recognised that these figures were lower than could be expected at the upland reservoir site and factored them upwards by 20%, he did not seem to appreciate that it was highly unlikely that an exceptional storm would have occurred in that particular short period. As a result he took "the greatest possible sudden burst of rain" as 2-inches of rain in 24 hours - in practice an event with a return period of about one year on site - and designed his overflow accordingly. When we compare his proposed 60-foot long overflow weir with the 200-foot long ones actually constructed forty years later (the upgrading of whose spillways is presently underway and the subject of A. Rowland's paper), it was fortunate that Mallet's dam was never built.

## BENEFITS OF EARLY IRISH DAMS

Whilst there is no doubt that Ireland's early dams have been of great benefit to the country they also had some negative social impacts at the time. Belfast Water Commissioners adopted a controversial policy of depopulating catchments, which required the displacement of over four hundred families in the Carrickfergus area alone. Dublin did not follow that policy but its Waterworks Acts are notable for the preferential treatment of wealthy landowners, with many special clauses included protecting individuals' interests. Following the Vartry dam leak of 1867, described in the previous section, a Mr Daniel Tighe took advantage of such a clause by suing the Corporation for depreciation because of the *possibility* of the dam failing. He had already received £5,000 compensation for losses due to reduction in flow of the river through his estate and through this action obtained a further £12,061. Contrast this with the treatment of his tenant miller who lost his means of livelihood and received £200 compensation. Even Dungannon Park Dam resulted in the demolition of a local hamlet and the dispersal of its inhabitants - it spoilt the view over the new lake! It is fair to say however that these negative impacts were relatively minor and short term when compared to the hundred and thirty year plus (and continuing) benefits provided by these reservoirs.

The Bann reservoirs, by increasing available river power fourfold, enabled the linen industry expand and this part of Ireland share in the prosperity brought with the Industrial Revolution. With the demise of water-power, Lough Island Reavy was converted in the 1970's to its present function as a water supply reservoir. Belfast's rapid growth in Victorian times as a major manufacturing centre would not have been possible without the series of reservoirs constructed to satisfy industrial and domestic water demands. Likewise, the provision of a clean, plentiful, high-pressure water supply made possible by Vartry Reservoir is regarded as the single most important action taken to improve public health in Dublin in the nineteenth century. In both cities these early reservoirs continue to be vital contributors to the water supply networks today. Irelands canals formed the first efficient transport network for heavy goods in the country and are now a widely used leisure and recreational amenity, whilst the lake at Dungannon Park, which was built for the privileged few, is today in public ownership giving pleasure to thousands.

## CONCLUSION

Although Ireland has very few pre-1870 dams, some of those built are noteworthy because of the well-documented construction problems encountered. On the Grand Canal techniques were developed for construction of canals across bogs using lightweight peat embankments, whilst Loch Island Reavy and Lower Vartry highlighted the potential hazards of buried culverts in embankments. These, together with Belfast's Woodburn reservoirs also illustrated the inadequacy of rainfall records of the time.

Most of these dams were crucial elements in the country's infrastructure and development and though some no longer serve their original function, all continue to provide beneficial service to society.

## ACKNOWLEDGEMENTS

I would like to acknowledge the generous assistance of the following in preparing this paper; John McKeown, Waterways Ireland, Ron Cox & Michael Gould, ICE Panel for Historical Engineering Works, Bertie McMinn, Dungannon District Council, Trevor Reeves-Smythe, Environment & Heritage Service.

## REFERENCES

Bateman J F (1848). Bann Reservoirs *Proc. Instn Civ Engrs*, 7, 251 - 274

Binnie A. (1913). Rainfall, Reservoirs & Water Supply, London

Binnie G M. (1981) Early Victorian Water Engineers, Thos. Telford, London

Cooper G A (1987). The Reservoir Safety Programme in Northern Ireland, *IWES Journal*, 1, 6

Cox R C & Gould M H (1998). Civil Engineering Heritage Ireland, Thos. Telford, London

Cox R C (1982). Robert Mallet 1810 - 1881, Royal Irish Academy, Dublin

Freemans Journal (1867). Reports on Vartry Leak 8th - 13th February, Reports on Tighe-v-Dublin Corporation October 1867

Given Anne (1988). Dungannon Park (Unpublished report for Environment & Heritage Service, DOENI)

Hassard R. (1861) On the River Vartry as a Source of Water Supply for Dublin, *Proc. Insnt Civ Engrs Ire.*

Hawkshaw J (1860). Royal Commission Report on Dublin Water Supply

Irish Times (1867) Reports on Vartry Leak 8th - 14th February

Loudan J (1940). In Search of Water, W M Mullan & Son, Belfast

Mallet R (1844). Proposed Formation of Reservoirs on the River Dodder, Alexander Thom, Dublin

McKeown J (2001) Heritage and Modern Technology The Irish Experience. World Canal Conference, Dublin

Mullins B & Mullins M B (1848). On the Construction of Canals in Bog, *Trans. Insnt Civ Engrs Ire.*, II, 37 - 41

Neville P (1874). On the Water Supply of the City of Dublin, *Proc. Instn Civ Engrs*, 38, 1 - 49

Neville P (1882). Presidents Address, *Trans. Insnt Civ Engrs Ire.*

Pigott P, Hanrahan E & Somers N (1992). Major canal reconstruction in peat areas *Proc. Instn Civil Engrs 1992, 96, Sept 1411 - 152*

Reeves-Smyth T (2001). Luttrellstown, Co. Dublin, An Outline Historical & Landscape Report (Unpublished report for An Taisce)

Rohan P K (1986). The Climate of Ireland, Stationary Office, Dublin

Smyth J (1869). Bann Reservoir at Lough Island Reavy A long existing leak and its repair. *Trans. Insnt Civ Engrs Ire.,* 9, 51 - 67

Walpole E H (1952). Mount Usher 1868 - 1952, Dundalgen Press, Dundalk

Walsh R (1867). Construction of Vartry Waterworks, MS 19814, National Library of Ireland, Dublin

# The Contribution to Society of Irish Hydro-Electric Dams

J D O'KEEFFE, Power Generation, Electricity Supply Board, Ireland

SYNOPSIS. A number of Hydropower projects were built in Ireland between 1925 and 1974. This paper briefly describes the various Schemes and gives the social and political impact they had on the development of the country. It also looks at the ongoing contribution to society of the Schemes.

## INTRODUCTION

When the Irish State was established in 1922, the Irish electricity industry had already been in existence for about forty years. However, progress in the development of the industry in Ireland, and indeed Britain, was much slower than in countries like Germany, Switzerland and the U.S. The industry was uncoordinated and inefficient, prices were high, the number of customers was small and there was a diversity of electrical systems within the country. The use of electricity was limited to lighting and heating. Little use had yet been found for electricity in either industry or agriculture. There was very little industry in Ireland with the exception of the North East of the country. The Government of the newly formed State was faced with an enormous task to build an economy capable of sustaining three to four million people, which up to this time was almost totally dependent on agriculture. The first major project undertaken by the State was the development of the Shannon Hydro-Electric Scheme. This was a major undertaking for such an underdeveloped country as Ireland was at that time. The project was successfully completed in a remarkably short period of time and this gave the new nation the confidence to tackle other major infrastructural projects such as those in the peat and sugar beet industries as well as a number of other hydro-electric projects. The first part of this paper gives a brief description of each of the Hydro-Electric Schemes built in Ireland between 1925 and 1974. The social and political impact of each of the Schemes on the developing economy and the contribution they made and are continuing to make in an ever changing Ireland are then outlined.

## DESCRIPTION OF SCHEMES

The following gives a brief description of each of the Schemes, in the order in which they were undertaken. Figure 1 shows the location of the rivers and their catchments for each development.

*Reservoirs in a Changing World*, Thomas Telford, London, 2002, 15–30

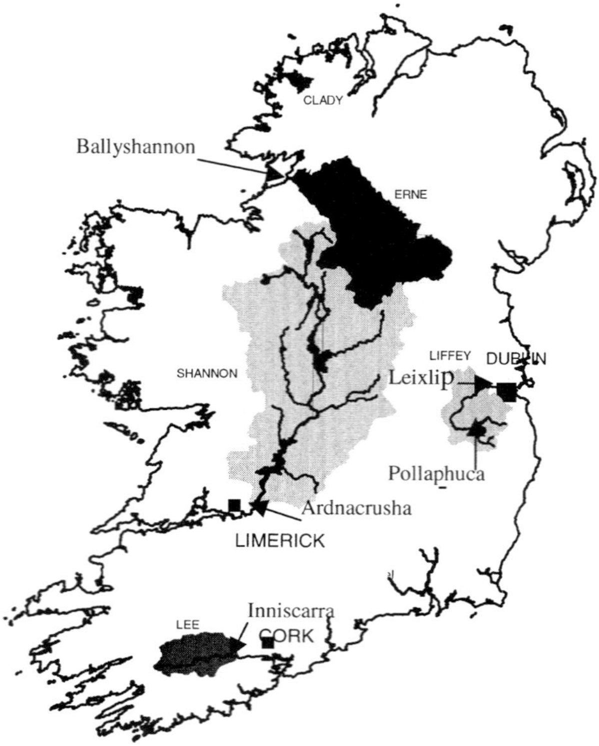


Figure 1 Map of Rivers and Catchments

The Shannon Scheme

The possibility of harnessing the River Shannon for hydropower was mooted as early as 1844 by Robert Kane. From the late nineteenth century until the formation of the Irish Free State in 1922 a number of hydro-electric schemes were proposed for the Shannon. These proposals varied in capacity from 11MW to as high as 70MW. However, for various reasons including the lack of hydrometric data, all of these proposals came to naught.

In 1922, Dr. Thomas Mclaughlin, a young Irish engineer, went to work for Siemens-Schuckert in Germany. Since his college days in Galway, he had been interested in the development of the Shannon for hydro-electric power. Shortly after joining Siemens, he developed a proposal for harnessing the River Shannon. He got support for his proposal within the company, and, through a friend from his college days, Mr. Patrick McGilligan, the Irish Government became interested.

In 1925 Siemens-Schuckert submitted detailed proposals for the Shannon Scheme and the electrification of the Irish Free State to the Irish Government. Despite a campaign of opposition to the Scheme, within six months of the proposal being submitted, a contract was signed in August 1925, between the Irish Government and Siemens-Schuckert, for the development of the Shannon Scheme and the electrification of the Irish Free State. Work began almost immediately as only three and a half years were allowed for completion of the contract.

The River Shannon has a very low gradient from its source as far as Killaloe, it then falls 30 metres between Killaloe and Limerick, a distance of approximately 20 kilometres. The Shannon Scheme utilises this fall for electricity generation at Ardnacrusha. The scheme involved the construction of a concrete Weir and intake structure at Parteen Villa, 8 kilometres south of Killaloe. The weir is fitted with six sluice gates which can discharge a range of flows, down the original river channel, from the minimum flow of 10 cubic metres per second to the maximum design flow of 900 cubic metres per second. A ladder or pool type fish pass forms part of the structure. This allows the passage of salmon past the weir. The intake structure has three large intake gates, which control the flow of water to the headrace and a smaller gate to allow the passage of boats. From the intake structure, a 12.5 km long headrace channel leads the water to the power station at Ardnacrusha. The headrace channel traverses nine natural valleys, eight of which are drained underneath the headrace. The headrace is formed by earthen embankments constructed from local materials. It varies in height up to a maximum of 20 metres above the original ground with a channel width of over a hundred metres at the top level of the embankments. The headrace ends at a 30 metre high concrete dam above the power station at Ardnacrusha. The dam incorporates a double boat lock to allow navigation past the dam, for boats of up to 150 tonnes. Water is taken from the headrace via four six metre diameter steel penstocks to the four turbine generator units before being returned to the river just upstream of Limerick, in a two kilometre long tailrace channel cut in rock.

The total catchment of the Shannon from its source to Ardnacrusha is 10,500 $km^2$. Storage for the scheme is provided in three existing natural lakes along the Shannon. The works involved the construction of three concrete dams, more than 50 kilometres of earthen embankments and the establishment of a number of drainage pumping stations at various points along the Shannon. A number of roads were re-routed. Three road bridges were erected across the headrace canal and one across the tailrace. Approximately 8.75 million cubic metres of material was excavated as part of the project, of which 1.25 million cubic metres was rock, and 0.25 million cubic metres of concrete was placed during the project.

The total generating capacity of the power station was 85 MW. Works began in late 1925 and the first generation was available before the end of 1929. The capacity of Ardnacrusha was increased to 92 MW in the last ten years, as part of a project to refurbish the turbines and generators.

The Liffey Scheme

Prior to the formation of the Irish State the development of the hydro potential of the River Liffey was under serious consideration. However, the development was dropped in favour of the larger Shannon Scheme soon after the establishment of the new state, on the basis that the Shannon Scheme was a national scheme whereas the Liffey Scheme was a regional scheme focused mainly on Dublin and the surrounding areas.

The demand for electricity grew rapidly during the early 1930s, and in addition there was an increasing need for an additional water supply source for Dublin City, so, soon after the completion of the Shannon Scheme, the development of the Liffey again began to be considered. In 1936 a decision was taken to harness the water of the Liffey for electricity and as a water supply to Dublin City. Construction work began in 1937 and the area upstream of Pollaphuca (the most upstream impoundment) began to be flooded in the autumn of 1940. The first power from the scheme was available in late 1943.

The scheme consisted of three concrete dams at Pollaphuca, Golden Falls and Leixlip respectively. Pollaphuca dam forms the main element of the project. It is a gravity dam 32 m high and 79 m long at a deep gorge on the river. It creates a reservoir of 20 $km^2$ with a large storage capacity of approximately 175 million cubic metres. The dams at Golden Falls, 2 km downstream from Pollaphuca, and Leixlip, a further 50 km downstream, are gravity type dams, 16 m high and 100 m long and 24 m high and 114 m long respectively, and are largely run of the river installations.

Other works associated with the Liffey Scheme were the clearing of 2,500 hectares of land before flooding, and the construction of five road bridges and a network of roads to replace the roads lost due to flooding. At Pollaphuca, a 5 m diameter tunnel 370 m long was cut in rock to take the water from the reservoir to the power station. A surge tank was built to protect the installations from a sudden shut down of the turbines. Each of the dams is fitted with three spillway gates to discharge excess floodwater. Leixlip dam has a lift type (Borland) fish pass to allow for migration of salmon. No fish passes were provided on the other two dams, as salmon fishing was not a major issue upstream of Golden Falls.

The Pollaphuca and Golden Falls elements of the project were carried out during the war and encountered many delays as a result. The full installations were not completed until 1947. Construction at Leixlip started in the late 1940s and was completed in 1952.

The total generation capacity of the scheme is 38 MW, 30 MW of which are at Pollaphuca. The initial agreement allowed for the abstraction of 100 million litres of water per day for Dublin City and its surrounds. The remaining water was available for power generation.

Erne Scheme

The Irish electricity market continued to grow during the 1940s. The River Erne was the next river to be considered for power generation. The River Erne rises in the Republic of Ireland and flows into Northern Ireland through Upper and Lower Lough Erne before flowing back into the Republic at Belleek. The river drops 45 m between Belleek and the sea at Ballyshannon, a distance of less than 10 km. A study of water resources in Ireland in the 1920s identified the Erne as second only to the Shannon in hydropower potential in an all Ireland context. However it was not considered at an earlier time because of the issues relating to the border between the Republic of Ireland and Northern Ireland

Agreement on the development of the Erne was reached between the ESB and the Ministry of Finance in Northern Ireland in the late forties. Two power stations were constructed in the Republic, 20 MW at Cliff and 45 MW at Ballyshannon. Northern Ireland benefited by way of significant drainage and flood protection in the areas around Upper and Lower Lough Erne. This was achieved by enlarging the 6 km long Belleek Channel and deepening of the 20 km long channel between Upper and Lower Lough Erne by excavation of 600,000 cubic metres of earth and rock material from these channels. These works reduced the flooding of approximately 1000 hectares of land mostly around Upper Lough Erne.

The River Erne has a catchment area of almost 4400 $km^2$ with an average annual precipitation of 1,050mm. Two gravity dams were constructed, one at Cliff, downstream from Belleek, 18 m high and 210 m long and the other at Cathleen's Fall, Ballyshannon, 27 m high and 257 m long. There is a small balancing reservoir between the two dams. Storage of 194 million cubic metres is available mainly in Upper and Lower Lough Erne. Floods are discharged through three spillway gates at each of the dams. There are pool type fish passes at each of the dams to allow for migration of salmon.

Work commenced on the construction of the Erne scheme in 1946 and the first unit was commissioned in 1950 and the final unit in 1955.

Lee and Clady Schemes

Two smaller hydro-electric schemes were developed during the 1950s, one on the River Lee, in the south of the country, with an installed capacity of 27 MW, and the other in north Donegal where a 4 MW unit was installed on the River Clady.

The River Lee rises in the mountains of west Cork and flows eastward to Cork City. Two dams were built on the river between Cork and Macroom.

A concrete buttress gravity dam, 45 m high and 245 m long, was built at Inniscarra and a conventional gravity dam, 22 m high and 107 m long, was built 20 km upstream at Carrigadrohid. The catchment area to Inniscarra is approximately 800 $km^2$. Two small reservoirs are formed by the dams, providing storage which is less than 5% of annual inflow. There are three flood discharge gates and a lift type fish pass in each dam. The works also involved the construction of several kilometres of roadway and three public road bridges. The development took place between 1952 and 1957.

The Clady River has a catchment area of less than 100 $km^2$. The scheme involved the construction of two small dams and a headrace canal, 3 km long. The water used in the power station discharges into the adjacent Crolly River. The river is self-regulating with excess water discharging over an overflow weir, on the headrace, into the Clady River.

Turlough Hill Pumped Storage Scheme

The final major hydro-electric development carried out in Ireland was the construction of a 292 MW Pumped Storage Scheme at Turlough Hill, in the Wicklow Mountains. This was completed in 1974. The Upper Reservoir is formed by a 1.5 km long 34 m high rock fill embankment dam with an asphaltic concrete lining. The lower reservoir was formed from a natural lake. The power station is underground and the transmission power lines from the station are taken underground for a short distance in order to reduce the visual impact of the development on the countryside in a very scenic area of the Wicklow Mountains at the Wicklow Gap.

## SOCIAL AND POLITICAL IMPACTS OF THE SCHEMES

At the time of the development of the Shannon Scheme the economy of the new state was primarily based on agriculture with little or no industrial base. Living conditions were very poor and any employment available was in agriculture or agricultural based products. Heavy emigration, which began around 1850, was still a significant factor in the Ireland of the 1920s. The main industrial base on the island of Ireland before the setting up of the State was in Northern Ireland, mainly in Belfast. When the Shannon Scheme was proposed, the new Irish Government saw it as a means of boosting the economy and indeed the morale of the nation, as well as a potential source of energy for other industries. It was determined that the project should succeed.

There was a lot of opposition to the proposed Scheme mainly from those who were already involved in local electricity operations in the cities and towns, many of whom were City and Town Corporations, Chambers of Commerce, business and professional people. The strongest opposition came from a body of people who argued that a development on the River Liffey would be more suitable. The opposition to the Scheme was strongly supported by The Irish Times as well as a number of other publications. A number of arguments, both technical and financial were raised, but the opposition never developed into any serious threat to the project. Despite reservations expressed by political representatives, the Shannon Electricity Bill was passed by both the Dail and Senate in June 1925 without a single dissenting vote. At a local level, the number of people whose property was directly affected by the Scheme was very small and by and large there were no great problems with the acquisition of land. The total area of land acquired for the scheme was less than 1500 hectares and this was mostly acquired in small parcels of land from any individual owner and only a small number of dwellings were directly affected. An ancient oratory on an island in the Shannon, due to be flooded as part of the scheme, was dismantled and re-erected in Killaloe where it still stands today. A school and a post office, which were on the line of the tailrace, were also relocated. The only real opposition came from the Abbey Salmon Fishermen on the lower Shannon. This opposition did not really materialise until the Scheme was completed and the fishermen experienced a negative effect on their livelihoods. The salmon were being attracted up the tailrace, as opposed to the river, by the flow of water from the power station, but fishing was prohibited on the tailrace near the power station for safety reasons. This lead to some bitter confrontations between the fishermen and the police in the early years of operation of the Scheme. In 1936 agreement was reached with the fishermen and they were paid compensation for loss of fishing rights and the Electricity Supply Board (ESB) then became the owners of all the fisheries on the River Shannon. A lift type fish pass was added to Ardnacrusha Dam in 1959 to allow the migration of fish in the headrace as well as in the river.

The Shannon Scheme was a massive undertaking, by any standards. But it is hard to imagine the magnitude of the undertaking, at that time, for a state in existence for just three years, and recovering from a civil war, which occurred in the first year of its existence, that left the people politically divided. It was the first major infrastructural undertaking since independence and the new government pursued the project with enthusiasm. Despite some Industrial relations problems and a very short construction programme the Scheme was completed on time and within the original contract price. The final price of £5.2 million included the construction of the transmission network for the whole country as well as the Shannon Scheme itself.

Rural areas of Ireland were experiencing severe economic depression in the 1920s and employment was very scarce. The Shannon Scheme provided a welcome boost to employment and workers came from all over Ireland to work on the project, although wages were relatively low. Up to 4,500 people worked on the project at its peak. Most would have been unskilled at the beginning of the project but many went on to get skilled jobs in the power station or on other construction projects in Ireland, England, and elsewhere after the project was completed.

Viewed from today the Shannon Scheme is seen by many as the foundation of modern Ireland. The scheme changed the social and economic life of the country. One of the many successes from the Scheme was the Electricity Supply Board (ESB) itself. The ESB was set up, as a semi-state company, (i.e. state owned but run by a Board appointed by the relevant Government Minister), in 1927 to manage the Shannon Scheme. ESB went on to design, construct and manage, from within its own resources, many hydro and thermal power plants as well as building the complete electricity infrastructure for the country. Today ESB is a successful international company having been involved, over the years, in projects in over eighty countries world wide, and is still the major Electricity Company in Ireland.

The Shannon Scheme has been well recorded in word and picture. However, the Scheme is unique in the fact that an official artist recorded the project. In 1926 a Limerick artist, Sean Keating, was commissioned to carry out a series of pictures of the Scheme. Keating completed twenty-six paintings and drawings of the project. He also produced some paintings of the Liffey Scheme.

The Liffey Scheme met with much more serious local opposition due to the fact that a much greater area of land and more homes and livelihoods were affected by the flooding. A total of 2,250 hectares of land was flooded upstream of Pollaphuca Dam. Seventy-six dwelling houses and fifty farms as well as a church and cemetery were in the inundated area and had to be evacuated. The graves from the cemetery were relocated to higher ground before flooding took place. Displaced residents were paid market value for their houses and land or were resettled locally. Problems arose because the acquired land was often of poor quality and the amount of compensation paid to dispossessed farmers around Blessington was not enough to buy viable holdings in the more expensive low lying areas of Kildare. In addition, no allowance, apart from compensation for loss of property, was made to people whose livelihoods were affected but who did not own land. The economic realities of the time, the war from 1939, the increasing national need for power and Dublin's need for water meant that farmers and others accepted the compensation offered and got on with life. The economic benefits from the construction of the Liffey Scheme and the spin off from power, water and leisure activities of the completed scheme resulted in a considerable improvement in the living standards of many people along the River Liffey.

There was similar opposition to the land acquisition for the Lee Scheme. However, the area flooded was smaller and consequently the numbers affected were fewer. The total area flooded is less than 1,500 hectares and 39 dwelling houses including three public houses were demolished before the flooding took place. Innisleena House, one of the larger houses demolished due to flooding of the Lee Valley, is shown in Figure 2 below. All those whose houses were affected were given equivalent houses in the area or alternatively were given compensation, whichever they chose. Some of the land acquired is only covered by water for short periods during the year. This land is leased back to the original owners or their families for grazing at a nominal rate. Upstream of Carrigadrohid Dam is an area known as 'The Gearagh' or Flooded Forest. This is a unique type of landscape and is the last of its type in Western Europe. Most of the Gearagh was damaged by the Lee Scheme but about 150 hectares remains intact and the area was declared a National Nature Reserve in 1987. The area is characterised by trees, mostly oak trees, growing between a network of streams and undergrowth rich in rare flora and fauna.

Figure 2. Innisleena House

Less than 300 hectares of land was acquired for the Erne Scheme. The area within the reservoir between the two dams is the only area flooded, approximately 2 $km^2$ in total. Five large 'estate' type houses and thirteen smaller dwellings were flooded or displaced. There was a significant visual impact on the Lower River Erne at Ballyshannon from the Scheme. The tailrace downstream of Cathleen's Fall dam was cut through the bed of the existing river. The water fall at Assaroe, where the river discharged into the sea was lost and what was a wide river flowing through the town of Ballyshannon was changed to a 15 m wide channel through the bed of the original river. This fall was a major tourist attraction for the town, prior to the Scheme. A significant area of land on the perimeter of the Erne Lakes was enhanced for farm purposes, due to the lowering of the water levels in the Erne Loughs and subsequent reduction in flooding. There was no significant local opposition to the Erne Scheme. However, some opposition was expressed in the Northern Ireland Parliament when the Bill of Agreement between the parties was being discussed. The Erne Scheme is remarkable for the ongoing level of co-operation between the bodies involved in the two parts of Ireland, given the level of mistrust that existed between North and South of Ireland in most other areas of activity. In fact, a joint interest in the drainage of the Erne predated the Scheme.

There was little or no national opposition to the Liffey and subsequent Schemes. By then it had been established that all electricity developments would be carried out by the ESB and consequently there were no other groups or individuals interested in developing these projects. These smaller Schemes did not have the same national impact as the Shannon Scheme. However, they each provided significant benefits to the local communities by way of employment and provision of services during the construction. They provided a number of permanent jobs in the power stations at each location subsequently. The projects also helped to improve the skills level in the work forces in the local communities and many went on to work on other large projects in Ireland or abroad. Many of those who were displaced by the Schemes benefited from the employment and other opportunities afforded by the Schemes.

## ELECTRICITY GENERATION

Generation output from the Shannon Scheme, in the early years, accounted for over ninety percent of total generation output for the country. At that, not all the available output was utilised, as the night time load demand was not enough to use all the available water during the winter months. Hydro continued to be the major source of generation up the 1960s accounting for approximately 40% of output in 1959. As the electricity market continued to grow in Ireland, hydro, as a proportion of the overall output, continued to decline to the point where hydro only accounted for approximately 3% of electricity generation in 2000.

The long-term average generation from hydro is approximately 725 Gwh per year over the last 45 years, but continues to decline as a proportion of the total electricity generation. Hydro still continues to provide a valuable source of peaking capacity for the national electricity system. The fore going does not include the impact of the Pumped Storage Scheme at Turlough Hill.

If the electricity produced from hydro was produced from coal it would result in the production of approximately 653,000 tonnes of Carbon Dioxide per year as well as the production of other gases harmful to the atmosphere.

WATER SUPPLY

The Liffey Scheme was the only one which was developed with a dual purpose of water supply and electricity generation. The Liffey Scheme was designed to provide 100 million litres per day for domestic water supply for Dublin City and surrounding towns. Over the years this quantity has increased gradually and approximately 400 Ml per day are now abstracted, approximately two thirds at Pollaphuca and the remainder at Leixlip. This is close to half the long term average annual inflow to the Liffey upstream of Pollaphuca. Water entering the river downstream of Pollaphuca cannot be stored and therefore cannot be depended on when estimating sustainable quantities for water supply. Five hundred and fifty million litres per day is considered to be close to the limit of the sustainable abstraction from the Liffey.

Since the early 1980s, water is also being abstracted from the Lee Scheme upstream of Inniscarra dam. Approximately 55 Ml per day is currently abstracted and arrangements are in place to increase this quantity to over 225 Ml per day in future years. The water is treated upstream of Inniscarra and is gravity fed to Cork City and Harbour areas.

Where the hydro schemes are being used for the abstraction of water, water supply is now given precedence over electricity generation. The Authority abstracting the water compensates the ESB for the loss of water available for electricity generation.

FISHING

All the above rivers are good salmon fishery rivers. Fish passes were constructed on all rivers as part of the Hydro Developments. Despite this the stocks of salmon reduced gradually on all of the rivers over the years. Salmon stocks also declined on rivers that are without dams, but not to the same extent. The exact reasons for the decline in stocks are not known but can be attributed to a number of factors including disease, netting, water quality and the dams, where they exist. The dams were accepted as being a factor in the decline in salmon stocks, from an early stage in their operation, so during the 1960s and 1970s, in an effort to redress the situation, restocking became a part of the management of the fisheries on the rivers with Hydro-Electric Developments. These measures helped to arrest the decline in the stock numbers, but the numbers are still well below what they were in the early part of the twentieth century. Restocking is carried out by rearing the fish to smolt stage, using brood stock from the particular river, and putting the smolt into the rivers downstream of the dams when they are ready to go to sea. This measure has been successful in increasing the numbers of adult fish returning to the rivers as far as the hatcheries or the point at which the smolt were put into the river. However, the numbers of fish that return to spawn in the upper tributaries is very limited. Efforts are now being made to address this problem and an experimental programme of putting ‘unfed fry’ into the rivers upstream of the dams has been carried out in recent years. This is apparently improving the rate of return of adult fish to the upper reaches of the catchments. This experimental work has yet to be conclusively proven. In the meantime the practice of planting hatchery smolt in the rivers is continuing.

Eel fishing is also a feature of the Shannon and Erne Rivers and to a lesser extent the Lee River. Eels continue to be fished commercially on these rivers. However, the young eels or elvers are artificially transported to the upper reaches of the rivers where they grow for many years before returning to sea as adults. The adult eels are caught for commercial purposes on their return journey to sea.

Coarse fish such as bream and roach have always been available in the natural lakes in Ireland. These coarse fish species are now also available on most of the artificial reservoirs formed by the Hydro-Electric Schemes. Coarse fishing has developed as a significant tourist attraction on many of these new reservoirs.

FLOODING

Only the Erne of the above Hydro-Electric Schemes was built with the express purpose of flood mitigation. However, significant flood mitigation benefits have also resulted on the Liffey, Lee and Clady rivers

The reservoir above Pollaphuca has a very large storage capacity relative to the catchment inflow. A large element of this storage is reserved for flood mitigation. This has allowed the vast majority of the inflow upstream of Pollaphuca, in the most severe floods, to be stored until the worst of the floods in the river downstream of Pollaphuca has abated. Since the dams were built, the maximum flow in the river down stream of Pollaphuca, during the largest floods has been reduced to less than 25% of what the estimated flow would have been otherwise. Even with this reduction the severe floods of 1954, 1993 and 2000 caused significant damage in the valley downstream of Pollaphuca. This damage would have been many times greater, on these occasion, but for the presence of Pollaphuca Dam. Figure 3 indicates the effect Pollaphuca Dam had on the flood of November 2000.

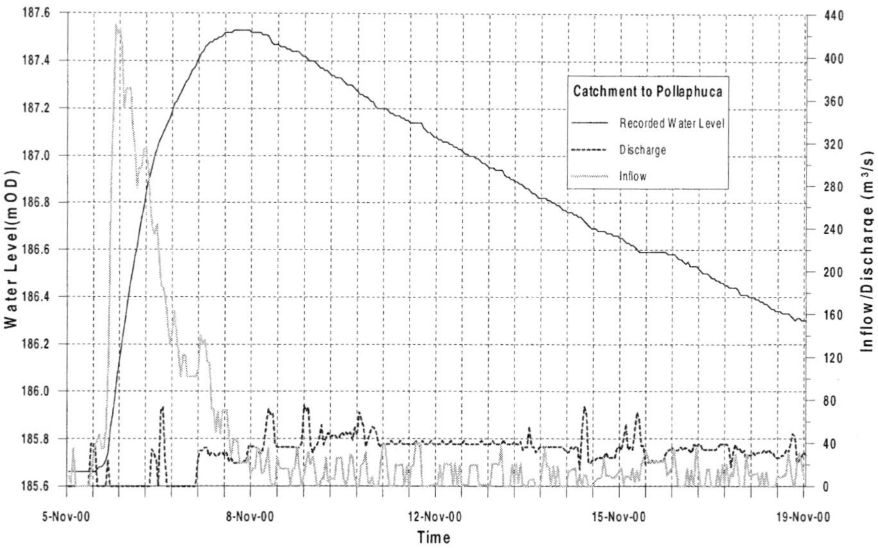


Figure 3. Pollaphuca Levels and Flows – November 2000 Flood

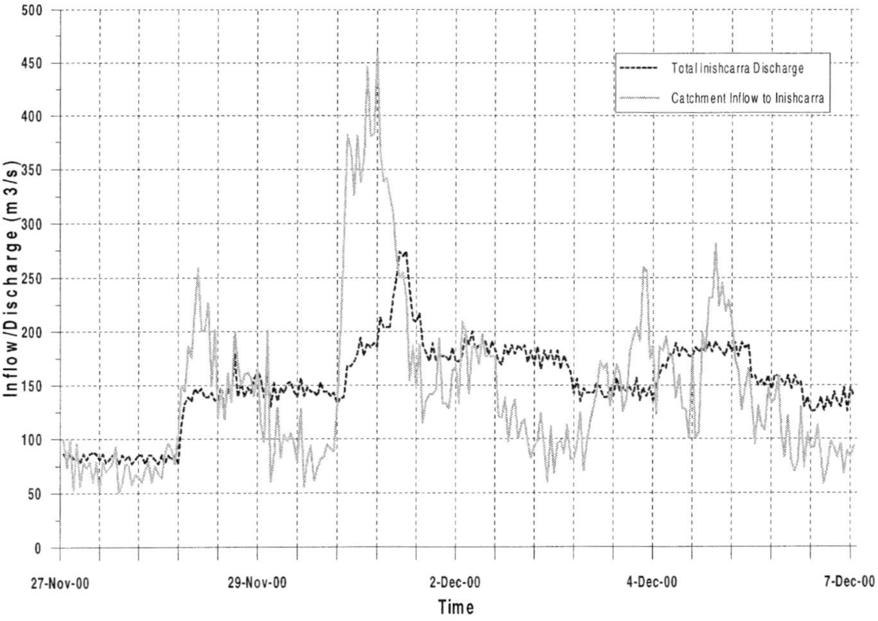


Figure 4. Inniscarra Inflows / Discharges – December 2000 Flood

The contribution of the dams to the reduction in flooding on the River Lee is far less dramatic than on the Liffey but is still significant, and because of the flashy nature of the river, floods occur here much more frequently. Inniscarra Dam attenuates flooding due to most short duration flood events on the Lee and it is possible to prevent the peak discharge reaching the peak inflow in all but the most severe events. Figure 4 above shows that the peak discharge at Inniscarra was approximately 200 $m^3$/sec less than peak inflow during a flood in December 2000. Yet this flood resulted in damage to property and considerable disruption to traffic downstream of Inniscarra dam despite the reduced discharge made possible by the dam. Careful management of flood situations helps to reduce the frequency and magnitude of flooding downstream of Inniscarra, but because of the limited storage available, the flashy nature of the river and the downstream topography, flooding can never be eliminated.

Control of flood waters on the River Clady is self regulating, but, because of the storage provided by the Scheme, all floods are attenuated as a result of the development.

In Ireland, in common with similar hydro developments world wide, downstream flooding has reduced in frequency and magnitude where ever dams have been built. This has resulted in human habitations encroaching on the flood plains downstream of the dams, where previously such habitation did not exist, due to the danger of flooding. The dams have reduced the risk of flooding but the risk has not been eliminated and occasional severe events result in flooding to these new settlements. Because of the control provided by the dams it is generally possible to provide warnings for people at risk of such events.

## RECREATIONAL BENEFITS

In the early years after the completion of the Shannon Scheme, the Irish Electricity System was very dependent on the Scheme to meet the Electrical Daily Load Demand. To meet this load full use had to be made of the considerable storage available in the three natural lakes on the Shannon, Lough Derg, Lough Ree, and Lough Allen. This generally meant that these lakes were full at the beginning of the summer with lake levels being low in late summer and autumn. In those years, power generation took precedence over all other activities on the lakes, including navigation. The Electricity System grew over the years and the importance of the Shannon to the larger system reduced gradually. Commercial navigation on the Shannon declined slowly in the first half of the twentieth century but from the mid 1950s navigation for leisure purposes began to grow, particularly during the summer months. These factors resulted in pressure from navigation and tourism interests to maintain the water levels on the Shannon Lakes high during the summer months. This has resulted in changes in how the Shannon is now operated for electricity generation. Little use is now made of the storage available along the Shannon for power generation. The Scheme is now used more as a ‘run of river type scheme’ as far as power generation is concerned. The overall power output from the Scheme has not been greatly affected. On the other hand the leisure tourism business on the Shannon is now a major part of the economy for all the towns along the Shannon and indeed, in recent years, into Northern Ireland, since the

refurbishment of the Shannon Erne Waterway in the 1990s, which connects the Shannon and Erne rivers.
The reservoirs on the Liffey, and the Lee, being in close proximity to the cities of Dublin and Cork respectively, are extensively used for water based leisure activities e.g. skiing, fishing, rowing, lake walks. These reservoirs also add to the visual amenity of the countryside.

The various uses of the reservoirs, such as power generation, water supply and the various leisure activities can easily coexist, once they are carefully managed. Thousands of people benefit from the amenities of these reservoirs annually. This in turn provides an economic spin off to local communities, in addition to the benefits derived from power, water supply and flood mitigation.

## FUTURE ROLE OF THE DAMS

All the dams were designed and constructed in accordance with the best practice at the time. However, there have been important developments in safety related aspects of dam and reservoir engineering since the dams were built. Major upgrading and refurbishment of the dams and associated structures and equipment was carried out over the last fifteen years, to bring them in line with best international practice ( O'Tuama & O'Mahoney, 1989). Ireland, unlike many other European countries, has no legislation covering the dams or their safety. However, a programme of inspection, monitoring and reporting has been put in place similar to those required, by legislation, in other countries. (O'Keeffe 1997).

All the installations at the above Schemes are now considered to meet safety standards in line with best practice. Provided the installations continue to be operated and maintained to a high standard, the Schemes can be expected to continue contributing to society in a variety of ways, as heretofore, for many years into the future.

## CONCLUSIONS

Hydropower now accounts for only a small portion of Ireland's electricity needs. However, the five Hydro-Electric Schemes discussed in this paper, have provided a huge contribution to the development of Ireland over the years, during their construction and operation. They continue to provide benefits to the economy and society, not only by power production and water supply but also in ways that were not fully appreciated at the time of their development.

There was relatively little opposition to the development of the Schemes and any opposition there was came from individuals directly affected by the Schemes. If these Schemes were being developed today it would be much more difficult to get the agreement of the various stakeholders for the projects. This can probably be attributed to the economic and employment realities in Ireland up to the 1970s as well as the propensity of modern society to oppose any developments that might have an effect on local or global environments. On the other hand any fair assessment of the Schemes would almost certainly come to the conclusion that the benefits far outweigh the negative impacts.

ACKNOWLEDGEMENTS
The author would like to thank ESB for permission to publish this paper. The assistance of ESB colleagues in the preparation of the paper is gratefully acknowledged.

REFERENCES

Annon. The SHANNON Hydro-Electric Scheme. Limerick. T.C. Carroll & Sons.

Duffy P (1987/1990). ARDNACRUSHA-Birthplace of the ESB. Dublin. Electricity Supply Board

Fitzpatrick J & Bree T (2001). Flood Risk Management through Reservoir Storage and Flow Control. IHP/ICID National Hydrological Conference. Tullamore.

Government Publication (1930). Lough Erne Drainage Report – upon Remedial Measures to Alleviate Flooding. Dublin. Government Publications Office.

Kenny L (1991). 50 Years on the Liffey, A History of Hydro-Electric Power on the River Liffey. Dublin. Electricity Supply Board

Lysaght W (1964). The Abbey Fishermen, A short History. Limerick. The Treaty Press Limited

Manning M & McDowell M (1985). Electricity Supply in Ireland, The History of the ESB. Dublin. Gill and Macmillan

O'Donoghue S (1996). The Flooding of the Lee Valley, The Lee Hydro-Electric Scheme. Cork. Tower Books

O'Keeffe J D (1997). Dam safety and risk analysis – Experience of ESB Ireland. 3rd International Conference on Hydropower. Trondheim, pp493-499.

O'Tuama F F & O'Mahony B (1989). Hydro Electric Dam Safety - Upgrading of ESB Dams. The Institution of Engineers of Ireland

# Design, construction and performance of Fullerton Pollan dam and reservoir, Co Donegal, Ireland

R C BRIDLE, Rodney Bridle Ltd, UK
J HOLOHAN, Donegal County Council, Ireland
D GILLESPIE, Nicholas O'Dwyer, Ireland
D A SMITH, MWH PM, UK
S FAWCETT, MWH PM, UK
S MCINERNEY, formerly Donegal County Council, Ireland
I C CARTER, MWH, UK
R EVANS, SMEC, Australia

SYNOPSIS. Fullerton Pollan Reservoir was completed in 1997 with a capacity of 4.6 $Mm^3$ and a yield of 33 Ml/d to municipal supply and compensation. It has a central conventional concrete gravity section flanked by earth core- rockfill embankments. The dam is 22.5 m high above lowest foundation. The concrete dam incorporates the spillway, fish lifts, draw-off and compensation release pipework. Thick deposits of till and peat deep blanket the site. The till was used for core material while rockfill for the shoulders was excavated from an on-site quarry. The rock foundation was grouted using the GIN method. The concrete mix designs for the dam made use of cement replacement to minimise thermal cracking. The performance of the dam over its first five years has followed expectations.

## PROJECT HISTORY

During the early 1980s Donegal CC identified a shortfall in water resources serving the Buncrana area. Feasibility studies identified the Crana River Basin as a potential source and a suitable site on the River Owennasop. Nicholas O'Dwyer & Partners were appointed as engineers with MWH as their specialist dam engineers. The detailed design of Fullerton Pollan Reservoir was prepared during the early 1990s with reservoir safety aspects being handled in the manner described in the UK's Reservoirs Act, 1975. Tenders were submitted in 1994 and a contractor, Ascon Ltd, of Co. Kildare was appointed in early 1995.

## HYDROLOGY AND CATCHMENT

The 14 $km^2$ catchment of Fullerton Pollan reservoir is a remote upland area, which is almost unpopulated. Land use is limited to peat workings and sheep grazing. The long term demand (year 2020) was estimated to be 20.5Mld. In addition to water supply requirements, river compensation flows were also necessary. The compensation flows at Fullerton Pollan were designed to maintain suitable conditions for salmon.

*Reservoirs in a Changing World*, Thomas Telford, London, 2002, 31–48

The compensation flow regime comprises:

- *base flow* (minimum of 0.025m$^3$/s) that is to be released at all times;
- *freshets*, which mimick natural floods to supplement river flows and encourage upstream migration of adult salmon to spawning grounds. These releases are made twice monthly, from June to December, with a daily hydrograph peaking at 5m$^3$/s;
- *flushing flows* required to replicate major flood events and flush fine sediments out of spawning gravels. These are to be timed to coincide with natural storm events and single releases will be made in October or November over a twenty four hour period with a hydrograph peaking at 8m$^3$/s;
- *spawning flows* to maintain adequate flow depth to support salmon spawning, generally during the period November to December. These flows to be achieved by supplementing the natural river flow to obtain a minimum flow value in the Crana River by a constant release from the dam, assessed on a daily basis, to a maximum of 0.75m$^3$/s;
- *smolt flows*, increased flow rates during April and May to attract and transport smolts migrating from upstream of the dam to the sea, at a constant rate of 0.3m$^3$/s during night-time hours.

The average demand for the various fish flows is 13Ml/d making the average yield of the project 33.5Ml/d. The required reservoir volume was assessed to be 4Mm$^3$ based on the target yield of 34Ml/d and 98% reliability (1 in 50 years). An extra 5% was added for uncertainty and 0.3Mm$^3$ for sedimentation and dead storage, giving a total reservoir volume of 4.6Mm$^3$. The corresponding Full Supply Level (FSL) of the reservoir was 143.8m AD. The surface area of the reservoir at FSL is 117 hectares and the height of the dam above river-bed level was about 13m.

## FLOOD ANALYSIS AND SPILLWAY FLOW

A flood peak inflow of 155m$^3$/s was derived as the Probable Maximum Flood (PMF) (Summer Event). The dam was placed in Category A, as a breach would put lives in a community at risk. The spillway was designed to pass the PMF outflow, which was determined as 146m$^3$/s. The flood lift at peak raise the reservoir level by 1.2m to 145m AD. The wave surcharge was estimated to be 1.0m.

## ENVIRONMENTAL ASSESSMENT AND MITIGATION

The project provides considerable benefits, notably industrial water supplies to expand the local economy, but there were several environmental concerns, particularly about the salmon fishery. An Environmental Impact Statement was prepared as required under Irish legislation (S I No. 25, 1990), the first to be carried out in Ireland following the EC Directive (S I No. 349, 1989), as described by Smith (1996a, 1996b).

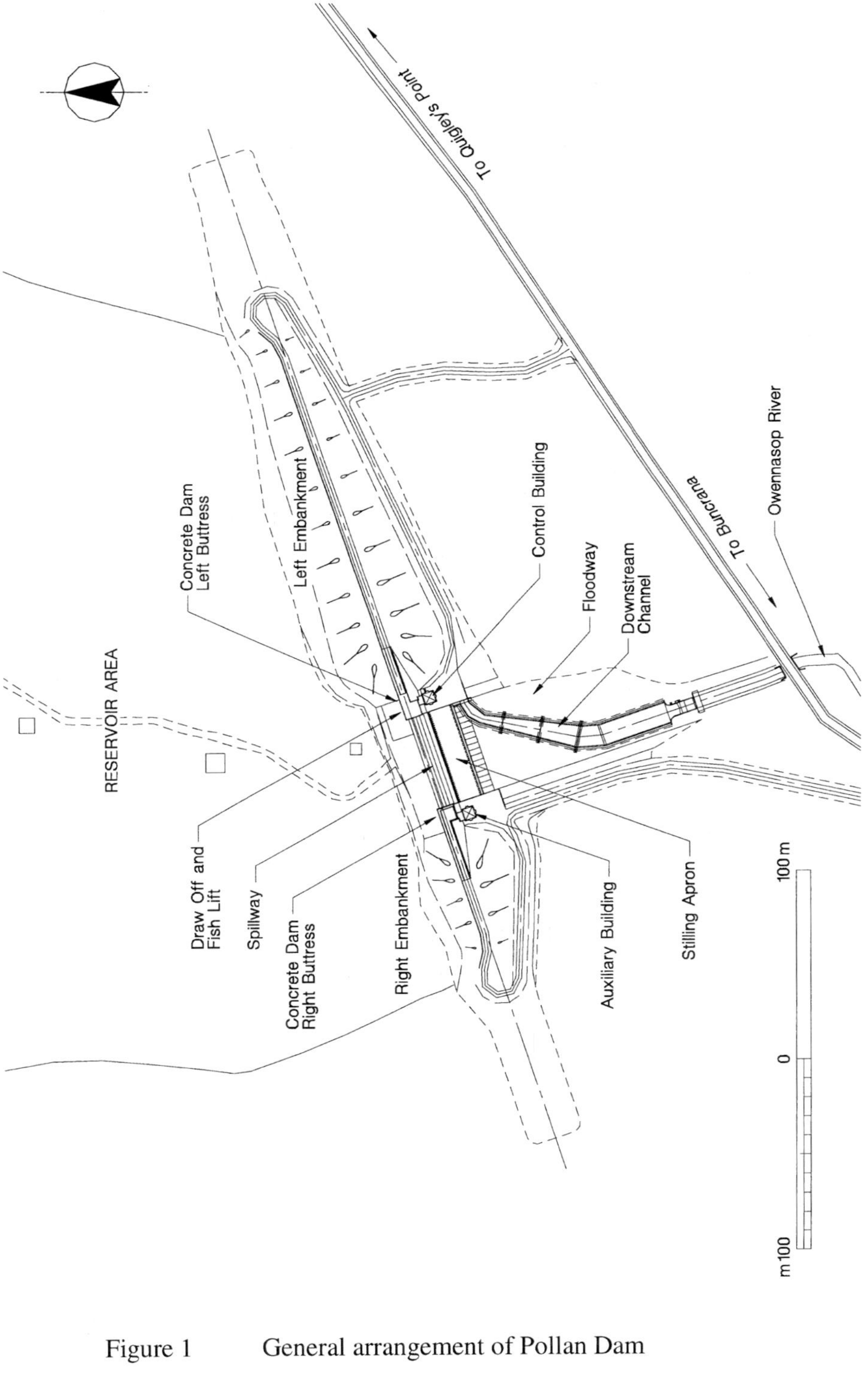


Figure 1 General arrangement of Pollan Dam

Consultations and preliminary field investigations were used to complete a scoping exercise defining the key issues and concerns to be addressed. A programme of fieldwork then collected the data needed for evaluation of the probable impact of dam construction and future operations. Fieldwork was undertaken in botany, ornithology, aquatic flora and fauna, water quality, sedimentology, fisheries, traffic, noise and archaeology. One of the main findings was that salmonids were present upstream of the dam site, which identified the need for the dam to incorporate fish passes.

The positive and negative effects of the scheme were derived and the following main factors identified as requiring mitigation:

- Interruption to natural river flows and effect on salmon life cycle
- Effects on river water quality from reservoir operation
- The dam as a barrier to spawning grounds and nurseries upstream
- Loss of area of blanket peat bog and river in the reservoir basin
- Traffic, noise, dust and pollution during construction.

Mitigation measures were incorporated into the design as it proceeded, and the selection of the dam type was influenced by the environmental considerations. The Buncrana Anglers Association was the major interest group affected by the project.

The fishery mitigation measures included:

- Compensation releases to maintain the salmonid environment
- Creation of a brown trout fishery in the reservoir to mitigate the loss of the reach inundated by the reservoir
- Incorporation of a fish passage device within the dam to mitigate against the barrier to fish movements
- Water quality (pollution control) restrictions during construction

During construction the contractor's anti-pollution measures were supervised and monitored by an Environmental Scientist working as a member of the Resident Engineer's staff. Watchers from the Buncrana Angling Club had free access to the site to satisfy themselves that the fishery was not being polluted. Following completion of the reservoir, it has been fenced and operated as a private fishery by the Buncrana Angling Club. No public access is allowed.

## GEOLOGY

Glacial tills and periglacial deposits cover the low-lying areas of the Innishowen Peninsula. The area was subject to two glaciations; during the latter phase the ground was subject to periglacial processes that resulted in solifluction deposits and related features. Post-glacial deposition includes alluvial spreads along rivers and their estuaries and the development of widespread peat deposits often as raised blanket bogs on the flatter ground.

The bedrock of the Innishowen Peninsula comprises a complex series of Dalradian meta-sediments dipping generally south-east. These are schists, slates, grits and sandstones with subordinate limestones, phyllites and intruded epidiorites.

The area has been subject to only a relatively low level of seismic activity throughout history. The seismic risk at Fullerton Pollan was assessed using the UK guidelines (Charles et al, 1991) and the peak ground acceleration (PGA) of Safety Evaluation Earthquake was deemed to be 0.10*g*.

The bedrock at the dam site was the Fahan Grit Formation, which consists mainly of medium grained, moderately strong, sandstone. The bedrock dips steeply to the south-east and has moderately developed foliation that has a similar dip and strike. The joints are mainly very closely spaced, tight and rough but occasionally were slightly open and oxidised. Some thin bands of moderately weak to moderately strong graphitic shale were also present, although these were impersistent and normally only present as lenses or irregular inclusions. Clay smears were observed close to rockhead where the rock was moderately–highly weathered.

The bedrock is entirely concealed beneath glacial till and peat. The glacial till is medium dense, blue-grey slightly clayey sandy silt with some gravel, cobbles and occasional boulders. The peat was widespread and covered most of the reservoir basin. It was up to about 4m deep at the dam site and up to 8m thick in some parts of the reservoir basin. It was very soft, mostly amorphous, locally fibrous, plastic material, highly humified (Von Post classification H7 – H8) and almost totally organic in composition.

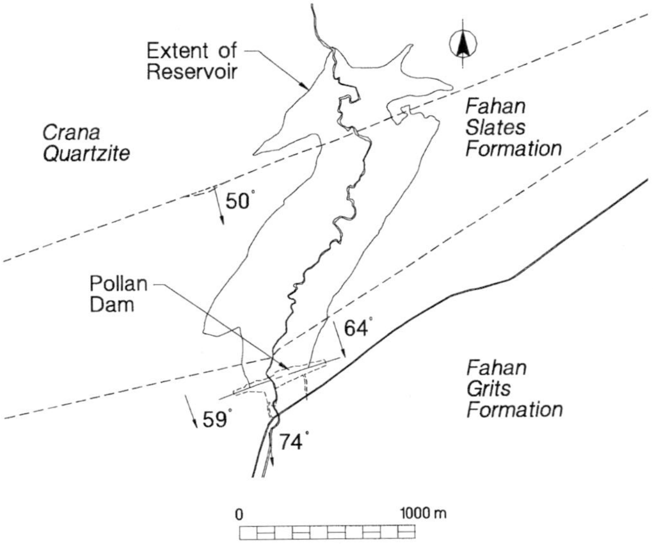


Figure 2 Geology of Fullerton Pollan Dam

## DAM FOUNDATIONS

The concrete dam was founded on bedrock. The shoulders of the embankment dams were founded on glacial till deposits while the core was taken down in a cut-off through to the underlying bedrock.

The till was dominated by a matrix of fines in which the coarse fragments appears as discrete particles. The matrix at Fullerton Pollan was predominantly non-cohesive (i.e. "rock flour"). The coarse fraction was sand, gravel and boulders. The till was silt of low plasticity with a low index usually in the range of 1% to 6% but non-plastic in some areas. The till was relatively homogenous with a permeability averaging 7.3 x $10^{-7}$ m/s.

The angle of shearing resistance ($ø'$) was estimated to be about 35° allowing embankment type dams with slopes up to 1 on 2 to be considered. Small settlements could be expected, as the coefficient of compressibility ($m_v$) in the dense till was less than 0.05$m^2$/MN. The average coefficient of consolidation ($c_v$), when measured in the laboratory on re-moulded samples, was relatively high at 22$m^2$/year. The glacial till was suitable for embankment dam fill and the estimated permeability of re-compacted glacial till was in the order of 3 x $10^{-8}$ m/s.

The bedrock is generally moderately strong to very strong with unconfined compressive strength in the range of 8 to 151MN/$m^2$, averaging 39MN/$m^2$. The elastic moduli show rock of good quality. Young's Modulus was between 10 and 80 GPa, averaging 36 GPa, and Poisson's Ratio between 0.1 and 0.2. These properties showed the rock to be suitable for the foundation of a concrete dam. The bedrock would also be suitable as the embankment foundation and as a source of rip-rap, filters and concrete aggregate.

The bedrock and the local groundwater regime were such that the reservoir would be watertight. Lugeon tests carried out in the bedrock showed Lugeon values between 1Lu and 39Lu, indicating low to moderate permeability, approximately 1 x $10^{-7}$ m/s to 4 x $10^{-6}$ m/s. Grouting was carried out to reduce the bulk permeability of the bedrock, check for local high permeability zones, and limit uplift beneath the dam by limiting the amount of seepage reaching the relief drains.

## DAM TYPES

A study was undertaken to determine the most suitable and economic dam to construct, as the dam site was suitable for both embankments and concrete dams. Embankment dams in earthfill, rockfill with glacial till cores, and concrete faced rockfill dams were considered. Conventional concrete and roller compacted concrete gravity dams were also investigated, as were composite dams comprising concrete and embankment sections.

Ultimately a tripartite dam was selected, comprising rockfill flank embankments with a conventional concrete gravity dam incorporating the spillway and draw-off works. The main criterion leading to this selection was that the concrete dam provided a convenient location for the spillway, the draw-off pipework, fish lifts and associated works. A further advantage of this option was that the concrete dam provided a convenient location for river diversions during construction, which were therefore less complex than they would have been with an embankment dam.

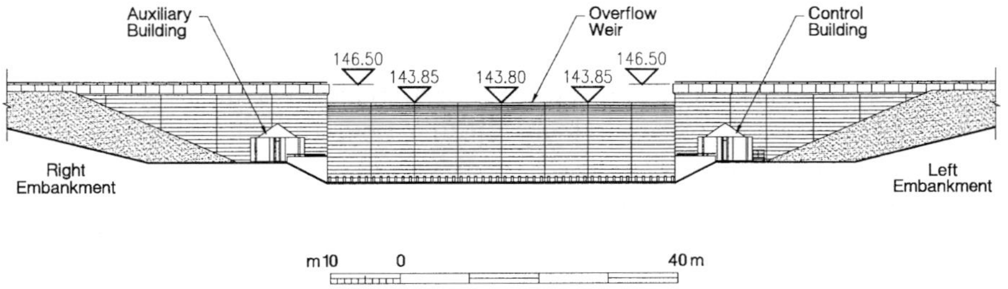


Figure 3 Elevation of Fullerton Pollan Dam

The size and complexity of the concrete dam precluded roller compaction, but the selected option was buildable and economic and despite the apparent complication of having two different types of dam, was readily shown to be the most favourable option. The dam would have a wavewall level of 146m AD, a top of core level of 145m AD, a crest length of 400m, a maximum dam height from crest to lowest foundation of 22.5 m and a maximum dam height from overflow level to river bed of 13 m. The estimated cost of the recommended dam (as at February 1992) was £6.1 million.

## DAM FOUNDATION TREATMENT AND GROUTING

The peat and till overburden and the upper weathered bedrock was removed over the width of the cut-off trench beneath the embankment dams, and over the entire foundation of the concrete dam. A 4 m wide, 500 mm thick grout cap was cast beneath the embankment dams. The grout cap continued around the upstream face of the concrete dam and a water bar formed a watertight connection between the concrete dam and grout.

The grout curtain comprised three rows of holes: two outer rows of vertical holes and a deeper central row of raked holes to intercept sub-vertical discontinuities in the rock. The outer rows were 6 m deep and the central grout curtain, which varied in depth depending on the depth of reservoir, reached a maximum depth of about 18m. The outer rows were grouted first followed by the central row. The primary holes were at 8 m centres and grouting was carried out in 3 m to 6 m deep stages, using the split spacing method. In many locations quaternary holes were needed.

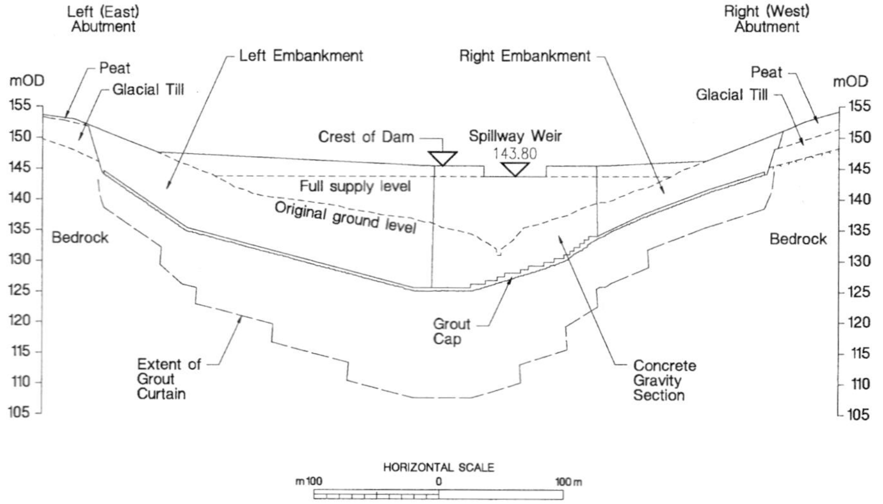


Figure 4 Foundation excavation and grouting at Fullerton Pollan Dam

Cement grout was used, injected following the Grouting Intensity Number (GIN) method developed by Lombardi (1993). The Grouting Intensity Number is the product of the grouting pressure and the grouting volume. The objective of the method is to control the grouting process by limiting either the pressure applied to the ground (in tight ground) or the rate at which grout is injected (in open ground) and by applying suitable pressures and injection rates in the ground between the extremes.

At Fullerton Pollan a Grouting Intensity Number limit of 1,000 bar.l/min was set. For holes less than 6 m deep, the Limiting Pressure was 10 bars and the Limiting Volume 150 l/m. For holes deeper than 6 m, the Limiting Pressure was 25 bars and the Limiting Volume 200 l/min. Initial water:cement ratios of 1:1 were used, which were thickened to 0.8:1 was permitted if the hole took grout freely. Absolute refusal was defined as acceptance of 3 l/min.

Much of the left bank grouting was carried out following the initially specified GIN and Limiting Pressures and Volumes listed above. The method was effective and reduced Lugeon values from a pre-grouting value of about 100 Lu to 1 Lu after grouting. However, there had been escapes of grout to the surface, and the high pressures were thought to be driving grout further upstream and downstream from the curtain than was necessary.

The grouting regime was therefore reviewed and modified. The GIN was reduced to 500 bar.l/min and the Limiting Pressures were reduced to 4 bars for holes up to 6 m deep, and to 10 bars for holes over 6 m deep. The Limiting Volume was maintained at 150 litres/min for holes to 6 m deep and reduced to 150 litres/min for holes over 6 m deep. These changes had little effect on the effectiveness of grouting and Lugeon values continued to reduce to the target values of 5 Lu or less, but did reduce the quantity of grout injected and the time taken to complete the grouting work.

Computers controlled the grouting operations and a full record of depths and spacings of holes and grout takes was kept. The 5,500 m of inclined grout holes in the central curtain took 143 t of grout at an average take of 26 kg/m. In the 2,700 m of outer holes, the average take was 46 kg/m.

## CONCRETE DAM

The crest carries a 50m wide ogee-crested overflow spillway with a stepped downstream profile. The remainder of the crest carries a cantilevered roadway. The draw-off pipework, the fish lifts and control building are incorporated to the left side of the spillway section. The building contains the valves and control equipment associated with water supply draw-off, fish compensation flows, fish lift controls and emergency emptying pipework. An auxiliary building is present on the opposite side from which there is access to the gallery, which is ventilated to prevent accumulation of carbon dioxide and other noxious gases.

The 130m long concrete dam was founded on slightly weathered, moderately strong rock or better. There are 75mm relief wells extending into the foundation. Discharge from the wells is measured and they ultimately drain through to the foot of the spillway steps. There are also upward drains from the gallery to assist in limiting uplift pressures on the joints in the blocks.

The concrete gravity dam was constructed in 18 blocks (monoliths) varying between 6.25 m and 13.2 m wide. The upstream face is vertical and the downstream face slopes at 0.75 (horizontal) to 1 (vertical).

The concrete mixes were designed for durability and to minimise the possibilities of thermal cracking. The principal means of achieving these objectives was to limit temperatures by varying lift depth, cement content, percentage of cement replacement, pour timing and aggregate type. The cement replacement was ground granulated blast furnace slag (GGBFS) from Castle Cement Works near Glasgow.

Trial mixes were carried out to design mixes to meet the following requirements for the concrete in the dam:

- Maximum peak hydration temperature in any pour to be less than 23°C.
- Maximum differential temperature between two parts of the same or subsequent pours was not to exceed 20°C.

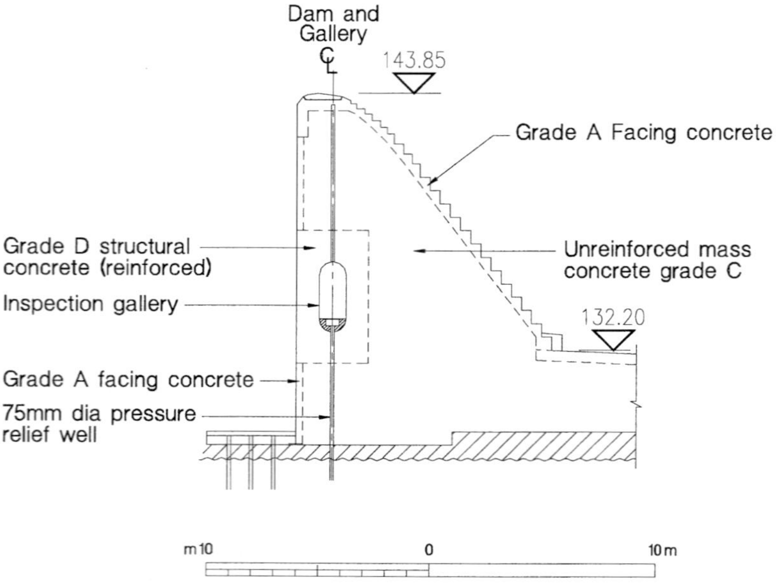


(a) Overflow Section

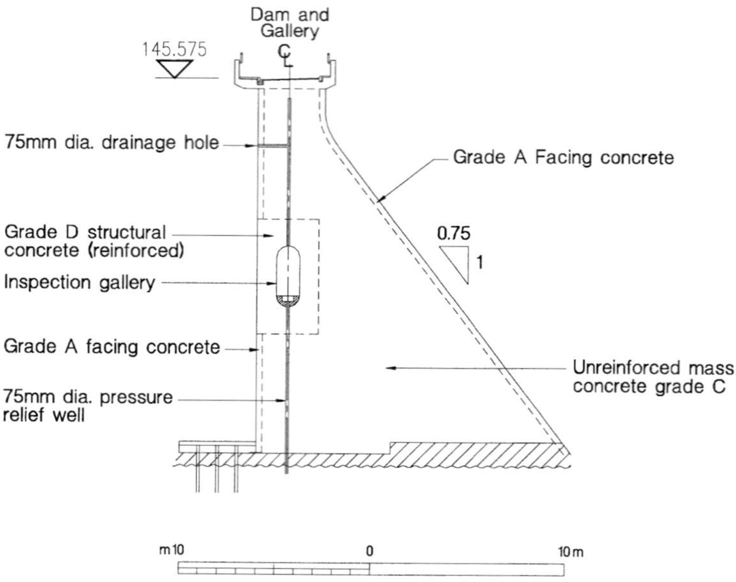


(b) Non-overflow section

Figure 5 Concrete gravity dam sections

- Maximum differential temperature between concrete surfaces and the daily ambient temperature on average prior to stripping shutters or removal of insulation was not to exceed 7°C, subject to not removing insulation or shutters on a falling thermometer.
- Subsequent lifts within the same block were to be placed within two days after achieving the peak hydration temperature in the present top lift. Lift depth was normally 1200mm, but in the event of non-compliance the next lift was not to exceed 600mm thickness.
- Concrete was not to be placed if the daily ambient temperature exceeded 23°C.

There were several concrete mixes designed for various situations in the dam. Mix A was the air-entrained reinforced and unreinforced 500 mm thick outer facing concrete on all exposed surfaces of the dam. To make the surfaces of exposed concrete denser, controlled permeability membranes were applied to all formwork. Mix B was placed in the foundations of the dam at the bedrock contact. Mix D concrete was reinforced concrete in unexposed locations, e.g. the structural concrete around the gallery.

The bulk of the concrete in the dam was the Mix C hearting concrete. To illustrate the controls applied, some details of how it was varied during the course of construction to meet the requirements as ambient conditions varied are given here. Mix C concrete was specified to contain:

- maximum size of aggregate of 40 mm.
- minimum cement content of 125 $kg/m^3$
- maximum cement content of 250 $kg/m^3$
- maximum free water:cement ratio of 0.8
- fresh concrete density of 2,350 $kg/m^3$

The maximum temperature requirement (i.e. absolute maximum temperature of 23°C) was difficult to achieve particularly as concreting extended into the summer months. During January, February and March 1996, cement replacement proportions were varied between 35% and 60%. After early experimentation when the cement content was varied from 160 $kg/m^3$ to 180$kg/m^3$, the higher replacement proportion was adopted for all subsequent Mix C concrete. Limestone aggregates from Dundaff Quarry were used initially, but after that quarry closed, sandstone aggregates were brought from Cassidy's Quarry in Buncrana.

As ambient temperatures increased during April and May 1996, it became increasingly difficult to restrict the peak temperature of the concrete. Measures such as cooling the mixing water and shading the aggregates were not sufficient to limit temperature rise, particularly as water volumes were low because cement content was low.

Limestone aggregate from the more distant Ballintra Quarry replaced the sandstone aggregate until completion of concrete in November 1996. The limestone has a lower coefficient of thermal expansion than sandstone and was able to sustain higher temperature gradients without cracking. The peak temperature requirement could therefore be safely increased to 32° C, thereby eliminating the difficulties in controlling peak temperatures that had been previously experienced. Some control remained necessary and cement was replaced with GGBFS up to a peak of 80% in August 1996, diminishing gradually to 20% in November.

A fully equipped on-site concrete laboratory was used to monitor concrete quality and record the temperature profile of at least one block per day of concreting. Temperatures were recorded using an array of thermocouples from which data was downloaded at intervals until the blocks were safely past their peak temperatures. Quality was monitored using rapid mix analysis during the hottest periods when the concrete was weakest (due to high cement replacement) because it was found to be more reliable than standard cube test results. It also enabled immediate adjustment to the mix.

The control measures requirements proved successful. Very few blocks show mid-width cracking on the downstream face. However slight “curling”, which gave rise to short, very narrow cracks at the outer edges of the upstream side of a few blocks did occur on occasions but these defects were successfully repaired by grouting.

Pre-cast concrete units were used for the spillway crest blocks and to form the inspection gallery in the dam.

## FISH LIFT

A Borland fish lift was selected to allow fish to pass at all times of the year and during a large range of water level in the reservoir. Two lifts have been provided to accommodate a water level range down to 139m OD from the Full Service Level of 143.8m AD. The lift channels are 1400mm square openings through the concrete. The fish enter along a low flow channel, through an entrance channel, where there is a fish counter and then collect in the downstream chamber. The downstream gate of the downstream chamber is closed, water accumulates in the channel and the fish are lifted and released above the upstream weirs.

## DRAW-OFF AND FISH FLOW PIPEWORK

There are two draw-off pipe systems through the dam, the water supply draw-off system and the combined fish compensation and emergency pipework.

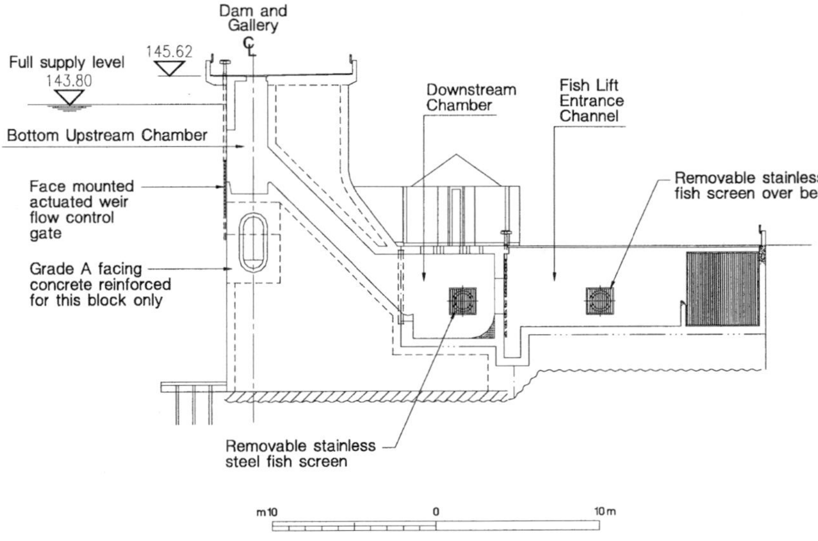


Figure 6 Fish lift arrangements

The water supply draw-off system takes water from two levels through screens into a common vertical draw-off pipe and then through a pipeline to the Illies Treatment Works about 4 km downstream. The draw-off pipework and valves are 600 mm in diameter. Upstream guard sluice valves are provided. Control is exercised mainly at the treatment works, or locally through a butterfly valve in the downstream chamber of the dam. There is also an electro-magnetic flow meter and a bypass to provide compensation flows, if the emergency draw-off pipe or the fish lift fail to do so.

The emergency draw-down pipe is 1200mm in diameter; it is also used for the larger fish compensation flows. Control is through a similar sized electrically actuated butterfly valve and discharges is measured through a 1200 mm diameter ultrasonic flow meter.

The 1000 mm diameter fish lift pipework releases water from the fish lifts and normally provides the compensation water flows. It is controlled by 700 mm diameter butterfly valve and there is a 200 mm diameter bypass to release the normal low compensation flows into the fish channel, where they provide a constant source of flow to attract the fish towards the fish lifts.

Emergency drawdown could be achieved effectively by opening a 1200 mm diameter butterfly valve, which releases up to about 17 $m^3/s$ of water downstream. Care would have to be taken in such situations to balance the risks of rapid emptying of the reservoir and of causing local flooding downstream.

SPILLWAY

A 50 m wide section of the concrete dam crest provides the overflow. The central 6.25m is set at 143.8m above datum and the remainder is 50 mm higher. The crest itself is ogee in profile while the downstream slope has 600 mm steps. A 13m wide apron is present at the foot of the steps with a 500 mm reinforced concrete slab forming the downstream section. At the downstream edge there is a 1 m wide drainage channel which releases water into the compensation channel downstream. However flow would discharge across the full width of the floodway during significant floods and this is protected by 'Reno' mattress and reinforced grass, which protects the surface from the erosive effects of flood flows.

The spillway is designed to accommodate PMF outflow of 146 $m^3/s$. During such events the outlet would be constricted at the Meenaharnish road bridge. A pond would form between the dam and bridge and water would be released under the bridge and over the road into the river channel downstream.

RIVER CHANNEL

A new river channel has been formed through the floodway with gabion side walls and a 'Reno' mattress floor. There are a series of pools retained by concrete weirs. The weirs have a 300mm wide notch with a crest 250mm below the general weir level to provide a simple passage for migrating fish. The channel gradually narrows as it approaches the fish lift entrance; this is to provide gradually increasing water flow velocities in order to draw the fish towards the fish lift.

At the downstream end of the river channel there is the River Owennasop gauging station, a 6m wide broad crested weir flume. There is a similar gauging station on the Camowen River, a left bank tributary of the Crana about a kilometre downstream of Meenaharnish.

EMBANKMENT DAMS

Earth core-rockfill embankment dams link the dam abutments to the central concrete gravity dam. The right flank dam is about 270m long and the left flank dam 140m long. These dams have rockfill shoulders and glacial till cores.

The core extends down into the cut-off excavated through the peat and glacial till into the bedrock below. The shoulders are founded on glacial till. The core is protected by a fine and coarse filter on the downstream side and this double filter extends across the surface of the exposed till under the downstream shoulder. There is a transition material zone on the upstream side of the core separating the core from selected finer Type B rockfill layer and the general coarser Type A rockfill.

Rip-rap wave protection is provided over the upper parts of the upstream slope and on the surface of the downstream slope there is underlayer and grassed topsoil. There is a precast concrete wavewall at the upstream side of the crest and a tarmac crest road. The upstream and downstream slopes are 1 on 2.5. The cut-off trench slopes are 1 on 1.5.

Most of the downstream foundation and part of the fill in the cut-off trench is protected by a fine filter layer with a 500 mm thick coarse filter layer over it to provide additional drainage capacity. This capacity is further enhanced by a finger drain at the base of the fine and coarse filter behind the core. On the upstream side a geotextile was laid as a separator between the glacial till foundation and the transition material below the standard rockfill.

The core fill was glacial till excavated from the cut-off trench and from borrow pits excavated in the reservoir basin. Organic materials and boulders greater than 250mm in size were excluded. The specification required that till placed as core had a fines content (i.e. materials smaller than 75 μm) of at least 15%. The as-placed permeability was less than $10^{-8}$ m/s. Selected glacial till was required at sensitive locations, such as the connection to the concrete dam and at the external edges of the core in the cut-off trench where the core material was laid directly against rock. All stones larger than 75mm had to be removed from the selected core material.

Allowance was made in the specification for placing the till by 'wet' and 'dry' methods. The wet placing method was to be used for till materials with a moisture content greater than about 2% above optimum. The 'wet' method used low ground pressure bulldozers to spread the till in layers not exceeding 500 mm thick and the same plant to compact it by tracking it in. The as-placed density was to be at least 100% of the standard (Proctor) density measured at the placement moisture content. The 'dry' method was to be used when the natural moisture content of the till was drier than optimum moisture plus 2%. The dry till was again spread by low ground pressure bulldozers in layers not exceeding 300mm deep, but compaction was by vibrating rollers to achieve the in-situ requirement outlined above. In the event most of the till was placed using the wet method.

The core fill was dilatant when placed and deformed in the characteristic constant volume 'cow-belly' manner. However, excess pore pressures quickly dissipated, as would be expected with a relatively permeable fill, and within a few days it was firm and stable.

The bulk of the rockfill (Type A) was placed as dug from the site quarry. It had a maximum size of 900 mm and a $D_{50}$ size of 20 mm to 300 mm. The finer Type B rockfill was used adjacent to the core zone as a transition material. It had a maximum size of 250 mm and a $D_{50}$ of 20 mm to 100 mm. Type B was processed on site from Type A rockfill using a bar grizzly. The coarse fraction was used as rip-rap. The maximum size was

1 m and the $D_{50}$ size was 400 mm to 600 mm.. Type A rockfill was placed in 1 m deep layers while Type B was placed in 500 mm thick layers. Vibrating steel drum rollers were used to compact the rockfill.

The quarry was excavated in fine grits and sandstones on the left flank of the reservoir by drilling and blasting. It provided very satisfactory rockfill and rip-rap for the dam. Afterwards spoil was placed in the deeper parts of the quarry and the face was covered with debris and a peaty layer placed to encourage vegetation to grow and disguise the rock scar. Safety fencing has been provided along the quarry edge to keep people away from the face.

Coarse and fine filter materials were imported from quarries off site.

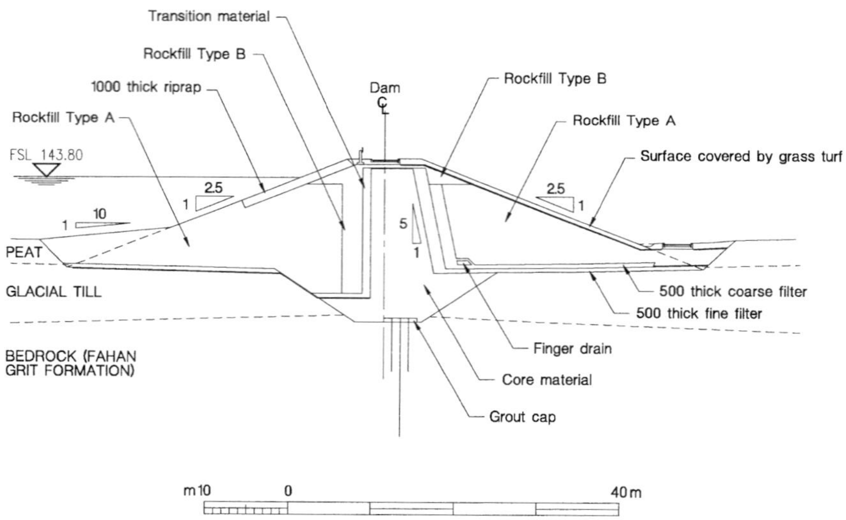


Figure 7 Typical section of embankment dams

RESERVOIR

Peat excavated from the dam foundations and the surface of the quarry and the borrow pits was spread on the left flank of the reservoir upstream of the borrow pit. There were some landslips in this material initially but during the construction period these were spread and stabilised. A rockfill bund was constructed at the toe of the peat spoil slope at Full Supply Water Level to contain the flows of peat spoil and limit wave erosion at its toe that might lead to further instability. Since construction was completed the peat spoil has been stable and now supports a good cover of vegetation.

Reservoir filling rates were controlled as far as possible to allow unsaturated peat to become gradually waterlogged, as there was concern about the possibility of peat floating up from the floor of the reservoir. A ramp was provided at the reservoir side as a landing point to which peat could be towed to and taken away for disposal. In addition, a boom was provided upstream of the dam, as a precautionary measure, to prevent possible clogging. However the boom was damaged during a flood event and was abandoned. Fortunately the prevailing wind is away from the dam and floating debris tends to be transported toward the head of the reservoir.

To aerate the water in the reservoir and combat stratification seven helixors were placed at intervals beside the river channel in the base of the reservoir.

## PERFORMANCE TO DATE

Inclinometers, levelling points, piezometers and seepage measuring devices have been provided at the dam to provide indications of dam performance.

Movement of the concrete dam has been barely measurable. There has been minor settlement of the embankment following first filling of the reservoir. Pore pressures have responded in an entirely predictable way in almost all cases. The upstream piezometers respond to changes in reservoir level promptly, as would be expected in relatively permeable fills. The core piezometers show no excess pore pressures and have gradually responded to new steady states as reservoir water level has changed.

Seepage quantities are low; at Full Supply Level the flow is about 0.6 litres/second from the left embankment, 0.45 litres/second from the right and 0.4 litres/second from the gravity dam relief wells. The flow rates lie within the predicted range. Suspended solids in the seepage flow are monitored and the levels have fallen to less than 0.5 mg/litre.

On one occasion there was an apparent increase in seepage baseflow but this was artificial and only reflected an accumulation of slime in the base of the V-notch measuring weirs. Since that time the weirs have received routine cleaning to guard against re-occurrence of misleading records.

Dissolved oxygen and turbidity levels in the reservoir are measured at the draw-off points. During first filling the dissolved oxygen reduced from 5 mg/l and 8 mg/l at the upper and lower draw-off levels to almost zero at both. Levels remain steady with little variation with season and reservoir level. The turbidity generally lies below 10 ntu, except when the reservoir is partly drawn down, at which time levels rise to 60 ntu.

The fish lifts and the fish flow regime have operated satisfactorily and water is drawn off and passed to the treatment works at Illies, prior to distribution to Buncrana and the surrounding district.

ACKNOWLEDGEMENTS

The authors would like to thank Donegal County Council for giving permission to publish this paper and for the help and encouragement they gave to the team responsible for the engineering and environmental protection measures. Many people played essential roles in the project and we would particularly like to mention the former Jack McInerney (County Surveyor and Engineer), Enda Killoran (Environmental Officer), Pauric McGeehan (Senior Technician) and Thomas Doherty (Clerk of Works), all of Donegal CC; Peter Horswill (MWH Geologist) and Brian Cunningham (Agent for Ascon Ltd).

REFERENCES

Charles J A, C P Abbiss, Gosschalk E M, Hinks J L (1991) An engineering guide to seismic risk to dams in the United Kingdom. Report BR 210, Building Research Establishment, Watford, 64pp.

Institution of Civil Engineers (1978) Floods and reservoir safety: an engineering guide. First edition, London, 58pp.

Lombardi G & D Deere (1993) Grouting design and control using the GIN principle. Water Power and Dam Construction, June, pp 15-22.

Smith D A (1996a) Pollan Dam – Incorporating environmental mitigation into dam design. Dams & Reservoirs, Vol 6, No 1, pp15-20.

Smith D A (1996b) Environmental mitigation at Pollan Dam – Design and construction implications. Institution of Engineers of Ireland, November, 10pp.

# Challenging values of dam builders

C S McCULLOCH, University of Oxford, UK.

SYNOPSIS. The work of the World Commission on Dams (WCD) reflects globalisation of concern over local problems arising from the construction of dams. Powerful support for local opposition has been gained by use of information technology to appeal globally to the values of people far removed from direct interest in the dam and reservoir sites. Increasing technological competence of dam builders has been accompanied by growing conflicts with people holding alternative views and beliefs. This war of values needs to be given more attention. Understanding of the values of dam builders is an important part of clarifying the arguments to help in future decision-making. Interviews with senior dam practitioners have been analysed in an attempt to discover the nature of the arguments used to justify dam construction and the philosophical bases of a technological culture.

## INTRODUCTION.

The WCD Report (2000) illustrates problems with dam schemes by selected case studies and recommends detailed processes of discussion with stakeholders. By focusing on results, the report avoided the main issues of the different values held by dam builders and opponents. The report has became exposed to criticism for selection of supposedly atypical dams as illustrations, for recommending talkshops without prospect of resolution and lack of attention to the benefits of dams in solving some urgent needs of humanity. Without further explicit examination of values, the extensive procedures are unlikely to lead to consensus in the resolution of conflict over dam construction. Such value examination is also needed to interpret the implications of recent, idealistic reviews of dams and reservoirs which produce long lists of the features required for sustainable reservoirs but without any costing or prioritisation (McCully 1996, Takeuchi et al 1998).

The aim of this paper is to make a start on filling some of the gaps left by the WCD report by interviewing experienced dam builders about their values and attitudes. A selection has been made of views critical to the ongoing debate. Direct quotations from recorded interviews with minor editorial changes only, or unpublished correspondence, are given in italics; other quotations from the literature are in normal text.

*Reservoirs in a Changing World*, Thomas Telford, London, 2002, 49–60

From the dam engineers the current controversy provokes four reactions: **disappointment:** *"Some of us entered into the dams field because we thought that it was a useful thing to be doing and benefiting mankind and now of course we are told by the detractors of dams that it is doing more harm than good."*
**aggression** dismissing the arguments of opponents as emotional, illogical or ill-informed, "*What I don't respect are the so-called ecologists because they pose as knowing everything and they know nothing." "I think the environmental anti-dams movement has got a lot to answer for in the sense that what sort of quality of life do they expect the ever-increasing population to have?"*
**engagement:** *"I guess I am more interested in the environmental component now these days than I am in the engineering. This is a huge dam but it is a bigger volume of the same thing whereas environmental issues are really challenging; every one is unique"*, or
**dissociation** from responsibility by attributing accountability to the clients, *" I really believe that the profession should not be held responsible for a government decision." "The engineer comes in when somebody says lets now have some new water works or treatment works, design it and we shall have it built. Then for anyone to say, this is wasteful, the engineer should not have done that, then you are choosing the wrong person." "Engineers are hired to work for the clients so your mission is to serve the client so when the client says to build a dam, normally you do not question it and also you want to make a profit". "I am providing a need that the society deems is there." "You have to deliver to the client what he wants not what you think that he should have.*"

These viewpoints illustrate a wide variation in attitudes. Without assuming uniformity, a selection of some value statements made by dam builders have been selected for further discussion. Alternative views, challenging those stated by the dam builders, have been given in each case to illustrate gaps in perceptions and beliefs, which need further discussion.

## METHODOLOGY AND PREVIOUS LITERATURE.

This research is largely based on transcriptions of fifteen, in-depth, semi-structured interviews with senior dam professionals, the majority trained as civil engineers but including one geologist. The dam dispute is not a simple one of engineers against ecologists/environmentalists/human rights activists. One of the main protest networks, the International Rivers Network, is led by engineers. Three of the engineers interviewed claimed to be environmentalists also. Human rights activists are drawn from a wide background. Yet, a distinctive technological view may be discerned.

Acknowledgement to those interviewed is given at the end of the article. Analysis is also based on further informal discussions, correspondence and reference to a growing literature on the philosophy of science and technology. Selection of declared values has been made to give priority to those, which are being actively challenged by opponents to dams. Six propositions have been chosen for discussion.

Key publications, which give bibliographies to relevant literatures, include Florman (1976,1994) who looks at the life experience of engineers and disputes with antitechnologists in general. Layton (1986) discusses the American civil engineering professional societies, the ideology of engineers and conflict between professional and commercial interests. Latour (1993) contributes critical science studies of the development of ideas from the premodern to modernity and the post modern. Hildyard & Goldsmith (1984), Pearce (1992), McCully (1996), Roy (1999) amongst others make the antidam protest case. Poff et al(1997) examines the ecological cost of river engineering. Schmuck-Widman (2001) takes anthropological view of the knowledge world of river engineers. Takeuchi et al(1998) is a collection of essays, which discuss sustainable reservoir development. The report of WCD (2000) also contains an extensive bibliography.

PROPOSITIONS

1. Water going to waste. The reservoir "*involves storing water, which is otherwise going to waste.*" "*At the borders of Iran and Iraq so much water ran to waste. We are collecting about 7000 million cubic metres of water. The water used to be wasted in border marshes and lagoons. Not only was there waste of water, there was waste of land which could be used for agriculture*". Viewing water purely as a quantity for the instrumental use of humans lies at the heart of antagonism between those concerned with the environment and the engineer. The concept of water going to waste implies little priority for, or even acknowledgement of, the services to both the human and natural environments downstream from a dam. The benefits of floods may be ignored; also the effects of changing water quality. Wetlands may be portrayed as wastelands yet these same wetland environments support the richest biodiversity with the highest turnover of nutrients of any habitat and may be home to people who have adapted to these special environments. Fish, birds both migratory and permanent residents, and a vast number of other life forms inhabit wetlands, which also provide links in the food chain for many marine species. To deny this life without regret for its loss is to diminish our own identity as part of a living interdependent ecosystem as well as the direct loss of many benefits provided by more natural regimes. Postel (1992) regrets that " in the quest for better living standards and economic gain, modern society has come to view water only as a resource that is there for the taking, rather then a living system that drives the working of a natural world." Pearce (1992) has even harsher views, "Modern engineers, in their hubris, have become the antirationalists, the mystics and the ideologues. Their view of a river as a piece of plumbing does not fit reality. It makes them as brutal towards the environment as any Stalinist hack towards the inhabitants of a people's republic. No wonder

these two groups, both perhaps destined to be dinosaurs from the fading age of modernism, have often got on so well together. They talk the same language of rationality and order, while creating disorder and tyranny."

2. Greatest good of the greatest number. Most of the interviewees claimed that they entered their profession to work "*for humanity*" whilst deferring to the majority or the state for legitimization of their work. " *They need the schemes for the greater good of the country*".

Reference to this utilitarian philosophy, expounded by Hutcheson (1738) and Bentham (1776), has so often been used to justify oppression by according power to the majority and neglecting the human rights of the minority. Bentham, himself, was opposed to the rights of man, which he characterised as "nonsense" (Russell, 1946). The alignment of dam builders with the power of the state in the execution of dam schemes has added to their power over people adversely affected by dam building. The equation of numbers giving might and right to a majority impedes equity and justice issues getting due consideration. One interviewee thought, however, that this philosophy gave comfort to displaced Chinese, "*They seem to have a nationalistic pride which gives them a little bit of comfort, like we are doing it for the good of the nation, like going to war.*"

Without recognition of the rights of minority groups for betterment from dam schemes, which interfere with, and possibly destroy, their traditional way of life, conflicts will continue. Some dam builders may even justify displacing people from the dam sites by portraying them almost as sub-human. "*The local population where our hydro projects are constructed are mostly tribals and socially backward communities who have been suffering from infections, diseases, ailments and other maladies arising out of unhygienic living conditions, malnutrition, contamination and animal and rodent bites.*" (Prasad, 2000). Such people should be grateful to be forced to join the mainstream and leave behind their "*humdrum and monotonous life*"! The appeal of anti-dam protest internationally is often founded on empathy with minority groups giving them the power to challenge large companies. Respect for traditional ways of life for their adaptation to their environment, their culture and society justifies the extensive consultation recommended by the WCD.

One dam builder thought there was no problem, not mentioning episodes of bloodshed over evacuation from a dam site in his country: " *the Government paid the price and the people were quite happy because before the land was not very good and did not have much price. But now they can get big prices, money, capital and they can invest elsewhere. So they are happy.*" Another claimed of displaced people "*They are more than looked after well. They play tricks on the companies.*" Most of the dam builders interviewed expressed some sympathy with the displaced, although they were prepared to follow the Government decisions, "*The people were living in the reservoir, which was designed, and that dam now is under construction. The reservoir was all the rice fields and this production; this land was belonging to them generation by generation. They were losing everything. This is very difficult psychologically and socially to remove these people...we have to follow because we are also a member of this society... seeing the future view of the Government for this area and maybe for 99% of the population, then you cool down and say OK. You have to think for the next generation.*" "*I am not trying to pass any value judgements, I am simply saying that if the Government believe that is what they need for the regeneration of the economy and they do this in terms of compensating everybody uniformly and far more generously than the value of the farms or the crofts that they would have to leave, then in the next generation everything is fine.*"

Unlike the anti-dam protests which portray straightforward David and Goliath struggles, the complexity of the disputes was emphasised with examples of delays causing previously-agreed compensation levels being diminished by inflation; companies being exploited by entrepreneurs buying land on the reservoir site just to get compensation; difficulties presented by lack of legal tenure and disagreements over whether compensation should be given to the landlord or to the tenant or to both. Resentment by local people of compensation given to refugees from across a national border was also mentioned.

Whilst the value of tradeoffs went unchallenged, "*Sure, it (the dam) changed what there was there previously and there are trade offs to everything*", it is to be noted that the WCD report assiduously avoids such a concept, which too easily appears to reduce costs and benefits to parity. Social injustice remains a crime even if the number of gainers exceeds the losers.

Despite a conviction that the majority view should carry the day otherwise "*nothing would get done*", several of the dam builders interviewed realised the value and importance of consultation and consideration for the sensitivities of other cultures and spiritualities. For example, two Australian dam builders emphasised their investment in aboriginal cultural history and extensive consultation with aborigines, whose land and sacred sites were affected.

Such progressive opening out of discussion predates the WCD recommendations. Priscoli (1983) advocates that water resource planners should not neglect public consultation, even if the state appears to be acting rationally, "The formula for survival is not power it is symbiosis". However, Layton (1971, 1986) found in his social history of the American engineering profession, values which might obstruct open discussion, "The most fundamental difference between engineers and progressive reformers was their attitude toward democracy. Engineers did not share the progressive's faith in democracy. The driving force moulding engineers into a cohesive unit and pushing them into politics was an essentially elitist concept, professionalism. Professionalism involved concepts of social hierarchy and the inequality of man, whether in the membership grades of technical societies or in society generally. Engineers insisted that one man's opinion was not as good as another's." One interviewee regarded consultation as valuable within limits, "*We consulted with the farmers. You did as information gathering. They come up with lots of suggestions, I mean, there is no way that someone who has not even been to school will see the overall plan.*"

3. Progress and the inevitability of change.

*Many of the interviewees adopted an uncritical view of progress and the inevitability of change, " We live in an environment where there is always change. The more people that there is living on the earth, the more rapid change occurs. We cannot put a little bit of the world in a box and leave it alone because it gets affected; if you put a fence around it, it would still get affected by what is going on outside." "The way I look at it you have to stick moving forward, any nationalistic groups that have become stationary and stagnant have perished in the past. Any historical group that has been evolved has disappeared when they have become stagnant." "People who are in favour of preserving today's ecology are denying everything that God wanted. God did make a world dynamically progressing. 250 million years ago 70% of all the fauna disappeared except those which could crawl, and from that the mammals evolved. Dinosaurs disappeared; all part of the dynamics of nature. Are we all of a sudden going to determine that the rate of change is becoming too fast, biologically too fast?"*

The progress of capitalism was seen as inevitable, even though costly investment in dams and other infrastructure has tied developing countries into debt and service to developed world, and subject to the vagaries of global markets. One interviewee felt the progress in the developing country had not lived up to expectations, although targets for the irrigation scheme had been met. Another had doubts "*that it is often the worst about the west that seems to get imported and makes you wonder if you are really doing the right thing.*"

One respondent thought that dams were an essential part of development "*The more dams built, the country's economic position is stronger, the people's living condition is better.*" The rich, developed countries alone could afford to give environmental concerns higher priority, having already developed their water resources.

These views are challenged not only by the ecocentric but also those who do not see the necessity of a homogeneous road to development which "leads to the imposition of programmes of structural adjustment, with all their horrendous and well-documented effects of polarisation, of yet-increasing hardship for the already-poor and especially for women."(Massey, 1999)

4. Production of new natures. " *You don't destroy the environment you create another one. The environment created by irrigating the desert is quite different from the one you had before. Both have their merits but the second is more productive; people can live on it.*"

Latour (1997) considers one of the features of modern life is our tolerance to allowing mediators to produce natures and societies. Nature becomes a relative product of history. In Europe and other densely settled parts of the world, centuries of profound human intervention in the environment have made the concept of nature relative rather than absolute. This lack of "purity" caused by domestication of nature may be a comfort; or sensed as a loss, or even greatly feared as a Frankenstein monster provoking the danger of retribution. The concept of new natures may be more acceptable where the contrast is between degrees of alteration rather than between pristine wild rivers and the controlled, dammed flow as in the American West, home of the protesting International Rivers Network.

The new natures brought about by dam builders has been challenged for:
a) **a limited vision of the criteria for success**. The instrumental value of nature constrains thoughts of ecological virtue to the survival of fish of commercial or sporting interest whilst neglecting other components of the ecosystem. Compensating flows from reservoirs may be geared solely to fishermen's demands rather than to maintaining the ecosystem. This vision may also encourage introduction of fish non-native to the environment, usurping native species from an ecological niche. Mitigation measures may be focused on a few species. For example, fish ladders for salmon have been developed but there has been relatively little research on the needs of smaller fish need to migrate upstream following floods.

b) **extinction of species**. Abramovitz (1996) claims that, "the rate of extinction is ..alarming -- a hundred to a thousand times the natural rate. By destroying species faster than nature can create new ones we are running a 'biodiversity deficit' that can never be recovered."

c) **current management approaches** which "often fail to recognise the fundamental scientific principle that the integrity of flowing water systems depends largely on their natural dynamic character." (Poff et al 1997). Maintaining a minimum quantity of flow in a river discharging from a dam is not enough for the integrity of the ecosystem. The following characteristics are vital: the magnitude, frequency, duration, timing and rate of change; water quality; temperature; seasonality; habitat diversity, and channel geomorphology. "The extreme daily variations below peaking power hydroelectric dams have no natural analogue in freshwater systems and represent, in an evolutionary sense, an extremely harsh environment of frequent, unpredictable flow disturbance" with consequent high mortality of aquatic populations.(Poff et al, op.cit)

Not all the ecological effects of dam building have been negative and some reservoirs have associated wildlife refuges, which have been colonised by many bird species. One interviewee claimed that a reservoir in the Amazon basin promoted an abundance of fish against all the negative prediction s of environmental opponents. Two South African dam builders quoted their new policies for the maintenance of the environment: " *Now in terms of our National Water Act, this principle is embedded in the Act, we have what is called the reserve components. One is the ecological reserve, where you have a certain water resource and you want to utilise it, you first have to slice off that quantity which is required to sustain the ecology in a given state".* Some of the dichotomies in this approach are discussed by Acreman (2001).

The dam builders resented the inability of ecologists to provide accurate predictions of the results of damming rivers and the high cost of environmental investigations. "*I asked the question to my marine scientist colleagues; how much water does an estuary need to keep in a reasonable state? They could not tell me.*" "*I started some of the very early fish ladder designs in the early 70s and the amount of scientific knowledge, which was available to design these fish ladders, was almost nil.*" Stimulation of research on these issues was seen as an important benefit.

However, the high cost of environmental investigations was resented. "*Of course, the way we have to do things in the States, in particular California, is to have environmental assessments done before the project even gets off the ground. In fact, they spend more money on the environmental inspection assessment reports than they do, many times, on the engineering on the project. From that standpoint, it seems sometimes to be out of hand*". Another quoted a case when the environmental assessment of upgrading a village track into a surfaced road would have cost more than the road itself!

5. Reversibility. "*Most, nearly all processes of that kind, which are incurred by man, are reversible. One thing you find is that nature is less reversible.*" " *No, we did the best we could, now let us recognise what we can do to correct that?... many of the things could be corrected.*"

Whilst the discourse of ecologists is often deeply pessimistic with talk of the killing of rivers, death of nature, destruction, irreversible damage to the environment, extinction of species including the human species, this optimism of the interviewees talking about the reversibility of their actions is striking.

The proposition seems to be that a series of technical fixes can reverse damage. Irrigation-induced salinisation can be controlled by increased drainage. Controlled flood releases can reverse much of the damage to ecosystems in flood plains downstream of dams. Controlled, channelled and straightened rivers may be restored to natural forms. This is a philosophy of attack rather than retreat and regret. A discipline of thinking about how engineering works may be undone if the circumstances change should be a useful exercise and may produce savings in the long run. But, even with some forethought, the cost of restoration is very high; the vitality of nature is not easy to recapture and the time is long for new environments to stabilise. If development and achievement of a wealthy society is a precondition before restoration or conservation of nature becomes feasible, it is doubtful whether nature can wait. Should nature be relegated to an optional luxury at some future date? Many would doubt whether resources could be mobilised throughout the world to increase consumption to that of the richest countries without producing great damage to the environment.

Although engineers are often accused of trying to control nature, the ideas expressed by the interviewees are more discriminating. Rather than the environmentalist assumption that anything wrong in nature is most likely to be the fault of humanity, their claim of reversibility of humanity's effects on nature was not thought to extend to domination of nature itself: *" You cannot really master nature. We are a grain of dust. All of sudden a volcano shows up or a huge earthquake that no one can predict. There are things that are beyond (our control). It is very much a case of education and assuming a position of co-participation with society. We are here to serve. To serve whom: society. By moulding nature, not by fighting. Nature is us, we are of nature."*

6. Beauty in artifice. Dam builders naturally take pride in the ingenuity and skill captured in the structures they have created and quote, as an important benefit, the numbers who visit the dams and the reservoir lakes for recreation. *"I am quite proud of what we have accomplished... I look back and I see a reservoir there and on weekends it is full of boats and people swimming and all sort s of recreation and fisheries that have developed and all that sort of stuff." "Beautiful lakes. I could not tell the difference between a reservoir like that and Loch Affric, which was one of the great beauty spots". "Do people like lakes or don't they? They like lakes. Cities that are beside lakes or rivers are much more pleasant". "Just imagine the land around this reservoir; it will be a fantastic holiday resort for the people. It will increase the land use, the landscaping, and gardening and probably be making hotels in beautiful places." "This gives me satisfaction that human beings are ingenious that I am producing 7000m cu m of water*

*in that reservoir for the people, not only for existing people but also for next generations and generations and generations. Producing electricity and making the environment nice and beautiful. Environmental beauty has enhanced, water is good use, electricity is good use."*

These views contrast with the scorn of some anti-dam writers. One describes as "tawdry" the recreational attributes of a reservoir "Fishing or boating on its drawn-down water, surrounded by miles of stinking mudbanks, is about as satisfying as playing basketball in a factory yard"(Vogt, 1949). Marchant (1983) writes "Mechanistic assumptions about nature push us increasingly in the direction of artificial environments, mechanised control over more and more aspects of human life, and a loss of the quality of Life itself". There are philosophical difficulties in imposing a hierarchy on other people's enjoyments and many poets have been inspired by harmonious man made landscapes. Yet Heidegger's (1955) regret at the denaturing of the Rhine in the interest of commerce and industrialisation has resonance. He (1955) advocated discernment: "we can say "yes" to the unavoidable use of technological objects, and we can at the same time say "no" in so far as we do not permit them to claim us exclusively and thus to warp, confuse, and finally lay waste to our essence"

It is not only aesthetics, which concern the opponents to dams and reservoirs but also the profound effect of water resource development on social relationships. Worster (1997) sees much more in the Hoover Dam than "the triumph of solid concrete over rambunctious river". In his view, dams both remake the face of the earth and alter the distribution of social and economic power on it. Power is given to those who profit most from domination of nature and power to the engineers who "offer the means of conquest". The death of the dominated river is linked to a reduced quality of life of the tourist, the average citizen, who has most of his or her life planned and administered by powerful external forces. For the benefit of cities in the desert, industry and large agribusinesses, the environmental services for the public from the river have been curtailed.

## CONCLUSION

The resourcefulness and problem-solving capabilities of engineers and dam-builders are now challenged as never before as definition of the problems and complexities widens. Dam engineers have a heavy and widening responsibility to ensure that their harnessing of water does not result in impoverishment of spiritual and cultural values whilst attending to the sustenance of humanity. Dam builders are faced not only with the novelty presented by each site and singularity of dam construction ("*every one a prototype*"); ecosystems which need understanding and care and responsibility with others for the social changes. Ashby (1959) claimed that for the civil engineer "the social consequences of his work are an integral part of his profession." Earlier thinkers have also recognised the pivotal role of engineers. Veblen (1921) thought the waste of capitalist production in salesmanship and advertising, raising the status of engineers and dispensing with the financial captains of industry could cure production of superfluities and spurious goods and a systematic dislocation! Such simplistic technocratic dreams have not been realised and we also now know how problematic and unsustainable societies have proved when concentration on technology has been at the expense of freedom of expression, cultural diversity and natural beauty of the environment. Progress in the development of water resources demands longer-term planning and commitment than is readily provided by market forces but restraint on the monopolistic power of the provider is needed to ensure representation of minorities, cultural heritage and ecological resources.

The WCD has played an important role in advocating participatory decision-making to open up discussion with a wide variety of people concerned with human rights and defence of environment. Now it is time for the dam builders to re-examine their values and augment their ambition for solution of human poverty with an increasing respect for cultural- and bio-diversity.

These preliminary findings from my research do not represent completion or closure. Reactions and response from the audience and readers are invited and part of a continuing research process.

ACKNOWLEDGEMENTS
The author is grateful to C. Faloudi, Alan Johnston, Victor de Mello, Nasser Tarkeshdooz, Alan O'Neill, Colin Gratwick, Jonathan Hinks, Jack O'Keefe, Robert Rangeley, Peter Riley, Paul Roberts, ST Su, Bob Wark, Michael Wilke and Peter Wolf, for giving so generously of their time to educating me on their perceptions of the dam debate. Extracts from their interviews are quoted verbatim with minor edits only. Also I thank Erik Swyngedouw for his stimulating discussions and advice. The responsibility for interpretation remains with the author.

REFERENCES

Abramovitz JN (1996). Imperilled waters, impoverished future: the decline of freshwater ecosystems. Worldwatch Paper 128. Worldwatch Institute, Washington, DC.

Acreman M (2001) Ethical aspects of water and ecosystems. Water Policy vol 3 Issue 3 pp257-65.

Ashby E (1959). Technology and academics. Macmillan, New York.

Bentham J (1776). A fragment on government. Payne, London.

Florman SC (1976, 1994). The existential pleasures of engineering. Souvenir Press, New York, London.

Heidegger M (1955). Discourse on thinking, quoted in Bate, J (2000) The song of the earth. Picador, London.

Hutcheson F (1738). An inquiry into the original of our ideas of beauty and virtue; in two treatises. I. Concerning beauty, order, harmony, design. II. Concerning moral good and evil Sec. 3 para 8.

Latour B (1993). We have never been modern p128 Harvard. USA.

Layton, Jr ET (1971, 1986). The revolt of the engineers. Social responsibility and the American engineering profession. The Johns Hopkins University Press, Baltimore, London.

Marchant C (1983). The death of nature: women, ecology and the scientific revolution. Harper, San Francisco.

Massey D (1999). Imagining globalisation; power-geometries of time-space in Brah A et al. Eds. Global future: migration, environment and globalisation pp 27-44 Macmillan, Basingstoke.

McCully P (1996). Silenced rivers. The ecology and politics of large dams pp 313-315 Zed, London, New Jersey, USA.

Pearce, F (1992). The dammed: rivers, dams and the coming world water crisis. The Bodley Head, London.

Poff LN, Allan JD, Bain MD, Karr JR, Prestegaard KL, Richter BD, Sparks RE & Stromberg JC (1997) The natural flow regime. A paradigm for river conservation and restoration. BioScience vol.47, No.11, pp 769-84.

Postel S (1992,1997). The last oasis. Facing water scarcity. Norton, NY.

Prasad Y (2000). Unpublished letter to the Secretariat of the WCD.

Priscoli JD (1983). Public involvement and social impact assessment: a union seeking marriage, pp 271-282. in Daneke, GA, Garcia, MW and Priscoli JD Public involvement and social impact assessments. Westview Press, Boulder, Colo.

Roy A (1999). The cost of living. Flamingo. London.

Russell B (1946). History of Western philosophy.p 695. Allen&Unwin, UK

Schmuck-Widmann H (2001). Facing the Jamuna River. Indigenous and engineering knowledge in Bangladesh. Bangladesh Resource Centre for Indigenous Knowledge, Bangladesh.

Takeuchi K et al (1998). Sustainable reservoir development and management. IAHS publication 251 pp 183-187. Wallingford, UK.

Veblen T (1921). The engineers and the price system.Viking Press, NY.

Vogt W (1948). Road to survival. Victor Gollancz, London.

World Commission on Dams (2000) Dams and development. A new framework for decision-making. Earthscan. London, Sterling USA.

Worster D (1997). Hoover Dam: a study in domination. in Jackson, DC (1997) Dams. Ashgate. Aldershot, Brookfield USA, Singapore, Sydney.

# Ghazi-Barotha hydropower project: social issues and engineering design

P E JONES, Embry-Riddle Aeronautical University, Prescott, Arizona, USA
(formerly Harza Engineering Co, Chicago)
J C ACKERS, Binnie Black & Veatch, Redhill, UK
A CHAUDHURY, ACE, Lahore, Pakistan

SYNOPSIS. The Ghazi-Barotha hydropower project in Pakistan is one of the largest currently under construction worldwide. The project demonstrates the capacity of engineers and social scientists to work together to avoid severe social impacts, with social considerations factored into the project design from the pre-feasibility stage. It was the first such project to integrate World Bank Operational Directives on resettlement and cultural properties into project design. The Resettlement Action Plan was regarded as the standard for 'world best practice' and continues to influence the design of projects elsewhere in the developing world. Well before the work of the World Commission on Dams, the design of the project covered many of the criteria and guidelines presented in their report, *Dams and development: a new framework for decision-making*.

## INTRODUCTION

The $2200M Ghazi-Barotha Hydropower Project (GBHP) in northern Pakistan is one of the largest currently under construction worldwide and is scheduled for completion in 2003. It will have a peak generating capacity of 1450 MW and provide an estimated annual energy output of 6600 GWh. The project, which is being built for the Water & Power Development Authority (WAPDA), includes three reservoirs:

- a pool created by a gated barrage on the Indus River at Ghazi (7km downstream from Tarbela dam) to divert up to 1600 $m^3/s$ into the 52km long power channel and provide additional operating head; and
- two headponds at the downstream end of the power channel, to provide storage to support diurnal peaking of power generation.

The project demonstrates the capacity of engineers and social scientists to work together to find creative engineering solutions to otherwise severe social impacts. The GBHP broke new ground in making the design of a massive infrastructure project responsive to carefully researched and defined social concerns. It was the first such project to integrate World Bank Operational Directives on resettlement and cultural properties into project design. Its Resettlement Action Plan was regarded as the standard for 'world best practice' and continues to influence the design of projects elsewhere in the developing world.

Social and economic analysis in the GBHP was based on extensive field appraisal, a programme of village scoping sessions that included the leaders and notables of every village directly impacted by the project, socio-economic surveys, and a baseline census of all affected households. Women social scientists held parallel scoping sessions with village women and carried out a socio-economic survey of the project area focused on women's issues. In addition, informational meetings were held at District Council level in Attock, Haripur and Swabi. Members of the Provincial Assemblies and the National Parliament representing communities in the project area were also briefed on the project.

## PROJECT REGION: LAND AND PEOPLE

The project makes use of the difference in the fall of the Indus River between Ghazi and Barotha and the more gradual descent of the land to provide a head of about more than 60m at Barotha, where the water is returned to the Indus through the powerhouse and the tailrace.

Given the size and nature of the GBHP, the region directly affected is extensive. Physical impacts (infrastructure components, spoil banks, drainage features, quarries, etc) cover an area approximately 60km by 8km, lying in two provinces of Pakistan: the North-West Frontier Province (NWFP) and the Punjab. Social and economic affects are, of course, greatest in the area of direct physical impacts, but are also felt over a considerably wider surrounding region (Figure 1).

For the purposes of social analysis, the project region was divided into three areas: Ghazi and Topi, the Chhachh, and Sarwala.

### Ghazi and Topi

The apex of the project is in the braided Indus floodplain above Ghazi town on the left bank and Galla village on the right bank. While there are effects on the right bank, the major impacts are on the left bank, where the power channel exits the barrage component and begins its long route of over 50km down to Barotha and the powerhouse.

Administratively part of Haripur District, Ghazi is located at the northern apex of the Chhachh Plain. The hilly area above the town is occupied by WAPDA's Tarbela Dam Colony. These hills are outliers of the main buttress of the Ghandghar Mountains, which farther upstream provides the left abutment of the great Tarbela Dam. Below Ghazi, the Ghandghar Ridge angles away from the river, opening a fertile plain about 11km long and 3km wide, the latter at the NWFP-Punjab border.

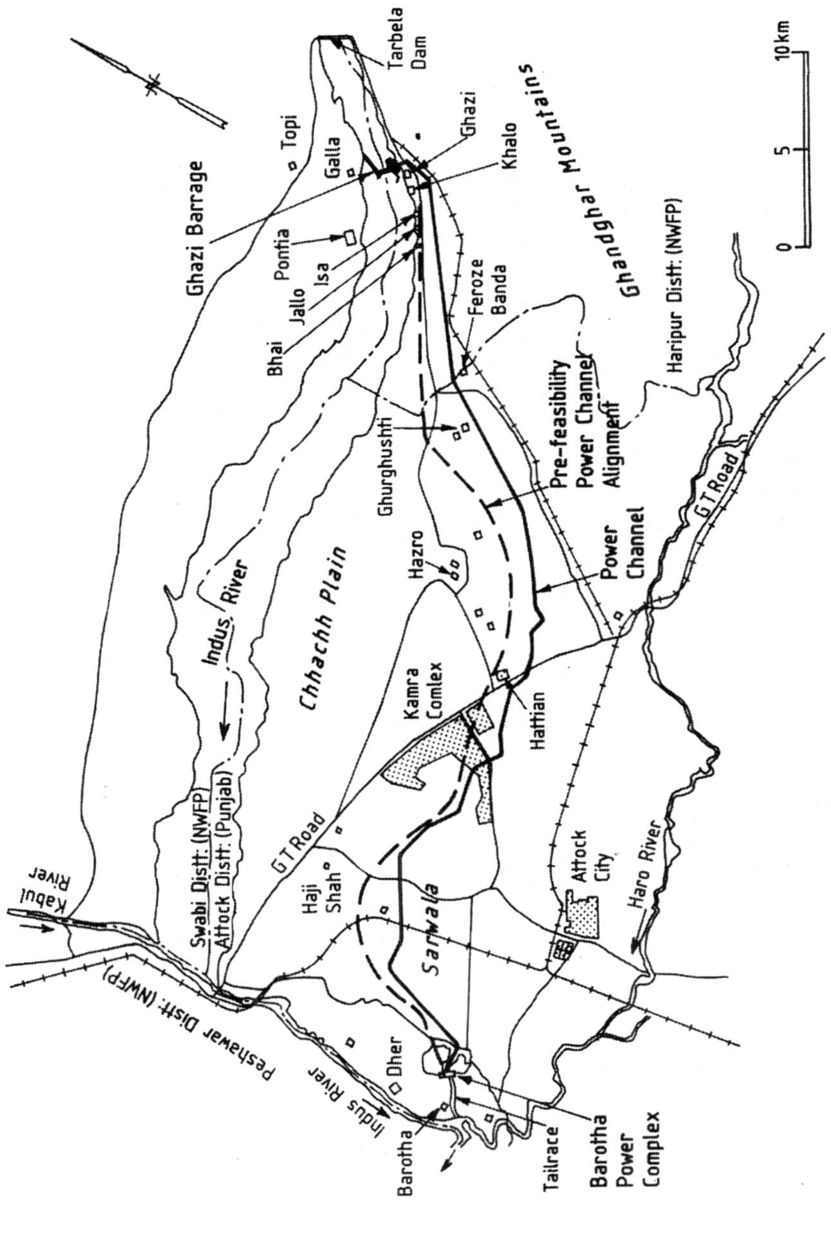


Figure 1 Layout of Ghazi-Barotha hydropower project

The plain is well cultivated, mostly with winter wheat, vegetables, orchards, and some tobacco. Although most cultivation remains dependent on rain, many farmers have put in tubewells. The dominant tribes in the area are Pakhtuns: Tahirkhelis along the Ghandghar and Ghazi plain, Utmanzai on the right bank of the Indus.

The Chhachh

Starting at the Punjab border (Attock District), the Chhachh plain widens as the Indus floodplain swings westward, gathers in the waters of the Kabul River and then turns south and enters the throat of the gorge at Attock. The Chhachh is a triangular plain, its sides bounded by the Indus left bank on the northwest (20km), the Ghandghar to the east (10km), and the hump of high ground along the Grand Trunk Road to the south (15km). Built up by alluvial deposition from the Ghandghar, the Chhachh is reputed the most fertile region in Punjab, providing high yields in wheat, high quality tobacco (for snuff), vegetables, fruit, fodder, wood for fuel and construction, and groundnuts (in the sandy soils closer to the mountains). Most landholdings are small for the region, averaging 1.5 ha. The population is a mixture of Pakhtun (Jadoon, Tareen) and Punjabi (Awan, Sayyid, Gujjar, Mughal).

The Chhachh long ago stopped supporting the population. Families typically keep a footing on their ancestral lands – a brother cultivates multiple small holdings – but send sons out to work in local defence industries (munitions factories and the Kamra aeronautical complex), or further afield to Lahore, Karachi and beyond. Ghurghushti, the largest village in the region, typically has from 30 to 40 percent of its population living and working in the Bradford region of the UK. The first generation of immigrants usually comes home to have the daughters married and then to retire. With fortunes made abroad, there is far greater demand for land than supply, making land prices abnormally high.

Sarwala

The third sub-region in the project area is south of the Grand Trunk Road (Rawalpindi to Peshawar section), which follows the watershed between the sluggish Sil Creek along the southern boundary of the Chhachh and the Haro River. The latter flows roughly along the southern perimeter of the project area and drains into the Indus River immediately below Barotha. This area comprises the Sarwala Tehsil of Attock District, a dry loess plain much gullied by erosion.

The population – mostly Awans and Khattars – is sparser and villages are located down in the lower gullies, where water is accessible through wells or from permanent springs The soils are dry and stony and completely dependent on rain (barani) in the winter and spring. The Grand Trunk Road; the bazaars at Hattian, Kamra Crossing, and Haji Shah; and the various defence establishments in the area provide employment for the people of this more impoverished region.

## BARRAGE ALIGNMENT, POOL AND RIGHT BANK EFFECTS

Five barrage sites along a 5km reach of the Indus River were examined during the pre-feasibility and feasibility studies. The farthest upstream site was chosen for engineering reasons, but it also had by far the least social impacts. The other alignments would have meant major disruptions to the villages of Bhai, Jallo, Isa and Khalo, and Ghazi town on the left bank, and Galla village on the right bank, forcing some involuntary resettlement and damage to water systems, roads, and cultivated areas. The necessity for embankments to prevent the barrage pool from flooding the lower sections of these inhabited areas would have blocked sewage outflows into the river and led to the pooling of polluted water between the embankments and residential areas.

**Ghazi barrage**

The barrage, situated near the small town of Ghazi, about 7km downstream of Tarbela dam, creates a raised storage pond confined largely within the normal width during floods of the original Indus River. The barrage has four principal purposes:

- forming a diversion structure in order to pass flows into the head of the power channel;
- raising the upstream water level and ultimately increasing the generating head available at the powerhouse at Barotha, at the downstream end of the project;
- providing sufficient storage to re-regulate outflows from Tarbela reservoir, which vary through the day to suit peaking operation of its powerhouse; and
- to act as a temporary settling basin for sediments passed downstream from Tarbela reservoir.

The barrage includes an eight-bay head regulator, with 18.3m wide radial gates controlling flows to the power channel, and 28 radial gate openings (also 18.3m wide), to convey downriver flows of up to 18 700 $m^3/s$, corresponding to the estimated 200-year flood. Eight of these gates are 'undersluices'. These are located adjacent to the head regulator and include facilities for flushing sediments deposited in the approach to the head regulator. The remaining 20 gates, when fully raised, provide the main open-flume spillway for passing downriver flows.

The current barrage site has the least upstream effects. The barrage pool will have a maximum surface area of 1140ha at a maximum retention level of 340m. The 7km reach of the Indus between the barrage and Tarbela dam lacks any permanent habitation and is owned by WAPDA. The braided floodplain, with its active creeks and pools, was used for fishing and as a source of grass and fuel wood, activities nominally administered by WAPDA. A few WAPDA assets in the colony area on the left bank – a horticultural nursery and a staff training facility – have been moved to higher ground. Permanent rim embankments on the right bank have been constructed to protect WAPDA facilities at Pehur. The right guide bank of the barrage is tied into the right bank about 0.75km upstream from Galla village.

The major social impacts of the barrage alignment and pool are:

- Loss of water along the riverfront at Galla. When in use during the summer high flow season, the standard bays on the barrage will channel water away from the right bank and the villages of Galla and Pontia. The riverfront at Galla was an important source of water, a place for washing clothes, and a daily meeting point for the village women – not an unimportant consideration in Pakhtun society, where women often are restricted to their homes. The recommendation was to put a pipe through the right guide bank to allow some flow of water down the old channel. Perhaps nervous at the idea of a permanent pipe through an impounding embankment, the engineers decided to provide Galla village with two tube wells, each to channel water down one side of the U-shaped ravine in which the village is set.

- The public road along the barrage will provide a much more direct link between Topi and Ghazi and their surrounding regions. Previously, traffic had to cross the river at Tarbela dam, an additional 18km. Although the people of Ghazi complain they will be at greater risk from smugglers and bandits from Swabi District, the road link will undoubtedly have major commercial benefits on both sides of the river.

- Protection for cultivation in the floodplain. Over the years, the villagers of Galla and Pontia have developed over 100ha for cultivation in the river bed under the line of higher ground on the right bank between the two villages. These are fertile alluvial fields, anchored by stone walls and belts of trees, and irrigated by artesian wells (26) and various pump wells (6). A temporary protection embankment was built during the construction of the barrage, as the main diversion channel was aligned along the right bank. At the time, the villagers favoured the permanent retention of the protective bund.

- Compensation for permanent loss of land in the floodplain islands (belas) and construction damage from access roads, mid-channel crushers and batch plants in borrow areas. Held as common land (shamilat), the belas were important sources of fodder, grazing, fuel wood, valuable shisham lumber, construction materials (sand and gravel), and gold dust from panning. Some 840ha of bela land was lost to the barrage and pool. Recognition of the resource value of the belas was an innovative aspect of social impact assessment, and the selection of the Ghazi site for the barrage greatly minimised the potential losses of the belas downstream.

- Commercial fishery is to be established in the barrage pool. This will have to be stocked from a hatchery, as the level of the pool could fluctuate as much as 5m per day, to the ruination of fish eggs in the shallows. The right to harvest fish on a franchise basis is to be offered first to households that previously depended on fishing in reach above the barrage site.

## MINIMISING RESETTLEMENT: REALIGNMENT OF THE POWER CHANNEL

The Ghazi-Barotha power channel is a truly massive undertaking, which must rank as one of the largest canals in the world. The choice of route for the power channel was a major and challenging part of its design and it was the largest element of the project which was significantly altered as a result of social considerations (Figure 1).

**Ghazi-Barotha power channel**

The Ghazi-Barotha power channel conveys the design discharge of 1600 $m^3/s$ from the barrage to the power intakes. It is of trapezoidal section, with maximum service depth of 9m, 1:2 sideslopes and water surface width of about 95m. The design velocity is 2.3 m/s and through most of its length the channel is formed entirely in cutting.

Concrete lining is required in order to achieve the requisite discharge capacity with the available hydraulic gradient (1:9600), whilst minimising the volume of excavation required. The concrete lining leads to the need to provide the power channel with an active underdrainage system.

The pre-feasibility alignment of the power channel followed the most economical 'engineering' route, that is, the route providing the best balance of cut and fill. Unfortunately, this route went directly through eight villages and one town (Hattian) and intercepted residential areas in another seven villages. It also went through the most fertile part of the Chhachh. Although a census of the potential affectees along this route was not made, the estimate from pre-feasibility field visits and discussions with village leaders suggested involuntary resettlement would involve approximately 2600 households, comprising some 14 300 individuals. This was considered unacceptable.

The adopted alignment avoids all villages by shifting the power channel to higher, less valuable ground, further away from the Indus River. This option significantly increased the requirement for excavation and therefore the overall cost of the project, but it markedly decreased the negative social (and environmental) impacts. Only 112 households have had to be relocated in the entire project area. Of these, 90 are due to the new power channel alignment. WAPDA has built two resettlement villages for 46 of these households. A third resettlement village for 22 households displaced by the power complex has been built near Barotha. The remaining 44 households were in scattered locations and were provided new homes close to their original dwellings on land purchased by WAPDA.

Rehabilitation of spoil banks

The overall volume of excavation material in the project is 110 $M.m^3$, of which 32 $M.m^3$ was required for structural components, such as bulk fill at the barrage site, embankments, filter materials, coarse aggregate and stone riprap. This left a balance of 78 $M.m^3$ of spoil for disposal.

A variety of options were considered, including large disposal in the Indus River, an unacceptable solution for environmental reasons. Some 10 $M.m^3$ was used to terrace wasteland and rugged areas along the eroded banks of the seasonal streams crossing the power channel from the Ghandghar. Another 4 $M.m^3$ was used to rebuild farmland along the left bank of the Indus in the Ghazi region, where scouring has destroyed several hundred hectares.

For the bulk of the remainder of the spoil, the engineers and social scientists developed a plan to concentrate the excess material in spoil banks on marginal land along the alignment of the power channel and to redevelop these spoil banks for cultivation.

A rolling construction methodology was developed that allowed the spoil banks to be constructed and stabilised, and then covered with topsoil originally stripped from beneath the spoil banks and from the land permanently lost to the power channel. Since the agricultural value of topsoil is in the organic activity concentrated in the top 300mm or so, the construction plan called for the stripping and separation of this matter, holding it for only a short time, and spreading it on the spoil banks while still in a fertile condition. The spoil banks required 6 $M.m^3$ of topsoil for rehabilitation out of the estimated 8 $M.m^3$ available.

The plan for spoil bank rehabilitation is a key feature of resettlement in the project, enabling farmers to remain in the region, rather than being forced to migrate 400km to marginal land available in the Thal Desert. A total of 3200ha of privately owned land, belonging to about 20 000 individuals was permanently acquired for the project. Most of these were highly fragmented holdings, of which 3.4%t was irrigated, 82% was rain-dependent (barani), and 15% uncultivable (ghair mumkin).

With a total surface area of 1640ha, the spoil banks were to be provided with tubewells and offered back to the farmers at low rates. As landholders had already been paid for the loss of their land, the buy-back provision was designed to attract those affectees still involved in cultivation. As irrigated land (chahi/nehri) is a little more than twice as productive as barani, the exchange of the latter for the former should enable the farmers to restore their previous livelihoods. In addition, the Resettlement Action Plan provided for a demonstration farm to be set up on a spoil bank, operated by agronomists, to show how this land could be developed.

Other realignment factors

The power channel was locally realigned to avoid cemeteries, shrines, archaeological sites and other cultural properties. The region is rich in such cultural properties, including an extensive necropolis east of the power channel, seven sufi shrines, and several stone age (Paleolithic), Gandharan Buddhist, Hindu Shahi, and Mughal period remains. In two instances – at Banda Feroze near Ghurghushti and near Kamra village – small numbers of graves had to be moved. These were moved by the affected families and in accordance with Islamic practice. Two lesser archaeological sites, Musa I and Musa II, could not be avoided and were excavated by the Department of Archaeology. Both turned out to be late Kushan Period (3rd to 5th century CE) Buddhist monastic sites.

Village crossings

Due to its length and width, the power channel will be a barrier to the daily activities and local movements of the rural population. In the majority of cases, village estates are shaped roughly as elongated rectangles with the long axis across the power channel. This means villagers will be cut off from a portion of their fields, virtually all of their grazing areas and wood lots, and, for many, access to graveyards, shrines, and common areas providing a variety of other resources. Furthermore, villages almost invariably are controlled by a tribe or clan (qaba'il, biraaderi) that maintains enmity with the tribe or clan of the village on either side. In most cases, the villagers are implacably opposed to sharing rights of way with their neighbours, hence the need for more than the normal number of crossing facilities. There crossings were tied into existing village rights of way.

The GBHP includes 45 cross-channel facilities, including bridges and cross-drainage 'superpassages'. Included in this number are 12 freestanding bridges for villagers, 14 roadways for tractor trolleys over superpassages, and another 12 pedestrian pathways and access ramps across superpassages. Access across the superpassages will be blocked for only a few days a year, when heavy monsoon rain in the Ghandghar Mountains floods down toward the Indus.

## THE POWER COMPLEX

The power complex is located within the village estates (mauza) of Barotha and Dher. The two villages are the most impacted of any in the entire project. Each lost far more land than any other village estate: Barotha 439ha, and Dher 608ha.

**Barotha power complex**

The power complex at Barotha comprises the following main components:

- the tail regulator, to maintain a constant flow depth at the downstream end of the power channel;
- the forebay, downstream of the tail regulator and upstream of the power intake, also connected to the headponds and the emergency spillway;
- the north and south headponds, impounded by embankment dams with a total length of about 8km and a maximum height of 60m, which provide sufficient storage in a 5m operational range of water levels to support four hours peaking operation of the powerhouse;
- the power intake, leading via surface penstocks to the five 290MW turbines;
- a switchyard and transmission lines;
- the emergency spillway, providing a bypass to the powerhouse when required for maintenance or during an unpredicted outage, for example because of a problem in the transmission system; and
- the 2km long tailrace, conveying flows from the turbines and spillway back to the Indus.

The headponds and their embankments, plus the associated facilities of the power complex are entirely located with these two villages. Most of the land taken was low grade barani or uncultivable and used for pasturage. The more valuable land on the river bench below the villages was not touched, but Barotha lost the lovely permanent spring, woods, and irrigated ground in the gully just south of the village – the seasonal haunt of the Paradise Flycatcher – to the tailrace. Both villages were provided tubewells to ensure water supplies. It is hoped that seepage from the headponds can be collected and used for irrigation on the cultivated benches along the Indus.

In addition to liberal compensation for land lost, a small resettlement village, construction period work, and the right to buy land on the nearest spoil bank, the most impacted householders in Barotha and Dher were promised permanent jobs once the project began operations.

PROJECT AREA ENHANCEMENTS

Mindful that such large-scale hydropower projects have area-wide impacts, the project design provided for the following social and economic components:

- Local training and employment to upgrade the local construction and industrial skills base in the project area.

- A requirement to give priority in hiring for work on the project, first to directly impacted households, then to the most affected villages, and finally to other area residents. This was bolstered by a work permit system and public information offices on both sides of the Indus River. The right to work preference was designed both as a compensatory entitlement and a strategy to prevent a large influx of outsiders.

- The establishment of the Ghazi-Barotha Tarraqqi Idara (GBTI), or Ghazi-Barotha Development Institute, an endowed organisation using community development planning for sustainable development. GBTI hired alumni from the highly regarded Agha Khan Rural Support Project and successfully implemented their participatory methodologies in the project area.

- An Integrated Rural Development Plan to upgrade agrarian technologies, agronomic practices, and agro-industrial investment.

- A Small Industries Estate to provide newly skilled workers with post-construction period jobs and reduce the impact of the inevitable project boom-bust cycle.

- Town Committees for Ghazi-Khalo and Topi. This was a recommendation to the Government of the NWFP. Town Committee status for both was long overdue, both having long outgrown their village origins. Both towns need the additional administrative infrastructure and revenue capacity to deal with problems of drainage, sewage, clean water, better market organisation, roads, and other municipal facilities.

## CONCLUSIONS

In conclusion, the social and economic impacts of the Ghazi-Barotha Hydropower Project on the local population were carefully identified and measured during the design phase. Throughout, social scientists and design engineers worked together to find innovative solutions to minimise and mitigate severe impacts. Not all social and economic impacts were susceptible of engineering solutions, but in some of these cases, the design of the GBHP broke new ground, as in the provision for an endowed local development agency.

In its approach to social issues in the project, the design teams covered many of the criteria and guidelines presented by the WCD final report. Field appraisal, scoping sessions, and the establishment of a project information office enabled designers to identify most stakeholder groups and draw them informally into the planning process. Impacts at the strategic and project level were assessed and great care was taken to minimise project impacts on cultural heritage sites. Social baseline data were collected both at the project level for the social impact assessment and at the household level for the Resettlement Action Plan. In some areas, WCD guidelines are more focused and refined than those used in the GBHP. The WCD's emphasis on Strategic Impact Assessment and Impoverishment Risk Assessment are innovative approaches to project design and welcome additions to the 'toolkit' of project design social scientists and engineers.

## ACKNOWLEDGEMENTS

The authors thank the Water & Power Development of Pakistan for permission to present this paper, and gratefully acknowledge the support of their colleagues in Pakistan Hydro Consultants, Lahore.

## REFERENCES

World Commission on Dams (2000). Dams and development: a new framework for decision-making. Earthscan Publications Ltd (London).

# RCC Construction at Tannur Dam

M AIREY, Mott Macdonald Ltd, UK.

SYNOPSIS. Tannur Dam was the first RCC dam to be constructed in Jordan. An extensive trial mix programme was undertaken to refine and optimize the high past RCC mix. In addition, various innovations were adopted during the construction to ensure that an RCC of a consistently high quality was produced and placed. These included the use of the slope layer method of placing instead of the more conventional horizontal layer system. The outer faces of the dam were formed from grout enriched RCC (GERCC). This was introduced instead of the slip-formed facing of conventional vibrated concrete that was originally specified.

INTRODUCTION.

The construction of Tannur Dam in Southern Jordan was undertaken between 1999 and 2001. The dam is situated approximately 150 km south of Amman on the Wadi al Hasa, which runs from east to west and outfalls at the Dead Sea. The dam has created a reservoir with a storage capacity of 16.8 million $m^3$ that will capture flash flood discharges arising in the desert catchment. The water stored will be used to supplement supplies to existing irrigation schemes in the Southern Ghors area.

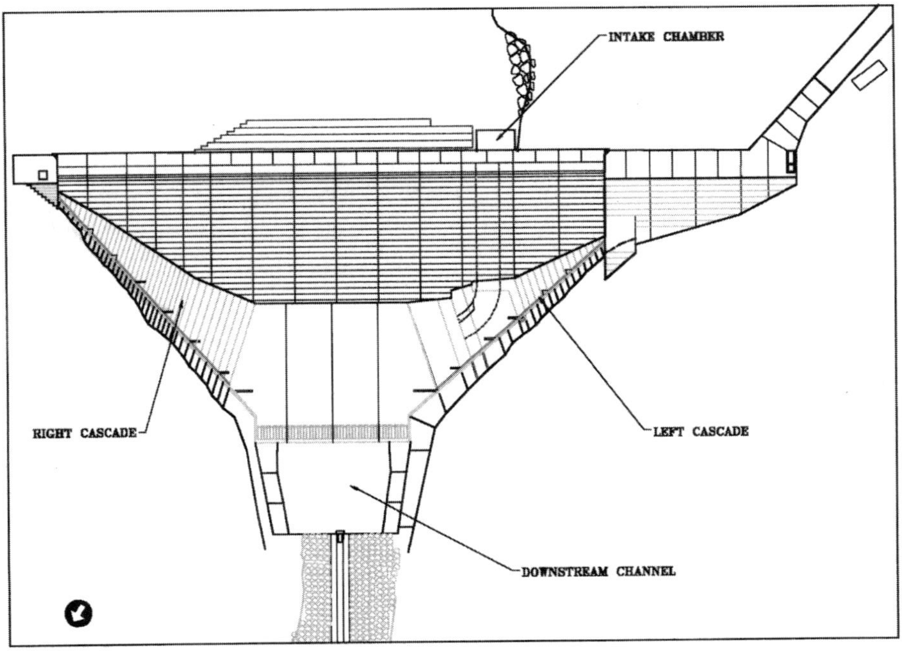


Fig. 1. General Arrangement of Tannur Dam

*Reservoirs in a Changing World*, Thomas Telford, London, 2002, 73–83

The dam is a mass gravity structure of roller compacted concrete (RCC), with a total volume of RCC of 210,000m$^3$. This was placed during the period January 2000 to November 2000 with a peak monthly production of around 45,000m$^3$. Tannur was one of three dams being constructed in Jordan during this period and it was the first time that RCC had been used in Jordan.

The cross section of the dam was changed during the design review stage. The tender design had been based upon an inclined upstream face and a relatively steep downstream face (1.0 : 0.5). However to improve the stability, to increase the energy dissipation characteristics of the downstream steps and to provide a design solution that was generally easier to construct, the profile was modified to a vertical upstream face with a flatter (1.0 : 1.8) downstream face. At the same time grout enriched RCC (GERCC) was substituted for the slip-formed immersion vibrated concrete that had been proposed for the upstream and downstream faces. These modifications represented a significant change and were introduced by means of a Variation Order that provided a saving of approximately JD 600,000.

The principal characteristics of the dam are given in Table 1 and Fig. 1 shows the general arrangement of the dam and its appurtenant works. Fig. 2 indicates the typical cross section.

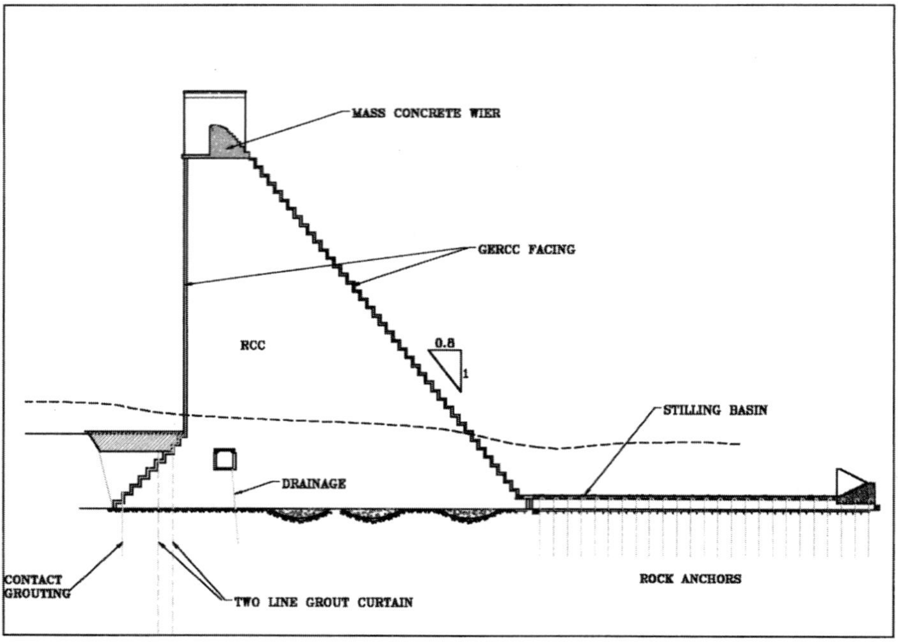


Fig. 2. Typical Cross Section of Dam

Table 1. Main Characteristics of Dam

| | |
|---|---|
| Type of Dam | Mass Gravity (RCC Construction) |
| Height of Dam | 60 metres |
| Length of Crest | 270 metres |
| Width of Spillway | 180 metres |
| Spillway Design Flood (PMF) | 3,357 $m^3/s$ |
| Type of Spillway | Free overflow weir discharging to cascades |
| Energy Dissipation | Stilling Basin and Cascades |

## RCC MIX DESIGN.

### General

The RCC specification was a high paste mix with a total cementitious content of greater than 150 $kg/m^3$. The high paste mix has an inherently lower permeability than medium or low paste mixes, whilst having superior compressive and tensile strengths. The main objective of the RCC mix design process was to obtain a product with the following properties: -

- A characteristic cube compressive strength of 20 MPa at 90 days, which would also ensure adequate tensile strength.
- A suitably workable mix specified by a target VeBe time of between 15 and 21 seconds.
- A robust mix that could be transported, placed, spread and compacted without segregation.
- A minimum cement content so as to reduce the heat of hydration, whilst still satisfying the strength and workability requirements.

It should be noted that the cementitious materials (cement and pozzolan) were measured and paid separately to the RCC itself. Hence any reduction in the amounts used provided a direct saving to the contract.

### Laboratory Trials

To achieve these objectives an extensive trial mix programme was undertaken in the initial phase of the construction period. The trials included laboratory trials that were commenced off-site, prior to the establishment of the site laboratory, and then completed on-site when the laboratory was operational, together with a full-scale placement trial. Both the cement (OPC) and pozzolan to be used were locally manufactured and were delivered to site in bulk tankers. During the laboratory trials a series of grout cubes were made with different proportions of cement and pozzolan at a constant water / cement ratio. The aim of this exercise was to ascertain the optimum proportion of the materials in terms of strength gain.

The results which are summarized in Table 2 indicated that a mix with up to around 35% of pozzolan could be used with no significant loss in long term strength when compared with a 100% OPC mix.

Table 2. Mortar Cube Compressive Strengths

| Mix Nr | Pozzolan Content (%) | Density @ 28 Days ($Kg/m^3$) | Cube Compressive Strength @ 28 Days (MPa) | Cube Compressive Strength @ 90 Days (MPa) | 90 Day Strength compared to OPC Only (%) |
|---|---|---|---|---|---|
| 1 | 0 | 2.284 | 30.70 | 36.93 | 100% |
| 2 | 30 | 2.280 | 29.18 | 35.95 | 97% |
| 3 | 40 | 2.320 | 27.94 | 28.37 | 79% |
| 4 | 50 | 2.280 | 25.63 | 25.41 | 71% |

External tests were also carried out to determine the heat of hydration developed for the cement and the cement / pozzolan blends. The results are given in Table 3 and it was demonstrated that with a 35% pozzolan content a significant benefit was achieved in terms of reduced heat development. This was extremely important in guarding against thermal cracking.

Table 3. Heat of Hydration Properties of OPC

| Time | 100% Cement (J/g) | 80% Cement to 20% Pozzolan (J/g) | 60% Cement to 40% Pozzolan (J/g) |
|---|---|---|---|
| 6 hours | 50 | 50 | 46 |
| 12 hours | 157 | 131 | 104 |
| 1 days | 253 | 210 | 151 |
| 2 days | 289 | 244 | 186 |
| 3 days | 299 | 257 | 199 |
| 4 days | 303 | 266 | 207 |
| 5 days | 307 | 271 | 213 |
| 21days | 309 | 272 | 213 |

To achieve the correct balance between strength and workability, another key aspect was the proportioning of the aggregates. The aggregates were obtained from the Naur limestone at a quarry that was established in the reservoir basin approximately 2km upstream of the dam. A dedicated crushing and screening plant was set up to produce all of the aggregates for the project. For the RCC, the aggregate sizes were 45 to 26mm, 26 to 5mm and 5 to 0mm. Extensive trials were carried out with the crushing plant, the screen arrangements and the re-circulation system to obtain satisfactory gradings and the right amount of each product. The average grading curves were as given in Table 4.

Table 4. Average Aggregate Gradings (Percentage Passing Standard Sieves)

| Aggregate Size | Sieve Size mm | | | | | | | | | | |
|---|---|---|---|---|---|---|---|---|---|---|---|
| | 53 | 7.5 | 19 | 9.5 | 4.75 | 2.36 | 1.18 | 0.6 | 0.3 | 0.15 | 0.075 |
| 45 to 26mm | 100 | 85.8 | 4.9 | 1.0 | 0.7 | 0.5 | 0.5 | 0.5 | 0.5 | 0.5 | 0.4 |
| 26 to 5mm | 100 | 100 | 92.7 | 32.1 | 6.6 | 1.3 | 1.3 | 1.3 | 1.3 | 1.3 | 0.7 |
| 5 to 0mm | 100 | 100 | 100 | 100 | 100 | 79.6 | 47.5 | 30.3 | 20.4 | 14.7 | 10.5 |

When combined in the proportions of around 40 : 20 : 40 %, this gave a combined grading curve that fell within the specified envelope. However one problem that persisted was a gap – grading of the fine material, with a deficiency of material in the 1 to 2mm range and excess "rock flour" material (> 10%) finer than 0.15mm.

The RCC that was produced with this aggregate was a harsh mix that was prone to segregation and which also required excessive quantities of cement and pozzolan to maintain workability. To solve the problem it was decided that a natural rounded sand should be imported and introduced into the mix. The laboratory tests showed that around 7% of the imported (Tafila) sand was sufficient to overcome the gap – grading effects.

Full – Scale Trial

A full scale trial embankment approximately 60m long by 12m wide was constructed in the reservoir basin prior to the placing of RCC in the dam wall. In total 10Nr x 300mm layers of RCC were placed using the most suitable mixes that had been identified from the laboratory tests. The trial embankment also served to test the plant and placing methods that were to be used on the dam. In particular it offered the first opportunity to develop techniques for the production and compaction of the GERCC facing under full scale conditions. It became apparent from these trials that penetration of thick grout mixes was difficult to achieve consistently under field conditions. The best results were obtained using 1:1 cement to water grout.

The RCC mix that was adopted from the trial mix programme had a total cementitious content of 200 kg/m$^3$ (125/75), and this mix was used in the first phase of dam construction. As more strength data become available it was possible to refine and optimize the mix still further. This resulted in a further reduction in cement and pozzolan contents and the final mix that was used for the majority of the dam had a cementitious content of 170 kg/m$^3$. Details of the final mix proportions are given in Table 5.

Table 5. RCC Mix Design Details (for $1m^3$)

| Material | Unit | Quantity |
|---|---|---|
| Cement | kg | 120 |
| Pozzolan | kg | 50 |
| Free Water | litre | 105.4 |
| Absorbed Water | litre | 66 |
| 45 to 26mm aggregate | kg | 676 (32.5%) |
| 26 to 5mm aggregate | kg | 634 (30.5%) |
| 5 to 0mm aggregate | kg | 624 (30.0%) |
| Tafila Sand | kg | 146 (7.0%) |
| Set Retarder | kg | 1 |
| Theroetical Air Free (TAF) Density | | 2422 kg/m$^3$ |

RCC MIXING AND PLACING.

Horizontal Layer Method

The RCC mixing plant was an SAE Type 70.275 twin shaft pugmill mixer rated at 300 m$^3$/hour. Batching was by continuous weighing on the conveyor belt feeds and the batching / mixing process was computer controlled. The mixing plant was located just upstream of the dam at an elevation equivalent to about the $^1/_3$ height. The RCC was transferred from the plant to the dam via a system of Rotec conveyors and discharged via a "swinger" unit. The conveyors were re-located and steepened progressively as the dam wall rose. The RCC was transported from the swinger to the dumping point using three 25 tonne Volvo A25C 6WD dump trucks. These were fitted with smooth profile types which helped to minimize damage to the fresh RCC surfaces.

The original placing system was based upon a horizontal layer method whereby a single 300mm thick layer was to be placed over the full area of the dam before commencing with the next layer. The layer was formed in a series of 15m wide strips that were placed from downstream to upstream and which were advanced across the width of the dam. One drawback of this system was that each lift joint between the layers had to be considered as a cold joint (greater than 36 hours old). This required extensive clean-up and preparation, initially specified as sand-blasting. The prepared joint was then covered with a 6mm thick layer of bedding mortar before the next layer could be placed. In the initial stages of construction there were various teething problems and plant breakdowns. In addition the tracking of the dump trucks and other plant across the previously prepared lift joint caused further contamination, which required joint clean-up activities to be repeated. As a consequence the whole operation was on a stop-start basis and not the continuous operation that is intended for efficient RCC production.

Slope Layer Placing Method

To overcome the problems the Contractor introduced a slope layer placing system. This system involved the placing and compaction of RCC in 300mm thick layers (as with the horizontal layers) but on a slope across the dam at a gradient of around 1 in 15. A 1.2m high lift was adopted (4Nr x 300mm layers) to correspond with the height of the steps on the downstream face.
The advantages of the slope layer method were: -

- Clean up operations were restricted to the horizontal surface on the top of each 1.2m high lift. This was a cold joint of fully hardened concrete and the clean-up could be executed by mechanical rotary brooming and high pressure air / water jetting.

- Bedding mortar requirements were reduced by around 60%. Bedding mortar was only used on joints where the underlying RCC was over 2 hours old, or 1 hour old in the case of a 4m wide strip adjacent to the upstream face.

- The sloping joints were generally fresh being less than 1 hour old. Because of the retarder, the underlying RCC had not achieved its initial set and a very good bond was provided between sloping layers.

- Overall production rates were increased by around 50%.

The principle of the slope layer placing method is illustrated in Fig. 3. In general the volume of each sloping layer was equal to about 1 hours RCC production. A pre-requisite for the introduction of this new method was the addition of a set retarder in the mix. Daratard P2 was added at a rate of 1 litre/$m^3$, which increased the initial set time to between 1½ and 2 hours.

With both placing methods the RCC was spread using a Caterpillar D6 bulldozer with a tilt blade. The bulldozer was fitted with low profile growser plates to minimize disturbance to the RCC surface. The D6 was found to be very effective, although perhaps a little too big for this application. This was particularly so in the latter stages of construction when the dam wall became narrower and the space was limited. The trimming of the layers was controlled by a laser with a target mounted on the blade of the dozer. For the slope layer method, the control system was supplemented by the painting of the layer profiles on the upstream and downstream formwork.

The RCC was compacted using 15 tonne Bomag BW 217 D single drum vibrating rollers with high frequency and low amplitude. These rollers had a much higher compactive effort than the smaller double drum rollers that are more commonly used. It was found that 6 to 8 passes were sufficient to achieve the required compaction of 98% TAF density. The desired result was a fully pasted smooth surface to the RCC. A "boney" surface usually indicated inadequate workability of the RCC or a lack of compaction. The smooth surface was a sign that the RCC had been fluidized under the action of the roller and it meant that the subsequent clean-up and preparation of the joint was considerably easier.

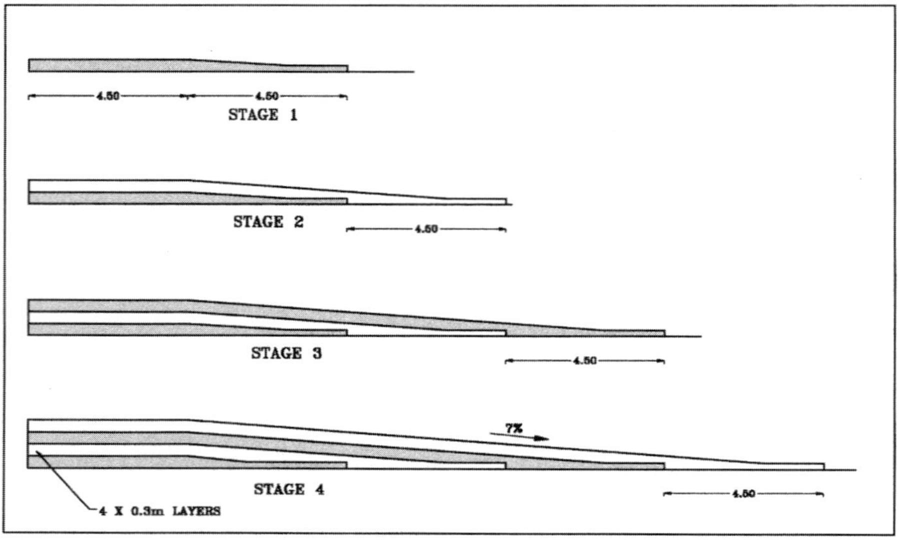


Fig. 3. Slope Layer Placing Method

OTHER RCC OPERATIONS.

GERCC Facings

The GERCC facing to the dam was produced by adding just sufficient 1 : 1 cement grout to the uncompacted RCC and then using conventional poker vibrators to compact the concrete so created. The grout was mixed nearby using hand-held mixers. The grout application rate was 8 litres/linear m for a 400mm width and 300mm thick layer. The grout was spread by hand using buckets. It was a relatively low-tech operation but a small team of workers were trained in the techniques of mixing, dosing and compacting so that the work was integrated into the overall RCC placing system. The width of the GERCC was initially specified as 400mm. However it was widened to 600mm for the slope layer method and locally in the vicinity of the contraction joint waterstops. The exposed horizontal surfaces of the downstream steps were also formed in GERCC and were given a wood float finish. GERCC was also used against the steep rock faces of abutment in lieu of the specified immersion vibrated concrete.

Transverse Contraction Joints

The contraction joints in the dam were located at a nominal spacing of 15m as indicated on Fig. 4. These joints were cut with a spade tool on a standard jackhammer. The open joint was filled with dry fine (Tafila) sand before it closed up. The surface was then re-rolled longitudinally using a small roller.

The object was to cut 200mm or $^2/_3$ of the depth of the contraction joint through each layer, leaving the balance to crack.

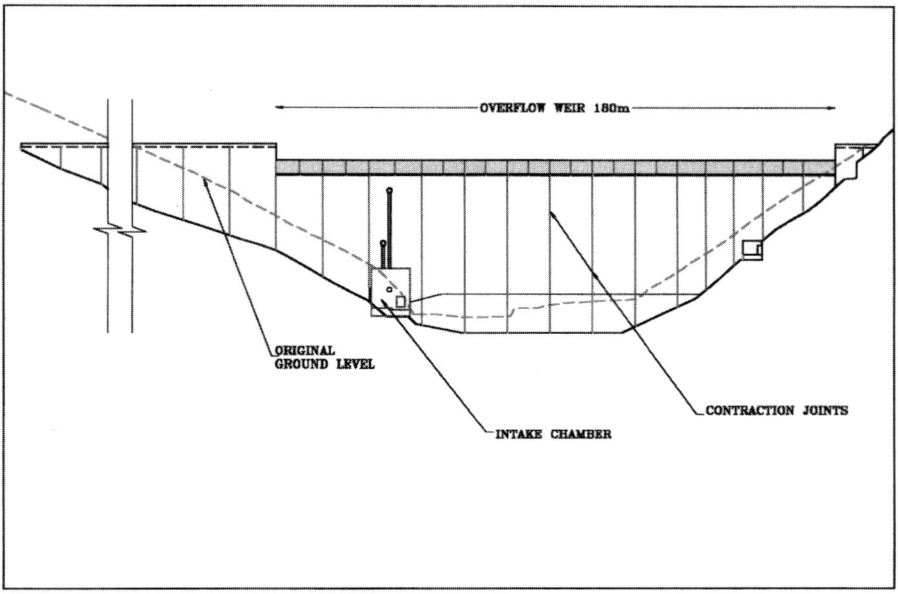


Fig. 4. Location of Contraction Joints

In the GERCC a thin sheet steel former was driven into the joint and left in place. In the vicinity of the waterstop, special steel formers were used to accurately locate the joint in relation to the bulb of the waterstop and to form the vertical drain behind the waterstop. For the horizontal layer method the steel formers were progressively withdrawn once the GERCC had taken its initial set, but these were left in place for the slope layer method.

Bedding Mortar

Various designs of a bedding mortar were tested in the site laboratory using a graded 5 mm material and aimed at achieving a 150mm slump. The best result was obtained with a mix containing 400kg cement and a blended aggregate containing 50% crusher sand and 50% Tafila sand. A mid range concrete plasticizer, Daracem SP4 was added to reduce the water content and improve workability whilst offering moderate set retardation.

The bedding mortar was normally produced in the conventional concrete batching plant and transported to the dam in truck mixers. Approximately 2m$^3$ of mortar was required per hour to keep pace with the RCC placement. The mortar was discharged from the truck via a chute. Towards the end of the project when access for the trucks was restructed, tower cranes were used to lift the bedding mortar to the point of placing. The bedding mortar layer of at least 6mm thickness was spread by hand using notched rubber blades that were manufactured at site. Two or three labourers were assigned to the bedding mortar spreading and worked just ahead of the RCC placing.

Other Uses of RCC

In addition to the dam wall, RCC was also used for the stilling basin floor slab and for the base of the downstream channel. In both cases it had been the intention that these works would be constructed from conventional concrete. However with the RCC operations under way it was concluded that the areas involved were sufficiently large to make RCC placing a viable and cost-effective alternative. In the case of the stilling basin, the original 2m thick reinforced concrete slab was reduced to a 0.5m thick slab that was anchored down to an underlying body of RCC. For the downstream channel, the base slab required approximately 3000m$^3$ of concrete with an average thickness of around 2m. In this area alone the use of RCC instead of normal mass concrete yielded a cost saving of almost JD 100,000.

RCC PLACING IN THE SUMMER.

During the early part of 2000 it became clear that the RCC placing would not be completed in a single winter season as had been the original plan. From an overall project point of view it did not make economic sense to stand down capital intensive plant and to delay the completion if this could be avoided. The possibility of placing RCC in the summer period was therefore examined. It was found that, provided the uptake of heat into the RCC surfaces during the heat of the day was prevented, then limitations on RCC placing temperatures alone would be sufficient to manage thermal stresses in the dam. Various measures were introduced to control RCC placing temperatures. As a result it was possible to continue placing during the summer although day-time placing was generally not possible, and in the hottest months the available window for placing was restricted to around 14 hours per day. The measures that were put in place to reduce RCC placing temperatures were as follows: -

- Stockpiling of aggregates in the winter period for subsequent use in the summer. The low thermal conductivity of the aggregate meant that the aggregate produced and stockpiled in the winter retained its cooler temperature into the summer.

- Evaporative cooling of the aggregate stockpiles by means of irrigation systems.

- Installation of a water chilling plant for the RCC mixing. However with only 105kg of water per m$^3$ this was of only limited benefit.

- The use of air/water mist sprays over the freshly placed RCC to provide evaporative cooling and to make up for surface water loss. In the low humidity environment it was found that the ambient temperature could be reduced by 5 to 10°C, so providing a cooler microclimate for RCC placing.

With these measures it was possible to reduce the temperature of the fresh RCC. The criteria that were established for placing to continue were that the instantaneous placing temperature should not exceed 27°C and that the rolling 3 day volumetric average RCC temperature should not exceed 26°C. In addition it was found that placing was not practical if the ambient temperature exceeded 35°C. A comprehensive monitoring regime was implemented to check both the fresh RCC temperature and the temperature build up in the recently placed RCC. For this purpose additional thermocouples were placed in the centre of each of the 1.2m thick lifts.

## CONCLUSIONS.

On the completion of placing of RCC to Tannur dam the following conclusions could be made: -

- The use of GERCC against formwork produced a durable outer surface to the dam at a lower cost than the slip formed system that had been specified.

- The use of the slope layer method for placing resulted in an improved final product with savings in the use of bedding mortar at lift joints and benefits in terms of increased production.

- The use of GERCC at the RCC / foundation interface proved to be an effective and economical way of achieving an intimate contact.

- The control measures that were introduced to allow RCC placing to continue into the summer were effective and were introduced without compromising the integrity of the structure.

- The use of RCC instead of conventional concrete in the floors of the stilling basin and the downstream channel provided further cost savings.

## ACKNOWLEDGEMENTS

The author is grateful to Mr Brian Forbes of Gutteridge, Haskins and Davey Pty Ltd, who were in joint venture with Mott Macdonald Ltd on the Tannur Dam project. He also expresses his thanks to the Project Manager, Mr Khalid Hazaymeh, and to other members of the Jordan Valley Authority.

# Seismic performance of dams

# Seismic assessment of Scottish dams

K J DEMPSTER, Scottish and Southern Energy, UK
A C MORISON, Halcrow Group, UK
S C GALLOCHER, Halcrow Group, UK
S BU, Halcrow Group, UK

SYNOPSIS. Halcrow Group are undertaking seismic assessment of the 90+ dams of various types owned by Scottish and Southern Energy. Initial work involved seismic assessment and ranking of all the dams and selection of key dams, for which a subjective 'seismic vulnerability index' was developed to complement the UK seismic risk classification. Assessment of the key dams was generally carried out using both dynamic finite element seismic analysis and a simplified method. A range of sensitivity cases were included in the key dam analyses. From these it was possible to assess the range of likely performance of the key dams under seismic conditions and, using simplified analysis, to extrapolate this to the majority of the other dams of similar type.

## INTRODUCTION

### Dam Stock

Scottish and Southern Energy's stock of dams is markedly different from any other UK dam owner. The dams are predominantly concrete structures with 45 gravity dams, 8 buttress, 4 arch, 1 pre-stressed, 14 earth embankment, 3 concrete faced rockfill, 9 composite concrete gravity and embankment and 9 small embankment or cut-off sections. One reservoir registered under the Reservoirs Act, 1975 is a 5 km long concrete lined open trapezoidal aqueduct. Fifty-six of the dams are included in the ICOLD world register, including one arch dam where the reservoir is marginally below the capacity for inclusion in the Act.

The buttress dam at Sloy is the highest at 56 m and Mullardoch Dam with a crest length of 727 m is the longest. Loch Quoich, impounded by a 38 m high concrete faced rockfill dam is the largest reservoir in the UK with a capacity of 382 x $10^6$ $m^3$. Other unique structures within the diverse range of concrete dams include Monar, a 35 m high arch dam and the only double curvature arch dam in the UK, and Allt-na-Lairage, a pre-stressed concrete dam.

Nine of the dams are more than fifty years old, most significantly those constructed by the Grampian Electricity Company as part of the Tummel Scheme in 1932. The oldest structure is Loch Mhor dam, which is of

*Reservoirs in a Changing World*, Thomas Telford, London, 2002, 87–99

concrete gravity construction with associated masonry and earthfill elements, all constructed in 1896. The remainder were generally constructed during the programme of intensive hydro-electric development between 1947 and 1963. The design of the majority of the dams is unlikely to have specifically included the effects of seismic loading.

Dam Management approach and the assessment project

Scottish and Southern Energy's approach to the development of risk based dam safety and asset management systems is well documented (Sandilands & Findlay, 2000). The Company policy on reservoir safety has always been aimed at compliance with UK legislation and with best dam engineering practice. As a consequence all dams comply with certain basic criteria: all dams are capable of passing design floods calculated in accordance with recognised UK standards; all dams are stable under static loads. To date assessment of static load has been based on the design flood level. This is currently being re-evaluated to ensure that dams have been assessed for stability during a probable maximum flood (PMF). As well as public safety risk, business risk also needs to be considered. Hydro generation assets have a very high replacement value and due to their maturity, the historic capital value is low and operating costs tend to also be low. Thus the Hydro Generation business is very important to Scottish and Southern Energy. Whilst no dam is business critical, loss of an individual dam would, of course be highly adverse to the business due to the high replacement value and the loss of income.

There are a number of factors, which give a significant degree of confidence in the safety of the dams:

- The majority of the dams are concrete dams of relatively modern design and construction, incorporating advances in technology since 1945.
- Almost all the dams are on competent rock foundations.
- Due to the shortage of suitable clays in the North of Scotland, concrete core walls are the most common form of impervious membrane at the non gravity sections.
- There are only two homogeneous earth embankments and both are relatively small.
- The reservoirs are operated under the Reservoirs Act 1975.
- Worldwide, the performance and safety of dams during earthquakes has been remarkably good.

However, although the risk of a serious failure of any of the dams is considered to be extremely small, the consequences of failure, should one occur, could result in extensive damage to property and infrastructure, and could put the downstream population at risk.

All of these issues and factors influence Scottish and Southern Energy's approach to risk awareness and management with a view to understanding, managing and mitigating the hazards posed by their dams. The current programme of seismic assessments, generally in accordance with the UK Seismic Guide (Charles et al, 1991) as modified by the application note (ICE, 1998) and the PMF load case check, being carried out by Halcrow, form an integral part in the risk management processes. When completed this will allow decisions to be taken on the need or otherwise for further action.

## INITIAL REVIEW

### Previous studies

Over the period from 1985 to 1999 Scottish & Southern Energy commissioned individual seismic stability assessments at a limited number of their higher risk dams from various consultants. Examination of the reports of these showed that, although each included an appropriate seismic analysis, a wide range of different approaches had been adopted, only some of which related to advances in the available methods of analysis over this period. Few made any comment on the sufficiency or likely applicability of the method of analysis adopted. None of these reports confirmed that the structures were unsafe under the assessed seismic loading to the extent of requiring protective remedial measures, although various degrees of cracking under seismic loads were identified.

Because of the different approaches used, it was not possible from the earlier studies to compare the seismic vulnerability of the individual structures. Discussion on this point confirmed the need for a coherent methodology for the project so that it would be possible to explore and compare the results from a variety of methods of analysis in context in assessing the need for and prioritising remedial actions.

Before the project started Scottish & Southern Energy had already addressed one major cause of variability by commissioning a report (EQE, 1998) to define the site-specific peak horizontal ground acceleration (PHGA) for each of the dams at different return periods. This data superseded the maps of PGA from the Seismic Guide. At most of the Scottish and Southern Energy sites the 1:10,000 year PHGA is in the range 0.24 to 0.2g, but in some areas of northern Scotland and the Outer Hebrides, this drops to below 0.1g.

### General approach to analyses

With over 90 dams to consider, the general approach adopted was to subdivide the dams by type, select key dams for detailed analysis by a range of methods and then to carry out simpler, less sophisticated, parametric analyses of all the dams in order to assess their performance relative to the key dams. Standardised load, material and foundation properties were adopted for the base case analyses of the key dams. The effects of possible variations of these were assessed as sensitivity analyses and taken into account in the parametric analyses of other dams. The methods of analysis used varied to suit the different types of dam, but ranged from simple 2-d pseudostatic analyses to 3-d dynamic analyses of the whole dam.

Selection of key dams

Key dams were intended to be among those where seismic loading was likely to be most critical but also, where possible, to be representative of a group of other dams of similar design. In selecting the key dams the "*seismic risk classification*" (SRC), as defined in the UK Seismic Guide (Charles et al, 1991), was taken as one of the primary factors, however the parameters used to define this relate principally to the downstream effect of failure and not to the vulnerability of the structure to failure.

ICOLD Bulletin 72 – "*Selecting Seismic Parameters for Large Dams*" lists the factors relevant for defining the parameters for seismic analysis of a dam as:

- The seismic hazard rating of the dam site
- The risk rating of the completed structure
- The type of dam and its possible modes of failure

It was therefore felt appropriate to take structural vulnerability into account in selecting the key dams, and a complementary "*structural vulnerability index*" (SVI) was developed to do this. The parameters adopted for this were:

- Dam height
- 1:10,000 year return period PHGA
- Type of dam, with additional adjustments for key risk factors of each type
- Dam foundation

There was some discussion over the need for a parameter relating to the existing condition of the dam. However, it was eventually agreed that only parameters easily derived from existing available data should be used.

Details of the index are too extensive to be included here, and are the subject of a separate paper, now being prepared. However, the index developed is entirely subjective and, as developed, relates only to the family of dams covered by this analysis. As it stands, it is probably not suitable for universal application.

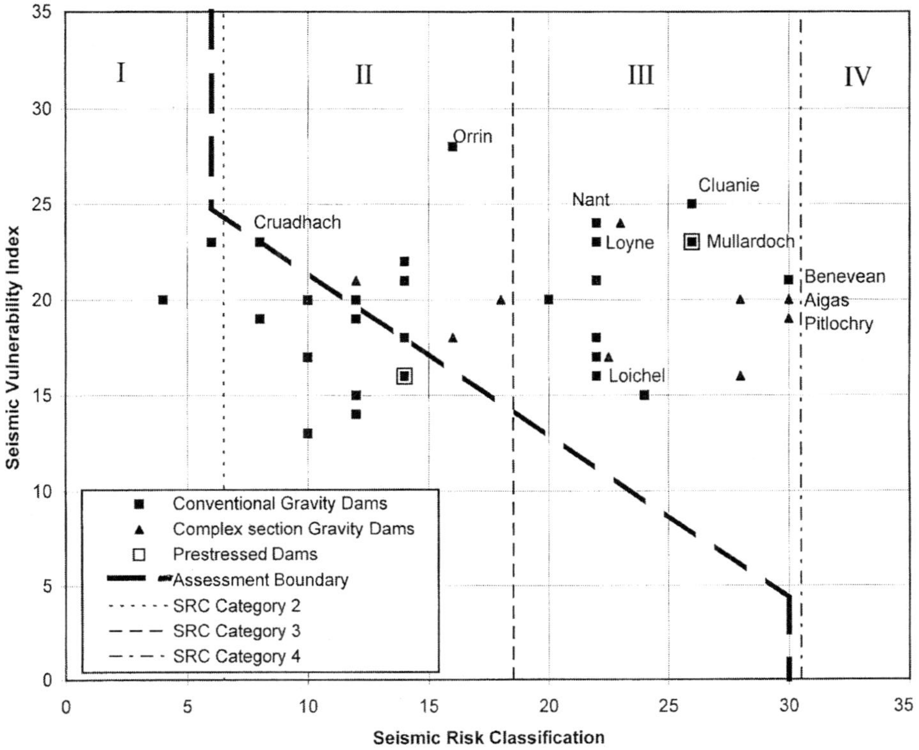


Figure 1. Plot of SVI against SRC for gravity dams

Table 1. Key Dams selected

| Key Dam | Type | Height (m) | PHGA (g) | SRC | SVI |
|---|---|---|---|---|---|
| Monar | Arch | 34.75 | 0.21 | 26 | 28 |
| Sloy | Buttress | 48.62 | 0.20 | 18 | 33 |
| Mullardoch | Gravity | 37.46 | 0.21 | 26 | 23 |
| Loichel | Gravity | 13.5 | 0.21 | 22 | 16 |
| Nant | Gravity | 23.49 | 0.21 | 22 | 24 |
| Glascarnoch | Embankment – Earthfill with central concrete core | 29.64 | 0.20 | 26 | 27 |
| Lairg | Embankment – Rockfill with central concrete core | 15.7 | 0.14 | 24 | 15 |
| Pitlochry | Complex Concrete Gravity, two drum gates, powerhouse and core wall embankment | 17.37 | 0.20 | 30 | 17 |

As an example of how it was used, Figure 1 shows a plot of seismic risk against the seismic vulnerability index for the gravity dams. It shows both the Seismic Guide classification boundaries, based on risk alone, and the combined boundary, beyond which more detailed assessment was considered appropriate. The key dams of each main type shown on Table 1 were generally selected from those in the upper right hand corner of the plots.

## ANALYSIS OF GRAVITY DAMS

### Approach and method

Concrete gravity dams are the most numerous type of dams for which analysis is considered necessary. These include both conventional spillway and non-overflow sections, and complex gravity sections incorporating gates, power stations and other structures that also act as gravity retaining walls. Many of the conventional gravity overflow sections are incorporated within dams with embankment section abutments. Initially only the conventional gravity sections have been analysed, but it is intended to return to the complex sections, which will require more individual attention, once the lessons from the generalised analyses have been learned. Most of those not requiring analysis are small spillway sections.

At the level of analysis conducted it was accepted that 2-dimensional analysis is appropriate for most gravity dam sections. Analyses were carried out using both the traditional rigid block gravity method with pseudostatic seismic loading and a 2-dimensional dynamic analysis using EAGD-SLIDE, (Chavez & Fenves, 1994), a well known public domain program incorporating the possibility of a slip plane developing at the base of the dam.

The seismic loads were all superimposed on gravity loads based on the dam geometry, hydrostatic loading and uplift with the reservoir at normal top water level. Sensitivity tests were carried out on the EAGD-SLIDE analyses for: dam height, PHGA, different seismic time histories, structural and foundation elastic modulus, structural damping, foundation contact slope, shear properties and uplift pressure distribution.

Review of the dam sections revealed that the designers of the Scottish and Southern Energy gravity dams used two separate design approaches, the first producing a steeper downstream face (typically 0.7:1), with a small upstream batter, incorporating dam body drainage and with the spillway ogee projecting downstream of the line of the downstream face. The second design typically had a downstream face of 0.75:1, with a vertical upstream face, no dam body drainage and the spillway ogees tangential with the downstream face but projecting upstream of the upstream face. Mullardoch and Nant dam sections, shown on Figure 2 are representative of these two approaches. Loichel is similar to Mullardoch dam in section, but less than half the height.

Results

Conventional rigid block gravity analysis with pseudostatic horizontal loading of the design ground acceleration and factored vertical acceleration showed that both Mullardoch and Nant Dam sections were satisfactory in compression, upstream face tensile stresses were within what could be expected of reasonable quality lift joints and shear-friction factors of safety were satisfactory provided expected values for cohesive shear strength were used. However, if cohesive shear strength

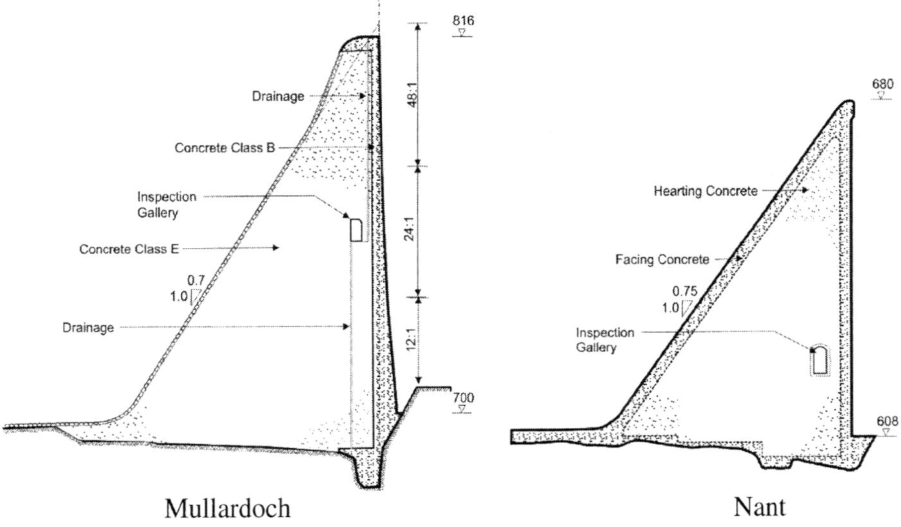


Figure 2. Comparison of Mullardoch and Nant Dam spillway sections.

was ignored, the sections would not be restrained from sliding by friction alone with tan $\phi$ = 1.0. Loichel was found to be stable in all respects.

The EAGD-SLIDE analysis of the highest spillway and non-spillway blocks at Mullardoch dam under the 1:10,000 year PHGA predicted sliding displacement of up to 7mm, with tan $\phi$ of 1.0 and c = 0, reducing to 1mm if tan $\phi$ is increased to 1.2. Peak effective tensile stresses at the upstream heel were generally less than 0.5MPa, compared with an allowable dynamic concrete tensile strength of 0.75MPa, but were high enough to cause cracking on the downstream face of the non-spillway section just below crest level, although the local stress concentration will be less in practice since the model did not include the radius between the dam crest and downstream face.

EAGD-SLIDE analysis of Nant Dam calculated 30mm sliding displacement of the non-spillway block and 15mm of the spillway block for c=0, tan $\phi$ = 1.0, reducing to 3 and 1mm for tan $\phi$ = 1.2. Peak effective tensile stresses at the upstream face were generally less than 0.3MPa. Nant dam is unusual in having a gallery but not drainage. Analysis showed that the addition of drainage would eliminate sliding.

The dynamic analysis of Loichel dam showed no sliding and effective tensile stresses at the upstream toe of less than 0.2MPa.

No significant correlation was found between the peak stresses calculated by the gravity method and EAGD-SLIDE using the same assumptions, however sliding distances calculated by EAGD-SLIDE did not exceed 1mm where the calculated shear-friction factor was greater than 1.0 from the gravity method.

The first natural frequency of the full height sections of both Mullardoch and Nant Dams both fell within the peak zone of the UK response spectrum between about 5 and 12 Hz, where maximum attenuation of seismic loads is expected. However lower sections have higher natural frequencies, and it was found that the lower blocks of the same sections had both lower stresses and were significantly less prone to sliding under the same seismic loading. Sliding of 1mm or less was found for all sections less than 15m high. Mullardoch and Nant dams are both the highest in their respective groups and are therefore likely to define the worst cases. However, Glascarnoch dam has a section similar to Nant and at 29.7m, is higher, but has a fully-drained section. A specific analysis of this section showed less than 1mm deflection and stresses within allowable limits.

Sliding distance from EAGD-SLIDE was found to be very sensitive to sliding parameters adopted, in particular the value of tan ϕ used, as shown by Figure 3.

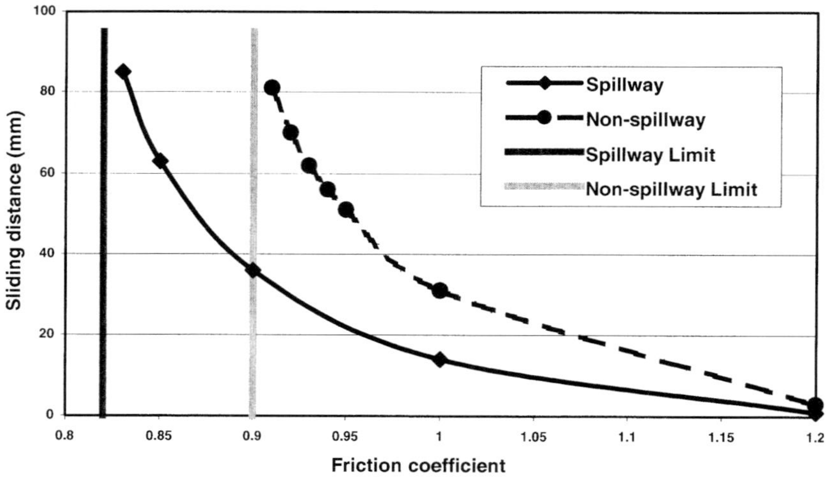


Figure 3. Calculated sliding distance for a range of tan ϕ values for Nant spillway.

## ANALYSIS OF BUTTRESS DAMS

### Approach and methods

Buttress dams were developed in the 1930s and the main concept of a buttress dam arose from considering where concrete could be left out of a gravity structure and how the area of uplift could be reduced. Clearly, because the dam comprises buttresses of varying height, analysing a buttress dam is more complex than a gravity dam, particularly under dynamic loading. The Scottish and Southern Energy buttress dams are mostly massive, smooth-faced designs, being essentially gravity dams with sections removed from the downstream face with the upstream face spanning as arches between the buttresses but also include round and diamond head buttresses where the upstream face is shaped to transfer lead to the buttresses, and each buttress acts as a separate structure.

For Sloy dam, a massive buttress dam, a 3D elastic SAP multi-buttress model was adopted to evaluate the stress distribution within the dam body. The dam-foundation rock interaction was modelled using non-linear link (NLLINK) elements. The earthquake-induced hydrodynamic pressures acting on the upstream face were simulated by the approximate method suggested by Zanger (1952). In addition, the earthquake-induced sliding behaviour of Sloy Dam was also conservatively assessed, using the two-dimensional EAGD-SLIDE package (Chavez and Fenves, 1994).

### Results

The peak earthquake-induced compressive stresses, tensile stress in the upstream -downstream direction, and shear stresses calculated within Sloy Dam body were generally less than the allowable strength of the dam concrete.

The calculated peak tensile stress found across the valley was 2.24MPa. Tensile cracking at construction joints from such stresses would cause leakage, but are unlikely to cause catastrophic failure. High tensile stress was also found at the base of the upstream heel and here the effective tensile stress significantly exceeds the tensile strength of the dam concrete. The buttress heads are therefore expected to crack horizontally at the upstream face under seismic loading. The cracked section was further investigated by a quasi-nonlinear analysis, carried out by releasing restraint at nodes in tension, which confirmed that the cracked buttress sections are stable after the earthquake under normal hydrostatic loading. However, assessment and remedial works may be required after an earthquake to provide suitable factors of safety for long-term stability.

The EAGD-SLIDE analysis showed that, under the 1:10,000 year PHGA, the higher buttresses might slide a short way downstream, although this is not certain to happen in practice. Displacement, should it occur, is expected to be significantly less than indicated by the numerical model as there is additional resistance provided by the sloping foundation surface, buttress embedment and lateral constraints not taken into account in the calculations. The combination of these and a high threshold value of $\phi$ before sliding is initiated are likely to provide sufficient restraint to prevent sliding movement under the 1:10,000 year seismic loading. The possibility of sliding was not supported by the calculated shear stresses from the stress model.

## ANALYSIS OF MONAR ARCH DAM

### Approach and method

Arch dams are designed to transfer most of the water load onto the sides of the valley. An arch dam can only fail because it is overstressed. Thus the assessment of arch dams generally centres on the magnitudes of stresses in the dam body. EACD-3D-96 (Tan and Chopra, 1997) was used to carry out a seismic analysis of Monar Dam.

### Results

The peak earthquake-induced compressive stress and shear stress at Monar Dam are within the allowable strength of the dam concrete. The peak upstream tensile stress, which is higher than the characteristic tensile strength of the dam concrete, occurs at the sides of the arch dam and, consequently, tensile cracking in this area can be expected. Such cracking at the upstream dam-foundation interface would not lead to catastrophic failure, as the corresponding downstream face is always in compression and the line of action of the loading remains within the structure without compressive overstress. Cracking in this region was predicted in the original design analysis. The peak downstream tensile stress occurs at the region at the centre of the dam crest, and is slightly higher than the characteristic tensile strength of the dam concrete. Minor tensile cracking may occur at the downstream face in this area. Inspection, assessment and remedial works at the deepest dam heel, the crest level of the crown, and the lower part of the sides of the dam may be required after an earthquake to provide suitable factors of safety for long term stability, but Monar Dam is considered capable of withstanding the 1:10,000 year PHGA without catastrophic failure.

## ANALYSIS OF EMBANKMENT DAMS

### Approach and method

Only the highest cross-section of the embankment dams were analysed as a two-dimensional case. Since the serviceability of a slope after an earthquake is controlled by deformations, analyses that predict slope displacements provide a more useful indication of seismic safety of embankment dams. The main approach adopted by the project was to undertake two-dimensional analysis using the program FLUSH (Lysmer, et al, 1975) to determine the earthquake-induced acceleration time histories throughout the dam. These accelerations were then used as the input to a sliding-block model at various slopes (Newmark, 1965) to assess the dam stability. The factor of safety against sliding varies with the ground acceleration throughout an earthquake, and displacements are calculated where this falls below 1.

An initial assessment of liquefaction (Kramer, 1996) has been carried out by addressing questions pertain to the liquefaction hazard susceptibility.

### Results

Under 1:10,000 year PGHA both Lairg Dam and Glascarnoch Dam were found to be stable with only limited permanent movements (2 and 11mm, respectively) in the upstream face. Displacements of these magnitudes, should they occur, would not lead to catastrophic failure. However, some cracking and surface sliding zones could develop at the upstream faces and assessment and remedial works may be required after an earthquake to provide suitable factors of safety for long-term stability.

A simplified method (Bray, 2001) was also used to assess the maximum earthquake-induced displacements of Lairg and Glascarnoch Dams. Comparative study reveals that the simplified method can predict reasonable results at a significantly lower calculation cost.

Based on the available information, an initial assessment of liquefaction shows that the dam materials of both Lairg Dam and Glascarnoch Dam are not susceptible to liquefaction. A FLAC analysis (Itasca, 1996) was carried out to verify this conclusion. However, there is insufficient geotechnical data available at present on the superficial foundation deposits left in place beneath the embankments to confirm absolutely that these could not be liable to liquefaction.

## CONCLUSIONS

The analyses carried out to date indicate that none of the Scottish and Southern Dams examined so far have a significant risk of catastrophic failure under the 1:10,000 year design seismic loading, which for the area has a peak horizontal ground acceleration of between about 0.2 and 0.24g. Localised tensile cracking of the higher concrete dams is expected at the upstream heel and at the change of section below heavy crest blocks, but the cracked sections remain stable under post seismic conditions. Some slight sliding displacement is also suggested by the base case calculations, which adopt inherently conservative assumptions, but from sensitivity analysis, this is considered unlikely in practice. The dams may require structural investigation, assessment and remedial works after such an earthquake.

The seismic vulnerability index proved to be a transparent method for helping to identify what indeed turned out to be the key dams, although it produced no major surprises for those already familiar with the Scottish and Southern Energy dams.

The potential for cracking and sliding of the concrete dams was found to be particularly sensitive to dam height and the sliding parameters adopted. All of the concrete dams less than 15m high studied had both stresses and sliding stability within acceptable limits, and are not considered susceptible to seismic damage under accelerations of up to 0.2g, provided they are initially free of structural cracks in critical locations, have well made lift joints and are adequately stable under normal conditions. This conclusion is also expected to apply to such dams elsewhere in the UK.

REFERENCES

Bray J D (2001). Methods of Analysis, Evaluation and Mitigation of Seismic Hazards, University of California, Berkeley.

Charles J A, Abbiss C P, Gosschalk E M & Hinks J L (1991). An Engineering Guide To Seismic Risk To Dams In The United Kingdom, BRE, London

Chavez J W & Fenves G L (1994). EAGD-SLIDE: A Computer Program For The Earthquake Analysis Of Concrete Gravity Dams Including Base Sliding, Department of Civil Engineering, University of California, Berkeley.

EQE International Ltd (1998). Seismic Hazard at Scottish Hydro-Electric Dams.

ICOLD (1989). Bulletin 72. Selecting Seismic Parameters for Large Dams.

Institution of Civil Engineers (ICE) (1998). An Application Note to "An Engineering Guide To Seismic Risk To Dams In The United Kingdom".

Itasca Consulting Group (1996), FLAC Version 3.3 User's Manual.

Lysmer J T, Udaka C F, Tsai & Seed H B (1975). FLUSH-- A Computer Program for Approximate 3-D Analysis of Soil-Structure Interaction Problems, Earthquake Engineering Research Center, Report No. UCB/EERC-75/30, University of California, Berkeley.

Kramer S L (1996). Geotechnical Earthquake Engineering, Prentice Hall, New York.

Newmark N M (1965). Effects of earthquakes on dams and embankments. *Geotechnique,* 15 (2) 139-159.

Sandilands N M, Findlay J W, (2000) Development of a Risk Based Approach to Dam Safety Management. 20th ICOLD Congress, Beijing, Q76, ICOLD.

Tan H & Chopra A K (1997). EACD-3D-96: A Computer Program For Three-Dimensional Earthquake Analysis Of Concrete Dams, Department of Civil Engineering, University of California, Berkeley.

Zanger C N (1952). Hydrodynamic pressures on dams due to horizontal earthquake effects. Engineering Monograph No 11, US Bureau of Reclamation.

# Assessing the Seismic Performance of UK Intake/Outlet Towers

W E DANIELL, Earthquake Engineering Research Centre, University of Bristol, UK
C A TAYLOR, Earthquake Engineering Research Centre, University of Bristol, UK

SYNOPSIS. The seismic assessment of a reinforced concrete intake tower in Scotland included a seismic "walk-down" inspection, ambient vibration tests and finite element analyses. The purpose of the inspection was to identify vulnerable features in the operation and structure of the tower. Ambient vibration tests on the tower were conducted to measure its modal properties. These were employed to validate a finite element model. A preliminary elastic seismic analysis of the model was run using synthetic accelerograms. The results indicated a non-linear response and the requirement for further detailed analysis to assess the seismic capacity of the tower.

## INTRODUCTION

Following publication of the 1991 BRE guide (Charles et al., 1991), many UK dams have now been assessed for their seismic risk. However, their appurtenant structures have received relatively little attention. Many of these structures may be vulnerable to earthquakes, as they were not designed for seismic loading. Depending on their function, appurtenant works can play an important role in reservoir safety, especially intake/outlet towers and gates as they regulate the release of water. Their continued operation could be required for emergency evacuation of the reservoir in the event of a damaged dam.

As part of their asset management (Sandilands et al., 1998), Scottish and Southern Energy (SSE) include seismic assessments of appurtenant works as well as their dams. Although there is ample published guidance on the seismic assessment of dams, internationally, there is little useful, similar guidance for appurtenant structures. To make up for this deficiency, the University of Bristol's Earthquake Engineering Research Centre developed a guide for appurtenant structures tailored specifically to SSE needs (Daniell and Taylor, 2000). It was based on a review of other guides and standards for dams and the seismic evaluation of structures generally, and also, publications on appurtenant structures, with regard to their seismic assessment and real earthquake experiences. As part of the development, two case studies were carried out on typical SSE structures. The work

*Reservoirs in a Changing World*, Thomas Telford, London, 2002, 100–111

included a preliminary seismic assessment of Loch Glascarnoch intake tower (Figure 1). The case study was not intended as a definitive seismic evaluation of the tower, but an investigation into the potential seismic risk to SSE towers and into an appropriate method for their numerical analysis. The study included a seismic "walk-down" inspection, ambient vibration tests and the development of a finite element model for its seismic analysis.

The seismic walk-down inspection was simply a visual inspection of the tower and its facilities. This type of inspection is used commonly for nuclear installations. The inspection requires judgement and knowledge of previous experiences, usually catalogued, to identify vulnerable features in the operation and structure of the tower.

The ambient vibration tests were conducted to measure modes and frequencies of vibration for the tower. It was particularly important to measure the fundamental mode of vibration, as this is the mode that participates most in the seismic response. Knowledge of its frequency establishes whether the tower will resonate and respond dynamically to a typical earthquake for the region, amplifying the ground accelerations over the height of the tower. A value for the fundamental frequency is required for equivalent-static seismic analyses where the acceleration of the structure is determined from a response spectrum. The acceleration and mass distribution of the structure are used to calculate equivalent static, and lateral, earthquake forces.

The results of the tests were also employed to validate a relatively straightforward finite element model of the tower developed for its preliminary seismic analysis. A time history analysis of the tower was run using an artificial earthquake generated from an appropriate UK response spectrum.

This paper discusses the walk-down inspection, the ambient vibration tests and the finite element study for Glascarnoch tower.

## DESCRIPTION OF GLASCARNOCH TOWER

The tower (Fig. 1) is constructed from reinforced concrete and sits on a relatively rigid base structure. The top of the tower and the base are connected to a covered inlet structure with trapezoidal wing-walls. Continuous reinforcement is provided across these connections, and, in between, there are vertical construction joints with continuous shear keys. The tower is approximately 6.4 metres square with a 4.9 metre diameter circular well. It supports a steel-framed gatehouse with reinforced concrete slabs and infill walls. There are three floors in the gatehouse, one at bridge level and two higher ones that support the hoisting gear for the intake gate and the trash rake. The overall height of the tower is 24.5 m.

Figure 1: Glascarnoch Intake Tower

A 50 metre long, four-span bridge provides access to the tower. It consists of reinforced concrete slabs supported on concrete encased steel edge beams. Pairs of 840 mm diameter reinforced concrete piers provide intermediate supports.

WALK-DOWN INSPECTION

The inspection included an examination of the structure of the access bridge and tower and of all the equipment, controls and power supply. The tower could only be accessed above bridge level, as the well below contained water. Details were identified that would need modifying to improve the seismic resistance of the tower.

The tower contained a number of cabinets containing electrical and electronic control and communications equipment. With regard to seismic shaking, the vulnerable features observed included:

- Loose cables and spare components inside cabinets that could be thrown around and damage other sensitive electronic components.
- Simple push fit circuit boards that would require retaining clips to secure them.
- A deep cabinet cantilevering out from a wall from which support motions could be amplified.
- Inadequate supports or fixings to the cabinets.

Cable support trays running between the cabinets and the operating gear were also observed to have inadequate fixings and to be relatively flimsy.

A cable running from the reservoir shore across the access bridge supplies the electrical power to the tower. The cable is fixed to one side of the bridge and runs taut across the joints between the spans. Relative movement of spans is not uncommon during earthquakes and was observed on access bridges affected by the Northridge earthquake in California in 1994 (Daniell et al., 1994). In these cases, service runs were not damaged as they had been designed to be flexible to accommodate large displacements. However, any significant displacement of individual bridge spans would sever the mains cable to Glascarnoch tower.

A rack of batteries provides the back-up power supply to the controls. However, they sit directly in a mounting bracket and could shake loose, breaking connections and being damaged.

In the event of a total loss of power, hoisting gear can be operated manually. However, this required the winch handle to be available. This was observed to be lying loose next to the equipment. In the event of an earthquake, it could be shaken into an inaccessible and hidden place. The same could happen to the keys for control cabinets for important equipment, which were hung from insubstantial hooks.

The intake gate is stored in the gatehouse when the intake is open. In its raised position, the gate rests on temporary supports to relieve the tension on the cables from which it is suspended. During shaking, the gate could rock on these supports and maybe even swing, damaging equipment around it.

The risk to the seismic integrity of the tower posed by all of these features could be mitigated by very simple measures. They could include restraints to important and heavy equipment, secure storage for keys, improvements to fixings and allowances for flexibility within cable runs.

## AMBIENT VIBRATION TESTS

Wind provided the excitation force for the ambient vibration tests that were conducted on the tower over a period of five days. Accelerometers were used to measure the responses, which were acquired, digitised and stored on computer. The stored data were processed, after the tests, using Fourier analysis to produce frequency response spectra from which natural frequencies of the tower were identified.

Figure 2 shows a typical frequency response spectrum for the tower for motion normal to the longitudinal axis of the bridge. Natural frequencies were identified from the peaks, which occur clearly at 4.4 and 16.7 Hz, with a double peak with close frequencies of 5.8 and 6.0 Hz.

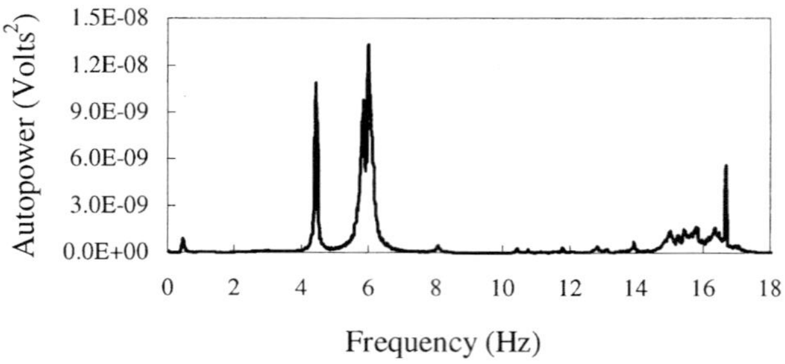


Figure 2: Autopower frequency response spectrum for E-W tower measurements

The frequency response spectra were used for further analysis to identify vibration mode shapes. To aid description of the mode shapes, modes of vibration in the direction of the longitudinal axis of the bridge will be referred to as N-S (north-south) modes. Similarly, modes normal to that direction will be called E-W (east-west) modes.

Modal analysis of the test data identified a number of important modes of vibration for Glascarnoch tower. They included the fundamental bending modes of the tower at 5.8-6.0 Hz for the E-W direction, and at 10.5-11.0 Hz for the N-S direction. For the bridge, two lateral (E-W) modes were identified at 4.4 Hz and 10.4 Hz, and a vertical mode at 8.3 Hz. The E-W mode shapes are shown in Figure 3 and natural frequencies are given in Table 1.

These results, together with a response spectrum for a typical earthquake at the site, provide useful information on the likely nature of the seismic response of the tower. The response spectrum shown in Figure 4 was used for the seismic analysis of Glascarnoch tower. It is a UK response spectrum for a site with "hard ground" and 5% damping for the structure. The spectrum is normalised for a peak ground acceleration (PGA) of 1.0 g. On the graph, the abscissa gives the natural frequency for a vibration mode of a structure and the ordinate the spectral response acceleration for that mode. It can be seen that the fundamental E-W mode (5.8-6.0 Hz) of the tower falls on the peak of the response spectrum. This demonstrates that the tower will respond dynamically with significant amplification of the ground motion. The selected return period for the Safety Evaluation Earthquake for the assessment of Glascarnoch tower was 10,000 years giving a peak ground acceleration of 0.2g (EQE, 1998). This indicates that at the top of the tower, the gatehouse and the equipment housed within it could be subjected to significant shaking with accelerations up to 0.6g.

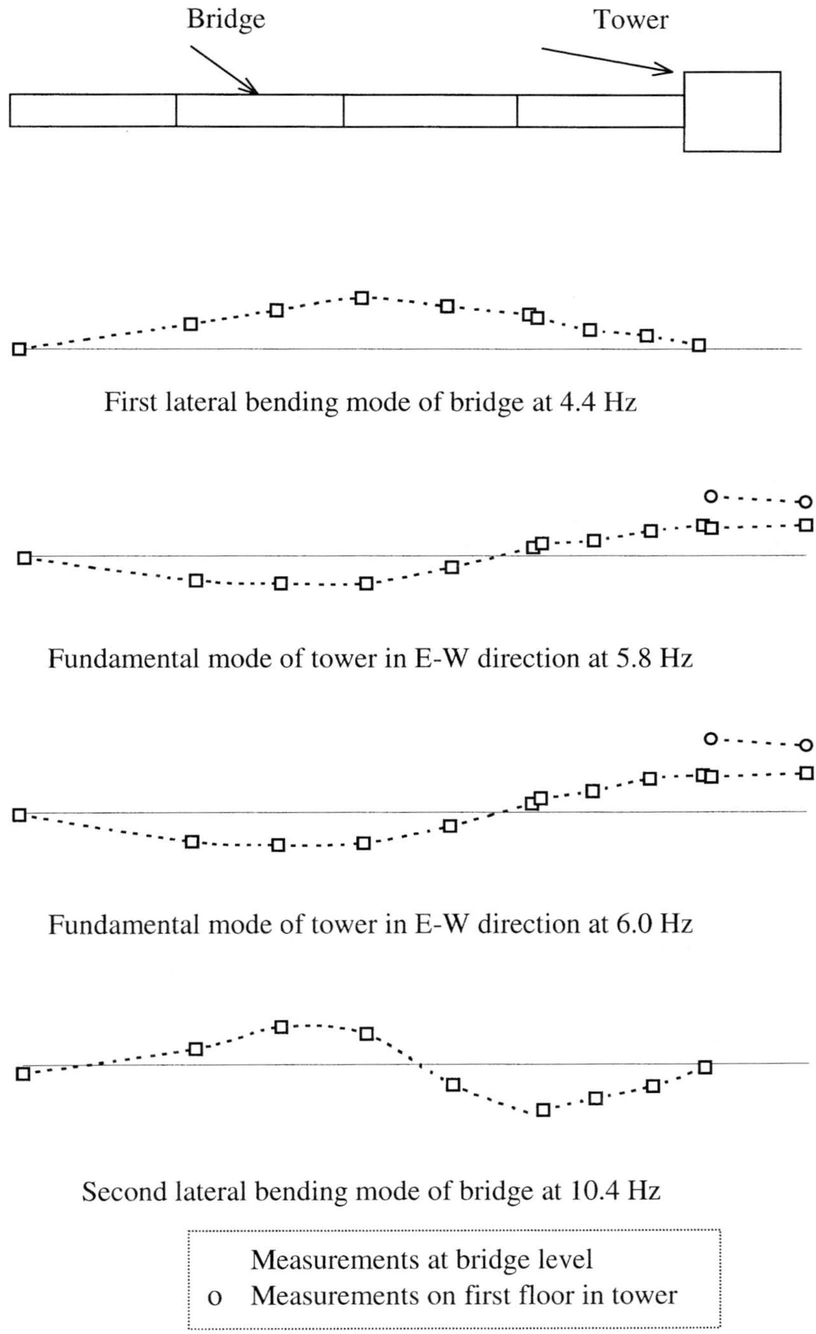


Figure 3: Modes shapes from tests

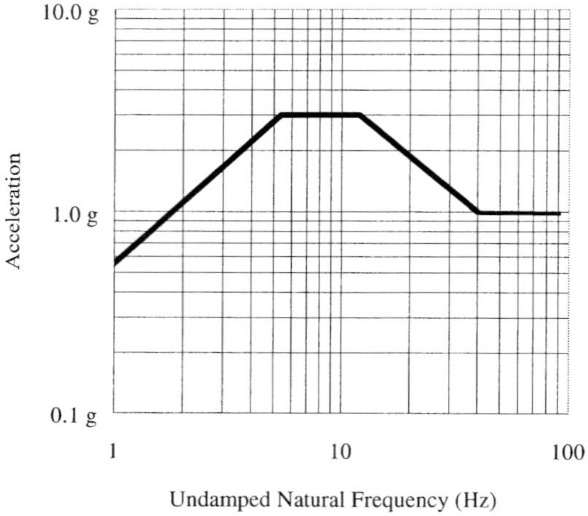


Figure 4: UK Response Spectrum for Hard Ground and 5% Damping Normalised to 1.0g PGA

## FINITE ELEMENT MODELLING

The finite element method is commonly employed for the numerical modelling and seismic analysis of intake/outlet towers and their access bridges. Tall, slender towers that respond mainly in bending can be represented by stick models using beam elements (e.g. Bureau, 1985), but solid models using 3-D brick elements are more appropriate for squat towers, which respond mainly in shear, or for towers with complex structural arrangements. Glascarnoch tower has a relatively squat and complex geometry, and therefore required a 3-D solid model.

The development and analysis of an accurate 3-D solid model can be time consuming and costly. However, such a model is not necessary for a preliminary seismic analysis of a tower where only estimates of its fundamental frequency and response are required.

As hand calculations of stick models are not suitable for complex towers, it would seem appropriate to develop a simple 3-D solid, finite element model for preliminary analysis. This can be achieved by simplifying the geometry of the tower so that less elements are required and employing a relatively coarse mesh. The results of the analysis of such a model will determine whether further analysis of a more refined and accurate model is even necessary. This was the approach used for Glascarnoch tower.

Details of Finite Element Model

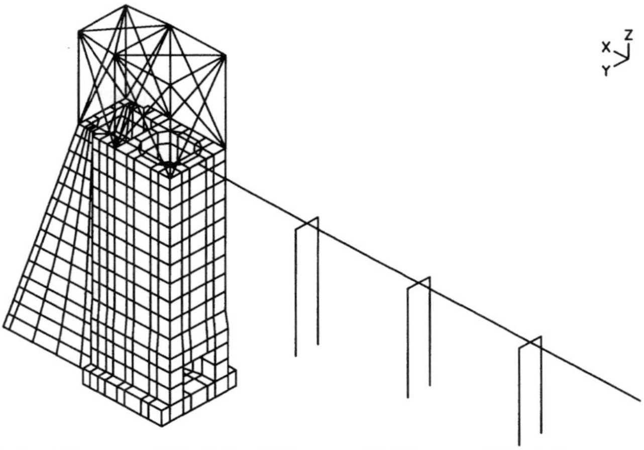


Figure 5: Finite Element Model of Tower, Bridge and Gatehouse

The finite element model of Glascarnoch tower (Fig. 5) included the tower, bridge, gate house and the dynamic effects of the reservoir. Soil-structure interaction was neglected, as the structures are founded on sound rock. The global X and Y axes for the model correspond respectively to the N-S and E-W directions for the real tower.

A linear-elastic model was developed for the seismic analysis of the tower even though the possibility of a non-linear response was recognised. This is a normal, if not essential, first step in a seismic assessment, identifying the dynamic characteristics of the structure, and indicating critically stressed regions, potential modes of failure and whether further more sophisticated analysis is required.

The tower and inlet structure were modelled with 3-D solid elements, and the base of the inlet structure was modelled as horizontal and level with the base of the tower, instead of sloping away from it. Only one element was used through the thickness of each wall. This provided an adequate representation of the distribution of the mass and stiffness of the tower for the modal analysis. Although this mesh was not sufficiently refined to evaluate seismic stresses accurately, it could identify the most highly stressed regions in the tower and be used to approximate the internal forces.

The vertical joints between the tower and inlet structure were modelled simply by altering the material properties of a column of elements in the walls of the inlet structure at each joint location. The modelled joints could transfer both tension and compression, whereas the real joints can transfer compression, but not tension. The inability of the joint to transfer vertical shear was modelled by providing a very low shear modulus, in the x-z plane, for the columns of elements.

The gatehouse was modelled as a rigid structure using very stiff beam elements, and the correct mass distribution was represented by appropriately placed lumped masses.

Single lines of beam elements represented the bridge deck and the piers. Equivalent section properties were employed, but approximations had to be made to represent the end conditions for each span.

The dynamic effects of the water surrounding the tower and bridge piers, and the water inside the tower, were modelled using the added mass concept. The hydrodynamic mass was calculated using Goyal and Chopra's method (1989) for intake/outlet towers and represented by appropriately distributed lumped mass elements.

Modal Analysis and Validation of Model

A modal analysis of the finite element model was run and resulting frequencies and mode shapes were compared with those measured from the tests. A comparison of the frequencies is given in Table 1.

Table 1: Natural frequencies in Hz from tests and finite element model

| **Description of Mode** | **Tests** | **FE Model** | **Difference %** |
|---|---|---|---|
| 1st lateral bending mode of bridge | 4.4 | 4.6 | +4.5 |
| 1st cantilever mode of tower in E-W direction | 5.8 or 6.0 | 5.5 | -5.2 or -8.3 |
| 1st vertical mode of bridge span next to tower | 8.3 | 7.6 | -8.4 |
| 1st cantilever mode of tower in N-S direction | 10.5 to11.0 | 10.3 | -1.9 to -6.4 |
| 2nd lateral bending mode of bridge | 10.4 | 12.4 | +19.2 |
| Torsional mode of tower | 16.7 | 17.3 | +3.5 |

The correlation between the theoretical and measured frequencies was very good with up to 9% difference between them, excepting the second, horizontal bending mode for the bridge, which showed a 19.2% difference. This larger difference probably resulted from the approximations used to model the end restraints of the bridge spans. Also, the finite element model did not yield two modes of a similar shape for the fundamental E-W mode of the tower, as measured from the tests. The reason for this was not investigated in depth, but possible explanations could include the modelling of the gatehouse as a rigid rather than flexible structure, or the approximate modelling of the dynamic effects of the reservoir. Mode shapes from the finite element model also corresponded well with those that were measured from the tests. Overall, these results confirmed that the relatively coarse finite element model of tower and bridge was a valid representation of the prototype structure, and that it could be used for preliminary seismic analysis and further development for a more accurate model, if required.

## SEISMIC ANALYSIS OF MODEL

The preliminary seismic analysis of the model included static and seismic loading. Three statistically independent accelerograms were generated from the response spectrum shown in Figure 4. Two were used as the orthogonal, horizontal ground motion both with PGAs of 0.2g, and one as the vertical ground motion with a PGA of 0.13g. The mode superposition method was employed for the time history analysis using all modes of vibration below 40 Hz, which was demonstrated to be more than adequate through prior response spectrum analysis.

The seismic analysis of the tower indicated that cracking of concrete could occur at the bottom of the tower where it connects to the base structure, possibly resulting in significant non-linear behaviour. At the critical cross-section of the tower, the ultimate moment capacity was exceeded by the demand calculated from the elastic analysis. Also, tensile stresses resulted at the interface between the tower and its foundation rock, indicating that separation and rocking could occur.

A more detailed non-linear analysis would be required to assess the seismic performance of the tower, allowing for a reduction in its stiffness caused by concrete cracking and energy absorption due to the ductility of the reinforcing steel. The reduced stiffness of the cracked tower will lower its natural frequencies of vibration, subsequently de-tuning the tower's response from the main frequency components of the earthquake. Furthermore, although not seismically designed, the tower may have some ductility that could be taken into account when estimating its capacity to resist a large earthquake.

There are essentially two methods available for a non-linear seismic analysis, a dynamic non-linear time history analysis of a finite element model, or a simpler static push-over analysis. Using the latter, the effective seismic force calculated from a linear-elastic analysis can be reduced by taking account of the available ductility of the structure. For this type of tower, a dynamic non-linear analysis requires the development of a large and complex 3-D finite element model of the tower representing the non-linear cyclic behaviour of concrete and reinforcing steel. Six time history analyses would be required with different earthquakes, as recommended in Eurocode 8 (DD ENV 1998-1-1:1994), to investigate a range of possible responses. This would be a very costly activity, and the validity of the model would be difficult to prove, as there has been little work done on valid methods for the non-linear modelling of typical intake tower structures. Although significantly simpler and much less time consuming, the push-over analysis will also include some uncertainty due to problems in assessing a reliable value for the available ductility of existing non-seismically designed towers. There has been some preliminary research on the ductility of lightly reinforced towers in the US (Dove, 1997), but further work in this area is needed.

Although non-linear analyses can be conducted for existing intake towers, further experimental research on their non-linear behaviour and ductile capacity is required to provide a measure of the reliability of their results.

## CONCLUSIONS

This case study on the seismic assessment of Glascarnoch intake tower indicates that it, and other similar SSE towers, may be vulnerable to large earthquakes in the region.

The fundamental frequency of the tower (the predominant mode in its seismic response), measured from ambient vibration tests, fell within the main range of frequencies likely to be excited by a typical UK earthquake. This means that significant amplification of the ground motion can occur in the gatehouse, which is a less robust structure than its supporting tower, and where the operational equipment is stored.

The walk-down inspection identified a number of vulnerable features that could result in damage to the equipment under severe shaking. However, relatively simple measures can be taken to mitigate this risk, such as providing proper restraints to equipment and allowances for flexibility within cable runs.

It has been demonstrated that a tower with a complex geometry can be represented adequately by a relatively coarse 3-D solid finite element model for its modal and preliminary seismic analysis. Frequencies and mode shapes measured from the ambient vibration tests correlated well with their corresponding theoretical modes, thus validating the finite element model.

The linear-elastic seismic analysis of the tower indicated a significant non-linear response for an extreme UK earthquake, and therefore the requirement for further non-linear analysis to determine more reliably the seismic demand and seismic capacity of the tower.

## ACKNOWLEDGEMENTS

The authors wish to acknowledge Scottish and Southern Energy for funding this project, and Dr A.K. Hughes for instigating it and for his advice.

REFERENCES

Bureau, G. (1985). Seismic Safety Analysis and Rehabilitation of Dam Inlet/Outlet Structures. 15th ICOLD Congress, Lausanne, Q.59, R.17, pp 297-319.

Charles, J.A., C.P. Abbiss, E.M. Gosschalk and J.L. Hinks (1991). An Engineering Guide to Seismic Risk to Dams in the United Kingdom. Building Research Establishment Report.

Daniell W. E. and C.A. Taylor (2000). Dam Appurtenant Structures: Seismic Assessment Guide, A University of Bristol Earthquake Engineering Research Centre Report for Scottish and Southern Energy plc.

Daniell, W.E., S.P.G. Madabhushi and A. Blakeborough (1994). Performance of Dams and Appurtenant Works, in The Northridge, California Earthquake of 17 January 1994. A Field Report by EEFIT, Institution of Structural Engineers.

DD ENV 1998-1-1:1994. Eurocode 8 : Design Provisions for Earthquake Resistance of Structures, Part 1.1: General Rules - Seismic Actions and General Requirements for Structures.

Dove, R.C. (1997). Performance of Lightly Reinforced Concrete Intake Towers under Selected Loading. Ph.D. Thesis, Mississippi State University.

EQE International Limited (1998). Seismic Hazard at Scottish Hydro-Electric Dams. EQE Report No 414-01-R-01.

Goyal, A. and A.K. Chopra (1989). Simplified Evaluation of Added Hydrodynamic Mass for Intake Towers. Proc. ASCE, Journ. of Eng. Mechanics, Vol. 115, No. 7.

Principia Mechanica (1981). Seismic Ground Motions for UK Design. A Principia Mechanica Report for British Nuclear Fuels Limited and Central Electricity Generating Board.

Sandilands N.M., M. Noble and J.W. Findlay (1998). Risk Assessment Strategies for Dam Based Hydro Schemes. The prospects for Reservoirs in the 21st century, British Dam Society, Bangor, pp 27-38, Thomas Telford, London.

# Seismic hazard in the UK – another look

C W SCOTT, Binnie Black & Veatch, UK
J J BOMMER, Imperial College of Science, Technology and Medicine, UK

---

SYNOPSIS This paper reviews the literature on the seismicity of the United Kingdom and discuss the options for assessing the seismic hazard to critical structures such as dams. The approaches to the assessment of seismic hazard are briefly described, together with parameters available for characterising seismic activity and ground motions. The vulnerability of dam structures are considered. A more rationally based approach for the assessment of the seismic hazard to dams in the UK is proposed.

## INTRODUCTION

The issue of seismic hazard and UK reservoirs has been a topic of discussion since the publication of the *Engineering Guide to the seismic risk to dams in the United Kingdom* (Charles et al., 1991), known hereafter as the BRE guide. Some of the recommendations in the BRE guide were controversial and in 1998 an Application Note was published (ICE, 1998) revising some of the guidance in the original document. The advice on the assessment of seismic hazard in the Application Note was based on Department of Environment funded research undertaken by AEA Technology.

## DEFINITIONS

### Seismic sources

Earthquakes are caused by ruptures in the crust of the Earth. These ruptures are the result of the release of stress built up within the Earth's crust arising from tectonic and other forces. For the purposes of assessing seismic hazard, zones of potential rupture are defined as seismic sources. There are three main types of seismic source:

- Faults. These are pre-existing fractures or rupture surfaces in the Earth that can be associated with continuing or possible seismic activity.
- Localising structures. These are identifiable geological structures assumed to generate or localise earthquakes.
- Seismotectonic province. These are geographical regions of seismological similarity assumed to possess uniform earthquake potential throughout. Recorded earthquakes may have occurred randomly in the region or be clustered in a preferred location but the seismologist has been unable to identify faults or geological structures allowing definition as either of the other source types.

*Reservoirs in a Changing World*, Thomas Telford, London, 2002, 112–125

Reiter (1990) gives a more detailed discussion of seismic sources.

Intensity

The *Intensity* of earthquake ground motion is a qualitative description of the felt effect of the earthquake at a particular location as evidenced by observed damage and human reactions. A variety of scales are available. Figure 1 shows the descriptions used by the Modified Mercalli Intensity Scale. A more detailed discussion of intensity is given in Kramer (1996) and Reiter (1990).

| | |
|---|---|
| I | Not felt except by a very few under especially favourable circumstances |
| II | Felt by only a few persons at rest, especially on upper floors of buildings; delicately suspended objects may swing. |
| III | Felt quite noticeably indoors, especially on upper floors of buildings, but many people do not recognise it as an earthquake; standing motor cars may rock slightly; vibration like passing of truck; duration estimated. |
| IV | During the day felt indoors, outdoors by few; at night some awakened; dishes, windows, doors disturbed; walls make cracking sound; sensation like heavy truck striking building. |
| V | Felt by nearly everyone, many awakened; some dishes, windows, etc broken; a few instances of cracked plaster; unstable objects overturned; disturbances of trees, piles, and other tall objects sometimes noticed; pendulum clocks may stop |
| VI | Felt by all, many frightened and run outdoors; some heavy furniture moved; a few instances of fallen plaster or damaged chimneys; damage slight. |
| VII | Everyone runs outdoors; damage negligible in buildings of good design an construction, slight to moderate in well-built ordinary structures, considerable in poorly built or badly designed structures, some chimneys broken, noticed by persons driving motor cars. |
| VIII | Damage slight in specially designed structures, considerable in ordinary substantial buildings, with partial collapse, great in poorly built structures; panel walls thrown out of frame structures; fall of chimneys, factory stacks, columns. monuments, walls; heavy furniture overturned; sand and mud ejected in small amounts; changes in well water; persons driving motor cars disturbed. |
| IX | Damage considerable in specially designed structures; well designed frame structures thrown out of plumb; great in substantial buildings, with partial collapse; buildings shifted off foundations; ground cracked conspicuously; underground pipes broken. |
| X | Some well built wooden structures destroyed; most masonry and frame structures destroyed with foundations; ground badly cracked; rails bent; landslides considerable from river banks and steep slopes; shifted sand and mud; water splashed over banks. |
| XI | Few (if any) masonry structures remain standing; bridges destroyed; broad fissures in ground; underground pipelines completely out of service; earth slumps and land slides in soft ground; rails bent greatly. |
| XII | Damage total; practically all works of construction are damaged greatly or destroyed; waves seen in ground surface; lines of sight and level are distorted; objects thrown into air. |

Figure 1 Modified Mercali Intensity Scale (Kramer, 1996)

Magnitude

Earthquake magnitude is a quantitative measure of the size of an earthquake.

**Magnitude Scales**

Local magnitude ($M_L$) This measure, also known as the Richter magnitude, is based on the maximum trace amplitude of a Wood-Anderson seismometer. It is appropriate for shallow local (epicentral distance of less than 1000 km) earthquakes in southern California.

The Richter magnitude is not based on the amplitude of any particular wave type. Other scales that are based on the amplitude of a particular wave have been developed.

Surface wave magnitude ($M_s$) At large epicentral distances the motion observed is dominated by surface waves. The surface wave magnitude is a worldwide scale based on the amplitude of the surface wave with a period of about 20 seconds. This scale is widely used far shallow earthquakes worldwide.

Body wave magnitude ($m_b$) Deep earthquakes do not produce sufficient surfac waves for reliable estimates of $M_s$ magnitude to be made. The body wave magnitude is a worldwide scale based on the amplitude of early P-wave or S-wave arrivals at the seismometer.

The different instrumental magnitude scales are unable to differentiate between very large earthquakes. This phenomenon, known as saturation, occurs at different points for different scales. Thus local and body wave magnitudes are unable to differentiate between earthquakes larger than about magnitude 7 and the surface wave magnitude saturates at about magnitude 8. The figure below illustrates this.

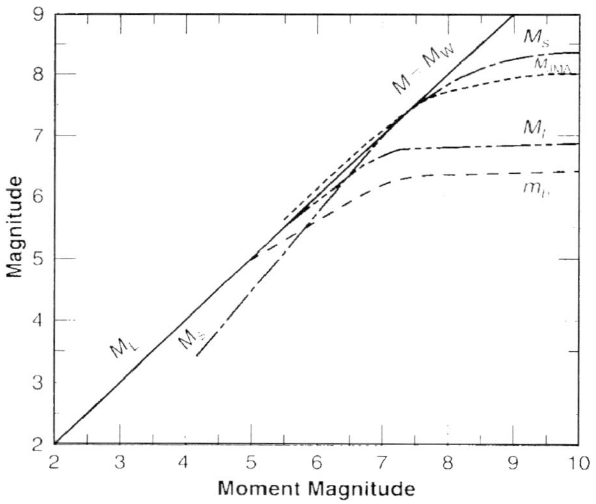


(Kramer, 1996)

Moment Magnitude ($M_w$) The only magnitude scale not subject to saturation is the moment magntitude which is based on the seismic moment, a direct measure of the fault rupture.

Essentially the earthquake magnitude seeks to indicate the amount of energy released by the earthquake. It is a measure of the absolute size of the event rather than its felt effect. Earthquake magnitude scales are derived from instruments. The differences between the most frequently adopted magnitude scales are outlined in the Box. A more detailed discussion of magnitude is given in Kramer (1996) and Reiter (1990).

## SEISMIC HAZARD ASSESSMENT

There are two distinct approaches to the assessment of the hazard posed by earthquake ground motions at a given site: deterministic hazard assessment and probabilistic hazard assessment. These two approaches are briefly outlined below.

### Deterministic hazard assessment

The process of a deterministic seismic hazard assessment can be summarised as follows (Kramer, 1996) and is shown diagrammatically on Figure 2.

1. Identification and characterisation of all earthquake sources capable of producing ground motion at the site.
2. Selection of a source-to-site distance parameter.
3. Selection of controlling earthquake.
4. Hazard defined, usually in terms of the ground motions produced at the site by the controlling earthquake.

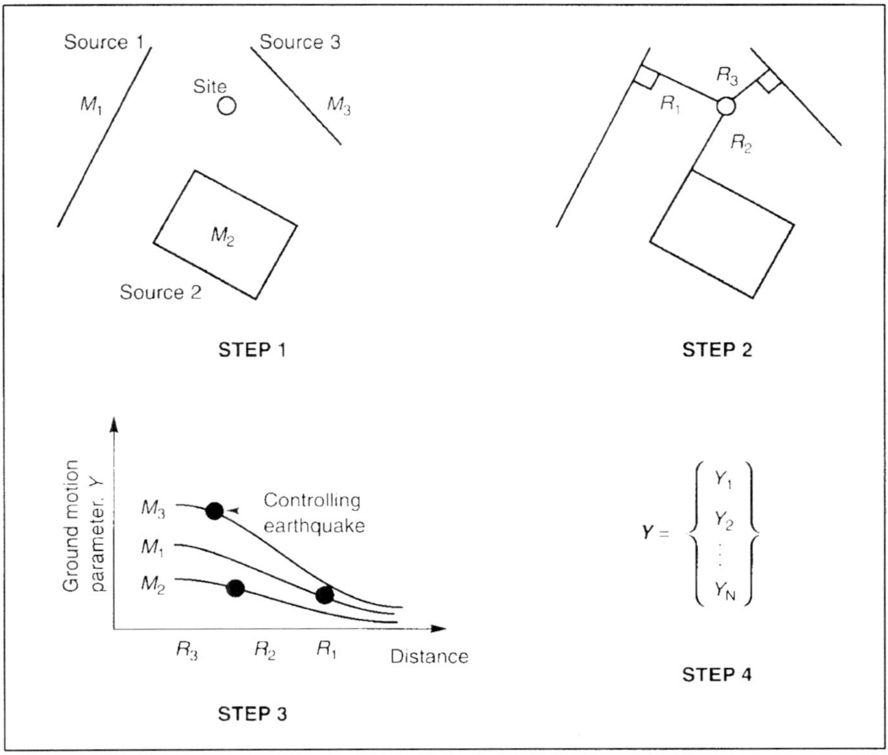


Figure 2 Four steps of a deterministic seismic hazard analysis (Kramer, 1996)

### Probabilistic hazard assessment

The process of a probabilistic seismic hazard assessment can be summarised as follows (Kramer, 1996) and is shown diagrammatically on Figure 3.

1. As step 1 for the deterministic approach, except that the probability distribution of potential rupture locations within the source must be characterised. A uniform distribution is usually used, implying that rupture is equally likely to occur at any location in the source.
2. The temporal distribution of earthquake recurrence must be characterised. This recurrence relationship is truncated at the maximum event for the source.
3. The ground motion produced at the site by any possible size of earthquake occurring at any possible location in each source is determined. The uncertainty inherent in the attenuation relationships used to make these predictions is also modelled.
4. The uncertainty in earthquake location, earthquake size and ground motion parameter prediction are combined to obtain the probability that the ground parameter will be exceeded during a particular time period.

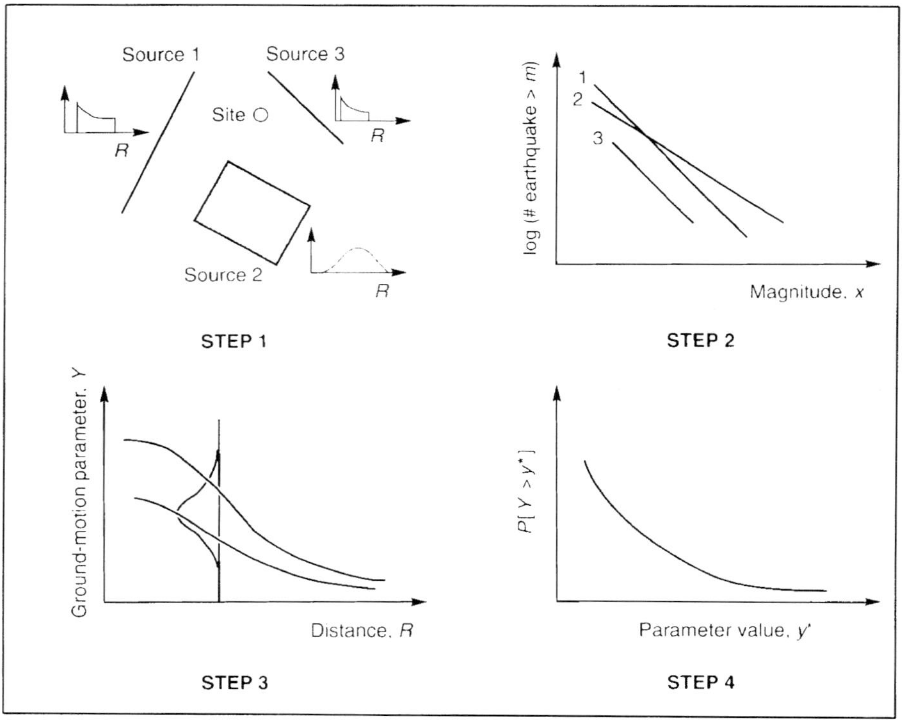


Figure 3 Four steps of a probabilistic seismic hazard (Kramer, 1996)

The selection of one or other hazard analysis approach has been presented as a stark choice between two diametrically opposed philosophies. This is gross simplification. All hazard assessment is a combination of judgement and probability (whether explicit or implicit) and the difference between the two approaches has been much exaggerated by the proponents of the two approaches. Bommer (2002) discusses the differences and similarities between various approaches and argues for the adoption of the combination of deterministic and probabilistic elements of hazard assessment according to the seismicity of the region, the purpose of the study, and the nature of the engineering project in terms of the consequences of failure.

## PUBLISHED GUIDANCE

Guidance on appropriate standards for safety assessment needs to address two separate issues:

1. The required safety standard, and
2. The means of estimating the input for that standard.

Flood standards provide an example of this separation. Table 1 in *Floods and Reservoir Safety* (ICE, 1996) defines the safety standard for different categories of reservoir and the proposed means of estimating the inputs is the rainfall runoff method as defined in the *Flood Studies Report* (NERC, 1975).

The BRE guide proposes a means of assessing the required safety standard based on the methodology in ICOLD Bulletin 72 (ICOLD, 1989). This approach categorises a reservoir on the basis of four factors:

- capacity
- height
- evacuation requirements, and
- potential downstream damage.

The Safety Evaluation Earthquake (SEE) for each of the four proposed categories is defined by return period as shown in the Table below. The level of safety assessment required is also defined on the basis of the reservoir categorisation.

Table 1 Dam categories for seismic assessment (Charles et al., 1991)

| **Dam Category** | **Return Period of SEE** |
|---|---|
| IV | 30,000 yrs |
| III | 10,000 yrs |
| II | 3,000 yrs |
| I | 1,000 yrs |

The BRE guide proposed Peak Ground Accelerations (PGA) for the four return periods for three geographical zones. The PGAs recommended ranged from 0.375g for the highest category reservoir in the most seismically active zone to 0.05g for the lowest category in the least active zone. (The third decimal place in these accelerations – and possibly even the second – convey a level of accuracy entirely unfounded in terms of the input data.)

The PGAs and the level of seismic assessment proposed by the BRE guide proved controversial. An ad hoc committee of the Reservoirs Committee of the

ICE undertook a review of the guide. The Application Note arose from this review (ICE, 1998). The Application Note supports most of the recommendations of the BRE guide. The level of safety assessment proposed is amended and the means of determining the SEE ground motion is changed. The safety standard itself is unchanged other than for Category IV where the SEE is changed to either the deterministically assessed Maximum Credible Earthquake (MCE) or the 10,000-year event.

In assessing the means of determining the input for the proposed safety standards, the Application Note took account of additional studies of earthquake ground motions undertaken after the publication of the BRE guide. The Application Note goes on to propose that the result of the study undertaken by Musson and Winter to produce a seismic hazard map for the UK be adopted in preference to the values suggested in the BRE guide. Musson and Winter's work was undertaken for the Department of the Environment. The authors have not be able to obtain a copy of the report but the approach is described elsewhere (Musson and Winter, 1997).

In discussing return period, the Application Note refers to the SEE rather than the PGA. In practice the term earthquake is often used interchangeably with ground motions at the subject site. That is the case here and the return period must refer to the ground motions as they are the critical input to the safety assessment not the earthquake itself for which the appropriate measure is recurrence interval.

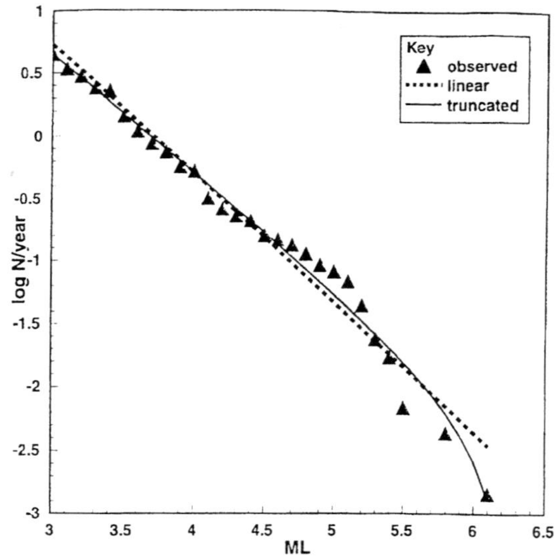


Figure 4 Recurrence relationship from Musson & Winter, 1997

## COMMENTS ON SEISMIC HAZARD MAPS IN PUBLISHED GUIDANCE

### Estimation of input parameters

In their 1997 paper, Musson and Winter describe the probabilistic hazard assessment methodology adopted to produce the hazard maps used in the Application Note. The steps of the assessment are as defined above. The next two sub-sections will briefly describe the assumptions made and their implications.

Spatial and temporal distribution (Source and recurrence models)

The source model used in the assessment largely followed the concentration of historic earthquakes in the UK. The source types are not defined but are generally seismotectonic provinces as there is little correlation between the historic record of UK earthquakes and underlying geological or tectonic features.

A common recurrence relationship is adopted for all source zones. The relationship is shown in Figure 4. Points to note are that earthquake magnitudes are defined using the local magnitude scale and the relationship is effectively truncated at $M_L$ 6.2, which has a recurrence interval of slightly less than 1,000 years. The hazard assessment adopted a logic-tree formulation with two larger maximum earthquakes also considered ($M_L$ 6.5 and 7), but with assigned weights lower than the $M_L$ 6.2 scenario.

The range of maximum magnitudes adopted by Musson and Winter does not appear to be widely supported in the literature on UK seismicity. Musson and Winter's own paper of 1996 stated that the maximum earthquake magnitude likely for the UK is $M_L$ 6.2. Ambraseys and Jackson in their 1985 paper proposed a maximum land-based earthquake for the UK of $M_S$ 5.5. Evidence has recently been presented for slightly larger events on a particular fault (Stewart et al., 2001). Musson and Winter justify the adoption of different values of maximum magnitude on the basis of that 7 represents a figure which intraplate earthquakes rarely exceed worldwide, it is not related to the context of UK seismicity. Indeed the $M_{max}$ of 7 implies a potential seismic rupture with a length in excess of 100km on a currently undetected fault! The 6.5 figure was selected as an intermediate value. Musson and Winter acknowledge that the general opinion among UK seismologists favours a low maximum magnitude on the basis of the historical record and the lack of tectonic reasons for supporting higher values. There is therefore a tacit acceptance that the source model adopted for this study overestimates the size of very rare events.

Ambraseys and Jackson also note that the focal depth of UK earthquakes increase with increasing magnitude thus most events larger than $M_S$ 5 are generally at depths greater than 10 km (Ambraseys and Jackson, 1985). Musson and Winter adopted a range of focal depths of earthquakes in their model including 5 km which was given a probability weighting of 20% (compared to 30% for 10 and 15 km and 20% for 20 km). The paper provides no evidence to support the adoption of a focal depth outside the observed range for large events. This actually represents a misuse of the logic-tree formulation since a fundamental principle of the tool is that at any node the different branches represent mutually exclusive possibilities (Abrahamson, 2000). Since Musson & Winter (1997) argue for a particular distribution of earthquakes with respect to focal depth, this focal depth distribution should have been incorporated into actual hazard calculations rather than assigning particular depths to branches of the logic-tree.

The above discussion illustrates one of the bizarre facets of the debate between probabilitistic and deterministic hazard analysis. Musson and Winter provide no justification for the adoption of the weights applied to the different maximum magnitudes or focal depths in the logic tree formulation. These figures have been selected on the basis of judgement (i.e. deterministically). Their selection and the values adopted will have a significant effect on the results of the probabilistic hazard analysis. It should be noted that seismic hazard assessments do entail these sorts of subjective judgement that are then treated as statistical values.

Attenuation relationship

Musson and Winter adopted an attenuation relationship derived by Dahle et al. (1990) for intra plate earthquakes. The relationship was one of the first derived for intra plate areas rather than more active inter plate zones. The relationship is not specifically for the UK and the database used is based on a very loose definition of intra-plate. It is particularly noteworthy that the authors of this particular relationship have long since abandoned its use (Bungum, 2001). The regression analysis was done using $M_S$ and that is the preferred magnitude scale for the relationship. The Musson and Winter work was done using the $M_L$ scale. As noted earlier, the different magnitude scales are not interchangeable. The relationship is a predictor for ground motions at rock sites.

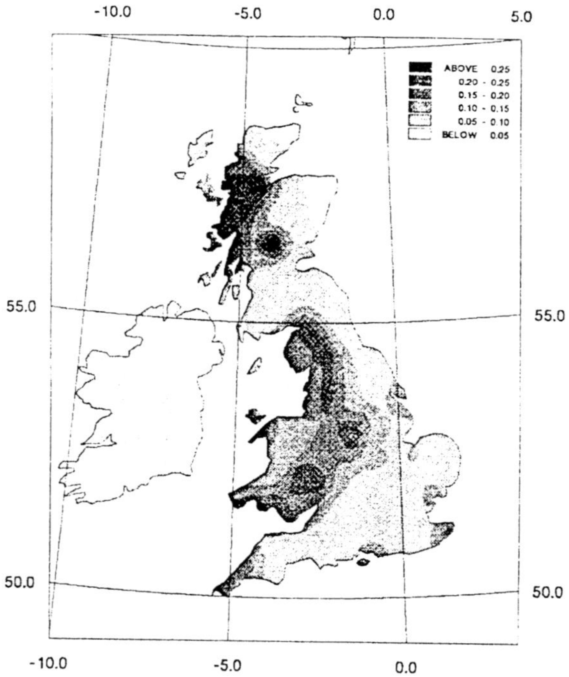


Figure 5 PGA with annual exceedance probability of $10^{-4}$ (Musson & Winter, 1997)

The hazard maps provide rock head PGAs and make no allowance for the amplifying or damping effect of any superficial deposits overlying rock at the dam site.

The contoured maps of PGA represent the result of the hazard analysis for given return periods. Musson and Winter acknowledge that at very low probabilities, such as for the 10,000 year return period shown on Figure 5, one of the most significant factors is the scatter in the attenuation relationship. The contours thus are more likely to represent a Magnitude 5 to 6 earthquake at short range producing unexpectedly large acceleration rather than the normal response to a large and extremely rare earthquake as demonstrated by Bommer et al (2000). Thus the uncertainty in the attenuation relationship has a significant effect on the estimated ground motions.

Attenuation relationships are generally developed by applying regression techniques to fit relationships to a set of earthquake ground motion data. The scatter of the data represents the uncertainty in the ground motion estimates derived from the relationship. This scatter is measured using the standard deviation of the data from the median line. Attenuation relationships are usually done in log log space, using either base 10 or e, these different bases make it invalid to merely compare standard deviations to compare the relative uncertainty in the attenuation relationships. Ambraseys and Bommer (1995) used the ratio of the PGA at the 85$^{th}$ and 50$^{th}$ percentiles of the distribution to compare the uncertainty inherent in attenuation relationships on a common basis.

Table 2 Comparison of uncertainty in three intra plate attenuation relationships

| **Attenuation Relationship** | **$PGA_{85}/PGA_{50}$** |
|---|---|
| Dahle et al, 1990 | 2.29 |
| Free et al., 1998 | 2.69 |
| Toro et al., 1997 | 1.36 |

Table 2 shows a comparison of the uncertainty in three intra plate attenuation relationships. The first two relationships use worldwide data while the third has been derived for central and western North America. As can be seen the worldwide relationships show significantly greater scatter than the regional equation. Figure 6 shows a comparison of the median PGAs predicted by the three attenuation relationships for a Magnitude 5.5 earthquake. The accelerations predicted by these attenuation relationships are quite similar at epicentral distances greater than about 10 km. Below that figure the Dahle et al. relationship predicts the highest PGAs and the Toro et al. the equation the lowest. The adoption of the Dahle et al. relationship in the hazard analysis (a deterministic decision) significantly affects the results of the hazard analysis because of the higher PGAs derived at short epicentral distances and the level of uncertainty in the relationship. No explanation for the adoption of the Dahle et al. relationship is given in Musson and Winter paper.

Ambraseys and Jackson (1985) indicate that the highest Intensity experienced in the UK over the past 700 years is VII. A crude correlation between Intensity and PGA (Wald et al., 1999) suggests that the highest PGA experienced in the UK in the past 700 years is of the order of 0.24g. (The data underlying the correlation at Intensity VII ranges between 0.07g and 0.40g.)

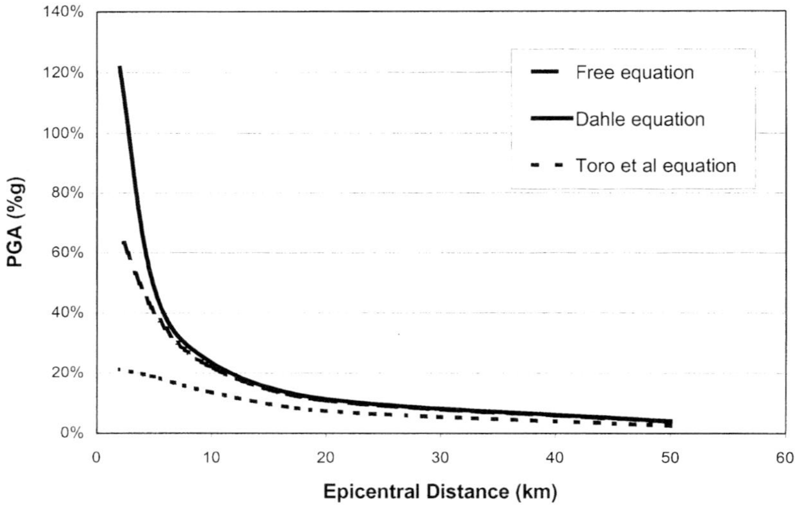


Figure 6 Comparison of median accelerations for M 5.5 earthquake

IMPLICATIONS OF COMMENTS

The discussion above has sought to highlight the very large uncertainties present in the prediction of ground motion characteristics for very rare events. The problem is a corollary of the difficulties in predicting very rare flood events and as engineers responsible for safety related decisions we need to be very aware of the uncertainties involved. The quality and quantity of data available is too sparse to provide 10,000 year PGA estimates with the spatial accuracy craved by the profession.

Using the median values of the attenuation relationship derived by Dahle et al., a Magnitude 6 earthquake would cause peak ground accelerations in excess of 0.1g within a 33 km radius of the epicentre. What is the probability of that happening? The recurrence relationship would suggest that the annual exceedence probability of such an earthquake happening anywhere in the UK is $10^{-3}$. So the probability of such an event happening within 22 km of a particular location must be much much less (The annual exceedence probability is about $1.4 \times 10^{-5}$ or 1 in 75,000. Using the Free et al equation this reduces to 1 to 160,000 demonstrating the sensitivity of the output to the choice of attenuation realtionship.). This simplistic view supports the contention that the ground motions derived from the charts produced by Musson and Winter are likely to be very conservative. Unfortunately the work of Musson and Winter remains the only UK-specific work to predict ground motions for such rare return periods. The contours on the published maps and the tables available quoting PGA to three decimal places for given site locations should be used with caution given the quality of the underlying data and the number of assumptions inherent in the methodology. Perhaps funding of an improved UK wide hazard study would be a useful target for reservoir related research.

It can reasonably be argued that the information presented in the Application Note provides a conservative means of assessing safety. If a dam can survive those values it is most certainly seismically safe. The results of a seismic assessment using the Musson and Winter charts would be a poor reason to spend any significant sums on complex dynamic analyses or improvement works. The money spent on the analysis could be more usefully invested in improving the estimates of the input parameters.

## CONCLUSIONS

This paper has described some of the terminology of seismic hazard analysis and has reviewed the background to the maps of seismic hazard proposed as input data to the seismic hazard assessment of UK dams. The paper has sought to underline the levels of uncertainty inherent in the analysis and the assumptions made by the hazard analysts.

The data presented in the Application Note provides a reasonable means of making a first pass safety assessment. The input parameters are however conservative and are a poor basis for investment in complex analyses or improvement works. Care needs to be used with the use of all seismic input data not derived from a site specific hazard assessment. Large scale hazard models are imperfect and their robustness reduces dramatically as return period increases.

An alternative approach would seek to pragmatically look at the risks presented by earthquakes in the UK. The fact is that no well built embankment dam has ever failed due to seismic action. The much quoted example of the Lower San Fernando dam in California refers to a hydraulic fill dam and the failure was the result of liquefaction of saturated cohesionless material. The authors are unaware of a similar dam in the UK. The UK is blessed with a stock of intrinsically seismically robust structures. Concerns can more validly be raised about appurtenant structures which are significantly more vulnerable than the dam structure.

REFERENCES

Ambrahamson, NA (2000). State of practice of seismic hazard evaluation. Proceedings of GeoEng 2000, November, Melbourne, Australia.

Ambraseys, NN and Jackson, JA (1985) Long-term seismicity of Britain, Proceedings of Conference on Earthquake Engineering in Britain, University of East Anglia, Norwich, 49-66, Thomas Telford, London.

Ambraseys, N.N. & Bommer, JJ (1995). Attenuation relations for the use in Europe: an overview. Proceedings, SECED Conference on European Seismic Design Practice, Chester, 67-74, Balkema.

Bommer, JJ, Scott, SG and Sarma, SK (2000). Hazard-consistent earthquake scenarios. Soil Dynamics and Earthquake Engineering, 19, 219-231.

Bommer, JJ (2002) Deterministic vs probabilistic seismic hazard assessment: an exaggerated and obstructive dichotomy. Journal of Earthquake Engineering, Vol 6, Special Issue No 2.

Bungum, H. (2001) Personal communication, October.

Charles, JA, Abbiss, CP, Gosschalk, EM and Hinks, JL (1991), An engineering guide to the seismic risk to dams in the United Kingdom, BRE, Watford.

Dahle, A, Bungum, H and Kvamme, LB (1990) Attenuation relationships inferred from intra-plate earthquake recordings, Earthquake Engineering and Structural Dynamics, 19, 1125-1141.

Douglas, J (2001) A comprehensive worldwide summary of strong-motion attenuation relationships for peak ground acceleration and spectral ordinates (1969 to 2000). ESEE Report No 01-1. Department of Civil Engineering, Imperial College, London.

Institution of Civil Engineers (1998), An application note to an engineering guide to the seismic risk to dams in the United Kingdom, Thomas Telford, London.

Institution of Civil Engineers (1996). Floods and reservoir safety. Third Edition. Thomas Telford, London.

Free, MW, Ambraseys, NN and Sarma, SK (1998) Earthquake ground-motion for stable continental intraplate regions. ESEE Report 98-1. Department of Civil Engineering, Imperial College, London.

Kramer, SL (1996) Geotechnical earthquake engineering, Prentice Hall, .

Musson, RMW (1996) The seismicity of the British Isles, Annali di Geofisica, 39, No 3, 463-469.

Musson, RMW and Winter, PW (1997) Seismic hazard maps for the UK, Natural Hazards, 14, 141-154.

Natural Environment Research Council (1975). Flood studies report. Natural Environment Research Council, London.

Reiter, L (1991) Earthquake hazard analysis: Issues and insight, Columbia University Press, New York.

Toro, G, Abrahamson, NA and Schneider, JF (1997). Model of strong ground motions from earthquakes in central and eastern North America: Best estimates and uncertainties, Seismological Research Letters, 68, No 1, 41-56.

Wald, DJ, Quitoriano, V, Heaton, TH and Kanamori, H (1999). Relationship between peak ground acceleration, peak ground velocity and Modified Mercali Intensity in California, Earthquake Spectra, 15, no 3, 557-564.

# A methodology for seismic investigation and analysis of dams in the UK

P RIGBY, Bechtel
S WALTHALL, Bechtel
K D GARDINER, United Utilities plc

SYNOPSIS. Soil parameters derived from existing data only have been found to give pessimistic results when used in seismic analyses. The paper discusses the design of ground investigations needed to obtain realistic parameters and the development of a methodology to determine the level of investigation required for dams in different seismic categories.

## INTRODUCTION

Following publication of the "Engineering Guide to seismic risk to dams in the UK" in 1991, recommendations for seismic risk assessments began to be made by Inspecting Engineers under the Reservoirs Act 1975.

Given the large number of analyses potentially required for its 160 reservoirs, North West Water (now United Utilities plc (UU)) commissioned Bechtel Water Technology (now Bechtel) to carry out a pilot study on five dams using information available from archive information. This would allow UU to estimate the likely cost of the analyses and establish the extent of additional Site Investigation that would be required. The dams chosen were at Arnfield, Belmont, Hurst, Rhodeswood and Rooden reservoirs.

The preliminary report's somewhat pessimistic main findings were that:

a) Four of the five dams had apparently lower factors of safety in the normal operating (static) condition than expected.
b) All 5 would theoretically suffer damage in a seismic event although this would not lead to immediate collapse.
c) The results were susceptible to small variations in the assumed parameters.

Following receipt of this report, UU established a steering group of three eminent dam Engineers, Dr J A Charles of BRE, Mr T A Johnston of Babtie Group and Mr D J Knight an independent consultant, to advise the company on how they should proceed in the light of the initial findings. The remit of the Steering Group was to review the Bechtel report and comment on the data used and methods of analysis, and to recommend and review further studies. The Steering Group's review concluded that:

a) The methods of analysis employed by Bechtel were correct but given the nature of the archive information the parameters used may have been over conservative.
b) The peak horizontal ground acceleration figure of 0.25g used in the seismic analysis was perhaps too high in the light of studies carried out since the publishing of the Engineering Guide and that 0.18g would be more appropriate.
c) The apparently low factors of safety for the static slope stability of the dams considered required verification.
d) The damage to ancillary structures predicted by the Bechtel's report were probably over conservative and were not likely to lead to failure of the dam
e) The Steering Group agreed with Bechtel that further studies were needed to obtain reliable parameters for use in the stability analysis.

The Steering Group proposed to UU that:

a) The dam cross sections should be verified by topographical survey.
b) Site Investigations should be carried out on the five dams studied to obtain representative parameters soils parameters
c) The dams be re-analysed using the new data.
d) A database of geotechnical information and analyses be established for all UU dams.

The views expressed are those of the authors of this paper and not necessarily those of the Steering Group.

## HISTORICAL BACKGROUND

The history of the dams and the development of construction techniques are an important consideration in the North West of England. Whilst the dams in this study were all built within a very short period of each other around 1850, the North West Water dams are known to range in age from the 1790's (Upper Chelburn) to 2000 (Audenshaw No3). They were built for a large number of different owners, usually the local town water company, by a very small number of Engineers, each with their own views on construction. It is therefore important to understand the age and origin of the dam before setting out to investigate its stability.

## METHODOLOGY

It was agreed with the Steering Group that the study would be carried out using conventional effective stress testing and classical soil mechanics theory for the development of slip surfaces. Whilst there are alternatives, it was considered that this would provide information suitable for long term use and for comparison with other studies and future work. The methodology followed the conventional route of obtaining driven tube samples from percussion boreholes, effective stress testing in the laboratory using standard methods, such as those described by Bishop and Henkel and given in BS 1377 Part 8 1990, followed by analysis with published limit equilibrium slip surface methods.

## DESK STUDY

The desk study proved to be essential to establish the history of the embankments. The construction record drawings were obtained from the various archives maintained by UU, which are an essential source of information. They need to be treated with care since they are often design drawings showing how the dam was intended to be built and a few 'as constructed' drawings. Where the dams have been modified by subsequent works these modifications can be of importance to the stability. Two cases from the study highlight this. At Belmont the current dam profile resulted from construction on top of an earlier dam, followed by the major stabilisation works 80 years later. Secondly, the construction at Hurst shows a blanket on the construction drawings but is referred to in the records as having a core. The 'core' is shown sketched in pencil on an old record drawing with no reference to its status.

## GROUND INVESTIGATION

### Topographical Surveys

A number of factors govern the stability of a slope particularly its slope inclination and internal geometry. Topographical surveys were undertaken on the downstream slopes with limited surveys on the upstream face. These were compared with sections through the embankments modeled by reference to the "as constructed" archive drawings for each dam. Discrepancies can occur between the recorded slope angles and those observed in-situ. Variations in the records are summarised in Table 1.

Table 1

| **Embankment** | **Slope angle** (from archive) | **Slope angle** (from survey) | **TWL** mOD | **Crest level** (archive) mOD | **Crest level** (survey) mOD |
|---|---|---|---|---|---|
| Arnfield | 22 | 18 | 164.77 | 165.5 | 165.8 |
| Belmont | 18 | 21 | 260.15 | 261.6 | 261.28 |
| Hurst | 25 | 26 | 215.46 | 216.7 | 216.8 |
| Rhodeswood | 31 | 26 | 175.16 | 176.15 | 176.6 |
| Rooden North | 22 | 21 | 325.53 | 326.62 | 326.8 |
| Rooden South | 28 | 21 | 325.53 | 326.67 | 326.8 |

### Field Investigations

The core was initially located by hand excavated trenching across the crest of each embankment. Core material can be difficult to locate in a trench and requires an experienced eye to differentiate between the various clay fills which can be found in the bank.

Boreholes

Three boreholes were required for each dam, one on the crest, one located one third and one half way down the embankment. In the planning stage, the crest boreholes to sample the core material were to be 6m in depth to prevent the borehole puncturing through any water-retaining layer. The other holes were to penetrate through the full depth of the embankment and 5m into the foundation materials (i.e. natural ground).

Evaluation of Field Investigations

A number of methods of investigation were used to evaluate the most cost effective methods for future investigations.

Conventional 'cable tool percussion' Conventional cable tool percussion (alternatively known as 'shell and auger') was used on all embankments. Good quality (Class 2, BS5930) samples were obtained and it was not necessary to re-drill using rotary coring methods which had been planned as a backup. Continuous sampling was carried out using driven U100 tube samples at 0.5m centres to gain a full picture of the fill materials and any structure such as layering and/or shear surfaces. SPT tests were not carried out as it was considered more appropriate to obtain driven Dynamic probe penetration tests (DPT) which were carried out alongside each borehole location to determine the relative density of the embankments.

Rotary drilling. Rotary drilling was carried out only to sample the rock foundations at Belmont as part of the subsidence investigations, but was not found necessary at other dams.

Light Dynamic rigs . These provide three investigation techniques, Window sampling, dynamic samples and DPT tests.

A Soil Sampling (Window) System (commonly referred to as window sampling) was used which uses a series of decreasing diameter open sampler tubes driven by a vibrating hammer. The samples are of small diameter, very poor quality, easily contaminated by material from a higher level in the hole and not suitable for testing other than for moisture content. They do provide a quick indication of the material types present in the embankment but tend to achieve only shallow penetrations where granular materials are present, particularly cobbles.

A dynamic sampling rig (known as the Archway Engineering AEC 150) was also used. This was able to take consecutive 90mm or 65mm diameter driven tube samples, each one metre long to a depth of 10m. This provided samples of at least an equivalent quality to the driven U100. The rig can also take SPT tests and install piezometers. Because of its limited depth capacity it was found necessary to plan for making three separate attempts at each location, the maximum depth achieved during this project being 9m.

Dynamic probe penetration tests (DPT) were carried out alongside each borehole location to determine the relative density of the embankments. These are of limited use when used alone but can be valuable when there are site specific samples available. DPT tests were used at each location to determine if this test could be used as a means of classifying the nature of the ground without resorting to a major drilling operation. There is insufficient sensitivity in the test to indicate local changes in the embankment materials, for example the presence of sandy or silty bands.

Access considerations

Impounding reservoirs are generally located in remote areas. The slopes of the embankments investigated in this study vary between $18^{\circ}$ and $26^{\circ}$ and therefore pose access problems. In order to complete boreholes safely the construction of extensive drilling platforms was required. These are expensive, the cost of each platform being greater than the cost of drilling each borehole.

The dynamic sampling rig (AEC150) used requires a much smaller, and therefore cheaper, platform than that required for cable tool and rotary rigs. The DPT and Window sampling rig could be placed directly onto the embankments requiring only a narrow berm cut into the slope.

Ground water monitoring

Conventional standpipe piezometers were installed in the cable tool percussion holes whilst driven Cambridge type were installed alongside the DPT holes. Standpipe piezometers in fill materials which are partially saturated are unlikely to be very effective at measuring low piezometric pressures in low permeability material. They did however prove adequate for measuring water levels in the saturated natural ground. Their use is however recommended throughout since if there is any water or piezometric pressure present then they are likely to reveal these conditions. Pneumatic, hydraulic or electrical piezometers should be reserved for special investigations.

## LABORATORY TESTING

Conventional laboratory tests

Laboratory testing consisted of routine classification tests and triaxial effective stress testing. The soil parameters for use in the analysis are derived from the effective stress tests. The results from the tests together with in-situ density, compaction tests, oedometers and other laboratory tests were examined to investigate whether information could be derived as to the history of the embankment. Of interest were the assessment of the degree of compaction applied at the time of construction, the amount and rate of consolidation since then, and the implications for the future of the embankment.

Effective stress testing

The laboratory test schedules and procedures were prepared after discussions with the Steering Group and the specialist testing laboratories. Effective stress testing was carried out to BS1377 Part 8 with the following protocols.

The soil samples were taken using driven 100mm diameter tubes and whilst some mechanical disturbance of these samples may occur it was considered that the moisture content would be preserved. All samples were tested as single stage tests at full sample diameter of 100mm. The consolidation and degree of saturation of these materials was unknown and therefore for the effective stress test it was agreed that the cell pressures required to achieve back pressure saturation should be kept to a minimum, of the order of $100kN/m^2$. However saturation often does not occur in the samples until a pressure in excess of $400kN/m^2$ has been applied. It was therefore agreed that the minimum cell pressure that achieved a 'B bar' of 85% would be an acceptable compromise rather than 95% as required by BS1377 5.3.2

Discussion

The s′/t′ plots have been examined in detail to see if the concept of c′>0 is valid for all or any of the materials or embankments. Low stress levels were specified in the laboratory testing to investigate this phenomenon. From the summary plots and statistical line fitting on the s′/t′ plots for each of the embankments it is suggested that c′=0 should always be used. The problem that using c′=0 creates in the stability analysis by producing shallow slips with a low factor of safety can be addressed by a suitable use of geometrical constraints and prudent choice of slip planes and circles. Shallow and deep slips are addressed separately in the analysis. The worst credible values are chosen from the detailed testing and as such should be expected to give a realistic factor of safety. A comparison of the parameters to put this conservatism in context has been carried out for Arnfield and a summary is given in Table 1. In addition to the calculation of $\phi'$ at maximum principal stress ratio it is possible to calculate $\phi'_{cv}$ at the critical state where sufficient samples have been tested.

Table 2 Arnfield

| **Material** | $\phi'_{\text{pk (statistical mean)}}$ | $\phi'_{\text{cv ( statistical mean)}}$ | $\phi'_{\text{design for this study}}$ |
|---|---|---|---|
| Core | 33 | 30 | 28 |
| Fill | 38 | 31 | 30 |
| Natural Ground | 27 | 22 | 24 |

This indicates that this study has used reasonable, though not excessively conservative parameters and design values for the construction of new embankments may be lower than those used in this analysis.

Correlation between soil parameters

At the start of the investigation it was suggested that for future analysis it might be possible to derive parameters from the classification tests with the following considered to be the most useful.

- moisture content,
- atterberg limits,
- grading,
- clay content.

So far the tests have not revealed any correlation which is sufficient for this purpose. This is an extensive topic and is outside the scope of this paper. The analysis of the factor of safety is very sensitive to the value of ($\phi'$). Correlations will only give ($\phi'$) to within five degrees and this is not accurate enough for meaningful analysis. It is therefore necessary to test each and every material.

## ANALYSIS

### Validation of the computer analysis programs

The programs used were STABLE by MZ Associates and SLOPE by Oasys, part of the GEO suite and both of these have been validated according to the AGS report on the Validation of Geotechnical Software. The validation checks that the programs carry out the analysis as described in their documentation and also that they are appropriate for the type of work being undertaken. Part of this is to ensure that the conditions modelled represent the conditions at the site and that the analyses of these models are realistic.

### Methods of analysis

The following methods of analysis were used for determination of the factor of safety against sliding.

Bishop (Stable and Slope)
Sarma (Stable).
Morgenstern and Price (Stable)
Janbu (Slope)

### Determination of the slip surface

Shallow slips

In order to ensure that the shallow slips analysed approach the critical slips, an estimate of the lowest anticipated factor of safety was calculated using the following equation: -

$$F = (\tan \phi'/\tan \beta)\ (1\text{-}r_u.\sec^2\beta).$$

The solutions are given in Figure 1 and are tabulated below together with the equivalent value for shallow slips determined by Bishop circular slips. This reasonable correlation can therefore be used for an initial assessment of the stability of an embankment.

Table 3. Shallow slips

| Dam | $\phi'$ embankment | $\beta$ (slope angle taken from survey) | (1) Factor of Safety $(\tan\phi'/\tan\beta)(1-r_u.\sec^2\beta^{\#})$ | (2) Bishop Shallow slips (lowest FoS) | Ratio (1)/(2) |
|---|---|---|---|---|---|
| Arnfield | 30 | 18 | 1.68 | 1.32 | 1.27 |
| Belmont | 34 | 21 | 1.66 | 1.50 | 1.10 |
| Hurst | 38 | 26 | 1.45 | 1.50 | 0.97 |
| Rooden South | 36 | 26 | 1.45 | 1.45 | 1.00 |
| Rooden North | 36 | 21 | 1.78 | 1.81 | 0.98 |
| Rhodeswood | 37 | 21 | 1.78 | 1.76 | 1.01 |

\# For $r_u = 0.05$

Shallow slips have lower factors of safety than deep slips. This has been demonstrated in a number of embankments where shallow slips have occurred usually as a result of some local disturbance such as water ingress due to leaking drains or the excavation of a trench which has permitted water to enter the slope. Shallow slips are more common in cut slopes than in the raised (embankment) slopes.

Deep Slips

Deep slips are uncommon but very serious when they occur. The results of the analyses carried out for this study show that they are less likely to occur than shallow slips but if they were to occur then the whole of the bank would be affected including part of the upstream face. The exit location of the slip surface is usually on the upstream face and is at or just below top water level. This means that, unlike shallow slips which can occur and be repaired without detriment to the reservoir, deeper slips would have very serious consequences. A guide to the effect of deep slips is given in Table 4.

Table 4. Deep slips

| **Slope** | | **1 in 2** | **1 in 2.5** | **1 in 3** | **>1 in 3.5** |
|---|---|---|---|---|---|
| Material | Likely $\phi'$ (degrees) | | | | |
| Rock fill | >35 | Case 2 | Case 3 | Case 3 | Case 3 |
| Stony fill | 30 to 35 | Case 1 | Case 2 | Case 3 | Case 3 |
| Sandy clay fill | 27 to 30 | Case 1 | Case 1 | Case 2 | Case 3 |
| Clay fill | <27 | Case 0 | Case 0 | Case 1 | Case 1 |

- **Case 3** has an acceptable factor from desk studies alone using Figures 1 or 2 or similar, a Factor of Safety of at least 1.7 for both shallow and deep slips is required.

- **Case 2** has an acceptable factor of safety after carrying out static stability analyses using assumed conservative parameters for both soil and water levels, a Factor of Safety of at least 1.6 is required.

- **Case 1** requires that sufficient field work and laboratory testing has been carried out by a preliminary investigation to produce acceptable Factor of safety of at least 1.5.

- **Case 0** It is suggested that remedial works are required if the factor of safety for deep slips is less than 1.3. However, if a similar value is obtained for shallow slips then the embankment may be acceptable. Any embankment with a factor of safety less than 1.2 requires urgent attention.

Seismic Events

A report specially commissioned by UU (EQE Report (draft March 2000)) indicates that the values of the seismic forces to be expected are not dissimilar from those given in the BRE guide and the ICE Application Note. These are summarised in Table 5.

Table 5.

| **Reservoir** | **SEE (Application Note)** | **10,000yr event (EQE)** | **30,000yr event (EQE)** |
|---|---|---|---|
| Arnfield | 0.24g | 0.207g | 0.281g |
| Belmont | 0.24g | 0.201g | 0.275g |
| Hurst | 0.24g | 0.208g | 0.282g |
| Rhodeswood | 0.24g | 0.208g | 0.283g |
| Rooden | 0.24g | 0.206g | 0.281g |

The pseudo static analysis has been described in the BRE guide and its use is described in the Application Note. These were followed in this study. It is of note that embankments with an adequate Factor of Safety in the static case are acceptable in the pseudo static case whilst those with a low static factor of safety generally have a low pseudo static stability and may be vulnerable to a seismic event.

## MOVEMENTS

Settlement

Settlement records are kept by UU for all the embankments. For those in this study the movements recorded are generally small and show no noticeable change over the last few years. Any bank showing significant movement should be treated with caution and subjected to a detailed investigation.

Seismic Events

Movements due to the effects of a seismic event have been assessed using the methods given by Ambraseys. These are qualitative calculations based on experience and are designed to give an indication of the likely deformation of an earth embankment after an event. The movements calculated are the horizontal movements due to an earthquake event in the North West of England. Whilst these calculations are very empirical for an embankment with an adequate factor of safety against static failure (greater than 1.3 for deep slips) the movements are very small and are of the order of a few centimetres. For embankments with a low factor of safety in the static case then the calculated movements can be significant.

Care should be taken in the interpretation of these results. Slip planes due to deep slips could exit on the upstream face, often at or below the water level. In these situations the presence of a failure plane within the water retaining part of the reservoir could cause progressive catastrophic failure. However, the UU embankments have significant freeboard for protection against floods, often of the order of a metre or more and hence the small movements associated with seismic events on embankments with an adequate factor of safety will not constitute a problem. This indicates that overtopping of the embankment is unlikely to be a problem after an earthquake event.

## FUTURE INVESTIGATIONS

A methodology is presented for future investigations, based on the lessons learnt during the two stability studies carried out to date. This takes the form of a decision tree which is described below.

Factors of safety

The choice of an acceptable Factor of Safety must be made on an individual basis with sound engineering judgement for each dam investigated. However it is possible to suggest some guidelines for use at each stage. It is suggested that it is necessary to differentiate between the effect of shallow and deep slips. For shallow slips which are unlikely to cause catastrophic failure of the embankment a lower factor of safety may be acceptable at each stage, whilst for deep slips which may affect the overall stability of the embankment a more onerous condition and higher factor of safety is required.

These factors apply to the investigation of an existing embankment, remedial measures will require a different approach.

These factors apply to the investigation of an existing embankment, remedial measures will require a different approach.

Pseudo static analysis

A pseudo static analysis should be carried out as described in the Application note to the Engineering Guide, but only after the static stability of the embankment has been established.

Ancillary structures

At the same time as the desk study is carried out an investigation should be made of the ancillary structures and their vulnerability to movements of the embankments, including those calculated for seismic events.

## CONCLUSION

This study has revealed that seismic analyses based on assumed data are likely to give conservative results that could lead to expensive remedial works. However dams assessed to have a sufficiently high factor of safety against failure using this technique are unlikely to require additional site investigation.

Should a site specific investigation be required the use of dynamic probing/window sampling techniques for sample gathering could be considered. This equipment can be manoeuvred to areas of difficult access on the embankment providing ground information cheaply to correlate with more conventional borehole data.

A decision tree (Figure 2) complied from the information and experience gained during this study, can be used as a starting point to determine whether a site specific ground investigation is required or if existing data can be used satisfactorily for dams in seismic categories II, III and IV.

This decision tree is currently being used as the basis for the investigation of embankment dams owned by UU.

## ACKNOWLEDGEMENT

The authors wish to thank United Utilities plc for permission to publish this paper.

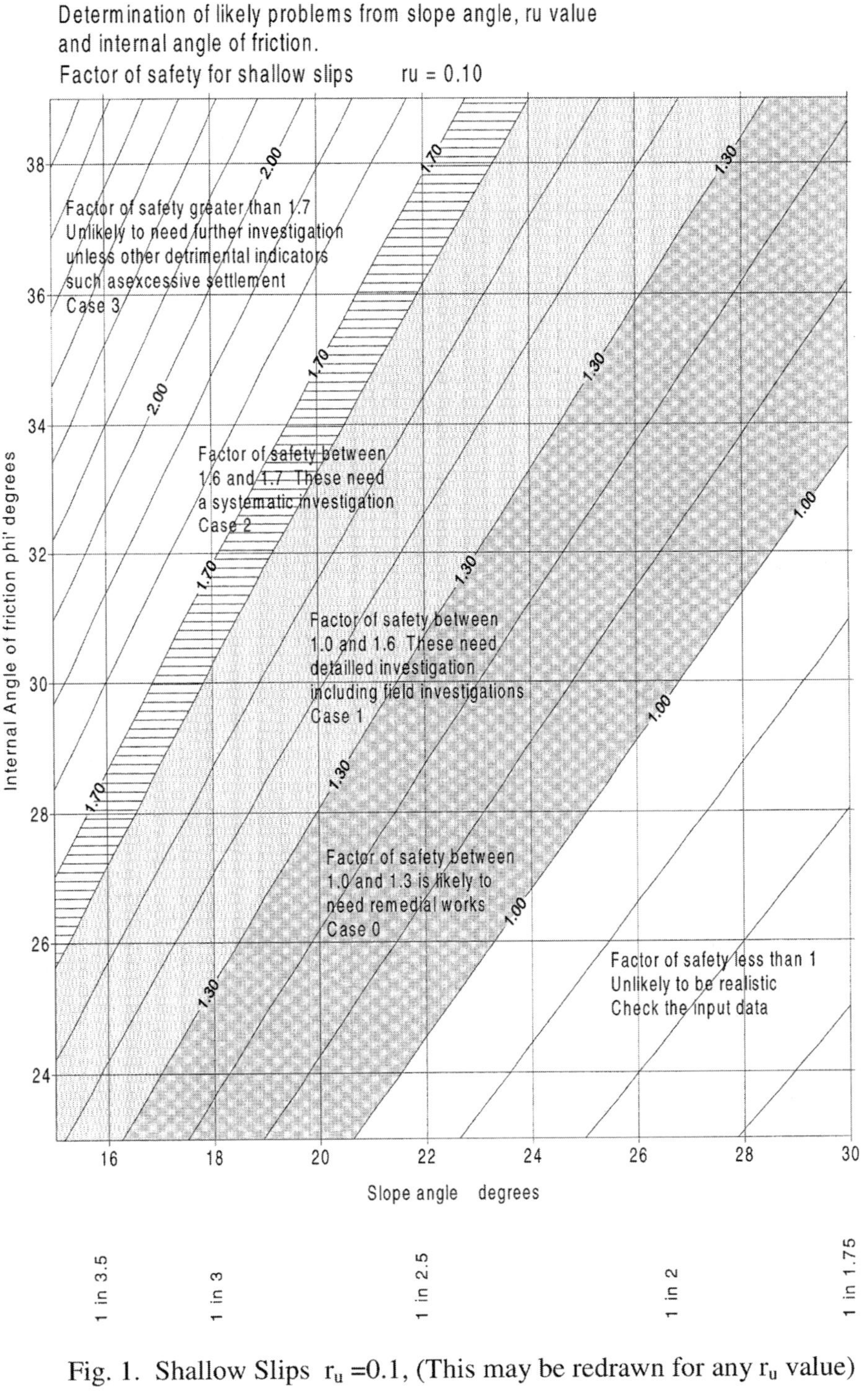


Fig. 1. Shallow Slips $r_u$ =0.1, (This may be redrawn for any $r_u$ value)

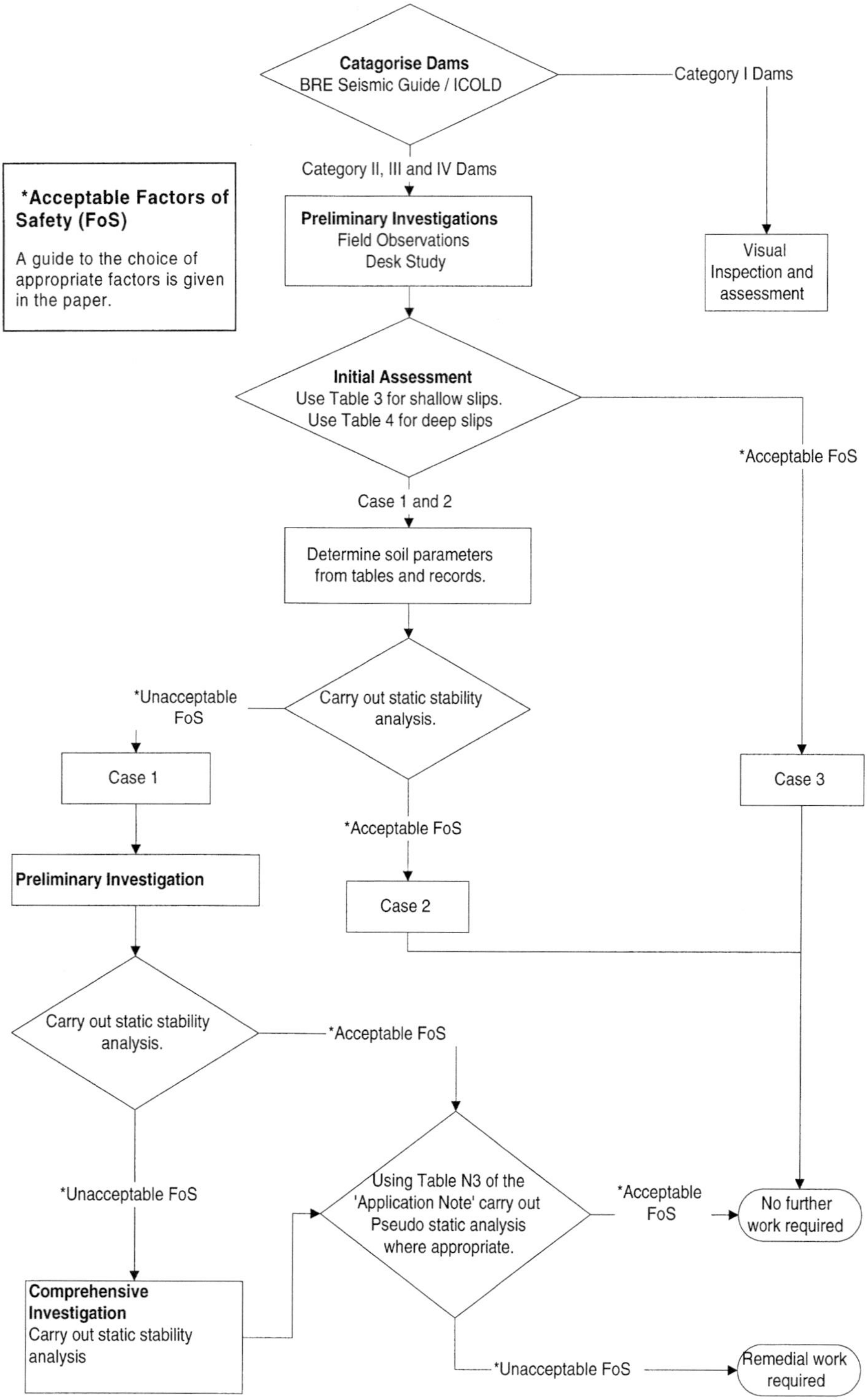


Fig. 2. Decision tree to determine whether a site specific ground investigation is required

REFERENCES

Ambraseys N N (1972). Behaviour of foundation materials during strong motion earthquakes. Proceedings 4th European Symposium on Earthquake Engineering London September 1972. Vol. 7 pp11-12 Sofia Bulgarian Academy of Sciences

A.G.S. Guide. (1994). Validation and use of Geotechnical Software, version 1.0. (Association of Geotechnical and Environmental Specialists)

Bateman J F La Trobe. (1884). History and Description of Manchester Waterworks.

Binnie A.R. (1925) Rainfall reservoirs and water supply. Constable London.

Binnie G M (1981) Early Victorian Water Engineers. Thomas Telford Ltd. 310pp

Binnie G M (1987). Early dam builders in Britain. Thomas Telford Ltd.

Bishop A W (19550. The use of the slip circle in the stability analysis of slopes. Geotechnique Vol. 5, N° 1.

Bolton M.D. and Steedman R.S. (1985) modelling the seismic resistance of retaining structures. 11th Int. Conf. on Soil Mechanics and Foundation Engineering, San Francisco, August.

Building Research Establishment. (1991) An engineering guide to seismic risk to dams in the United Kingdom.

Building Research Establishment. (1996) Investigating Embankment Dams. A guide to the identification and repair of defects.

Burns J.C.O. (1932) Some examples of the raising of reservoir embankments. Transactions of the Institute of Water Engineers, vol. 37, pp 259-283

Charles J A, Abbiss C P, Gosschalk E M and Hinks J L (1991). An Engineering Guide to Seismic Risk to Dams in the United Kingdom. Building Research Establishment Report Publ. BRE, Watford.

Crabb G I and Atkinson J H (1991) Determination of soil strength parameters for the analysis of highway slope failures. Slope Stability Engineering developments and applications (edited by R J Chandler) Thomas Telford

Failures in Earthworks. (1985) Proc. conf. Institution of Civil Engineers .

Hoyle N. (1987). Reservoirs from Rivington to Rossendale.

ICE (1998). An Application Note To An Engineering Guide To Seismic Risk To Dams In The United Kingdom

International Commission on Large Dams. (1989) Selecting seismic parameters for large dams. Bulletin 72. Paris, ICOLD,.

Leech A H and Son. (1988) Mining Appraisal on Certain Reservoirs in the North West. (For North West Water Authority).

Morgenstern N R and Price V E. The Analysis of the Stability of General Slip Surfaces. Geotechnique, Vol 8, 3

Penetration Testing in the U.K. Proceeding of the Geotechnology Conference.– I.C.E Birmingham 1988, Thomas Telford.

Proceedings of the 2nd European Symposium on penetration testing (ESOPT 11) Amsterdam May 1982. Pgs 191-386.

Quayle T (1988). Reservoirs in the Hills.

Rigby P J (1990). Dynamic Probing Systems. MSc Thesis, Leeds University (Unpublished)

Sarma S K (1979). Stability Analysis of Embankments and Slopes. Proc. Am. Soc. Civ. Eng. J. Geotech. Eng. Div.

Skempton A W (1989). Historical Development of British Embankment Dams to 1960. Clay Barriers for Embankment Dams. Proceedings of the Institution of Civil Engineers Conference, October 1989, pp15-52. Thomas Telford, London.

Swales J K (1926) Repair works in connection with the Belmont Reservoir of the Bolton Corporation. Transactions of Institute of Water Engineers, Vol. 31, pp 203-240.

Swannell N G (1994). Simplified Seismic Safety Evaluation of Embankment Dams. Dams and Reservoirs.

Tang W H, Stark T D, Angulo M. (1999) Reliability in back analysis of slope failures. Soils and Foundations Japanese Geotechnical Society Vol 39 No5 pp73-80 Oct

Watts W (1896). Notes on sinking, timbering and refilling concrete and puddle trenches for reservoir embankments. Transactions of British Association of Waterworks Engineers, vol. 1 pp 82-117.

EQE International Ltd (2000) Proposal for the Seismic Hazard Evaluation at Dams in North West England. Confidential report for NWW.

# Remedial works to concrete and masonry dams

# Stability Reassessment and Remedial Works at Leixlip Dam

B O'MAHONY, ESB International, Ireland
B HAUGH, ESB International, Ireland

SYNOPSIS. Leixlip Dam is located on the River Liffey, 20km upstream of Dublin City. Concerns regarding the stability of the dam, due to the presence of weak layers of gouge material interspersed with the limestone layers in its rock foundation, have lead to a series of stability studies being carried out over a number of years. This paper provides a brief history of these studies and describes in more detail the most recent reassessment of the stability of Leixlip Dam. It further describes the remedial works, principally involving the installation of prestressed rock anchors, which were carried out as a result of the most recent reassessment of the dam's stability.

## INTRODUCTION

The River Liffey rises in the Wicklow Mountains near the Sally Gap and enters the sea at Dublin. Leixlip Dam, owned by the Electricity Supply Board (ESB), is the lowermost of three hydro-electric dams on the River Liffey and is located approximately 20km upstream of Dublin. The dam is of the mass concrete gravity type, founded on stratified limestone rock interbedded with weak layers of gouge material (possibly resulting from a mylonitic-type process whereby extensive movement in the bedrock may have occurred in geological time). The limestone beds are extensively jointed and dip from the right bank at angles of $15^{\circ}$ to $30^{\circ}$ in a direction parallel to the dam centre-line.

Construction of the dam was completed in 1952. The crest length of the dam is 114m and its maximum height above foundation level is approximately 23.5m. Short earth embankments extend upstream at each end of the dam. The concrete dam comprises nine separate blocks of varying width and height which are keyed together, i.e. intake block, fish pass block, two spillway blocks, comprising three spillways, and five conventional gravity blocks. Fig. 1 shows a plan of Leixlip Dam, indicating the dam blocks.

## ORIGINAL DESIGN OF LEIXLIP DAM

The original design of the dam was based on the "Gravity Method of Stress and Stability Analysis", as described in the handbook "Engineering for Dams Volume 2" (Hinds, Creager and Justin, 1946). This method of design did not allow tensile stresses within the body of the dam under any loading

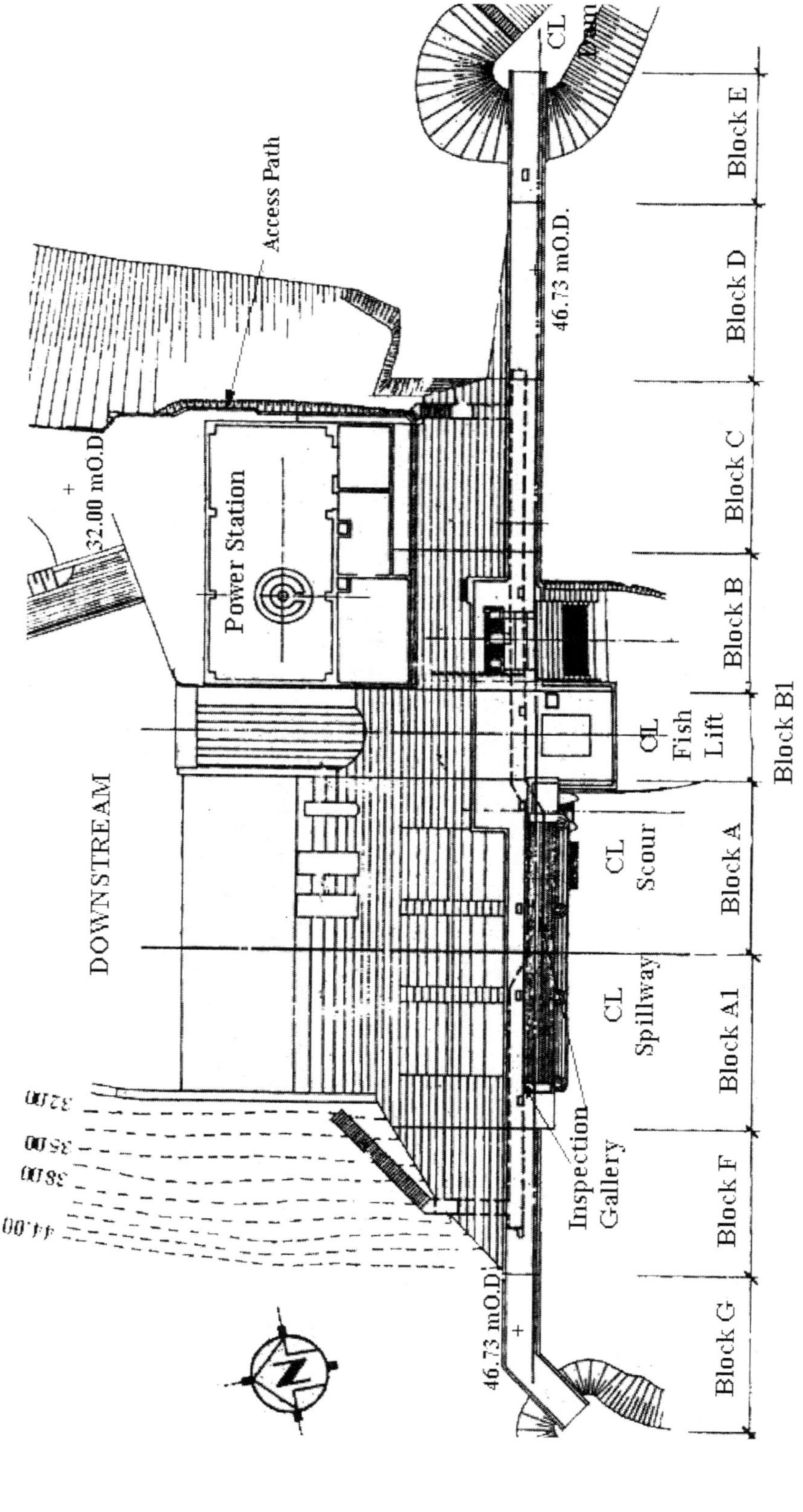


Fig. 1 Plan of Leixlip Dam

condition. It was assumed that the water level was approximately one metre below crest level. Uplift pressure of 67% of the reservoir head was assumed to act at foundation level and at the horizontal construction joints of the dam.

The foundation was not considered as an integral part of the dam. Furthermore, stresses arising from seismic loadings were not considered in the design because of the prevailing view at the time that such events were not relevant in an Irish context.

The foundation of the dam was designed as a series of steps, rising from the excavation in the river bed. The depth of the foundation generally depended on the depth of a suitable rock stratum. Anticipated locations for steps were line-drilled and the excavation worked towards the line-drilling. In many cases it was necessary to redrill these lines in order to obtain a suitable rock foundation. In particular, for Block C, located immediately upstream of the power station, excavated steps failed on the weak layers of gouge material and slid into the excavation. Therefore, under this block the rock was pinned with 50mm diameter bars grouted into holes drilled 6m into rock below the excavation level.

## DESIGN REVIEWS AND REMEDIAL WORKS (1984 –1990)

In 1984, as part of an overall review of the safety of all of its dams, the ESB undertook a design review of its three dams on the Liffey (Brogan, Hayes and O'Mahony, 1986). Part of this overall review involved the setting of new design standards for ESB dams. Under these new standards, ESB dams are required to withstand both the 10,000-year flood event, which for Leixlip involves reservoir levels up to crest level, and the 10,000-year seismic event.

During this review, the structural stability of Leixlip Dam was examined using the information available at the time. This examination, carried out on typical blocks, comprised a static analysis and also finite element analyses of stresses resulting from both static and seismic loads. For the purposes of these studies typical metre wide sections of the dam blocks were analysed. Due to concerns raised by the available geological information relating to the weak seams, a sensitivity study was carried out, involving the inclusion of weak layers with varying properties into the finite element models. These studies concluded that tensile stresses could occur in the foundation rock under some loading conditions. Due to the unfavourable orientation of bedding planes and the presence of weak layers, the occurrence of tensile stresses could facilitate the ingress of water to the weak layers, further weakening and lubricating them. To prevent this and to improve the stability of Leixlip Dam, it was recommended that the installation of prestressed rock anchors should be considered. This requirement for prestressed anchors was confirmed by an independent consultant employed by ESB in 1986, Vorarlberger Illwerke AG (VIW) of Austria.

Following the recommendations of the ESB and VIW studies, further stability studies were carried out on Leixlip Dam in 1989. These studies used additional new geological information obtained at this time and concluded that a total of 26 prestressed rock anchors (1,000kN working load) were required in five blocks of the dam. These anchors were installed in 1989/1990. Fig. 2 shows the installation of one of these anchors using a crane located in the tailrace. At this time other improvement works were also carried out on Leixlip Dam, i.e. foundation grouting, construction of a concrete apron and guide wall in the river downstream of the dam to protect its toe from erosion during floods, raising of essential spillway gate control equipment above the 10,000-year tailrace level, installation of movement monitoring systems and drainage relief holes and installation of a floating boom to protect the spillway gates from floating trees and other debris (O'Tuama and O'Mahony, 1989).

Fig. 2 Installation of Anchor Using a Crane Located in the Tailrace

Towards the end of this review period a number of organisational changes relating to dam safety took place within ESB. Civil Works Department, ESB, which had carried out the original design and subsequent stability studies on ESB dams, was incorporated into ESB International, which operates as an independent consultancy company. In addition, the current

ESB dam safety organisational structure was instigated. This included the appointment of an External Dam Safety Committee (EDSC), which reports directly to the Chief Executive of ESB. This committee of international experts is an independent board that oversees the safety evaluation of ESB dams.

EXTERNAL DAM SAFETY COMMITTEE INSPECTIONS (1991/1996)

The appointment of the EDSC led to further investigation of the weak layers of gouge material between the limestone beds at Leixlip. The first EDSC 10-year Inspection of Leixlip Dam took place in 1991. Following this inspection samples of the gouge material were retrieved from cores and laboratory tests were carried out. These tests included large displacement shear box tests to simulate possible mylonitic type action. These tests indicated that the strength of this material was less than originally considered and provided the following revised strength parameters:

$$\varphi = 25^{\circ} \text{ and } C = 0, \text{ where } \sigma_n^1 < 192 \text{ kN/m}^2$$
$$\varphi = 6^{\circ} \text{ and } C = 70 \text{ kN/m}^2, \text{ where } \sigma_n^1 > 192 \text{ kN/m}^2$$

Preliminary studies using these revised strength parameters for the gouge material were carried out and further detailed discussions on the issue took place during the EDSC 5-year Review of Leixlip Dam in 1996.

One of the problems with the weak layers is that they can be very thin and can be difficult to identify from cores. Therefore, it was agreed during the 5-year Review that it should be assumed that the weak seams could occur everywhere under the foundation of the dam, with a nominal spacing of one metre between seams.

STABILITY REASSESSMENT OF DAM

In 1997/1998, following the discussions with the EDSC, ESB International undertook a reassessment of the stability of Leixlip Dam. Unlike previous investigations, where typical metre-wide cross-sections of dam blocks were considered, this study treated each block of the dam as a three-dimensional body. In addition, the rock wedge method of analysis was used. This method assumes that the plane on which the dam might slide has a saw-tooth shape, formed by the weak seams at one metre centres and the orthogonal joint planes (J1) in the limestone beds. The following strength parameters were applied for the J1 joints:

$$\varphi = 67^{\circ} \text{ and } C = 20 \text{ kN/m}^2, \text{ where } \sigma_n^1 < 40 \text{ kN/m}^2$$
$$\varphi = 48^{\circ} \text{ and } C = 70 \text{ kN/m}^2, \text{ where } \sigma_n^1 > 40 \text{ kN/m}^2$$

This saw-tooth shaped surface was considered to be the most likely sliding failure surface. Because of the conservative assumption that the weak seams could occur at one-metre centres everywhere under the foundation of the dam, it was considered acceptable that support from downstream structures and the passive resistance of downstream rock masses could be included to achieve acceptable factors of safety against sliding if necessary. Fig. 3 provides a typical representation of the sliding failure surface assumed under a dam block.

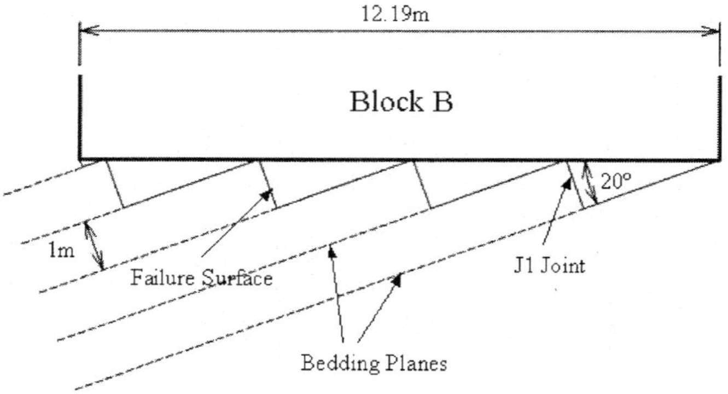


Fig. 3 Typical Representation of Sliding Failure Surface Considered Under a Dam Block

For this reassessment uplift pressures were based on the full reservoir head. In addition, seismic loads based on the findings of the report "Seismic Design Criteria for Dams in Ireland" (Principia Mechanica, 1986) were included in the analyses. Seismic loads due to the Maximum Credible Earthquake were calculated, using the pseudo-static method, based on 0.15g horizontally, in the upstream/downstream direction, and 0.075g vertically.

LOAD CASES

Generally each block was assessed for four loading conditions, with factors of safety being calculated. In accordance with international practice, the following factors of safety were required for both sliding and overturning:

<u>Load Case 1</u>
Maximum Normal Operating Level (M.N.O.L.)
"Sliding plane" at the dam/rock interface
Required Factor of Safety = 1.5

<u>Load Case 2</u>
10,000-year Flood Level (Crest Level)
"Sliding plane" at the dam/rock interface
Required Factor of Safety = 1.3

Load Case 3
Maximum Normal Operating Level + Maximum Credible Earthquake
"Sliding plane" at the dam/rock interface
Required Factor of Safety = 1.1

Load Case 4
Maximum Normal Operating Level + Maximum Credible Earthquake
"Sliding plane" at a deeper level, taken as 15% of the average block height below the dam/rock interface
Required Factor of Safety = 1.05

Where the required factors of safety could not be achieved for the existing conditions at a particular block, the number of anchors needed to achieve the required factors of safety for that block was also calculated.

## GENERAL DESCRIPTION OF ANALYSIS PROCEDURE

All the loads acting on a block for a particular load case were calculated and applied as appropriate. The vertical loads considered, where appropriate for each block, were the self-weights of the block, rock and backfill, the weights of water on the upstream and downstream faces of the block, the weights of water in structures (e.g. upstream of the intake gate), vertical earthquake loads due to the masses of the block, rock, backfill and water, uplift and the vertical component of the existing anchor loads. The horizontal loads considered, where appropriate for each block, were the hydrostatic loads from the reservoir and tailrace, the hydrodynamic load from the reservoir, the horizontal earthquake loads due to the masses of the block, rock and backfill and the horizontal component of the existing anchor loads in the upstream/downstream direction.

The distribution of stresses on the "sliding plane" was calculated for each load case, taking all relevant vertical and horizontal loads into account. Where negative stresses were calculated at the heel of a block, a crack propagation analysis was carried out to determine the maximum length of crack to ensure that the negative stresses did not have a destabilising effect on the block. Generally, negative stresses only occurred for earthquake loads cases. The resistance to sliding in these instances was calculated based on the area of the foundation subject to positive stresses calculated following the crack propagation analysis. Stresses normal to the bedding planes and J1 joints were calculated using Mohr's Circle. Fig. 4 indicates a typical Mohr's circle calculation for one of the dam blocks.

The resistance to sliding due to friction and cohesion on the bedding planes and J1 joints was calculated for each load case, using the revised values for the strength parameters noted above. The parameters for the gouge material were used to calculate the sliding resistance on the bedding planes. This allowed the factor of safety against sliding for each block to be calculated.

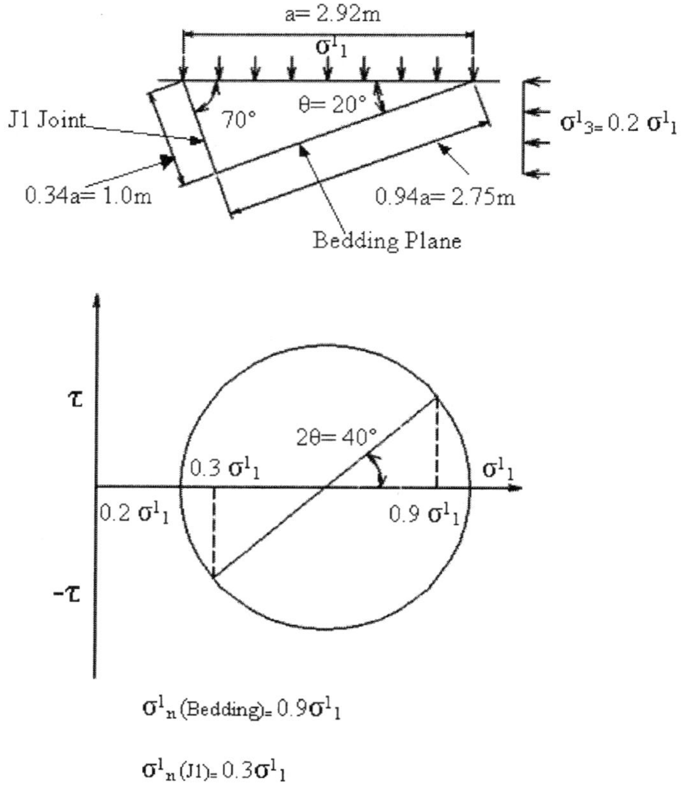


Fig. 4 Mohr's Circle Calculation for a Dam Block

Where the factor of safety against sliding for a number of dam blocks was calculated to be less than that required for a particular load case, additional support from downstream structures or the passive resistance from downstream rock masses was considered. The downstream structures included the power house (Block B), fish pass barrel (Block B1), stilling basin (Blocks A and A1) and a mass concrete retaining wall (Block A1). The additional resistance to sliding due to these structures was calculated in a similar manner to the block itself. The factor of safety against sliding was then recalculated.

For Load Cases 1, 2 and 3, for each dam block, the factor of safety against overturning was also calculated.

Where adequate factors of safety could not be achieved, even when the support from downstream structures and passive resistance from the downstream rock mass were taken into account, the number of additional anchors required was calculated for each load case. The calculations for additional anchors were carried out in a similar manner to the procedures outlined above.

RESULTS OF STABILITY REASSESSMENT OF LEIXLIP DAM

The reassessment of Leixlip Dam indicated that Blocks F, A1, B1, B, C and E could achieve adequate factors of safety for all load cases, without the need for further remedial measures. Due to favourable results from previous more simplified analysis, it was also concluded that the stability of Block G is adequate.

For one of the spillway blocks (Block A), adequate factors of safety could be achieved for all load cases under the most favourable assumptions. However, there was some uncertainty regarding the dip of the bedding planes under this block and it was considered prudent to consider the implications of this uncertainty. It was concluded than, even under the worst assumption regarding the dip of the bedding planes under this block, i.e. a dip of $20^{o}$ (instead of $30^{o}$) across the entire width of the block, the installation of four additional 1,000kN anchors would provide adequate factors of safety.

In the case of one of the smaller gravity blocks (Block D), near the right abutment of the dam, factors of safety less than those required were obtained for all load cases. It was calculated that the installation of five 1,000kN anchors in this block would provide adequate factors of safety for all load cases.

The stability of a rock mass immediately downstream of the right abutment, which is adjacent to a quarry and the excavation for the power station, was also assessed. The integrity of this rock mass was adversely affected by joints and seams. Roots from trees and shrubs growing on this rock mass were tending to widen the joints and seams. There were concerns that this rock mass was a danger to personnel and also to the structure of the power station. The results of this assessment also indicated factors of safety less than those required. It was calculated that the installation of one 1,000kN anchor in this rock mass, at an angle of $30^{0}$ to the horizontal would provide adequate factors of safety for the assessed load cases. However, to achieve the conditions assumed in the analyses, drainage relief holes were required in the vertical face of the rock mass, to ensure relief of uplift and hydrostatic pressures for a distance of 3 metres from the face.

Therefore, the main conclusion of the stability reassessment of Leixlip Dam was that ten additional 1,000kN rock anchors should be installed, nine in the dam and one in the rock mass downstream of the right abutment.

INSTALLATION OF ANCHORS

In 1998/1999 P.J. Edwards & Co. Ltd. of Dublin undertook a contract to install the required additional anchors.

The anchors installed were manufactured from non-alloy 7-wire dyform strand to BS 5896 Part 3 (1980). The bare strands are doubly protected by cement grout and corrugated plastic sheathing over a 6m fixed length in the limestone rock, with cement grout outside the sheath. Over the free length each strand is protected by a grease filled polypropylene sleeve, with cement grout both inside and outside a smooth plastic sheath. Fig. 5 shows typical details of the rock anchors used.

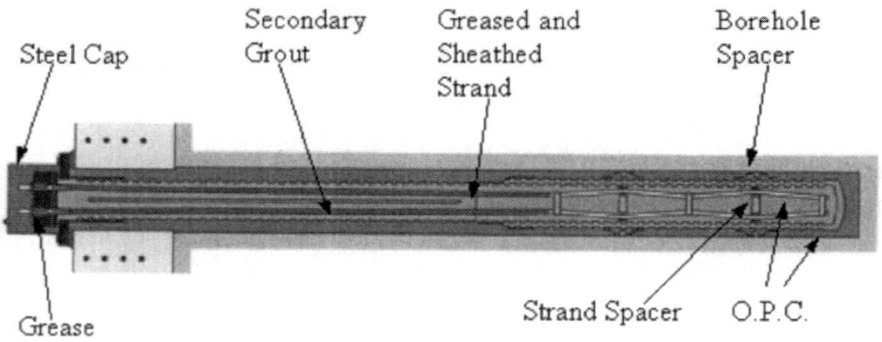


Fig. 5 Typical Rock Anchor Detail

The anchors were designed in accordance with BS 8081 using the following assumptions:

- Cement Grout/Rock – Ultimate Bond Stress = 1.75N/mm$^2$
- Strand/Cement Grout – Working Bond Stress = 1.3N/mm$^2$
- Minimum Embedment of Anchors in Rock = 10m
  (Actual minimum embedment = 12m)
- Included Angle for Rock Cone Resisting Uplift = 60$^\circ$
- Compressive Strength of Cement Grout = 45N/mm$^2$ @ 28 days
- Strand Breaking Load = 300kN

Boxouts were formed by stitch drilling on the downstream face of the two dam blocks at the locations where anchors were to be installed. 150mm diameter holes were then percussion drilled through the concrete dam blocks and into the limestone rock. These anchor holes were drilled in the same orientation as the original 26 anchors installed in 1990. This orientation was based on the requirement to intersect bedding planes as close as possible to perpendicular while still providing sufficient resistance to sliding. The length of drilling in concrete varied from 4m to 13m, while the length in rock varied from 12m to 14.5m. The hole downstream of the right abutment was also drilled 15m into the rock at 30° to the horizontal.

Each hole was pressure grouted with a sulphate resistant cement grout to fill any voids in the bedrock. The grout mix has a water cement ratio by weight varying between 1.3:1.0 and 1.0:1.0. Grouting pressures were generally low to reduce the risk of grouting pressures causing uplift. There was generally very little grout take in the holes. During grouting operations all existing movement monitoring instruments on the dam in the vicinity of the grouting were monitored regularly. In addition, an extra inclinometer tube was temporarily installed in the particular block where grouting was taking place. No significant movements were detected during the grouting operations.

The anchors were installed in the dam using mobile cranes located on the right bank. Prior to anchor installation, the holes were redrilled and tremie grouted with cementitious grout. Subsequent to the anchors being installed the void inside the corrugated sheathing was also grouted with a cementitious grout. Following installation of the anchors, concrete blocks were cast into the boxouts on the downstream face using Grade C40 concrete.

Fig. 6 300t Jack in Position During Stressing of a Rock Anchor

Stressing of the anchors was carried out in accordance with BS 8081 using a 300t hydraulic jack. Fig. 6 shows the 300t jack in position during the stressing of one of the anchors. Subsequent load tests indicate that the anchors are operating satisfactorily.

Prior to stressing the new anchors, existing anchors in the relevant blocks were load tested. They were again tested following stressing of the new anchors, with no loss of load capacity being detected.

The installation of the 1,000kN anchor in the rock mass downstream of the right abutment was carried out in a similar manner to those in the dam. In addition, during the anchor installation contract, to improve the integrity of the vertical face of this rock mass, some trimming of the faces of the rock was carried out to remove loose blocks, to increase personnel safety and to reduce the risk of blocks of rock falling on the power house. Where the rock was not considered to have been adequately stabilised by this trimming, a series of rock bolts was also installed.

## CONCLUSIONS

The stability reassessment of Leixlip Dam was carried out using the available information and considered all relevant load cases. While the analyses were based on geomechanical models and simplified failure modes, the assumptions are considered to range from realistic to conservative. The implementation of the measures recommended as a result of this reassessment improves the stability of Leixlip Dam to a level which is in general accordance with modern standards. However, due to the nature of its foundation, Leixlip Dam must be kept under continuous surveillance. This will ensure that potential decreases in factors of safety, resulting from decreases in anchor loads with time, increases in uplift pressures, development of seepages or other deterioration, are identified at an early stage.

## ACKNOWLEDGEMENTS

The author is grateful to the Electricity Supply Board for permission to publish this paper.

## REFERENCES

Hinds J, Creager W P and Justin J D (1946). Engineering for Dams, Volume 2. John Wiley & Sons, New York, USA.

Brogan O, Hayes T & O'Mahony J (1986). A Review of the River Liffey Dams Spillway Capacity and Structural Stability. The British National Committee on Large Dams, The BNCOLD Prize.

O'Tuama F F & O'Mahony B (1989). Hydro Electric Dam Safety – Upgrading of ESB Dams. The Institution of Engineers of Ireland.

Principia Mechanica (1986). Seismic Design Criteria for Dams in Ireland. Unpublished document for Electricity Supply Board.

# Rehabilitation of old masonry dams at full reservoir level - A comparison of successful rehabilitation projects

V BETTZIECHE, Ruhr River Association, GE
C HEITEFUSS, Ruhr River Association, GE

SYNOPSIS. At the beginning of the last century about 40 gravity dams were built in Germany. During the last years extensive rehabilitation measures were carried out at several old masonry dams. Rehabilitation concepts which called for the construction of inspection/drainage galleries inside the dams turned out to be rather cost-efficient. Different methods for the construction of these inspection galleries were used – from manual driving of the tunnel to the drill & blast method and the use of a tunnel boring machine. The specific costs of these galleries varied, depending on the construction method.

## INTRODUCTION

During the first 20 years of the last century about 40 gravity dams were built in Germany (Figure 1). These structures were designed as so-called „Intze-type" masonry dams with a curved base and without any joints.

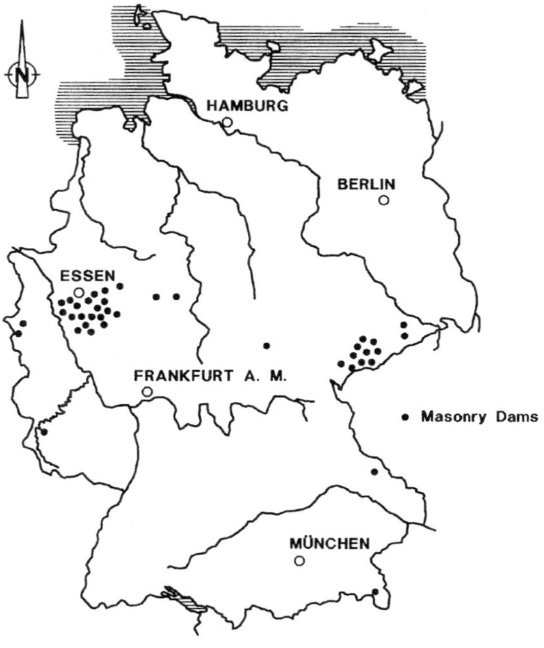


Figure 1. Masonry Dams in Germany

*Reservoirs in a Changing World*, Thomas Telford, London, 2002, 155–166

Hydraulic structures and dams are subject to ageing - like any other structure. The enormous importance of a reservoir for the infrastructure of the supply area and its damage potential require a continuing adaptation to the established technical standards. During the last years extensive rehabilitation measures were carried out at several old masonry dams. The specific costs for rehabilitation varied from 40 € to 600 € per cubic metre of dam volume.

Rehabilitation concepts which called for the construction of inspection galleries inside the dams were cost-efficient. These galleries serve as means for the inspection of the condition of the foundation joints, as well as drainage system for increased water pressures and for the installation of monitoring equipment inside the dams. Different methods for the construction of these inspection galleries were used – from manual driving of the tunnel to the drill & blast method and the use of a tunnel boring machine. The specific costs of these galleries varied from 1.100 € per m³ to 2.200 € per m³ of gallery volume, depending on the construction method. Three rehabilitations at the Fuerwigge Dam, the Gloer Dam and the Ennepe Dam are described.

Table 1: Dams data

| | Fuerwigge Dam | Gloer Dam | Ennepe Dam |
|---|---|---|---|
| Year of completion | 1904 | 1904 | 1904/1912 |
| Height of Dam [m] | 29 | 32 | 51 |
| Length of Dam [m] | 166 | 168 | 320 |
| Volume of Dam [1000 m³] | 26 | 35 | 106 |
| Storage Capacity [1000 m³] | 1670 | 2100 | 12600 |

THE DAMS

The reason for the construction of the Fuerwigge- and Gloer-Dam was the expansion the activities of the steel manufacturing industry in the Sauerland mountain region during the late 19$^{th}$ century and its growing demand for a sufficient and reliable water supply for hydro energy purposes. Due to lack of water, the water mills and the affiliated manufacturing plants had to be shut off partly or completely for longer periods every year. The Fuerwigge- and Gloer-Dam were built between 1904 and 1906 based upon the design of Prof. Intze. With dam heights of about 30 m and storage capacities of about 2 hm³ they can be considered as small reservoirs.

The Ennepe Dam is a masonry dam with a length of 320 m and a height of 51 m and was built between 1902 and 1904 by the former owner, the Ennepe Water Association. Its main purpose was to stabilise the discharge of the Ennepe River, thus being a reliable source for the generation of hydropower for the factories at the lower reaches of the river even in periods of drought.

Initially the dam was only 41.4 m high, which resulted in a storage capacity of 10.3 million m³. Between 1910 and 1912 a masonry block with a height 10 m was added to the crest of the dam. This enabled the former owner to raise the maximum water level by 2.5 m and created a storage capacity of 12.6 million m³.

Problems with these old masonry dams

Originally these three dams were not equipped with an inspection- and drainage gallery. As usual at many old masonry dams, a drainage system consisting of vertical stoneware pipes had been installed right behind the upstream face of the dams. Unintentionally these drainage pipes had been filled with grouting material during several repair works in the1950s. Thus during the last decades there was no effective drainage system available both in the dams and in their bedrock.

These old masonry dams were designed without taking the pore pressure and the uplift into account, based on the basic design principles Prof. Intze applied at the early masonry dams [1]. Therefore the entire structure proved to be rather slender. At the beginning of the 1980s this problem was detected by the Reservoir Supervision Authority. According to the current view of the physical effects of the uplift phenomenon the authorities demanded the immediate adaptation of the dams to the established technical standards. The maximum storage level had to be reduced for safety reasons at many old dams in Germany.

For different reasons the required adaptation of some dams was not carried out until the 1990s. In June 1997 the Ruhr River Association took over the Ennepe Dam from the former owner, the Ennepe Water Association under the obligation to adapt the dam to the established technical standards and to carry out rehabilitation measures for the long term safety of the structure.
The Fuerwigge Dam has been owned by the Ruhr River Association since 1933. Its refurbishment was postponed for the time being. The same goes for the Gloer-Dam, which is currently refurbished by the Ruhr River Association on behalf of the owner, the City of Luedenscheid municipal water works.

Specially the Ennepe Dam - as mentioned above - was built for the water supply of 170.000 consumers in the Ennepe-Ruhr District. Therefore during the rehabilitation process the reservoir could not be emptied without causing major problems. Former investigations show that a temporary conversion of the water supply system to other sources would cost about 13 million US-$ and had only a slight chance for realisation therefore. The basic principles for the adaptation of the dam to the established technical rules had to take this into account.

Different concepts were worked out, to adapt the dam to the established technical standards. As at many other old masonry dams in the neighbourhood the first idea was to build a concrete diaphragm wall at the upstream side of the Ennepe dam. First calculations showed that this would take about 40 Mio. €.

REHABILITATION CONCEPT "DRAINING THE DAM"

A concept for the rehabilitation of dams has been developed further by the Ruhr River Association, being used first time in 1965 at the rebuilding of the Lister Dam: to stabilise the entire structure by reducing the uplift.

The most important elements of this concept were:

- the construction of a drainage gallery close to the upstream face at normal reservoir level and
- to drain masonry and bedrock with fans of drainage borings.

Additional rehabilitation measures were carried out at all reservoirs e.g.:

- the replacement of the intake gates and conduits
- the rehabilitation of the gate towers
- a new layout for the water supply intakes.

This concept was developed so far, that there was no doubt about the feasibility and then submitted to the district authorities for permission and funding.

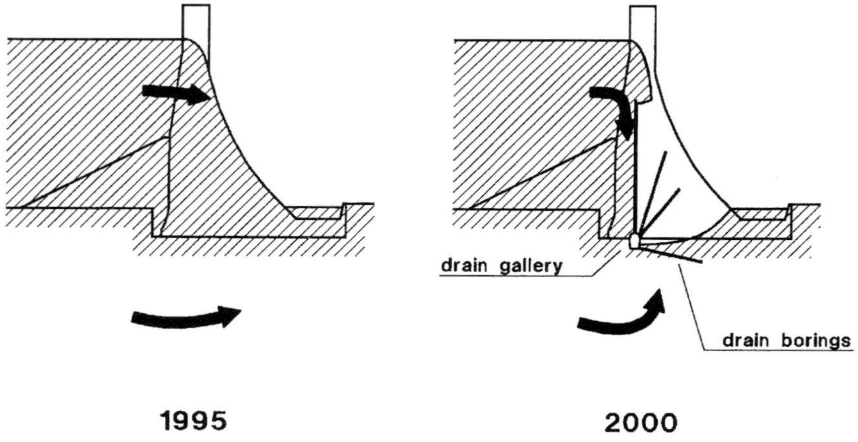


Figure 2. Realised concept of rehabilitation of the Ennepe Dam, using draining

The Reservoir Supervision Authority agreed upon the entire rehabilitation concept, under the reservation, that measurements had to prove the success of the rehabilitation.

FEM-Model

The rehabilitation concept was based upon a detailed feasibility study, applying different numerical simulation methods. On the basis of these simulations the Reservoir Supervision Authority agreed in the concepts of rehabilitation.

Three numerical models, using the Finite-Element-Method (FEM) were used:

- a fluid-FEM-model to analyse the seepage inside the dam and the effect of the internal waterforces (porepresure)
- a FEM-model of temperature distribution for the quantification of the influence of the seasonal temperatures and from this resulting the internal stresses in the dam (Bettzieche, V. 2000b)
- a FEM-model of crack propagation to prove the stability and the occurrence of cracks, essentially affected by the stresses, determined by the first two models (Bettzieche, V. 2000a)

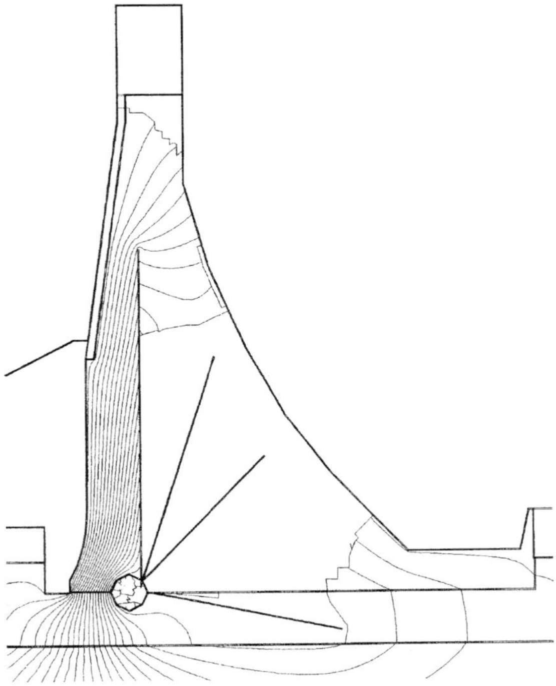

Figure 3: FEM-calculated field of porepressure

The layout scheme of the drainage borings, as one of the most important assumptions, had to be checked. The numerical calculations led to a provisional distance of 4 m between the drainage fans at the Ennepe Dam and 3 m at the Glör Dam. It had to be examined, if this distance was sufficient for a reliable reduction of the uplift pressure inside and underneath the dam.

Table 2: Data of the Rehabilitation

| | Fürwigge Dam (Concept) | Glör Dam | Ennepe Dam |
|---|---|---|---|
| Rehabilitation Concept | only borings | blasting a tunnel | tunnel boring machine |
| Length of the Gallery | 0 m | 45 m | 430 m |
| Profile of the Gallery | no gallery | 2,20 m x 3,00 m | ∅ 3,00 m |
| Total Length of Drainage borings | 900 m | 580 m | 1350 m |
| Costs of Drainage Gallery | - € | 330.000 € | 4.000.000 € |
| Costs of Borings | 205.000 € | 150.000 € | 250.000 € |
| Costs of Injections | 95.000 € | 65.000 € | - € |

The investigations at the Fuerwigge Dam indicate that a drainage gallery is not necessary. The rehabilitation concept calls for the installation of a fan of drainage borings only. These borings can be driven from the bottom outlet galleries, which cross the dam (s. Figure 4).

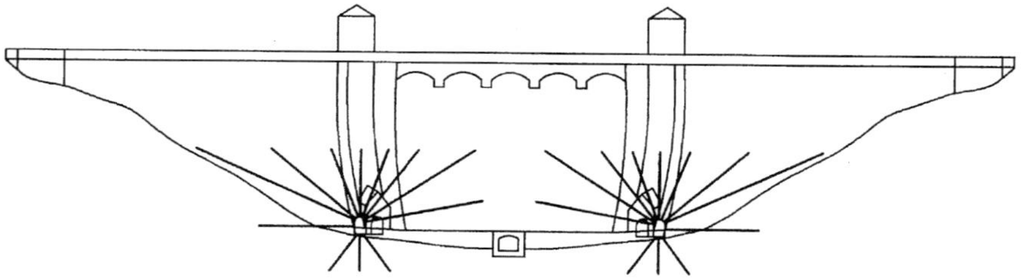

Figure 4. Drain Borings inside the Fürwigge Dam (Scheme)

Measurements

It has been mentioned, that before the execution of final stability calculations the effects of the drainage measures on the pressure conditions inside the dam and the bedrock had to be investigated by experimental measurings (Bettzieche, V. & Heitefuss, C. 2001).

At the Ennepe Dam i.e. the following measuring devices have been installed, according to the German Guidelines (ATV/DVWK 1991):

- 2 plumblines, l = 50 m (from the crest to the gallery),
- 2 invert plumblines, l = 25 m (in continuation to the plumblines),

- 2 inclinometers for monitoring of possible movements of the crest
- 2 measuring sections with 9 piezometers each, in order to monitor the piezometric pressures from the upstream to the downstream face of the dam.
- 2 measuring sections with 40 temperature gauges together and an additional fibreoptical sensor (Bettzieche, V. 1997b).

Since the Ennepe Dam was supposed to be run without a permanent operating crew, all relevant data of the structure are provided for external monitoring via a data transmission system.

## REHABILITATION

General solutions for the rehabilitation of dams do not exist, but a wide variety of technical concepts. The choice of a rehabilitation concept requires the adaptation of the optimal technical and economical solution to the specific hydraulic structure. During the last years some gravity dams were refurbished by the installation of an concrete diaphragm wall at the upstream face. This requires the complete drawdown of the reservoir, allowing a restricted water supply. Additionally the sealing of the bedrock is necessary in order to prevent uplift pressures.

In many cases dams have been refurbished by a combination of grouting and drainage measures. For this purpose cement is injected through boreholes into the dam and the bedrock, reducing the permeability to a tolerable level. The sealing of the dam and the drainage of the bedrock reduces the uplift pressure. The drainage borings collect the seepage water in order to guarantee the prevention of excessive uplift pressures.
The construction of the boreholes requires the driving of a drainage- and inspection gallery at the upstream foot of the dam. These galleries with a width of 2 to 3 m and a height of about 2.5 m run through the dam longitudinally in the foundation joint.
These inspection galleries have been driven with various methods, for instance the manual driving, the use of a tunnel boring machine and the drill & blast-method.

### Manual Driving of the Gallery

For the construction of the inspection gallery at the Unteren Herbringhaeuser Dam near the City of Wuppertal
the manual driving was chosen (Aberle, B. & Hellmann, H. 2000). This method can be adapted to the local conditions very good. The precise contours and varying courses of the gallery can easily be driven. Masonry made of greywacke or material of similar strength allow only very limited rates of advance - often less than 0.1 m per work shift. The excavation of the masonry can be supported by cracking equipment and hydraulic jacks, even though the use of machinery is limited by the small cross sections of the galleries. The irritation by dust and noise in the narrow galleries in

combination with the heavy work during many months puts an enormous strain on the tunneling workers.
The strain on the tunneling crews can be reduced by the use of overlapping core drills. Due to the very limited rates of advance this method can be economically used only in very specific situations or at very short tunneling stretches like break-throughs.

The Drill & Blast - Method
The drill & blast - method has successfully been used by the Ruhr River Association in the 1970s for the driving of a longitudinal inspection gallery at full reservoir level at the Moehne Dam. During the last years inspection galleries were driven into several dams using the drill & blast-method. All projects showed, that this method can be adapted to every necessary geometry on site (Aberle. & Hellmann, 2000). Very steep slopes at the abutments, right-angled junctions and even shafts can be driven precisely. The very small distance of the inspection gallery to the upstream foot of the dam requires a very careful and rock-protecting blasting method. By protective blasting the vibration load and the weakening of the surrounding masonry can be limited.
Due to precise blasting these inspection galleries could be constructed economically with little additional work due to over- or underbreak. Almost every inspection gallery was constructed without final shotcrete or concrete linings.

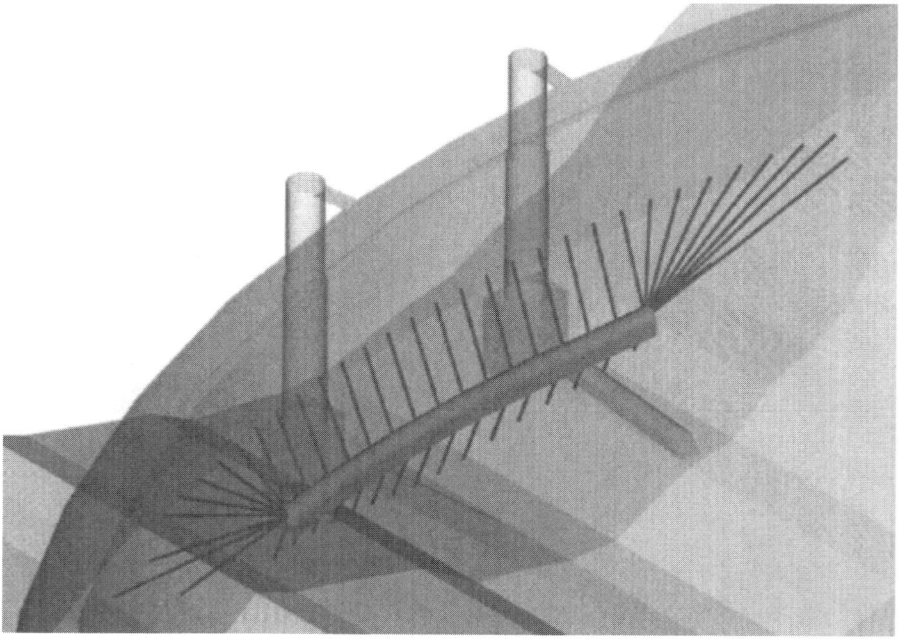

Figure 5. 3-D-CAD-View: Drainage gallery inside of the Glör Dam

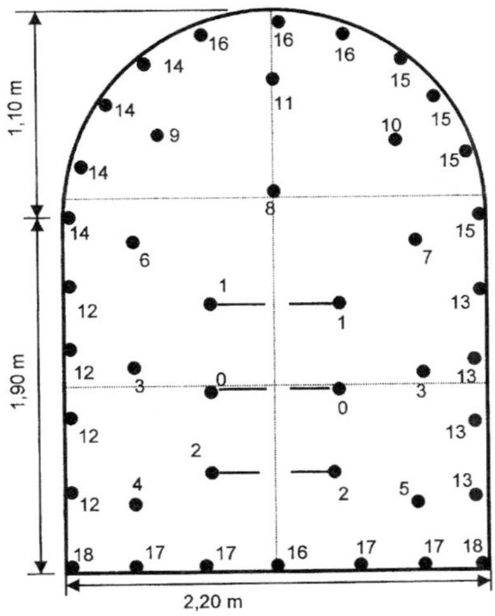


Figure 6. Blasting scheme of the drainage gallery of the Glör Dam

Use of a Tunnel Boring Machine

At the Ennepe Dam the Ruhr River Association suggested the construction of the drainage gallery with a tunnel boring machine (TBM). This construction method was accepted by the Reservoir Supervision Authority. Even though there was no specific experience with the use of a TBM under these conditions, there seemed to be big advantages concerning the quality of the tunnel. The lack of structural disturbance of the bedrock and the masonry surrounding the tunnel opening would make any kind of lining unnecessary, turning the gallery into a large scale drainage boring (Heitefuss & Rissler, 1999).

Figure 7: Tunnel boring machine

In the beginning there seemed to be some problems associated with the use of a tunnel boring machine,

- the curved axis of the gallery with a radius of 150 m,
- the very steep curve of the gallery at the abutments,
- the length of the gallery of only 370 m, being unfavourable for the economical use of a TBM

This demanded the use of a small and manoeuvrable tunnel boring machine like the Robbins 81-113-2 TBM by the Murer AG from Switzerland. This TBM is equipped with only one pair of grippers. Therefore this TBM is comparatively manoeuvrable.
The TBM started on the 24. October 1997 and reached the left end of the gallery on May 14, 1998. Seven weeks later, on August 18, 1998 the TBM appeared at the target shaft at the right abutment. The average rate of advance had been 6.7 m per day, the peak performance was 20 m per day.

It can be stated that the TBM has driven a mostly smooth and circular gallery 90-95 % of the gallery can remain unlined with no additional support. In the bottom reach, the upper half of the gallery runs through the masonry of the dam. Since this part is virtually unlined, the visitor has a remarkable view into the interior of the masonry, which is almost 100 years old.

## PROOF OF THE SUCCESSFUL REHABILITATION

Additionally to the measurements of the described measuring instruments the seepage was measured, which flowed out from each individual drainage drilling.
A comparison of these measurements with the values expected on the basis the seepage model is possible by averaging the measured outflow of the drillings. The quantities measured at the drainage in the masonry dam are clearly below the predictions of the model, while the quantity of the rock drainage reaches these. Also the values of the surface of the gallery are from same order.
At all dams the measured values prove the tightness of the masonry dam, which is substantially higher than assumed. The permeability of the rock corresponds to the assumption.

## COMPARISON OF THE CONSTRUCTION METHODS

Table 3 compares the described methods. The given values are based on the evaluation and on the experiences from the rehabilitation work at a number of dams. All given numbers are average values and have to be considered as a rough estimates.
It can be shown that the rate of advance per workshift and the unit price for the driving of one cubic-metre correlate. In case a TBM is used, major investments and deductions have more importance.

Table 3. Comparison of the construction methods

| Method | TBM | manual*) | core drilling*) | blasting |
|---|---|---|---|---|
| Performance per worker and shift (WS) | 1 m³/ WS | 0,4 m³/ WS | 0,2 m³/ WS | 0,7 m³/ WS |
| Costs | 1.300 € / m³ | 1.800 € / m³ | 2.200 € / m³ | 1.100 € / m³ |
| Advantages | rapidly, less damage to the rock | flexible minor expenditure | flexible, minor vibrations | flexible, minor expenditure |
| Dis-advantanges | only large diametres | dust and noise conditions for the workers | time consuming, expensive | accurate supervision needed |

*) s. Aberle & Hellmann, 2000

COSTS AND CONCLUSIONS

The 100 years old Fürwigge, Glör and Ennepe dam had to be adapted to the established technical standards. By numeric simulations and measurements as well as new procedures for the propulsion of the drainage gallery the costs of rehabilitation could be reduced. Former solutions, as the use of a diaphragm wall (s. Figure 8) were neglected as rather expensive.

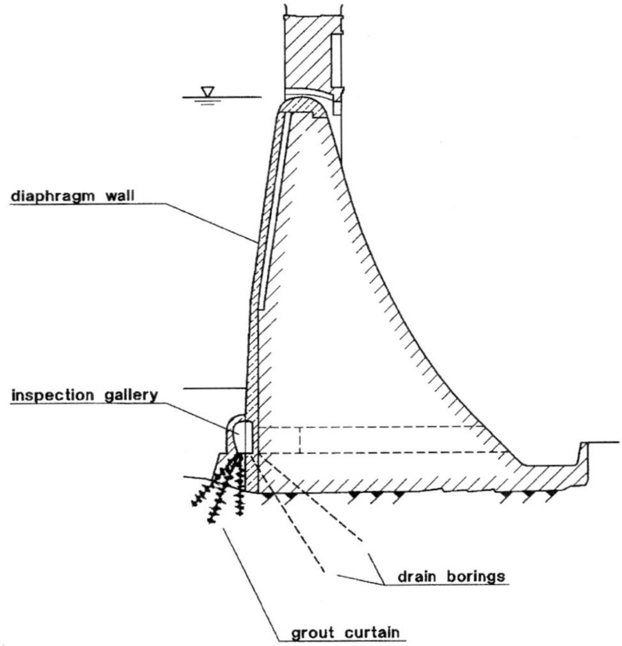


Figure 8. Earlier expensive concept, using a concrete diaphragm wall

At the Ennepe Dam the total costs of the rehabilitation could be bisected of 40 millions € to 20 millions €. This factor is also confirmed by the comparison of the rehabilitation of the Fürwigge, Glör and Ennepe Dam with other rehabilitations based on diaphragm walls, as shown in Table 4

Table 4. Comparison of realised Rehabilitations (Total Costs)

| Dam | Year of comple-tation | Volume of Dam [1000 m³] | Storage Capacity [1000 m³] | Rehabilitation Concept | Costs per Dam Volume [€ / m³] | Costs per Storage Capacity [€ / 1000 m³] |
|---|---|---|---|---|---|---|
| Fürwigge (Concept) | 1904 | 26 | 1670 | only borings | 62 € | 0,96 € |
| Glör | 1904 | 35 | 2100 | blasting a tunnel | 63 € | 1,05 € |
| Ennepe | 1904/12 | 106 | 12600 | tunnel boring machine | 189 € | 1,59 € |
| Dreiläger | 1912 | 85 | 4280 | concrete dia-phragm wall | 177 € | 3,50 € |
| Brucher | 1913 | 27 | 3300 | | 334 € | 2,73 € |
| Fuelbecke | 1896 | 17 | 700 | | 350 € | 8,57 € |
| Jubach | 1906 | 27 | 1050 | | 291 € | 7,62 € |
| Hasper | 1904 | 59 | 2050 | | 304 € | 8,78 € |

REFERENCES

Aberle B & Hellmann H (2000). Sprengen in Gewichtsstaumauern unter Vollstau, Proccedeings of the 12th Bohr- und sprengtechnisches Kolloquium, Clausthal

Bettzieche V et al. (2000a). Simulation of hydrofracturing in masonry dam structures. XIII International Conference on Computational Methods in Water Resources. Calgary

Bettzieche V (2000b). Temperature Measurement in a Masonry Dam by Means of Fibreoptical Sensors. Commission Internationale Des Grands Barrages. Beijing

Bettzieche V. & Heitefuss C (2001). Monitoring as a basis of cost-effective rehabilitation of an old masonry dam.

German Association for Water Resources and Land Improvement (ATV/DVWK). 1991. Measuring devices for the check of the stability of gravity dams and embankment dams. ATV/DVWK-Merkblätter zur Wasserwirtschaft (ATV/DVWK-Guidelines). Volume 222

Heitefuss C. & Rissler P (1999). Upgrading of the Ennepe dam. Hydropower & Dams Issue Two. Volume six

# Underwater work as a means for the rehabilitation of large hydraulic structures under full operation and unrestricted water supply

C HEITEFUSS, Ruhr River Association, GE
H J KNY, Ruhr River Association, GE

SYNOPSIS. The layout of many hydraulic structures built during the beginning of the 20th century did not make provisions for emergency gates, because the complete draw down of the reservoirs for repair purposes was considered possible. However, a complete draw down would restrict the water supply and cause severe ecological problems as well. Therefore rehabilitation works had to be done at reservoir levels which allowed an unrestricted water supply. That required underwater work at the intake structures and the use of emergency gates for safe working conditions. Extensive underwater work has been done in order to adapt the existing hydraulic structures to new operational needs.

## INTRODUCTION

The reservoirs of the Ruhr River Association ( in German: Ruhrverband ) facilitate the water supply of about 5 million people in the Ruhr area. The reservoirs cover a design and construction period from the beginning of the 20th century until 1966, when the largest reservoir, the Bigge Dam was put into operation. The Moehne Dam, known for its severe damage during World War II, was the first dam which had to undergo extensive rehabilitation works at its 4 bottom outlets in the early 1990s. The Verse Dam followed in the years 1995 – 1997. The bottom outlets of the Ennepe Dam were refurbished from 1997 until 2002.

Like most hydraulic structures of that period the original layout of the intake structures did not make provisions for emergency gates, since the complete draw down of the reservoirs for repair purposes was considered possible when the dams were designed. It is now known that a complete draw down of a reservoir has to be considered as impossible, due to possible restrictions of the water supply and severe ecological repercussions as well.

If a certain reservoir level is maintained during the rehabilitation works a nearly unrestricted water supply is possible. The partial draw down requires underwater work at the intake structures and the use of emergency gates for safe working conditions under atmospheric pressure inside the penstocks and the bottom outlet galleries. Thus, an important part of all the projects has been the construction of support structures for emergency gates.

*Reservoirs in a Changing World*, Thomas Telford, London, 2002, 167–178

The placing of emergency gates is also a prerequisite for the subsequent installation of guard valves and improved water supply facilities inside the bottom outlet galleries.

The paper will present the experiences from the underwater rehabilitation activities of the Ruhr River Association of the last 10 years. Some screen shots of animated underwater video sequences will illustrate the construction methods.

## PROJECTS

### The Moehne Dam Rehabilitation Project from 1992 -2002

The Moehne Dam (Klein, Harder, Klahn 2002) was built from 1908 - 1912 as a curved masonry dam with a height of up to 40 m, a length of 650 m and a maximum storage capacity of 134.5 million $m^3$. The Moehne Dam was considered by far the largest dam in Europe of that time. The Moehne Dam shows the characteristic signs of an Intze-type dam, based on design principles of Professor Otto Intze from the end of the 19th century. One of the typical signs is the so-called Intze-wedge, a wedge-shaped clay embankment at the upstream foot of dam. This wedge reaches up to one half of the structure and was originally supposed to serve as upstream sealing. It is now known that the sealing effect of this clay wedge can not be taken into account.

Another typical sign of Intze-type dams are the circular gate towers on the upstream side of the dam. The four gate towers of Moehne Dam were equipped with two gate valves each, which serve as closure gates for the bottom outlets. The discharged water is drawn through a bottom outlet tunnel underneath the Intze-wedge. After the second gate valve the pipework with a diameter of 1.4 m penetrates a circular sealing plug made of brick and continues into the bottom outlet gallery. Finally the water is released into the compensation basin at the downstream foot of the dam. The four bottom outlets have a discharge capacity of about 23 $m^3/s$ each. The valves at the bottom of the gate towers have been installed in the beginning of the $20^{th}$ century and showed strong symptoms of deterioration after almost one century of operation.

The gate towers itself proved to be in very bad condition as well. Due to major leakages any inspection or maintenance work at the bottom of the towers had to be considered as impossible. The concept for the rehabilitation of the entire bottom outlet structures is based upon the following ideas:

- to move the new intake trumpets about 30 m to the upstream side outside the Intze-wedge
- to use the new intake trumpets as support structures for emergency gates
- to replace the old intake gates

A prerequisite for the installation of new closure gates is the placing of emergency gates. The original layout of the entire hydraulic structure at Moehne Dam did not make provisions for emergency gates. The intake tunnel was able to carry only the dead weight of the Intze-wedge but not the hydrostatic pressure of a full reservoir. Therefore it had to be strengthened by a steel relining.

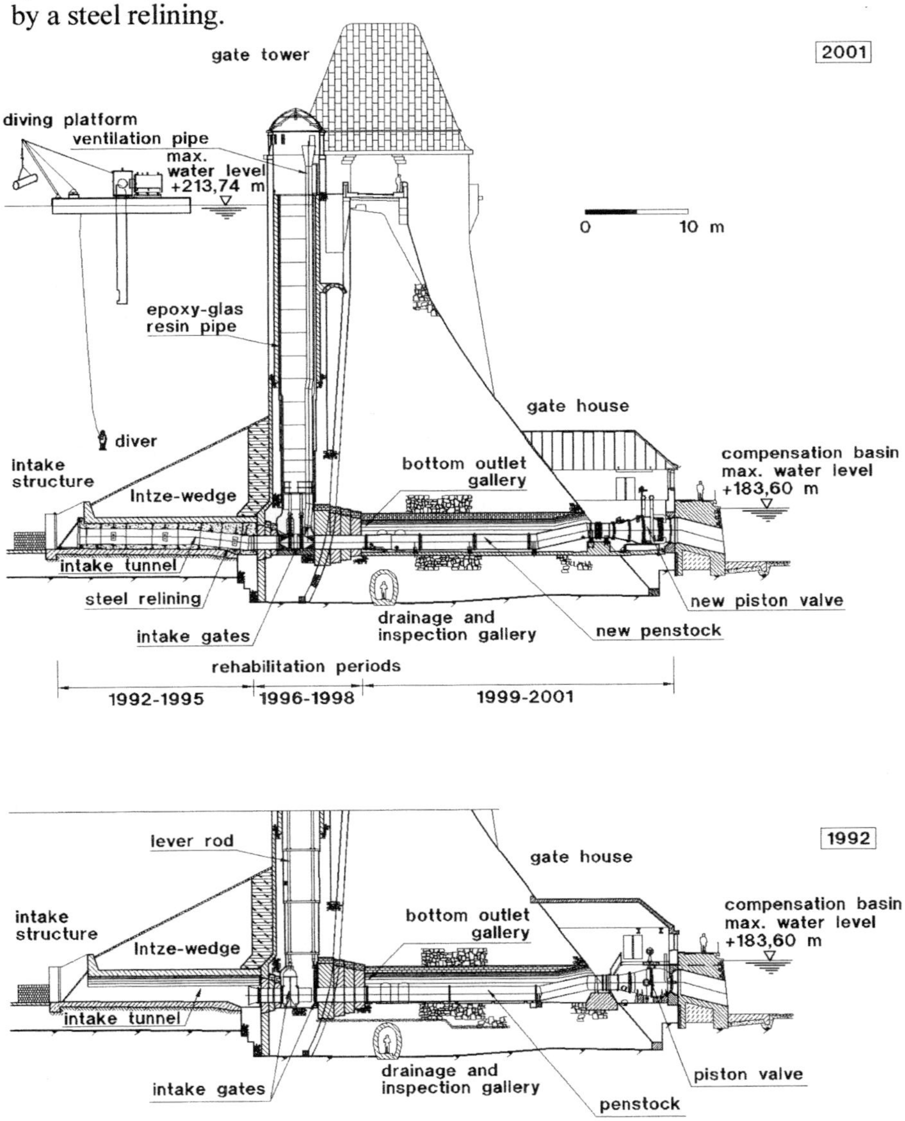


Fig. 1. Cross sections of Moehne Dam before and after rehabilitation

The work at the Moehne Dam (Figure 1) started in 1992. The diving contractor was equipped with a special diving platform which has been developed in cooperation with the Ruhr River Association (Mantwill & Campen 1993).

The diving platform is equipped with a decompression chamber which is connected directly to an entry pipe. This pipe reaches 9 m under the water surface and can be filled with compressed air. Thus, the diver can enter the pipe under dry conditions and under pressure. With an elevator he is lifted to a pressure lock, where he is undressed by an assistant. Then he enters the decompression chamber for the reduction of pressure. This equipment allows the controlled decompression under dry conditions and improves the safety. The platform is also equipped with a mobile pressure chamber, which can be connected quickly to the pressure lock. In case of a diving accident there are two options. The diver is either placed in the mobile pressure chamber and brought to a hospital via airlift within 30 minutes or medical assistance can be brought into the decompression chamber via the pressure lock.

The components of the steel relining of the bottom outlets were produced and then test-assembled by the diving crews in order to avoid problems during the underwater installation. Then the components of the relining pipes were installed. The dimensions of these pipes where chosen in such a way, that one diver is able to handle each component easily with lift-bags and push it into the right position inside the bottom outlet. When the steel pipe of the bottom outlet is in its final position inside the tunnel, the so-called end shield is brought into position at the intake of the bottom outlet tunnel.

The end shield serves as support for both the intake trumpet and the emergency gate. It is also used as sacrifice formwork for the underwater concreting. The end shield has to be supported by a steel framework during the pouring of the concrete.
One of the last steps of the underwater work after installation and fastening of the end shield is the concreting. After the hardening of the concrete the bottom outlet tunnel is capable of bearing the full hydrostatic pressure.
The final step of the underwater work in the bottom outlet tunnel was placing of the newly built emergency gate.
After the placing of the emergency gate the rehabilitation work inside the gate towers and the bottom outlet galleries started.
The gate valves at the bottom of the gate towers were then removed and replaced by new ones. The gate towers were relined with epoxy-glas resin pipes with a diameter of 3.4 m, in order to provide safe working conditions at the bottom of each tower. The 4 bottom outlets of Moehne Dam went back to full operation in the year 2002. The cost of the diving work was about 1.6 Mio. EURO.

The Verse Dam Rehabilitation Project from 1995 -1997
Based upon the experiences from Moehne Dam the bottom outlets of the Verse Dam underwent rehabilitation (Heitefuss & Kny 1997) from 1995 - 1997.

The Verse Dam is an earthfill dam with a concrete cut off and a height of 52 m. The Verse Reservoir can be considered as a multi-purpose reservoir. Two bottom outlets (Ø 800 mm) serve as penstocks for a small hydropower station with an installed capacity of 400 kW. The Verse Reservoir has a storage capacity of 32 million $m^3$ and provides about 8 million $m^3$ of drinking water per year for the adjacent cities. In combination with other reservoirs the Verse Reservoir also guarantees a minimum flow in the Ruhr River.

The design of the Verse Dam began by the end of the 1920s. Construction started in 1932. The intake structure of the bottom outlet except valves and penstocks was completed in 1944. The construction work was interrupted from 1944 to 1948 due to World War II. After the completion of the dam the impounding of the reservoir began in 1952. In 1988/89 the penstocks and valves except the intake gates were replaced. Each penstock of the bottom outlet of the Verse Dam was equipped with the following components:

- intake gate (wedge gate valve) operated from a gate chamber at the end of the penstock gallery
- bulkhead with a self-closing steel gate at the concrete cut off
- spherical valve in the penstock gallery
- needle valve at the outlet of the penstock

A major part of the rehabilitation project at the Verse Dam was the installation of two guard valves (butterfly valves) inside the penstock galleries in order to eliminate a safety deficiency in case of a pipe rupture between the bulkheads and the spherical valves.

A prerequisite for the installation of guard valves is the placing of emergency gates. The original design of the intake structure did allow the installation of emergency gates, since (like at many other dams) the complete draw down of the reservoir for repair purposes was considered possible when the dam was designed.

Since the Verse Reservoir is important for both the local and the regional water supply, a complete draw down of the reservoir had to be ruled out. The rehabilitation work at the Verse Dam had to be done at a reservoir level which allowed an unrestricted water supply. This partial draw down allowed underwater work at the intake structures with reasonable ground and decompression times for the diving crews. The use of emergency gates allowed safer and more economical work under atmospheric conditions inside the penstocks and the penstock galleries. Thus, as at Moehne Dam a few years before, an important part of the project was the construction of support structures for emergency gates. Theoretically it would have been possible to use the existing intake gates as emergency gates, but for several reasons this was ruled out:

- the intake gates were about 55 years old
- the material (cast iron) was showing signs of spongiosis
- one of the intake gates was leaking
- the maintenance and repair of the gates was not possible, since there was no access to the gates in case they are stuck and no access to the bearings and stuffing boxes in the gate chambers.

Since there was no reasonable way to rehabilitate the intake gates, they had to be removed by divers.

Before the underwater work began, extensive studies were done concerning the optimal scheduling of the work and the necessary reservoir volume during construction. A minimum reservoir level had to be found, which could guarantee the safety of water supply for the adjacent cities and allow reasonable and economical diving and decompression times for the diving crews. The calculations showed that in order to start the underwater construction work in May, a minimum volume of 13 million m³ had to be stored in the reservoir.

The concept of the rehabilitation was based upon the ideas of the Moehne Dam rehabilitation but adapted to the different design of the intake structure at Verse Dam. The idea was:

- to move the new intake trumpet about 2 m to the upstream side
- to use the new intake trumpet as support for a new emergency gate
- to remove the old intake gate and replace it by a new guard valve inside the penstock gallery

The rehabilitation work at the intake structures had to be scheduled for two consecutive years, since at least one bottom outlet had to be in full operation.

The underwater work at the Verse Dam (Figure 2) started by the end of May 1996 using the same diving platform like at the Moehne project.

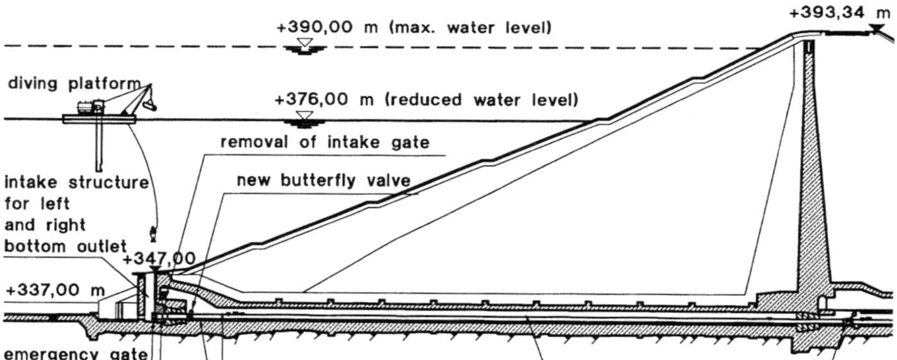


Fig. 2: Cross section of Verse Dam during rehabilitation

While the discharge for water supply and hydropower generation was shifted to the left bottom outlet, the first step of the rehabilitation work was to remove the old intake gate at the right bottom outlet. The guide rails were dismantled and the spindle of the gate valve was cut through. At the bottom of the intake structure, a layer about 40 cm thick had to be removed in order to provide sufficient space and good securing for the end shield. The end

shield serves as support for both the intake trumpet and the emergency gate. It is also used as sacrifice formwork for the underwater concreting. The dimensions of the end shield were chosen in such a way that it could be lowered from the diving platform through the access shaft of the intake structure to the new position. Counterweights were attached to the end shield in order to make the underwater handling easier.
One of the last steps of the underwater work after installation and fastening of the end shield at the right bottom outlet was the concreting. The back of the end shield was equipped with injection hoses in order to obtain good quality concrete. Joints and cracks due to shrinkage were grouted.

The final step of the underwater work in the bottom outlets was to place the newly built emergency gates. After completion of the underwater work in the bottom outlets the work in the penstock gallery started. In order to create space for the new guard valves about 30 m³ of concrete and about 7 m of the old penstock at the end of the gate chambers were removed. Under the protection of the emergency gates self-closing butterfly valves were installed inside the penstock gallery.
The rehabilitation work at Verse Dam ended in 1997, leaving this large hydraulic structure with two bottom outlets which meet the latest safety and operational standards. The cost of the entire rehabilitation project amounted to nearly 2 Mio. EURO.

The Ennepe Dam Rehabilitation Project from 1997 -2002

The Ennepe Dam is a curved masonry dam with a height of 51 m and a length of 350 m. With its max. storage capacity of 12.6 million m³ the Ennepe Reservoir is providing about 9 million m³ of drinking water per year for the adjacent cities. The Ennepe Dam is owned by the Ruhr River Association since 1997. In order to be adapted to the established technical standards and safety regulations, the Ennepe Dam has gone through a rehabilitation program since then, as requested by the Reservoir Supervision Authority of the state of North Rhine - Westphalia.
Like the Moehne Dam (built 4 years later) the Ennepe Dam is a characteristic Intze-type dam. A typical sign of Intze-type dams (besides the Intze-wedge) are the circular gate towers on the upstream side of the dam. The two gate towers were equipped with one gate valve each, which served as closure gate for the bottom outlet. These cast iron gates were operated manually by means of a rod lever from the top of the tower. Before reaching the intake trumpet of the gate tower and the first closure gate inside, the discharged water is drawn through a horse-shoe-shaped bottom outlet tunnel underneath the Intze-wedge. These tunnels are slightly curved, have a length of about 30 m and a height/width of 2 m. After the first gate valve the bottom outlet pipe penetrates a circular sealing plug made of brick and continues into the bottom outlet gallery. Finally the water is released via a needle valve into the stilling basin at the downstream foot of the dam. Both bottom outlets have a discharge capacity of about 6 m³/s each. They also serve as intakes for the local water works.

The gate valves at the bottom of the gate tower have been installed in the year 1902 and showed strong symptoms of deterioration like spongiosis and cracks. Due to a lack of financial resources the former owner had tried to repair the valves by sealing them with a thick layer of concrete.
The gate towers itself proved to be in very bad condition as well. Due to major leakages both intake towers where lined with prefabricated circular concrete liners 40 years ago. Thus the space inside the towers was substantially reduced.

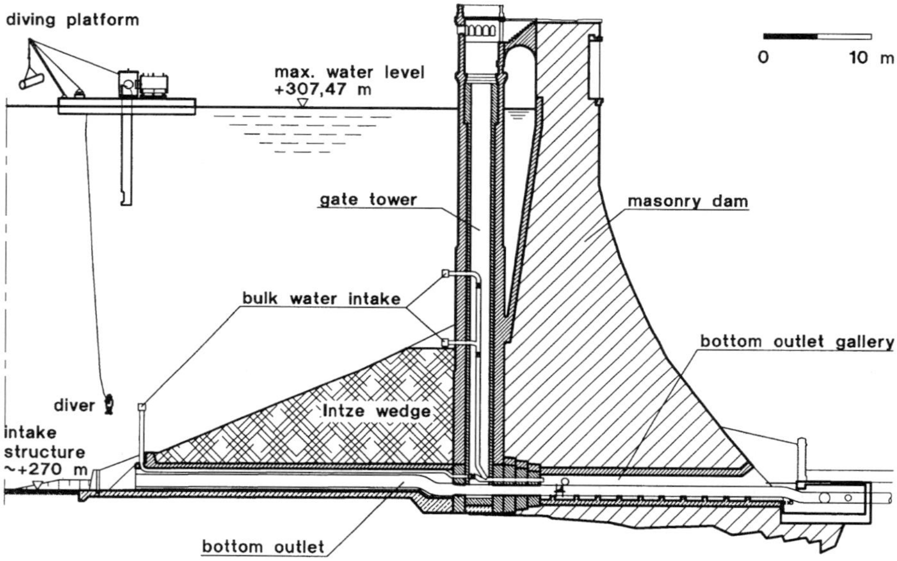


Fig. 3. Cross section of Ennepe Dam after rehabilitation

Another problem associated with the basic layout of this hydraulic structure was the fact, that the bulk water could be drawn only from the bottom of the reservoir via the bottom outlet. It was not possible to draw bulk water from different water levels. This caused severe treatment problems for the water works, since the sediment content of the bulk water used to be rather high during rain storms respectively flood periods..
The concept for the rehabilitation (Figure 3) of the entire bottom outlet structure is based upon the following ideas, which had already proven their feasibility at the Moehne and Verse Dam:

- to move the new intake trumpet about 30 m to the upstream side outside the Intze-wedge
- to use the new intake trumpet as support structure for an emergency gate
- to remove the old intake gates and replace them by new guard valves inside the bottom outlet galleries
- to create new intakes for bulk water at different water levels, which can be operated independently from the bottom outlet, thus allowing the optimization of the water quality.

The major difference to the rehabilitation projects at Moehne and Verse Dam is the consistent use of stainless steel. In order to assure a long term durability of the newly built bottom outlets and water intake pipes it was decided to use stainless steel for all components.
This required special conditions during production and installation of the components in the plant and on the construction site (Heitefuss & Kny 2001).

The rehabilitation of the upstream side of the bottom outlet was done very similar to the work at Moehne Dam. The intake tunnel had to be relined too, since it was in bad condition and not designed for the water pressure at maximum reservoir level.
The underwater work at the Ennepe Dam started in 1998. This was the third consecutive project for the special diving platform, once developed for the Moehne project.
While the discharge for water supply purposes was shifted to the right bottom outlet, the first step of the rehabilitation work was to penetrate the walls of the gate towers for 3 bulk water intake pipes and the bottom outlet pipe plus additional holes for concreting. This was done with core drilling equipment with diameters up to 600 mm (Figure 4) and jackhammers. The gate tower remained flooded in order to guarantee a balanced pressure.

Fig. 4. Penetration of the gate tower with 600 mm core drills
(sequence of video animation)

Meanwhile the components of the relining of the bottom outlet were manufactured. Based on experiences made at the Moehne project the test assembling of the components was carried out by the diving crews in order to avoid problems during the underwater installation.
The first step was the installation of the steel relining pipes in the horse-shoe-shaped bottom outlet tunnel. Based of the experiences from the Moehne Project, these pipes (Figure 5) could be handled easily by one diver. When both pipes of the bottom outlet were in their final positions inside the tunnel, the end shield was brought into its position at the intake of the bottom outlet tunnel like at the Moehne and Verse project.

Fig. 5. Handling of the steel relining pipes (sequence of video animation)

The last step of the underwater work after the pouring of the concrete was the installation of the emergency gate, then the rehabilitation work inside the gate tower and the bottom outlet gallery started.
The cast iron gate valve (installed in 1902) was removed in order to create space for the new bottom outlet pipe . The water supply pipes are also getting installed inside the gate tower. In order to connect the upstream side of the bottom outlet with the gallery on the downstream side, the circular sealing plug made out of brick has to be penetrated. This is done with core drillings, positioned around the old pipe. After the removal of the old pipe there is access to the bottom of the gate tower. In order to guarantee the safety inside the tower, additional support columns are necessary, which are designed to support the concrete lining of the inner wall of the tower. This

support structure had to be made from stainless steel in order to avoid electro-chemical corrosion. After the closure of the penetration the guard valves (type: butterfly-valve) were installed, in order to protect both the bottom outlet and the water supply pipes against leakages or ruptures. Finally the connection to the existing needle-valve at the downstream face of the dam is reconstructed.
In order to avoid electr :hemical corrosion between the stainless steel of the new pipes and the regular steel of the old pipework and the needle valve, pipes made of glass-fibre reinforced resin were installed in the bottom outlet gallery.

After completion of the rehabilitation work at the left bottom outlet in the beginning of the year 2000 the work at the right bottom outlet is now in progress. Surprisingly at the right bottom outlet the contractors had to face much more difficulties due to heavy leakages. The end of the rehabilitation work at the Ennepe Dam is scheduled for the year 2002.

Fig. 6. Schematic view of rehabilitation work with diving platform (Sequence of video animation)

By then the Ennepe Dam will be equipped with two bottom outlets which meet the latest safety standards and allow optimized intake strategies for the drinking water production. 4 million EURO is the estimated cost of the entire bottom outlet rehabilitation project.

## CONCLUSIONS

During the last 10 years three major underwater rehabilitation projects for large hydraulic structures have been carried out by the Ruhr River Association. It has been possible to perfect not only the diving equipment like the platform but also the manufacturing and installation procedures in order to guarantee safe and economical conditions. Thus the underwater rehabilitation work has become an important alternative to conventional rehabilitation techniques for large hydraulic structures.

The experience of three large projects shows, that particularly in underwater construction experienced contractors are essential for the success of these projects. Therefore the specifications and tender documents need the maximum attention of the engineers in charge of the planning and supervision of the work. Nevertheless it has to be stated, that every major underwater rehabilitation project has its surprises, which can be very costly. The reason is, that the status of very old hydraulic structures can be assessed only to a certain degree.

All newly built or adapted hydraulic structures have performed well so far, the stainless steel structures have to prove their long term reliability. Therefore the question, if the use of stainless steel with all its difficulties is <u>the</u> economical solution in underwater rehabilitation, can not be answered yet.

## REFERENCES

Klein, P., Harder, L. & J. Klahn 2002. Möhnetalsperre - Sanierung der Grundablässe, der Schiebertürme und des luftseitigen Mauerwerks, Moehne Dam – Rehabilitation of the bottom outlets, the gate towers and the downstream masonry face, to be published in "Die Wasserwirtschaft" in 2002

Mantwill, H. & F. Campen 1993. Rehabilitation of the bottom outlet works of the Moehne Dam. Ruhrwassermenge 1994 (Annual Report on Water Resources Management 1994). Ruhrverband (Ruhr River Association) Essen, Germany

Heitefuss, C. & Kny, H.-J. 1997. Rehabilitation of the intake structures at the Verse Dam, Germany. In E. Broch, D.K: Lysne, N. Flatabo, E. Helland-Hansen (eds), Hydropower ´97, Proc. 3rd intern. conf. on hydropower, Trondheim, Norway, 30 June–2 July 1997. Rotterdam: Balkema: 405-411

Heitefuss, C. & Kny, H.-J. 2001. Rehabilitation of the bottom outlets of Ennepe Dam / Germany with stainless steel pipes. In G. Midttomme et al (eds.), Dams in a European Context, Proc. of the ICOLD eur. symp., Geiranger, Norway, 25–27 June 2001. Lisse: Balkema: 429-434

# Hydraulic structures – hydrology

# Sluiceway isolation for gate replacement at Kotri Barrage in Pakistan

I E PADGETT, WS Atkins Consultants Ltd, formerly of WSP International, UK
K F MORRISON, WSP International, UK

SYNOPSIS. The 910m wide Kotri Barrage on the River Indus was constructed in 1955. In 1989, a feasibility study concluded that the 44 barrage gates had experienced serious loss of metal and recommended their replacement. Kotri barrage had been constructed without the means to isolate individual sluiceways. A concept was selected for sluiceway isolation comprising a floating bulkhead gate, to be upended, ballasted and positioned between the barrage piers. This paper describes the successful development of the bulkhead gate and associated equipment, and its operation in the replacement of the gates of the Kotri barrage.

## INTRODUCTION

### Background

Kotri Barrage is the lowest of the barrages on the Indus River in Pakistan. Operated and maintained by the Irrigation and Power Department of the Government of Sindh, the barrage was commissioned in 1955 to irrigate the command areas of Lower Sindh and supply water to the cities of Karachi and Hyderabad.

Throughout the 1980s, concerns were expressed at the rate of loss of metal thickness in the gates due to corrosion and the erosive action of the silt laden Indus waters. A feasibility report completed by Coode Blizard in 1989 recommended a programme for rehabilitation of the barrage which included the replacement of all of the main barrage gates. In common with many barrages constructed at the time, Kotri barrage had been designed and constructed without the means to isolate individual sluiceways by stoplogs or emergency gates (Padgett et al, 1997). The concept of floating bulkhead gates was developed and finally adopted for isolation of barrage sluiceways. This enabled the safe and economic replacement of existing barrage gates in each sluiceway, in the dry, and allowed inspection and refurbishment of embedded parts.

### Barrage Design and Construction

It will be helpful to describe the design of the sluiceways to be isolated. The barrage comprises 44 gate bays of 18.3m width. All barrage piers and divide walls are constructed of mass concrete with stone masonry facing. The total

*Reservoirs in a Changing World*, Thomas Telford, London, 2002, 181–192

width of the barrage between abutments is 910m. The general layout of the barrage is shown in Fig. 1.

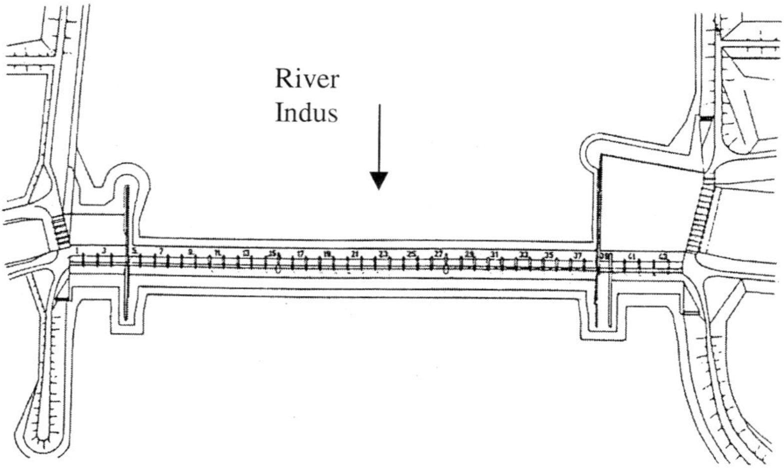


Fig. 1. Plan of Kotri Barrage

Each sluiceway has a raised concrete cill 2m wide. On the upstream side, a 600mm thick concrete glacis slopes down at 1 in 5 to meet the natural river bed level, which is protected immediately upstream of the barrage with a stone apron. Downstream, a 2.4m thick concrete glacis slopes at 1 in 3 to a horizontal concrete pavement extending 20m downstream. All concrete is un-reinforced. A cross-section through a sluiceway is shown in Fig.2.

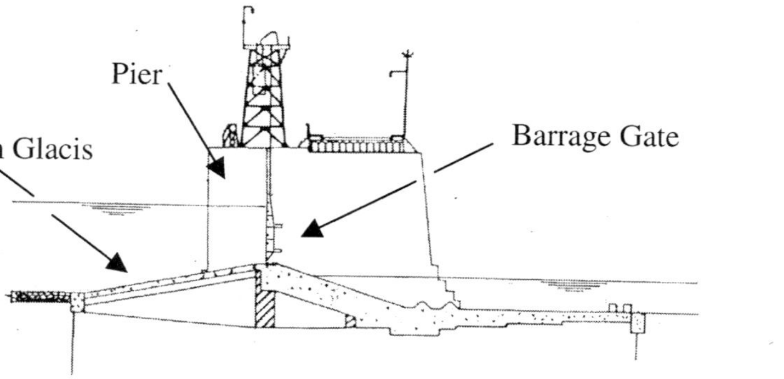


Fig. 2. Section through Barrage Sluiceway

The original barrage gates were of the counterweighted, vertical lift type, 6.4m high and of riveted steel construction. Gate operation is by manually turned rope hoists located on a high level operating platform.

The gates run in vertical cast iron built in parts set into and extending the full height of each pier. Water pressure on the gates is transferred to the built in parts by Stoney rollers.

## REQUIREMENT TO ISOLATE THE SLUICEWAYS

The barrage had been constructed without the means to isolate individual sluiceways by stoplogs or emergency gates. To remove the barrage gates and install new gates in the dry, safely and economically, it was necessary to devise a method of isolating and dewatering each sluiceway.

There were several problems specific to the barrage which had to be overcome in devising a suitable method of isolation.

- The isolating gate had to be manoeuvred into position from upstream because access for heavy craneage was not possible from the road bridge.
- Hydrostatic loads had to be carried by the barrage piers.
- The isolating gate must be operable for the full range of winter season pond levels.
- The roughness of the masonry was such that special attention would be needed to make an effective seal.
- The bottom seal had to be on the narrow sluiceway cill because the thin upstream glacis was not capable of resisting unbalanced uplift pressures.

## OPTIONS FOR SLUICEWAY ISOLATION

### Leading Options

The four leading options for isolation of the sluiceways considered by the Consultant were :

- The barrage had been constructed inside three separate ring bunds. The use of similar techniques, or of smaller cofferdams located on the upstream glacis, was considered for sluiceway isolation. However, these methods would have been costly and would have subjected the upstream glacis to unbalanced uplift pressure, for which it had not been designed.

- The possibility of cutting slots into the pier sides for the introduction of stoplogs was considered. There were practical difficulties associated with slotting of the piers, and with placement of the stoplogs from the road bridge due to the presence of existing gas pipes spanning the barrage piers on the upstream side.

- Curved caisson gates had been used for the replacement of the gates at Sukkur Barrage on the River Indus, some years previously (Sir M. MacDonald & Partners, 1989). The use of these at Kotri was investigated. However, the major modifications required to compensate for the curved face of the caisson, for the different pier length at Sukkur, to enable replacement of the barrage gate from upstream, and to avoid unbalanced uplift pressure on the glacis, ruled it out.

- It was concluded that the concept of a Bulkhead Gate, which could be floated into position from upstream, upended and ballasted down into position between the barrage piers, would provide the most cost effective solution. Additionally, the gate and associated equipment would be available to the Barrage operator after gate replacement, for seasonal maintenance work.

Bulkhead Gate

Key features of the selected option of the bulkhead gate are described below.

The bulkhead gate comprised a buoyant rectangular section compartmented into tanks which could be individually filled or emptied to ballast the gate from a horizontal floating position to an upright position spanning between the barrage piers, thus isolating the sluiceway. Figure 3 shows the concept of the bulkhead gate, with the gate positioned between the sluiceway piers.

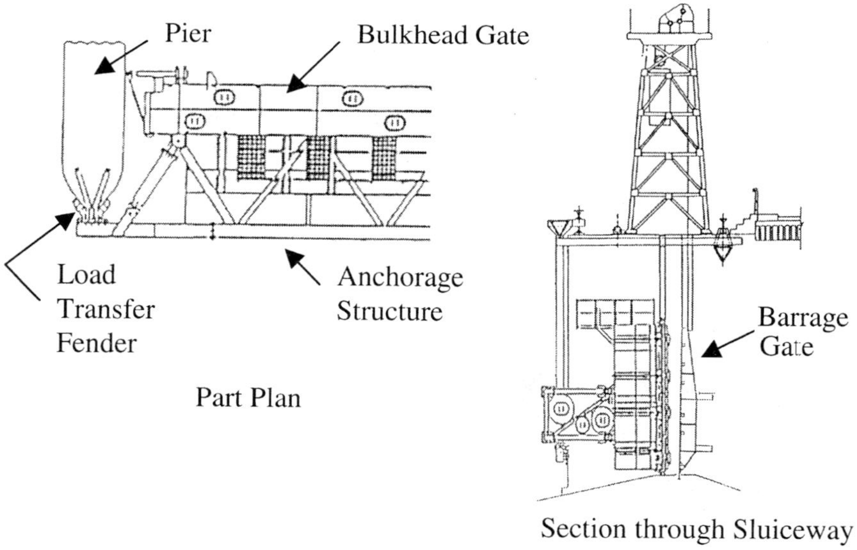


Fig. 3. Concept of Bulkhead Gate positioned within Sluiceway

The hydrostatic loading was carried by an anchorage structure projecting from the upstream face of the rectangular gate with arms projecting on either side. These arms would engage on the flanking pier noses so that the load on the finally positioned gate could be transferred into both piers. Each pier nose had a steel load transfer fender on which the arms of the gate anchorage structure bear.

Bottom sealing would be effected by a seal projecting from the downstream bottom lip of the gate body, which was landed on the sluiceway cill. Side seal assemblies would be capable of extension both outwards and downstream, so that the seal would land onto the secondary concrete embedding the barrage gate built in parts, which was significantly smoother than the adjacent masonry.

The concept did not include any machinery or power to be incorporated into the gate itself other than the hydraulic rams required to extend and retract the moveable side seals. A purpose-built pontoon would carry all necessary water pumps, hydraulic power packs, compressors, generators and cranes required to ballast and position the bulkhead gate. The pontoon was also equipped with winches and anchors to enable accurate manoeuvring and positioning of the gate in the sluiceway. Motive power to manoeuvre the whole arrangement was provided by a tug.

## CONTRACTUAL STRATEGY

In order to satisfy the requirements of the agencies funding the project, separate contracts were established for gate replacement (K2A) and sluiceway isolation (K2B). Contract K2B was to include operation and maintenance of the bulkhead gate and associated equipment for the period required for barrage replacement.

There were clear advantages in working closely with a contractor in the detailed design of the bulkhead gate. In common with other British consultants, the Consultant did not have the in-house, specialist expertise but it was known that number of specialist international contractors had such skills. The merits of entrusting the K2B Contractor with the detailed design, or of engaging another contractor to assist the Consultant with his detailed design were reviewed. It was concluded that cost savings could be expected if the Contractor were free to design these structures to suit his operational strategy. Accordingly, Contract K2B was let to Whessoe Projects Ltd, for design, fabrication, and operation and maintenance of the bulkhead gate and associated equipment. Detailed designs would follow the concept given in the Tender Documents and would be subject to the final approval of the Engineer, Coode Blizard Ltd.

It was decided to provide three bulkhead gates to allow replacement of barrage gates in three sluiceways concurrently. This, it was expected, would enable barrage gate replacement in possibly three, but no more than four working seasons. A single pontoon and a tug would service all three bulkheads. The three bulkheads and the pontoon were to be removed from the river at the end of each working season and re-launched next season from a slipway, purpose built on the right bank.

The pontoon and three bulkhead gates were to be fabricated adjacent to the slipway. This was also used as a storage area for equipment during the flood season. The tug was to be built in the UK and shipped to Pakistan.

## DESIGN OF THE GATE AND ASSOCIATED EQUIPMENT

### Bulkhead Gate

The final design of the gate comprised a self contained, rectangular, buoyant structure, fabricated from steel plate. The gate had 24 compartments positioned in four layers with three upstream and three downstream compartments at each level. Pipework was incorporated so that each tank or compartment could be individually filled or emptied by the same pump. Sumps were incorporated to allow the tanks to be fully emptied.

Three cylindrical tanks were nested within the double lattice girder anchorage structure. One of these could be ballasted to provide fine adjustment while the other two provided additional stability to the gate in the upright position. This was especially useful in the final stages of uprighting and before final ballasting down on the bottom seal, when there was a danger of the gate "flipping" over and capsizing. This tendency to capsize was further controlled by attachment of the two knuckleboom

Fig. 4. Bulkhead Gate on Slipway, ready for launching

cranes, fitted to the forward end of the pontoon. Figure 4 shows the fabricated bulkhead gate awaiting launching into the river.

<u>Gate Arms</u>
Arms would project from each side of the anchorage structure, for engagement on load transfer fenders accurately positioned on the flanking pier noses. Thus the load on the finally positioned gate would be transferred into both piers.

Different arms were required to cater for different pier configurations. "Normal" arms would be used for mid-river sluiceways, and different "abnormal" arms would cater for anchorages on, respectively, right and left abutments, right and left divide walls, and the wide piers. Figure 4 shows normal arms. These different arms were secured to the anchorage structure by bolted connections. Although it was intended that arms could be changed on the river, in practice this proved too difficult an operation, and arm changes were made after the bulkhead had been landed on the maintenance area on the right bank.

<u>Seals</u>
The Contractor followed the Consultant's basic concept for sealing except that he opted to effect a side seal against the masonry face of the pier instead of the secondary concrete. This meant that the side seal mechanism was achieved more simply. The Contractor decided that dressing of the irregular masonry face of the pier, in the area where the side seal would "land", could be accomplished.

<u>Load Transfer</u>
Each pier nose was fitted with a steel load transfer fender. The fenders were constructed from universal columns fitted with saddles which straddled the

Fig. 5. Upstream View of Barrage showing Load Transfer Fenders

pier nosings. They were anchored to the piers by anchors drilled through the masonry and grouted into the hearting concrete.

Load transfer fenders were positioned accurately, to suit the load pads provided on the extremity of the anchorage arms on the bulkhead gate. Figure 5 shows the load transfer fenders installed on the upstream pier noses of the barrage. This figure also shows the gas pipes that were important in the selection of the bulkhead gate option.

Pontoon

The pontoon performed as a service barge for the bulkhead gate and was constructed of steel, with overall dimensions of 14m x 17m x 2m deep.

Fig. 6. Pontoon at foot of Slipway

Two 100kVA diesel generators provide an electrical power source, primarily to an hydraulic power pack, and for supply to ballast water pumps, a compressor, and lighting and small power. A wind generator provides emergency power. Two knuckleboom cranes with a lift capacity of 7t were installed at the forward end of the pontoon, for fitting to the top of the bulkhead gate in the final stages of ballasting and lowering. Two further cranes of 1t capacity were fitted to the aft end of the pontoon for handling and positioning anchors. Other equipment included various winches, anchors, switchboard and control desk.

The pontoon was fitted with docking arms to facilitate central positioning of the bulkhead between piers and to ensure engagement of the bulkhead gate anchorage arms with the load transfer fenders, as the floating gate enters the sluiceway. Figure 6 shows the Pontoon at the slipway.

Tug

The tug provides motive power to enable the pontoon and bulkhead gate to be moved on the river, into and out of a barrage sluiceway. The tug is a steel hulled ship with twin diesel engines and a pusher bow. Winch and stern sheaves are provided and wheel and engine controls are located in a wheelhouse.

## OPERATIONAL DESIGN

After launching into the river by means of the slipway, the tug was used to propel the pontoon and a bulkhead gate downstream to the barrage. The gate was floated in a horizontal attitude with the anchorage structure uppermost until it reached the vicinity of the sluiceway to be isolated. Ballasting of the gate into a near upright and stable position was then carried out using the pumps on the pontoon. The knuckleboom cranes were attached to the top of the gate before final manoeuvring. The pumps on the pontoon were also used to ballast the pontoon, which was itself compartmented, to provide resistance to the downward forces exerted on the cranes at the forward end.

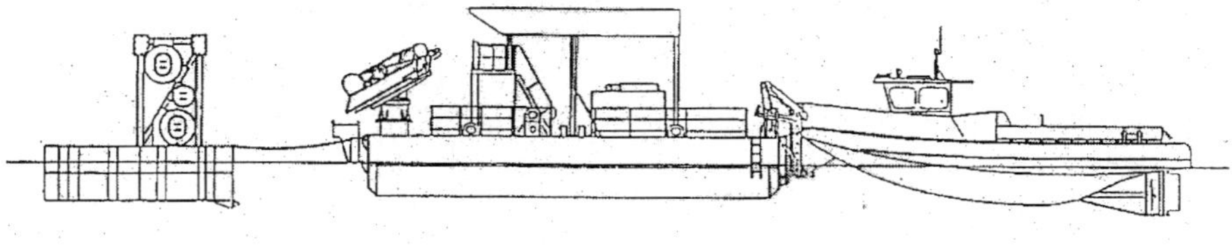

Fig. 7. Bulkhead Gate and Pontoon, manoeuvred on river by the Tug

Figure 7 shows the concept of the bulkhead gate and pontoon being manoeuvred towards a sluiceway by the tug.

Final positioning of the bulkhead gate into the sluiceway was carried out using two aft anchors on the pontoon. These are controlled by winches, also on the pontoon, and attached to adjacent pier tops. Side spacers fitted to the pontoon facilitated accuracy in placing the gate centrally in the sluiceway. This also ensured engagement of the anchorage structure arms on the load transfer fenders.

Once located in the correct position, final uprighting was completed and the side seals extended to make contact with the dressed masonry pier face. Ballasting was continued while the knuckleboom cranes were continuously lowered until the bottom seal made contact with the sluiceway cill. The side seals were pushed, using 40 bars pressure, firmly against the pier sides using the rams with hydraulic power from the pontoon. The gate was then further ballasted to give the required compression of the bottom seal.

After positioning, the space between the bulkhead gate and the barrage gate could be dewatered to allow removal and replacement of the barrage gate.

Upon completion of gate replacement, the procedure was reversed and the bulkhead gate was floated into the next sluiceway to be isolated.

OPERATING EXPERIENCE

All equipment performed well and generally as intended, although the Contractor made some adjustments to operating procedures and early unsuccessful attempts to seal the gate in the sluiceway required changes to the seal design.

The Contractor undertook laboratory tests of seal configurations before final design. The initial seal used on both the side and bottom seals was made by gluing strips of expanded neoprene together to form a rectangular shaped seal which, in cross-section, resembled fingers extending at right angles to the sealing surface. This worked at the top of side seals but tended to be pushed out by the water pressure towards the bottom of the seal, that is under the greatest pressure. Also, the force of compressing the seal against the masonry tended to de-bond the fingers and, therefore, rendered it useless for repeated isolations. Unless the seal was compressed by 70 mm it was not sufficiently rigid to withstand the hydraulic pressure and was simply pushed through the gap between the underside of the steel housing and the concrete cill. The initial attempt to isolate a sluiceway was, therefore, unsuccessful.

The seals were then wrapped in a sheet of neoprene which enclosed the rectangular cross-section of the seal. This was successful for the side seals and was used for the remainder of the project. The wrapping also kept the "fingers" together and made the seals more durable. However, although giving greater rigidity there were still areas on the bottom cill where sealing was not achieved because the seal had not been adequately compressed, either due to variations in the level of the cill concrete or where the concrete had been damaged.

After detailed surveys of the cills, it was decided to carry out underwater repairs to the cill concrete in 17 of the 44 bays, using proprietary underwater repair materials. This solved the problem and thereafter, all bays were successfully isolated. However, because the initial bottom seal was not considered adequate for long term use it was replaced by a Linatex proprietary seal.

The Engineer's original concept provided for "reverse upending" on the river in which the pontoon's knuckleboom cranes would be used to upend the gate so that the anchorage structure was below water and the side and bottom seals were exposed above water. The purpose was to allow seal maintenance and replacement on the river. In the event, this operation proved unsafe and was not pursued. As a result, all seal maintenance was undertaken on the bank, after landing the gate via the slipway.

Upon completion of barrage gate replacement, the Contractor prepared an Operation and Maintenance Manual for the Employer (Irrigation & Power Dept, 1999) and undertook training of the Employer's staff. This ensured that the Employer could use the bulkhead gates and associated equipment to isolate, inspect and maintain the barrage sluiceways in future winter seasons.

## PROJECT IMPLEMENTATION AND PROGRESS

Contracts K2B and K2A were let in July and November 1993 respectively. Overall contract completion, that is replacement of all 44 barrage gates, was set for January 1998. However working in the river was not possible during the flood season from June to the end of September of each year.

Initially the Contractor programmed for a 30 day cycle to remove the existing gate and erect the new gate. This meant that with 3 bulkhead gates, 18 gates per season could be changed. With 44 barrage gates to replace this would mean completion in the 1996/97 working season, leaving the period from October 1997 until January 1998, at the end of the Contract period, to complete any outstanding work.

In the event, the contract got off to a slow start and delays in completion of the slipway and construction of the bulkhead gates, including early sealing difficulties, meant that only one gate was replaced by the end of the 1994/95 working season. The K2A Contractor introduced a night shift to recoup delays and this was continued in the 1995/96 season. In this way the Contractor managed to cut down the cycle to 15 days which enabled all gates to be replaced by February 1997, some 10 months before the end of the contract period (WSP International, 2000).

## CONCLUSIONS

The Kotri Barrage Rehabilitation Project, of which the use of bulkhead gates for sluiceway isolation was a key component, was completed ahead of programme and within budget.

The development of the concept of a bulkhead gate at Kotri, using a pontoon for provision of services and a tug for propulsion on the river, proved to be practical, safe and cost effective. Technically, the equipment performed as expected although modifications were made to the side and bottom seal arrangements. These modifications highlight the need for detailed surveys of the seal contact areas beforehand and for exhaustive trials of the chosen seal design.

The decision to entrust the Contractor with the detailed design and with the responsibility for operation was vindicated. Following handover of this equipment and training of operatives, the Government of Sindh now has the facility to inspect barrage gates during the winter season. This avoids the need for an annual raising of the barrage gates and closure of the canal gates and the attendant loss of water downstream.

This concept for sluiceway isolation is applicable to similar river and maritime structures. Other barrages on the Indus River in Pakistan are known to be in need of rehabilitation similar to that undertaken at Kotri. These were also constructed without stoplogs and the use of the bulkhead gate concept for sluiceway isolation would be appropriate for these barrages.

ACKNOWLEDGEMENTS

Acknowledgements are due to the Irrigation and Power Department, Government of Sindh, Department for International Development, Asian Development Bank and WSP International for their permission to publish this paper. Acknowledgements are also due to Messrs Lewin Fryer who, as part of the Coode Blizard team, contributed towards the development of the original concept of the bulkhead gate.

REFERENCES

Coode Blizard Ltd (1989). Feasibility Report on Rehabilitation of Kotri Barrage.

Irrigation & Power Department, Government of Sindh (1999). Kotri Barrage Operation and Maintenance Manual vol 6.

Padgett I E & Dahar B A(1997). Removal of the Head Limitation at Kotri Barrage, Pakistan. 19th ICOLD Congress, Q75, R19.

Sir M Macdonald & Partners (1984). Sukkur Barrage Gate Study, Final Report.

WSP International Ltd (2000). Kotri Barrage Rehabilitation Project Phase II Completion Report.

# Maintaining the Thames tidal defences in a century of climate change

J LEWIN, Consultant, UK
S LAVERY, Environment Agency

"There was last night the greatest tide that ever was remembered in England to have been in this river: all Whitehall having been drowned".

Pepys in his Diary for December 7, 1663

SYNOPSIS. The paper briefly reviews the origin and nature of the Thames Tidal Defences. It considers the reliability of the current defences and the actions which may be required in coming years to adapt, maintain and operate the defence structures under changing conditions, particularly the world's changing climate. This will result in a continuous rise in sea level and increased storminess, causing higher storm surges and greater frequency of operation of the movable tidal defence equipment.

## INTRODUCTION

The Thames Estuary is one of the great natural assets of the UK, a highly complex and rich wildlife habitat linking London, Kent and Essex to the North Sea. It provides a home for some 119 species of fish, over 300 species of invertebrate and each year welcomes hundreds of thousands of visiting wildfowl and waders. It is also the most densely urbanised estuary in the country with a complex system of tidal defences providing protection to property valued at over £80 bn. This property is protected to a high standard with tidal defences designed and built in response to the catastrophic floods of 1953.

During the 1953 flood, the tide was elevated 1.98 m above predicted level, causing loss of life and catastrophic damage at the Thames Estuary. The surge diminished as it approached London's centre and only minor damage occurred in the capital. Following the recommendation of the Waverley Committee and the report by Professor Bondi on the cost/benefit aspects of a barrier to protect London (Bondi, 1967), construction of the Thames Tidal Defences began. Bank raising was completed and the movable defences became fully operational during the early 1980s.

Of course tidal flooding in London and the Thames Estuary corridor is not new. The earliest surviving record of flooding is the Anglo Saxon Chronicle of 1099. Since then, numerous floods have been documented each recording "the greatest tide" ever seen.

*Reservoirs in a Changing World*, Thomas Telford, London, 2002, 193–208

However, the continuing rise in sea level and the lowering of the land mass of South East England, together with longer term effects reported by the International Panel on Climate Change (IPCC), pose problems which were not envisaged when the Thames Tidal Defences were conceived.

These issues have prompted the Environment Agency to initiate reports leading to a strategy plan for flood risk management in the Thames Estuary for the next 100 years. This plan, which is due for completion and submission to government (DEFRA – Department for Environment, Food and Rural Affairs) in the year 2006, will consider the long term requirements for management of flood risk against the conflicting pressures of urban development and environmental protection in the inner Thames Estuary (from Teddington in West London to Sheerness/Shoeburyness). All relevant factors will be taken into account, including environment, navigation, commerce, biodiversity, urban development, heritage and amenity considerations. Striking a balance between these issues and the flood defence needs of the Thames Estuary will form the subject of future papers and reports.

However, the principal concern of this paper is to clarify a variety of ways in which climate change will impact on the operation and reliability of the Thames Barrier, and outline some of the proposed technical and operational solutions put forward to solve the resulting problems, in the context of increasingly frequent flood defence closures and ageing of plant.

## THE THAMES TIDAL DEFENCES

London and the Thames Estuary corridor are protected from tidal flooding by nine major barriers and tidal defence structures owned and operated by the Environment Agency. In addition, 33 floodgates protect vulnerable riverside industry downstream of the Thames Barrier. Upstream of the Thames Barrier there are over 380 smaller movable defences operated by their owners. These must be closed when the tide forecast is high enough to cause localised flooding but is not sufficiently high to trigger closure of the Thames Barrier.

Between Teddington and Sheerness/Shoeburyness, there are over 300 km of sea walls and embankments owned and maintained by various individuals and authorities, including the Environment Agency. The sea walls and embankments were upgraded in the 1980s, with crest levels downstream of the Thames Barrier raised by up to 2.5 m. Many lengths of defence were of new construction in the 1980s but others, particularly upstream of the Thames and other barriers, are based on the historic legacy of defence dating back several hundred years. The oldest functioning tidal defence is at the Tower of London where the medieval river wall is over 700 years old.

The major moving structure is the Thames Barrier, Fig. 1, spanning the Thames at Woolwich Reach West. It comprises four 61 m clear width main navigational openings, two 31.5 m subsidiary navigational openings all controlled by rising sector gates and four 31 m span falling radial gates.

Fig. 1. Thames Barrier at Woolwich

The concept and design of rising sector gates which control the navigation openings are dealt with in papers by Holloway et al (1977) and Clark & Tappin (1977). Figure. 2 shows the 61 m span rising sector gate.

The Barking Creek Barrier is located on the North bank of the Thames and controls the navigation opening by a 38.6 m span vertical lift gate which is normally stored at high level. There are three subsidiary gates of 12.0 m span, also of the vertical lift type.

The Dartford Barrier comprises a double leaf vertical lift gate, also normally stored at high level, Fig. 3. The barrier is located on the South bank of the Thames.

The King George V basin is protected by a top hinged flap gate which is moved across the navigation opening on a carriage, and the same function is carried out by a mitre lock gate at Gallions Reach, the entrance to the Royal Albert Dock. Further downstream, the Tilbury barrier, similar in design to the King George V gate, protects Tilbury Docks and there are three further structures, Fobbing Horse, Benfleet and East Haven, protecting the tidal creeks around Canvey Island.

## CRITERIA OF DESIGN AND OPERATION – THAMES BARRIER

The high water level recorded at London Bridge during the 1953 flood was +5.40 m AOD and the trend line of high water levels between 1780 and 1950 was 0.72 m increase per century. This suggested that high water levels of +5.7 m in the year 2000 and +6.1 m in the year 2050 could occur. The design load of the gates was a surge reaching a downriver water level of +6.9 m with an upriver water level of -1.5 m, and an extreme surge level of +7.2 m with a lower upriver water level of -2.7 m (Holloway et al 1977). The design conditions for extreme surge level assumed that the gate crest level was at +6.9 m, to ensure that in such an event overtopping occurs first over the barrier gates, rather than over the more vulnerable downriver embankments. The sill level at the 61 m span rising sector gates is -9.25 m. The present operating level of the gates when they are elevated to the near vertical position is with the tip of the gates at +7.65 m, a margin of 0.45 m over and above the extreme surge level assumed during the design.

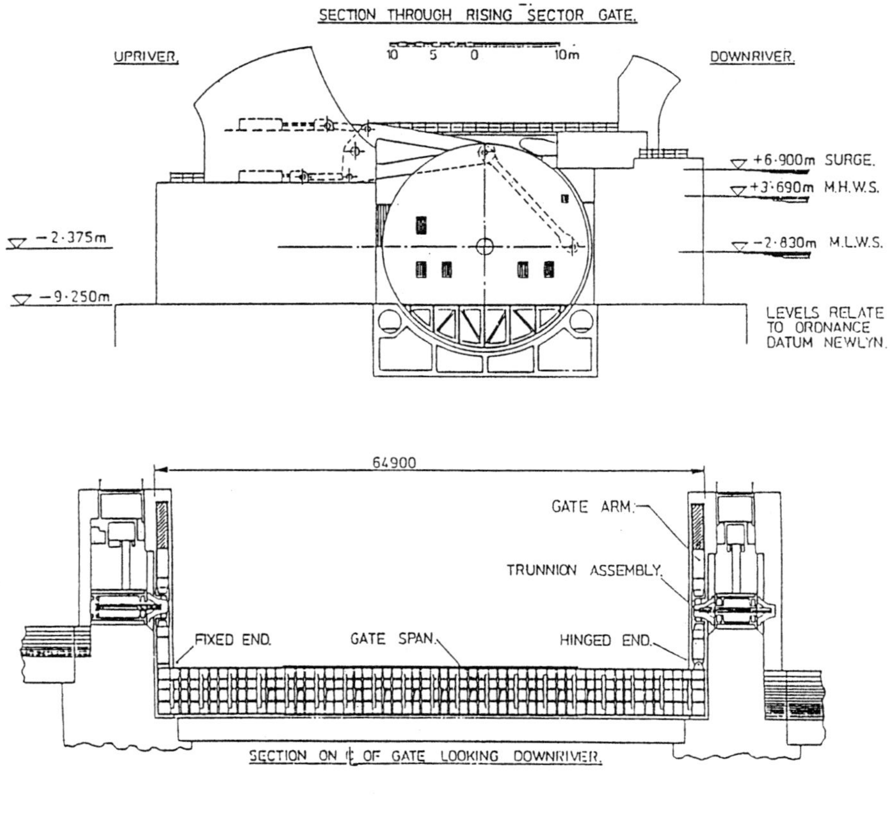


Fig. 2. Rising sector gate of the Thames Barrier

Fig. 3. The Dartford barrier

Currently, Barrier closure is determined by a model predicting open-river Tower Pier level, based on Southend level and Thames flow. For example, a Southend high water level of 3.8 m OD combined with low river flows of 100 $m^3/s$ triggers Barrier closure; a Southend high water level of 3.6 m and a river flow of 360 $m^3/s$ also initiates Barrier closure. At a flood flow of 600$m^3/s$ the model indicates Barrier closure at 3.15 m high water level at Southend.

Between the first event in February 1983 and September 2001 there have been 64 closures of the Thames Barrier to prevent flooding. Fig.4 plots annual closures. Even if recent closures to prevent fluvial flooding are disregarded, a significant upward trend is apparent after 1997. (The experience of high fluvial flows in 2000/2001 suggests that the Barrier closure model should be extended to include an upriver trigger point, for example at Richmond.)

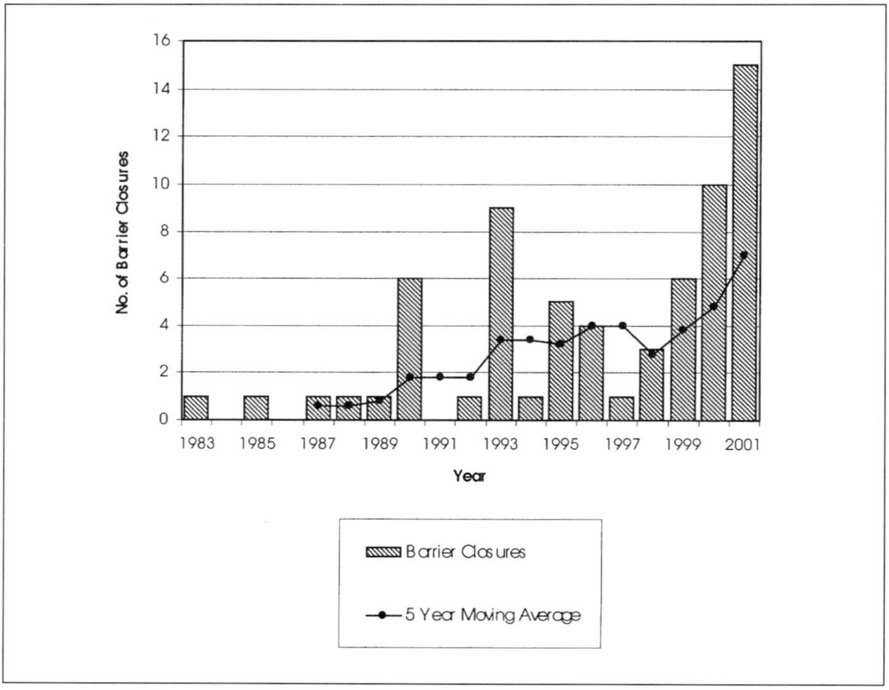


Fig. 4. Annual barrier closures

Records of water levels at the barrier gates during closure only go back to 1999. The maximum load during a surge event depends on the difference between the downriver level (the surge level) and the upriver level, which is largely determined by the river flow. The maximum downriver levels, Fig. 5, show a steep increase similar to that seen in the number of closures. The increase during the year 2001 was equivalent to 70 years of the trend line of sea level rise between 1780 to 1950. The average loading of the gates was about 35% (15% above and 20% below) of the design conditions, and shows a less steep increase compared with the downriver water level and frequency of operation of the barrier.

Since the predicted change in sea level is a gradual process, the recent higher surge levels at the closed Barrier may be a consequence of the increase in frequency and magnitude of storminess which was envisaged in the IPCC (International Panel on Climate Change) Reports of 1996 and 2001. The number of closures in 2001 may prove to be exceptional, but there is a probability that high numbers of annual closures will continue. If this remains the case, even to a lesser extent, when sea levels have risen, it will have serious consequences.

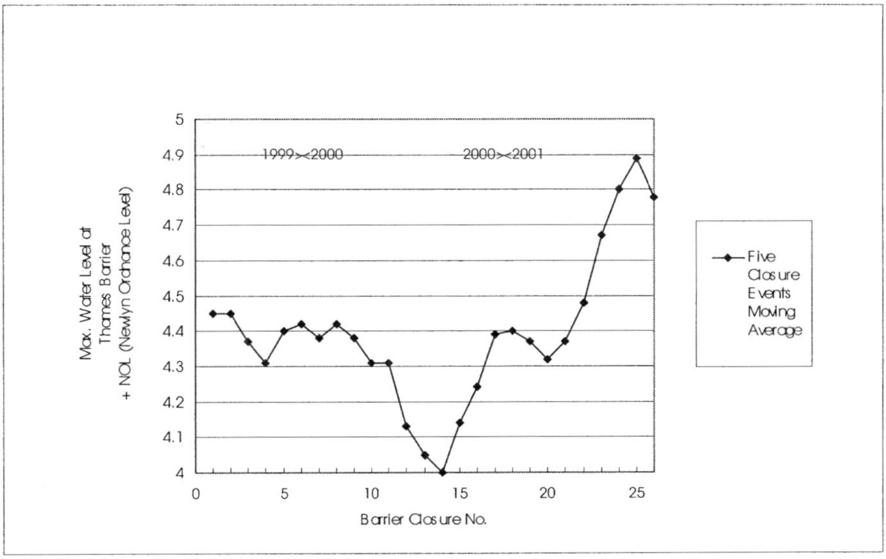


Fig. 5. Water levels downriver of the Thames Barrier (surge levels), moving average

RELIABILITY OF THE THAMES BARRIER

The design of the Thames Barrier and associated defences allowed for a continuing rise in high water levels, at the historic rate, up to the year 2030, when the Thames Barrier will have been in operation for 50 years. The resulting defence levels were based on an event which would have a 1 in 1000 probability of occurrence in any one year, under the estimated conditions in the year 2030.

After the Thames Water Authority took over the management of the Thames Barrier from the Greater London Council, they instructed the Systems Reliability Directorate (SRD) of the United Kingdom Atomic Energy Authority to carry out a Reliability Assessment of the Thames Tidal Defences (SRD 1988). For the Thames Barrier this was determined at 1.55 x $10^{-4}$ per gate per demand. Expressed differently, there is a chance that a single gate will fail to close on one in 560 closure demands, and that two of the 10 gates will fail to close at one full closure in approximately 6000 closure demands. A good industrial system standard is 1 x $10^{-4}$ per demand. Subsequently, the control of the Thames Barrier was upgraded by replacing the relay system, which had been prone to repeated failures, with programmable logic controllers (PLCs).

In 1997 the legal firm Barlow Lyde & Gilbert was commissioned to undertake a “disaster enquiry – without the disaster”, to test the robustness of the overall management system of the Thames Barrier and associated defences. This included consideration of later (post 1988 SRD report) developments such as the existence of London City Airport in the proximity of the Barrier. In 1999, Lewin & Ballard undertook a post project appraisal of the newly installed PLC system.

The potential for accidents was recognised in the 1988 SRD report. An event of this type occurred in 1999 when the dredger Sandkite collided with one of the piers and partially sank, discharging its cargo of sand and gravel onto one of the main rising sector gates. The accident occurred in October during neap tides and there was little structural damage to the gate. Circumstances might have been different if it had happened in the peak flood season and the gate had been damaged severely enough to render it inoperable.

The SRD assessment of reliability was time specific and was carried out when the frequency of barrier closures per annum was below one. By 2001 closures had risen to an average of 3.4 per annum, and if the last 4 years only are considered the annual rate has increased to 8.5. The 1988 reliability assessment of failure per demand could be equated to failure per annum. If the present probability of failure per annum is considered, there is a quantitative deterioration in reliability.

The reliability of the Thames Barrier will have to be improved if it is to remain consistent with the original reliability criterion. The Thames Barrier and Associated Gates Strategy currently being undertaken, takes into account improvements made since 1988 and also the effects of ageing of the plant and movable structures.

## THE PROBLEMS OF AGEING AND WEAR

Counteracting past improvements is the ageing process of an installation which is now over 20 years old. A planned preventative maintenance (PPM) system for the Thames Barrier and associated Gates is under constant review to take account of changes to critical elements of the installation. The tidal walls and embankments which form part of the Thames tidal defence system are subject to the same ageing process. Designed and maintained to a high standard, the 1980s defences nevertheless present a particular challenge of strategic planning in that they will all come to the end of their design life at approximately the same time. Irrespective of any environmental factors, it is likely the defences will cease to provide adequate protection at some point between 2030 and 2050 if improvements are not made. These improvements will be costly – preliminary estimates indicate capital investment of the order of £4 bn may be required – and major municipal works of this type could take up to 30 years to plan, design and obtain approvals.

The Barrier also has a number of components with a large installed population. All of these are potentially vulnerable to faults and early detection of wear or age-related problems is vital. There are over 300 limit switches, several hundred oil hydraulic valves, check valves, directional control valves, off loader and relief valves. Bearings, crossheads and operating cylinders are duplicated at each pier. In some cases components are no longer available, or were purpose made for the Barrier.

To anticipate failures requires estimating the residual life of components and forward planning for refurbishment and replacement of parts and, in the longer term, whole systems.

In the case of flood defences there are also practical problems due to the restricted outage time available to carry out remedial work or the need to test replacement components. Certain tasks may have to be carried out over an extended timescale and require careful planning. It may be necessary to replace some components before the end of their residual life, otherwise potential increased demand frequency, the consequent shortening of available maintenance time and the amount of time required to effect replacement may combine to make this task difficult.

## THE CAUSE OF STORM SURGES IN THE NORTH SEA

Surges are caused by depressions which have originated in the Atlantic, passing the North of Scotland in an easterly direction and then turning south easterly across Southern Scandinavia towards Germany. The track and speed of movement of the depression affects the height of the surge in the Thames Estuary. It is amplified by the shallower waters of the Continental Shelf. A surge is more dangerous if it occurs on a high spring tide than on a much lower neap tide.

## CLIMATE CHANGE

An extensive investigation was carried out in 1997/98 to assess sea level rise, possible increase in the height of storm surges and the consequences for the number of barrier closures, and to consider what work might have to be carried out to the movable and permanent defence structures to enable them to withstand higher water levels and greater forces (RKL-Arup 1998).

An Intergovernmental Panel on Climate Change (IPCC) was jointly established by the World Meteorological Organisation and the United Nations Environment Programme to assess scientific information on climate change, the impact of such change and to formulate response strategies. The first IPCC scientific assessment of climate change, published in 1990, included estimates of possible changes in sea levels, given a range of scenarios.

The National Rivers Authority (NRA, now the Environment Agency) accepted the IPCC predictions and issued a Policy Implementation Guidance Note (NRA 1992) which summarised the allowances to be made for sea level rise and for changes in land levels. For the Thames Estuary area they resulted in a total relative rise of 240 mm for the period 1990–2030, and for 2030–2100, 560 mm.

The second IPCC report was published in 1996. The main conclusion was that the sea level would rise at a somewhat slower rate than previously estimated and that the global average sea level rise between 1990 and 2100 would be 500 mm, with a wide range of uncertainty from 200 to 860 mm. The estimate of total relative rise for 1990–2100, including land level change for South East England, amounted to 640 mm (including a 140 mm rise relative to land level).

Taking 1996 as the baseline for comparison, Table 1 was prepared. It shows that from 2030 the operational demands on the Barrier and the resulting restriction of navigation may require mitigating measures, such as raising the upriver permanent defences.

The third IPCC report was published in 2001. The main relevant conclusions are that confidence in the ability of models to project future climate has increased and that global average temperature and sea level rises are projected under all IPCC Special Reports on Emission Scenarios (SRES). A few scenarios are illustrated in Fig. 6:

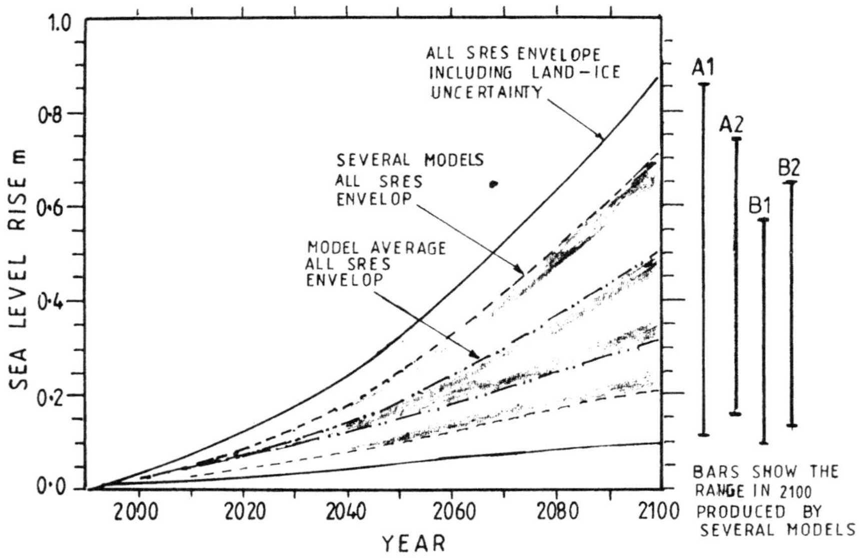


Fig. 6. Global sea level rise. All special reports on emission scenarios (SRES ) (after IPCC 2001)

A1 storyline and scenario family describes a future world of rapid economic growth, global population which peaks in mid century and then declines, and the rapid introduction of new and more efficient technologies.

A2 describes a heterogeneous world in which the underlying theme is self-reliance and preservation of local identities.

B1 describes a convergent world with rapid change in economic structures towards a service information economy.

B2 describes a world in which the emphasis is on local solutions to economic, social and environmental sustainability.

The scenarios do not include initiatives that implement the United Nations Framework Convention on Climate Change or the emission targets of the Kyoto Protocol. The graph, Fig. 6, shows the range of sea level rise and the dependence on factors such as global population, political, social and technical developments.

The summary of climate change for the 20th Century was (Alcock 2001):

- global mean sea level rose by 0.15 ± 0.05 m
- there were no significant trends in storm frequency or severity
- there were no significant trends in storm levels
- there were no significant trends in extreme sea levels, other than that associated with mean sea level change

The projections for 1990–2100 were:

- global mean sea level will rise by between 0.09 and 0.88 m, with a best estimate of 0.47 m
- tidal high water will increase in the British Channel and the East Irish Sea
- extreme sea levels will occur with increased frequency as a result of mean sea level rise
- extreme sea levels will be further increased if more frequent or severe storms occur, but changes cannot be predicted with confidence

A study by the UK Met Office of climate change and storminess predictions (McDonald 2001) concluded that the effect of changing meteorology (resulting from increased greenhouse gas forcing) is to increase the 50 year return height at most locations, but the pattern of rise is spatially inhomogeneous. At some locations, the changes in surge height are statistically significant and increasing mean sea level by 0.5 m results in a 50 year surge height of approximately 0.5 m at most locations.

The Met Office graph reproduced in Fig. 7 shows that the present day return period of the storm design surge level for the Thames Tidal Defences will become substantially shorter in future climate.

The expected effect of climate change on the frequency of Barrier closures is shown in Table 1.

Table 1: Effect of climate change on the frequency of Barrier closures

| **Year** | **No. of Test Closures** per annum | **No. of Defence Closures** per annum | | **Total Closures** per annum |
|---|---|---|---|---|
| | | 0.4 m uncertainty allowance | uncertainty reduced after 2030† | |
| 1996 | 12 | 4* | 4 | 16 |
| 2000 | 12 | 5* | 5 | 17 |
| 2010 | 12 | 10 | 10 | 22 |
| 2030 | 6 | 21 | 19 | 27–25 |
| 2050 | 6 | 81 | 50 | 87–56 |

*Notes:*

* 5 year moving average; allowance is for uncertainty in forecasts of water level

† other mitigating measures to reduce the frequency of closures will be investigated

Met forcing, that is the effect of more intensive storms and the direction of the storms, is not taken into account

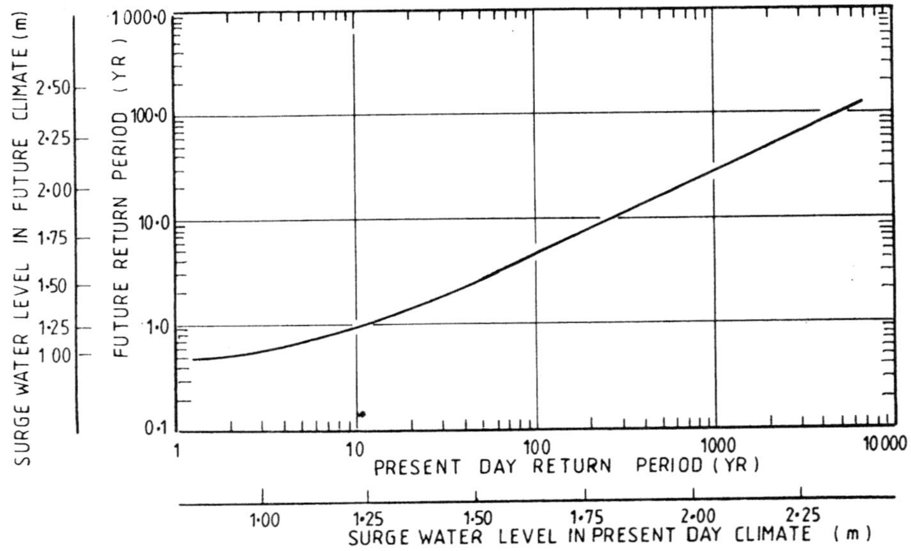


Fig. 7 Change in frequency of extreme water level (after McDonald, 2001)

## POSSIBLE ACTIONS TO MAINTAIN THE EFFECTIVENESS OF THE THAMES BARRIER AGAINST HIGHER STORM SURGES

The 1998 report by RKL-Arup investigated the structural consequences of the movable gates operated by the Environment Agency, Thames Region, being subjected to higher storm surges over a range of increased downriver flood levels.

Structural alterations to withstand higher loads would require isolation of gates while work is carried out and unavailability for protracted periods. Structural alterations would also result in increased demands on the operating machinery, which might have to be replaced, and also increased loads on the civil engineering structures. The proposed solutions were based on avoiding outage of the gates for structural alterations and to minimise increase of load on the operating equipment.

The suggested solution for the rising sector gates of the Thames Barrier was to raise the gates above their present crest level. This reduces the overlap of the gates and the sill beam in the gate elevated position, without increasing the hydrostatic load on the gates. The falling radial gates can be adapted to withstand higher storm surges by raising the sill. If the hydraulics of river flow are affected when the tide level is above 0.0m AOL, a movable sill beam can be introduced which is lowered or withdrawn by hydraulic cylinders. Protection of exposed areas at lower level at the piers will be required.

At the Barking Creek Barrier, an independent sill beam would enable the main navigation gate to withstand higher storm surge levels. Navigational depth at this location is already reduced by silt in the channel so a sill beam need not pose further restrictions. Other solutions have been devised, such as balancing the surge pressure by increasing the water level in the entrance to the upriver basins for the surge protection gates at the King George V lock and at Gallions Reach, both on the North bank of the Thames.

## EFFECTS OF CLIMATE CHANGE ON THAMES BARRIER GATES AND MACHINERY

The operating and control systems of the Barrier have exceptional redundancy, provided in the form of standby plant and alternative methods of operation. At many components and systems this reduces wear because they are used alternately. Where there is no redundancy the design is very robust.

Wear of the machinery and parts does not depend on gate loading, which so far has been below 50% of the design criteria, but solely on frequency of operation. Apart from routine test operations, the closure of the Barrier to withstand surges is increasing and is projected to continue increasing at an exponential rate.

Other factors such as deterioration of the paint protection system and structural loading are also influenced by climate considerations. The paint is affected by extended submersion and the presence of silt. Structural loading will increase from its current low base relative to the design criteria as sea level – amplified by storminess – rises. Operational changes will be made, as previously outlined, when surge loading could exceed the design stresses. (Periodic selective inspection of fracture critical elements will be introduced as gate loads increase or when sample inspections show this to be required.) Non-metallic parts, particularly elastomeric components such as seals, break down over time due to exposure to ultra-violet light and molecular changes. The end of their usual assumed safe life is now close and is monitored by the planned preventative maintenance system.

The anticipation of failures due to wear – a function of climate change – and structural load increase – also a consequence of climate change – is paramount and underlines the necessity for a long term flood risk management plan.

## CONCLUSIONS

Climate change will have a direct impact on the future workings of the Thames Barrier in the form of higher storm surges. Additional consequences will arise from increasingly frequent Barrier closures, such as accelerated wear and operational changes. The increased number of closures since 1997 and the recent rapid increase in surge levels at the Barrier may prove to be atypical, but moving averages show a distinct progression.

If sea level rise and greater intensity and frequency of storms act together, the time available to implement measures to maintain the reliability of the barriers and the permanent tidal defences may be appreciably shorter than assumed solely on the basis of projected sea level rise.

The robust design of the Barrier and its good performance to date does not allay concern or diminish the significance of the problems which lie ahead.

A framework for a comprehensive and systematic study has been set up by the Environment Agency to address the concerns and produce data for the future. This project, Planning for Flood Risk Management in the Thames Estuary, is an initiative from Anglian, Southern and Thames Regions of the Environment Agency, and covers a strategy area from Teddington in West London to Sheerness/Shoeburyness at the outer Thames Estuary. An important part of this initiative is the planning of a long term strategy for the Thames Barrier and associated gates. Following a pre-feasibility study (Lewin & Ballard 2001), the Agency has commissioned WS Atkins to produce a strategy plan to the end of the century for these important elements of the tidal defence system in order that there "will be increased certainty with which investment decisions on subsequent defences can be made".

## ACKNOWLEDGEMENTS

The authors gratefully acknowledge permission by the Environment Agency to publish this paper. Thanks are due to Mr. Andrew Batchelor, Tidal Defence manager, Thames Tidal Defences for his support.

Thanks are also due to RKL-Arup, now Arup Water, for permission to use some material from their 1988 report, specifically from the section in which the first author collaborated.

## REFERENCES

Alcock G, (2001). Intergovernmental Panel on Climate Change (IPCC) Third Assessment Report 2001. Storminess and Sea Level Rise, Conference Instn Civ. Engrs Water & Mar. Engng. June 2001

Bondi Professor H (1967). London Flood Barrier, Report to the Ministry of Housing and Local Government

Holloway BGR, Miller Richards G, Draper RC (1977). Basic concept of the rising sector gate. In Thames Barrier Design. Proc. Conference, London Oct 1977, Instn. Civ. Engrs, 1978

Clark PJ & Tappin RG (1977). Final design of Thames Barrier gate structure. In Thames Barrier Design. Proc. Conference, London Oct 21977, Instn. Civ. Engrs, 1978

IPCC (1990). Climate Change: The IPCC Scientific Assessment. Report of the IPCC Scientific Assessment Working Group

IPCC (1996). Climate Change1995: The Science of Climate Change. Contribution of Working Group 1 to the Second Assessment Report of the Intergovernmental Panel on Climate Change. Cambridge University Press, 1996

IPCC (2001). Summary for Policy Makers of the IPCC WG III Assessment Report

Lewin J & Ballard GM (1999). Thames Barrier Pro-Logic Controllers, Post Project Appraisal (not published)

Lewin J & Ballard GM (2001). Strategy Plan for the Environment Agency (not published)

NRA (1992). Policy Implementation Guidance Note – Climate Change, Ref: TE/FD/001, March 1992

RKL–Arup (1998). LondonTidal Defences, the development of a strategy for the 21st Century (not published)

McDonald R (2001). Climate Change and storminess predictions. Storminess and Sea Level Rise, Conference Intn Civ. Engrs Water & Mar Engng, June 2001

SRD (1988). A reliability assessment of the Thames Tidal Defences. Safety and Reliability Directorate, United Kingdom Atomic Energy Authority, SRS/ASG/31447 (not published)

# Flood control using the automatic tops spillway gates: A case study of the Avis Dam, Namibia

P D TOWNSHEND, Flowgate Projects (Pty) Ltd, RSA
K A LUND, Lund Consulting Engineers, Namibia

SYNOPSIS. The Windhoek City Council in Namibia required the maximum discharge from the Avis dam for the 1 in 100 year flood event to be limited to almost half the inflow peak to prevent undue flooding through the city. The original dam spillway and basin could not achieve that. Further, electrically actuated gates were not a safe option considering the remoteness of the site, vandalism and the harsh environment. The TOPS automatic spillway gate was selected as the only viable and safe option to achieve the performance criteria. Two 3,5m high by 11m long TOPS gates were employed using a unique level control device to release water from the dam without an increase in water level up to the maximum flow required and thereafter the flood would be attenuated. The gates are automatic and self actuating and provide a unique and interesting solution to this problem of flood control. This paper presents a case study of the problem, solution, design, model study, fabrication and commissioning of the TOPS Spillway gates for this dam.

## INTRODUCTION.

Namibia is situated on the West Coast of Southern Africa adjacent to a cold ocean current. It consequently has an average mean annual precipitation of less that 250mm per year. Although mostly dry, thunderstorms do occur during summer, sometimes resulting in flooding. The German colonialists built the Avis dam on the outskirts of Windhoek as the city's water supply dam. The dam is an earth embankment dam with a concrete facing slab and a side channel spillway. As the city grew, alternative water supply was provided and Avis dam reverted to a popular recreational dam.

## THE PROBLEM

The watercourse downstream of the Avis dam runs through the city of Windhoek. With increasing development adjacent to the watercourse, the City Council needed to ensure that a maximum flow of 350m$^3$/sec for a 1 in 100 year recurrence interval flood would not be exceeded. The peak inflow rate into the dam was determined by the project consultants as 600m$^3$/sec.

The dam with its uncontrolled rock spillway and its limited storage capacity, was not adequate to restrict the peak discharge to the required flow.

*Reservoirs in a Changing World,* Thomas Telford, London, 2002, 209–217

The initial solution to the problem was to excavate a 5m deep by 20m wide channel through the spillway to utilize the available storage to achieve the required peak outflow. However, this ran foul with the environmentalists and the public who could not accept a 5m drop in the FSL, particularly in such a dry country. Electro-mechanical gates were not an option for many reasons including cost, environmentally adverse impact of power cables, gantries etc, as well as vandalism.

THE SOLUTION

The TOPS automatic spillway gate was selected by the project consultants as it could meet the performance requirements. The TOPS gates are automatic, self actuating, flexible in usage, reliable and safe. The gate was also ideally suited for the flat invert of the excavated spillway channel. Two TOPS gates were employed, each 11m long and 3,5m high. The gates were modified to incorporate a level control valve to activate the gates' motion.

A further less obvious advantage provided by the TOPS gates was the provision of a walkway on the gates for pedestrians and horses, thereby eliminating the need for a relatively expensive and unsightly bridge over the spillway channel.

DESCRIPTION OF THE TOPS GATES

The Tops gate is so named because it is a top hung gate supported from two trunnions situated above the water level and up stream of the gate. The gate consists of an upstream closure plate which seals against vertical sides and along the sill. The closure plate forms part of a ballast tank attached to the downstream side of the closure plate. The ballast tank is connected to the dam by four openings through the closure plate sothat the water levels in the dam and the ballast tank are in equilibrium.

The mass of water in the ballast tank creates a moment about the trunnion which is greater than the opening moment induced by the upstream water in the dam, and so the gate remains closed. The gate remains closed to allow an overflow in the order of 300mm so that regular instream flows can be discharges without the gates opening.

For these particular TOPS gates, a telescopic arm connects the ballast tank to a float controlled discharge valve. When the water level rises above 300mm over the gate, the discharge valve will open automatically to release water from the ballast tank thereby reducing the closing moment which then causes the gate to open.

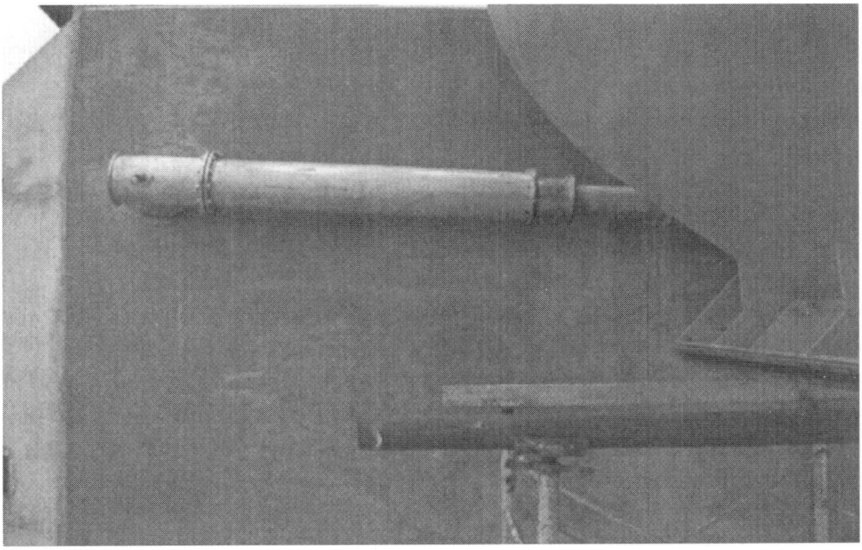

Fig. 1. Telescopic arm connecting the ballast tank to the float controlled valve.

As the water level in the dam drops as a result of the discharge through the gates, the discharge valve closes automatically, water flows into the ballast tank through the 4 openings to the dam to add mass to the ballast tank thereby causing the gate to close. The TOPS gate therefore opens and closes automatically in response to the water level in the dam.

Fig. 2. Float controlled discharge valve discharging water from the ballast tank

The TOPS gate also incorporates other safety features such as an emergency discharge valve in the ballast tank as well as a manually controlled valve to ensure the gate opens in the unlikely event of the primary actuating system not working. The TOPS gates are therefore failsafe to open and close to meet dam safety requirements.

DESIGN

The level controlled gate has a significant influence on the flood routing through the dam as indicated in the diagram below. Once the water level reaches 300mm over the gate, the gate will open due to the discharge of water out of the ballast tank, and will close down as the water recedes. In effect the gate will maintain a constant water level in the dam within a 30mm variation. There will therefore be no increase in storage in the dam and the inflow and outflow will essentially be the same for the initial part of the rising limb of the inflow hydrograph.

The gates are set with a 50mm incremental level difference so that one gate will operate before the other. The gates are restricted from opening more than 2,6m under the gate to give a total discharge of $305m^3/sec$ at the water level at which the gates open.

Thereafter, the water level in the dam will rise with increasing inflow, thereby using the available storage in the dam to attenuate the peak inflow. In this case an additional 1,5m rise in water level is required. In this way the two TOPS gates are able to restrict the discharge to the required maximum of $350^3/sec$ while attenuating the peak in flow of $600m^3/sec$.

Further it satisfies the environmental aspects in that

- It achieves the highest possible water level in the dam
- It is aesthetically pleasing
- No H.T. cables or dangerous mechanisms are required
- It provides a safe walkway over the spillway channel

From a structural aspect, the large ballast tank provides a large moment of inertia which results in low stresses and deflections.

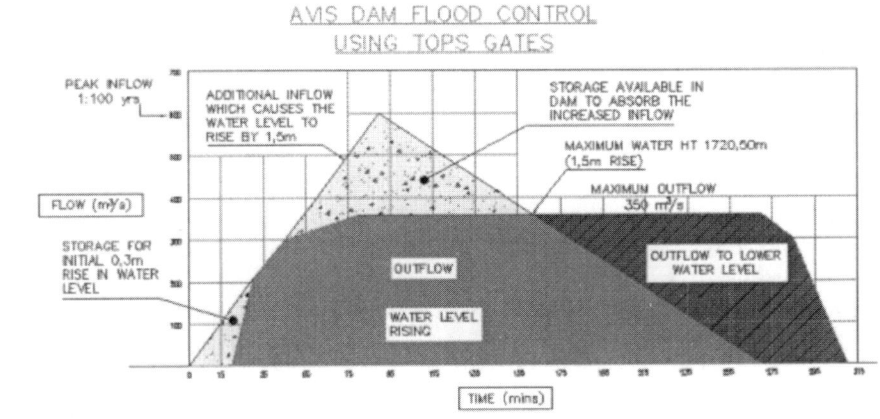


| | | | | | | |
|---|---|---|---|---|---|---|
| STATUS OF GATES | CLOSED | CLOSED | OPENING | FULLY OPEN | FULLY OPEN | GATES CLOSING |
| STATUS OF DAM | DAM FULL | SMALL OVER-FLOW | | WATER LEVEL RISING TO MAXIMUM | WATER LEVEL DROPPING | WATER LEVEL DROPPING |

## MODEL STUDY

A 1 in 20 scale perspex model was built of the gate and spillway. It was tested in the Stellenbosch University Hydraulics laboratories by independent researchers. The angular displacement of the gate was measured for different upstream water levels and flows and the discharge characteristics of the prototype determined. It was determined that the gates would rotate to 55° to the horizontal to pass the required flow.

Fig. 4. Model

It was also noted that the gates could open considerably more to pass larger flows but in this case the movement of the gate was stopped at 55° rotation to give the required discharge. The model was also tested with large debris and floating logs and found to have no adverse effect.

## FABRICATION

The gates were fabricated by Concor Engineering, a member of the Hochtief group.Normal workshop practice for low pressure vessels was adopted. The gate is manufactured out of carbon steel. The trunnion shaft as well as the seal plates are stainless steel grade 304L. The seals are double stem bulb seals which are hydraulically pressurized to assist sealing. The front upstream closure plate was fitted with studs to receive a 80mm thick nylon fibre reinforced concrete facing on site. This is to provide both effective corrosion protection to the closure plate as well as impact resistance. The corrosion protection to the remainder of the gate consisted of 500 microns inside the ballast tank and 200 microns external twin pack polyamine system and a 35 micron external coat of twin pack polyurethane weather resistant final coat. The gates were transported from Johannesburg to Windhoek, a distance of almost 1500km.

## INSTALLATION

The gates were loaded from the transporters and placed on tressles where the trunnion arms and seals fitted. The gates were then lifted into position by a 35 tonne crane and placed over holding down bolts to locate and secure the trunnion brackets. The gates were then rotated down into position where the seal plates were adjusted to ensure a 2mm pre-compression on the seals. The gate was tracked up and down to ensure a free movement. Second stage concreting to the seal plates and trunnions was the final operation together with casting the concrete skin to the closure plates.

Fig. 5. Gate lifted complete into position

Fig. 6. Gates installed ,one partially open

COMMISSIONING

An earth coffer dam was constructed 20m upstream and water then pumped in behind the gates over two days to reach the final depth of 3,5m above sill level. The gates did not show any sign of distress.

Fig. 7. Gates retaining full 3,5m water

The various components such as the level control discharge valve and the manual butterfly valve were wet tested.

Fig. 8. Gates ready for wet commissioning

With 60 interested observers and dignitaries in attendance the float controlled valve and manual valves were opened to release water from the ballast tank. At a drop of water level in the order 0,9m in the ballast tank, one gate opened and then the other.

Fig. 9. One gate opening

Unfortunately here was insufficient volume of water behind the gates to maintain an almost constant water level which is required to open the gates fully. The first gate opened partially which rapidly drew the water level down behind the gates so that full opening could not be achieved. The gate was closed down and the water level restored behind the gates by pumping. The second gate was then opened in the same manner. Despite not being able to open the gates fully, wet commissioning was nevertheless successfully achieved.

## CONCLUSION

This installation demonstrated the effectiveness and flexibility of operation of the TOPS spillway gates to maximize available storage in a dam to reduce flooding downstream as well as providing additional safe storage above the spillway level.

# The release of large diameter draw-off and control valves.

R P ENSTON, Hydra-Ject, UK
D C F LATHAM, Hydra-Ject, UK

SYNOPSIS. A patented new technology to release in situ stuck water valves without interrupting control or flow processes is explained. The technology can be applied to any type and size of valve, gate, butterfly, ball plug check etc. The release of both a 48-inch and 21-inch diameter gate valves and two 48-inch diameter butterfly valves at two raw water storage reservoirs belonging to Anglian Water Services in December 2001 is described.

## INTRODUCTION

General principles behind the technology

The majority of seized valves are caused when scale or sediment deposits accumulate between the internal moving parts (e.g. the gate) of a valve and the internal static components (e.g. guides, sealing faces etc.). The problem is further compounded by the fact that some valves cannot be operated on a frequent basis and the build up of the deposits increases with time. When this occurs the risk of an internal failure such as a broken spindle is greatly increased when the need to operate the valve is considered imperative and excessive torques are applied in an effort to release the valve.

(Less frequently, seizure can occur when a valve is over-torqued at either extremity of the open or closed positions.)

The end result is that the valves are broken or remain seized and either become redundant to the system, - impairing design capability of a dam, plant or pipeline, or have to be replaced or repaired. Frequently this requires a costly shutdown and expensive capital project costs.

This problem is common to most process systems and in many industries the traditional solution has been to implement a maintenance programme that enables process valves to be lubricated on a regular basis. However this is only possible if the valves are specified with lubrication capability at the design stage of a plant.

Historically, valves that are used throughout the water industry cannot be lubricated - such design features add significantly to the cost of manufacturing - and until recently a lubricant that would not have a detrimental effect on Water Quality was not available.

*Reservoirs in a Changing World*, Thomas Telford, London, 2002, 218–223

The Hydra-Ject technology has addressed these problems and now offers an inexpensive and novel solution that enables significant cost benefits and enhanced operational control to the Water Industry as a whole.

Covenham Reservoir and Cadney Carrs Reservoir

Both these reservoirs come under the 1975 Reservoirs Act and Anglian Water is responsible for ensuring that the inspection and maintenance is fully compliant with the legislation.

Covenham reservoir is an 11M cubic metre pumped storage reservoir, which had one stuck 21-inch scour control gate valve. This valve was fitted with a manually operated gearbox. Cadney Carrs Reservoir is a 910 thousand cubic metre bank side storage reservoir with a stuck 48-inch diameter inlet control gate valve and two stuck 48-inch diameter top intake and top draw-off butterfly valves. All three of these valves were capable of remote operation with Rotork electric actuators.

At Cadney the Inspecting Engineer had instructed the client to release all four valves and to ensure that they were in operating condition

The options available to the client were to effect an in situ repair or to remove and repair or replace. Removing the valves would have put in jeopardy the continuous operation of the reservoirs and would have been very costly. Alternative solutions were therefore sought.

INVESTIGATION

The first stage in the application of the technology required that the exact position of the valves be determined. This could not be reliably ascertained from a visual inspection alone. (In previous instances valves that were originally thought to be open were subsequently found to be closed and vice versa.)

A small hole was drilled and threaded - (in the valve bonnets) to receive an adapter that was used to insert a small diameter fibre optic endoscope. The scope enabled a visual inspection of the general condition inside the valve bodies – the extent and location of scale deposits and the accurate determination of the position of the shut off mechanisms (e.g. gate or discs). The Cadney Carrs 48-inch diameter inlet control valve was found to be approximately 50 per cent closed, and the 48-inch diameter intake and off take butterfly valves were found to be fully closed and approximately 90 per cent closed respectively. In these three cases the Rotork position indicators did not match the observations made from the endoscope inspection.

At Covenham, the 21-inch diameter scour valve is operated as part of the inspection routine under the 1975 Reservoir Act. Over the past two years increased difficulty in opening the valve had been evident at the inspections. The Supervising Engineer then asked that action was taken to free the valve. The endoscope inspection confirmed that the valve was 100 per cent closed. This was in agreement with the reading observed on the mechanical position indicator fitted to the valve.

The endoscope was also used on both the upstream and downstream pipeline sections adjacent to the valves. This provided valuable data on the general condition of the pipelines in that location.

The Cadney Carrs 48-inch diameter inlet control valve, as well as the pipeline upstream of the valve were thought to contain fresh water Zebra mussel deposits – a local environmental condition prevalent at the abstraction point on the river concerned.

In both cases the endoscope inspections revealed both locations to be comparatively unaffected. This enabled the opening of the scour valve to be undertaken with greater confidence that water quality would not be impaired at the discharge point in the river

METHOD OF WORKING

Following the endoscope surveys, a pattern of injection ports were then designed for each valve. The location of these ports was decided based on the data obtained from the surveys, essentially the ports are formed by drilling and threading holes around the areas where the scale deposits are located.

The number of ports required is largely dependent on two factors,

- the size of the valve
- the extent of scale deposits within the valve.

At Cadney Carrs the 48-inch diameter inlet control gate valve required 14 ports (ref Fig 1) - whereas the 48-inch diameter intake and off take butterfly valves required eight each (ref Fig 2). At Covenham, the 21-inch diameter scour valve required a total of six.

As a precaution an injectivity test was then carried out on each port using potable water to ensure the injection flow path was clear and pressure build up within the valve casing would not occur.

Following this, a small amount of the WRAS approved lubricant was then injected into each port using a pneumatic injection pump. The objective was to achieve a high nozzle velocity during the injection process this has the effect of displacing the scale deposits and permitting the lubricant to make contact with the metal surfaces of the moving and static parts (ref Fig 3).

This is a physical process and does not involve any chemical reaction between the lubricant and the scale deposits.

The lubricant used in the process is a specially formulated synthetic compound and has been tested and approved for use with cold potable water by the WRAS (The Water Regulatory Advisory Service - formerly the Water Bye-Laws). The Drinking Water Inspectorate has advised that for this application the provisions of Regulation 25 (1) (b) apply to the lubricant. This is for products with a small surface area of contact with potable water.

Once the injection process had been completed the valves were then cycled. The electrically actuated valves at Cadney Carrs were switched to manual override and once the initial moving torque had been applied, all the valves were able to move and progressively became easier to operate on each cycle as the lubricant came increasingly into contact with the internal surfaces of the valves. However, it has not yet been possible to fully cycle the 48-inch diameter inlet control gate valve at Cadney Carrs for the reasons that follow.

The reservoir at Cadney Carrs is critical to the water supply of industrial and residential consumers for large parts of South Humberside. Non-availability from this source of supply for any extended period could have major consequences. For this reason it was considered prudent that the inlet valve should only be cycled to the halfway closed position in the unlikely event that once the valve had been fully closed, it became stuck and could not be opened. As a precaution, a contingency plan is to be put into effect at a later date such that a rapid valve replacement can be effected for this scenario.

## CONCLUSION

All four valves were successfully released to the satisfaction of the client. The injection ports have been left in situ which will enable easy lubrication of the valve faces in future years. This will ensure that the valves continue to be operational.

The solution adopted was considerably less expensive than the alternatives, was speedy and did not interrupt ongoing operations.

Valuable ancillary data is made available from the initial investigations vis-à-vis pipeline condition monitoring and the upstream and downstream valve environment

The technology described can be applied to any type of valve and can be used over a wide range of diameters from 100 mm to 2m.

## ACKNOWLEDGEMENTS

The authors are grateful to Anglian Water for their kind consent to the publication of this paper.

**Fig 1 Cadney – Carrs 48 inch Gate Valve**

**Fig 2 Cadney - Carrs 48 inch Butterfly Valve**

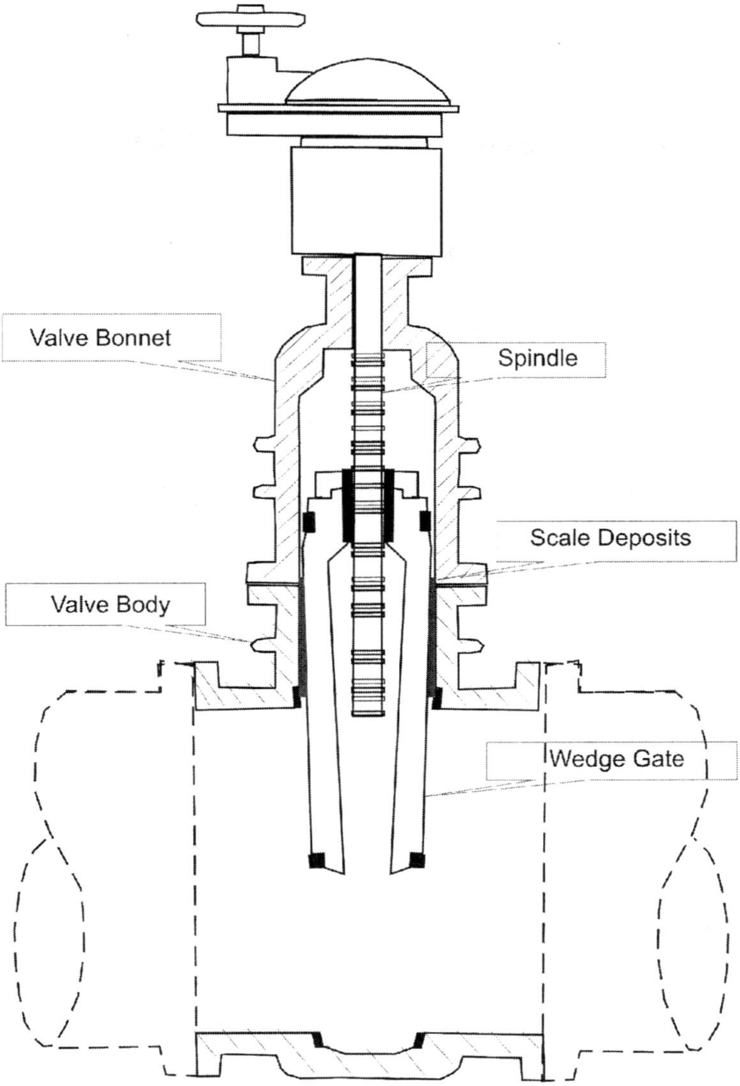


Fig. 3. Cadney – Carrs 48 inch Gate Valve Section

# Remedial works at Brent Reservoir to address leaking sluice gates

R A N HUGHES, Binnie Black & Veatch, UK
P KELLY, British Waterways, UK

SYNOPSIS. Brent reservoir is impounded by a 170 years old embankment dam with a puddle clay core and was constructed as a feeder reservoir for the Grand Union Canal. The reservoir has a chequered history; problems with leakage through the embankment are well documented. The reservoir is now mainly used for recreational purposes and the area upstream is a designated Site of Special Scientific Interest.

Two culverts pass through the dam at low level which discharge flood water into the horseshoe shaped stilling basin approximately midway along the dam. The culverts are controlled separately by cast iron sluice gates at their upstream end. During operation of the right hand sluice gate in 2001 severe vibration was experienced in the lifting mechanism; the degree of vibration persuaded the reservoir keeper to close the gate. The two sluices are used as part of the overflow works and are operated in conjunction with the four original fixed weirs, syphon spillways and emergency overflow over the dam crest.

Inspection of the gate by divers showed there to be substantial leakage through the top and bottom seals. Binnie Black & Veatch have been appointed to undertake a review of feasible options for stopping leakage through the culvert without affecting the SSSI site.

This paper will describe the options considered and detail the recommended solution for stopping leakage through the two sluice gates.

## INTRODUCTION

### Background

Brent Dam is a 9m high earth embankment dam with a puddle clay core. The dam was constructed in 1835 to supply water to the Grand Union Canal (Paddington Arm). In 1841 the dam partially collapsed following a seven-day period of heavy rain and was subsequently raised in 1852 to its present height.

The dam is a category A dam as defined by the Institution of Civil Engineer's publication "Floods and Reservoir Safety, 3rd Edition". Remedial works were carried out to the dam in 1984, which included

*Reservoirs in a Changing World*, Thomas Telford, London, 2002, 224–235

provisions to enable it to withstand overtopping during a 10,000 year flood. Before being overtopped the other overflow facilities would be in operation, those being two low-level culverts, a relatively short section of fixed weir spillway and five syphon spillways.

Discharge through the low-level culverts is controlled by sluice gates at their upstream ends. The gates are operated automatically from a gatehouse above, commencing opening when the reservoir level is 3 feet 6½ inches (1.077 m) below weir crest level.

During operation of the right hand sluice gate in March 2001 the reservoir keeper noted it was not opening correctly and during closure, movement was slow and there was heavy vibration.

Inspection of the sluice gate was carried out on 14 March 2001 by a diving team, Info-Tech Marine following the problems with its operation. The sluice gates were inspected from the upstream and downstream sides. From upstream the divers noted that there was leakage occurring through the right hand gate whilst it was in the closed position. Visibility was poor due to the quality of the water and the exact cause of the problem could not be determined.

The downstream inspection involved a diver walking up the culvert from the downstream end of the spillway. A video recording of the inspection showed that heavy leakage was occurring around the top and bottom seals.

Following the inspection, British Waterways (BW) decided to keep the gate in the closed position thus reducing the capacity of the low-level culverts by 50%. Binnie Black & Veatch (BBV) was appointed to undertake a review of technically feasible solutions and advise BW as to the best option for addressing the leakage around the sluice gate.

## OPTIONS STUDY

### Constraints

The options considered in study were limited to those that would allow the reservoir to remain full during the repair works. The constraint was stipulated by BW for the following reasons:

- The reservoir is a designated Site of Special Scientific Interest (SSSI). Emptying the reservoir could harm the site and generate unfavourable public opinion.

- The reservoir is heavily silted. Emptying the reservoir to carry out works to the sluice gates could potentially release large quantities of sediment into the downstream watercourse.

- The reservoir is used for recreational purposes; there are at least five different Sailing Clubs situated near to the right abutment of the dam. Emptying the reservoir to carry out repairs to the gates would disrupt its use as a significant amenity facility.

During the diving inspection carried out in March 2001, the divers were unable to get close to the right hand gate as the flow of water through the gate was sucking the divers towards the gate. To permit an underwater repair of the gate to be safely completed the flow through the gate would need to be reduced to a level that would allow the divers to approach the gate without danger.

Gate repair

The actual condition of the gates was unknown when the options report was being written. Several possible causes of the leakage were considered ranging from the failure of the rubbing strip to the warping of the gate frame. The severity of the problem would dictate whether the gate could be repaired in situ or would require off site repairs.

Both insitu and off site repair would require the gate to be opened or removed to enable the repair to be carried out. To prevent emptying of the reservoir a means of controlling flow through the culverts would be required.

Brainstorming

Early in the options study a brainstorming session was conducted to identify possible solutions to the problem. BBV's dams specialists were present in the session and numerous solutions tabled which can be grouped as follows:

1) Repairing the gate in-situ using either an upstream cofferdam or stoppers in the downstream culvert. This option could be viewed as a permanent solution or the first stage for option 2 or 3.
2) Line the culvert and install a penstock at the downstream end of the culvert. The upstream gate could then be repaired or abandoned.
3) Install two pipes within each culvert with valves at their downstream ends.
4) Modify existing weir levels & install smaller penstocks
5) Install inflatable valve in the culvert.

Temporary works

To safely repair the gate in-situ, the flow of water through the gate would need to be reduced. Two technically feasible options were considered: the installation of an inflatable stopper in the cast iron culvert, and the construction of a cofferdam within the reservoir.

A stopper system was viewed as the simplest means of stemming the flow through the culvert. The cast iron lined section of the culverts is lozenge shaped, 1800 mm high by 990 mm wide. The liner is constructed in units approximately 1.2m long joined together by bolted flanges. The unconventional shape of the culvert meant an off the shelf stopper was unavailable; the preferred solution involved a twin stopper system with an interstitial space that would be filled by water when inflated. The water pressure would be such as to be half that exerted on the upstream face of the first stopper. The twin bag system provided additional protection against failure of the stopper. A stopper would necessitate the use of divers to carry out the repair to the gates.

Also considered as a means of stemming the flow through the sluice gates was the construction of a cofferdam within the reservoir. The cofferdam would allow the reservoir to remain full whilst permitting works to the gates to be carried out in the dry therefore obviating the need for divers to undertake the repair works.

Numerous difficulties existed for this option including the connection detail between the cofferdam and the clay core of the dam, dealing with inflow to the reservoir whilst the cofferdam was dewatered and the bracing arrangement around the gatehouse tower.

Cost estimates

For each of the technically feasible options an estimate of the likely construction cost was prepared. Of the five options considered cost estimates were prepared for only three of them.

Modifying the existing weir to allow the low level culverts to be abandoned would result in the reduction of the reservoir's top water level. Given that a prime constraint on the study was the maintenance of the reservoir water level, the option was rejected.

The installation of an inflatable valve in the culvert was also rejected pre costing due to the logistical difficulties with installation, operation and safety.

Recommendations

The most economical solution was found to be the installation of a temporary stopper in the downstream culvert with the repair of the gate being carried out by divers. The cost of this option was estimated to be approximately 25% of the cost of the next cheapest option.

The draft options report recommended that this option be used to address the leakage through the sluice gates.

Culvert inspection

Following submission of the draft options report, the authors carried out an inspection of the two culverts in order to ascertain the condition of the cast iron lined and masonry sections. The inspection was used to determine whether the use of an inflatable stopper was feasible.

Overall the cast iron lining was in fair condition but the surface was heavily pitted and there was slight water ingress through the joints. Ultrasonic measurement of the lining was attempted at three locations without success. The unevenness of the cast iron surface meant the probe could not achieve a satisfactory contact for a measurement to be taken. The masonry-lined section of the culvert was in good condition with no cracks visible but a few of the joints were unpointed.

There were cracks present in the sidewalls of the cast iron lining in both culverts. The cracks were at roughly the same location immediately upstream of where the culverts enter the masonry wall of the spillway. The cracks began at the invert of the lining and disappeared approximately three quarters the way up the side walls. The locations of the cracks in the left culvert are recorded on Figure 1.

There are few record drawings of the dam, the ones detailing a section along the low level culverts indicate, by virtue of the style of hatching, that the area around the culvert where it enters the masonry wall is of a material other than masonry. The nature of the cracks in the cast iron liners suggests that the culverts had settled which would only be possible if the surrounding material was compressible. The position of the cracks coincided with the point where the height of the material above the culvert is greatest.

The inspection also identified that the left sluice gate was leaking through the top seal. The degree of leakage was much less than through the right hand gate. The amount of leakage through the right sluice gate made it impossible to get sufficiently close to the gate for a thorough inspection.

After discussions with the Inspecting Engineer, who carried out the last statutory inspection, it was determined that reliance could not be placed on the continued long-term structural integrity of the culvert.

Revised recommendation

Following the findings of the inspection, the original recommendation of using an inflatable stopper was rejected due to concerns about the ability of the cast iron lining to withstand the pressure that would be exerted by the stopper. All options were revisited and following a meeting at which Barhale Construction (BC), a term contractor with BW, were present it was decided to abandon all options which relied on the use of a stopper.

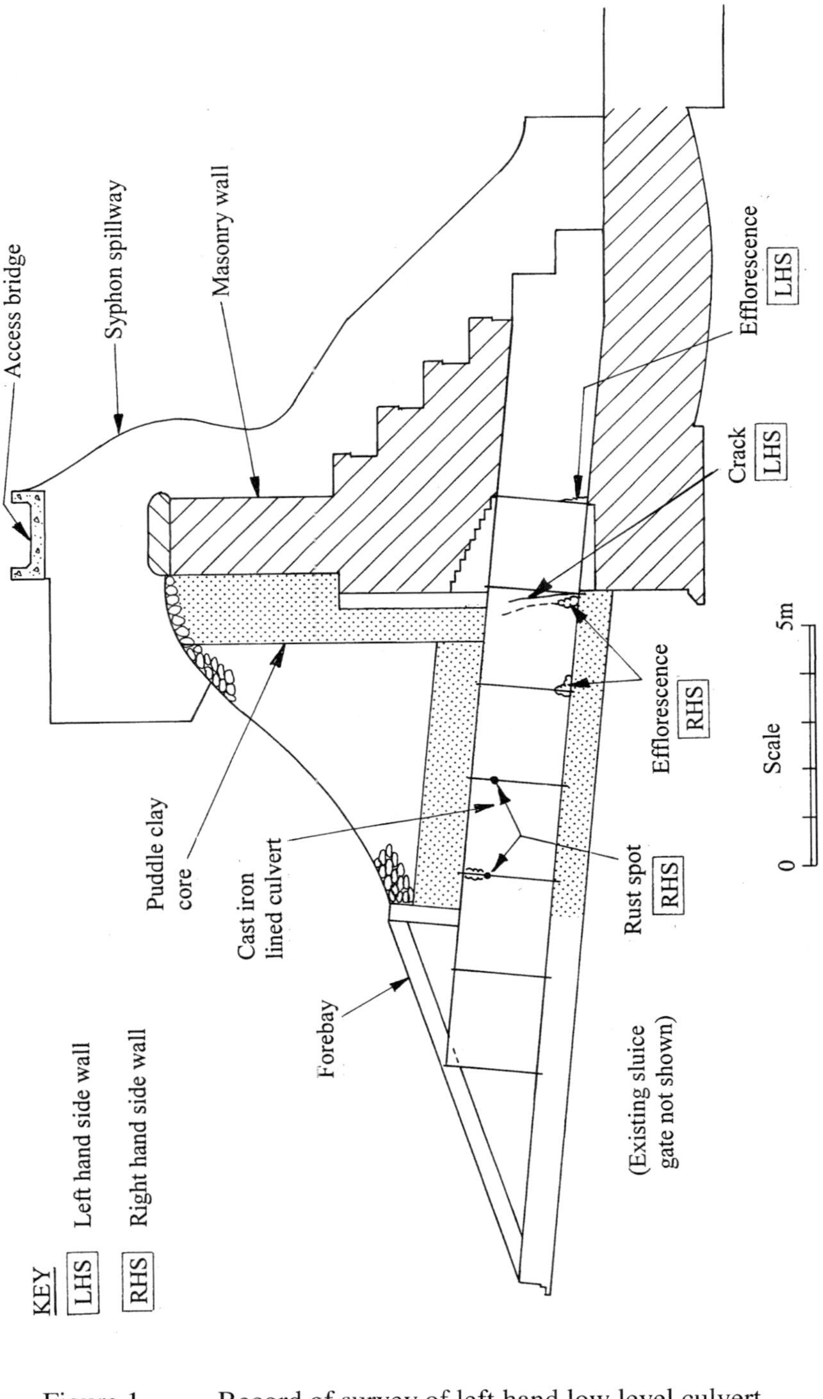


Figure 1 Record of survey of left hand low level culvert

The revised report recommended that the cofferdam option should be adopted, but due to the high cost, BW were uncomfortable with progressing with the option. The only options that remained involved stemming the leakage from downstream which would means the culverts would be full of water and therefore pressurised. Concerns regarding the ability of the cast iron liner to withstand internal pressure meant that some form of lining would be required. Barhale Construction proposed the use of Glassfibre Reinforced Plastic (GRP) liners within the existing culverts, the annulus between the cast iron and GRP liners would then be grouted to secure the liners in place and seal the culverts. Downstream control of the culvert would be provided by means of penstocks bolted to the downstream face of the lined culverts.

Following discussions to discuss the practicalities of installation of the liner under flowing water conditions it was agreed that the method used to address the gate leakage would be the lining of the existing culverts with GRP and the installation of automated penstocks at the downstream end. The recommended solution is shown in Figure 2.

## DETAILED DESIGN

Following the meeting, BW instructed BBV to carry out the detailed design of the remedial works. Barhale Construction was also appointed to construct the works, the early appointment of a contractor was intended to speed up the overall construction programme by obviating the need for competitive tendering.

### Culvert lining

The design of the liners was to be in accordance with Water Research Centre publication No. 4-34-02, Specification for Glassfibre Reinforced Plastics (GRP) Sewer Linings, Type II design. The liner would be designed to withstand internal and external pressures that would be present in a 10,000 year flood event.

The unusual shape of the culvert would require the GRP units to be of a non-standard profile. Channeline Sewer Systems Limited were approached to prepare the liner units since it specialises in the production of non-standard profile GRP products.

Due to the restricted access and confined space working environment the liners would be constructed in sections, approximately 1.5m long. During the detailed design stage it was envisaged that there would be some flow passing through the culverts whilst the lining works were ongoing. To ensure that the annulus between the GRP and cast iron liner remained dry for the grouting operation the uppermost liner would be formed with a thickening at its upstream end which would be modified on site to form a tight fit between the old and new liners.

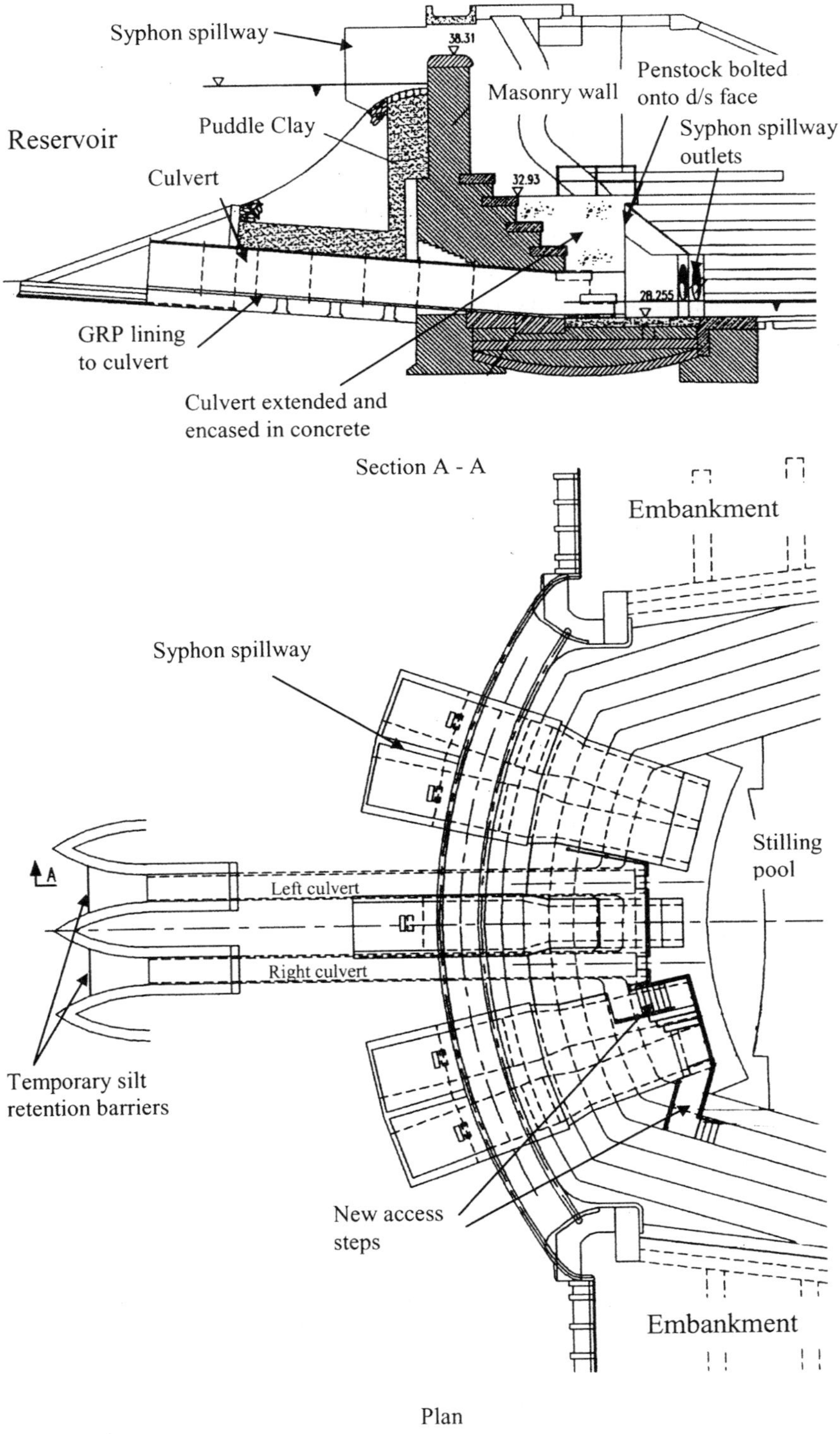


Figure 2: GRP liner & downstream penstocks solution

At the upstream end of the lined culvert a transition unit would be required to minimise disruption to the flow. Originally the transition unit was to be preformed off site but due to uncertainties with the profile of the cast iron lining it was decided to form the transition in situ following the installation of the GRP liner.

Upon award of the contract to produce the liners, a representative of Channeline visited site and carried out a detailed survey of the culverts. Based on the survey a wooden template was prepared which was for a single unit of the GRP liner. The template was brought to site and taken up the culverts to check whether it would fit. Following minor adjustments on site to the template the final profile of the liner was agreed.

Penstocks

Four companies were approached for the design and supply of the downstream penstocks. Several materials were offered, the contract to supply the penstocks was awarded to Aquatic Control Engineering Limited who offered to supply two penstocks manufactured from stainless steel with High Density Polyethylene.

The purpose of the penstocks is to isolate a culvert. Their closure will allow divers to safely access the upstream sluice gates to carry out any repairs. Should the existing gates need to be removed from site for repair, the downstream penstocks will be able to be operated temporarily as flow control penstocks until the repaired gates are returned to service. The prolonged operation of the downstream penstocks for flow control is undesirable, as the impermeable zone of the dam will be different from the existing arrangement, which may affect the watertightness of the embankment.

The penstocks will be automated and controlled from the gate house. Concrete steps will be constructed to provide access to the penstocks in case of failure of the automated systems.

Overflow rating curve

The installation of the GRP liners within the existing culverts will result in the reduction of the effective waterway. Analysis of the culverts concluded that the hydraulic capacity of the culverts would be reduced by approximately 10%. Table 1 summarises the findings of the analysis.

The above values are for the culverts flowing full; this scenario is considered to be unlikely since air will become trapped at the upstream end of the culvert causing them to operate with entrance control.

For entrance control, the hydraulic capacity of the culverts equates to approximately to 25 $m^3/s$ at Top Water Level. The rating curve used in the flood routing study adopted this value.

| | Hydraulic capacity ($m^3/s$) | |
|---|---|---|
| Reservoir water level | 38.01 mOD (Normal Retention Level) | 38.31 mOD (Top Water Level) |
| Unlined culverts (Total) | 29.5 | 30.1 |
| Lined culverts (Total) | 26.6 | 27.1 |

Table 1: Hydraulic capacity of the lined and unlined culverts.

Given that the capacity of the lined culverts is greater than that used in the rating curve, the ability of the dam to safely pass a 10,000 year flood will not be compromised.

Health & Safety

All of the work within the existing culverts will require stringent safety precautions as it is classed as a confined space. It is envisaged that a permit to work system will be adopted by Barhale Construction. The formation of the transition piece in situ at the upstream end of the GRP liners will result in fumes being present within the culvert, it is envisaged that breathing apparatus will be required for the personnel working in the culvert at that time.

The culverts are inclined at a gradient of 1 in 12. The GRP liners have a low coefficient of friction so in order to minimise the potential for falls and slips, a non-slip coating will be applied to the invert of the liner units in the factory. The effect of the coating on the overall hydraulic capacity of the culvert will be negligible.

During the course of the detailed design, leakage through the right hand sluice gate increased significantly. The increased was caused by the sluice gate lifting slightly on its own. During the works, the relevant sluice gate would be isolated to prevent their accidental operation. However this would not prevent the gates from lifting slightly as was demonstrated above. A site meeting between BW, BBV and BC concluded that working in the culvert whilst the reservoir was full would be too dangerous. A decision was made to approach affected parties and request that the reservoir be drawn-down for the duration of the lining works.

The Environment Agency, English Nature and the affected sailing clubs were approached to seek approval for emptying the reservoir. Following numerous discussions approval to empty the reservoir was given, on condition that the reservoir was refilled by 7$^{th}$ March to allow for the bird breeding season. Before the reservoir was totally de-watered, a fish rescue was undertaken which involved the removal of 25,000 lbs of fish to a temporary location within a length of canal 10 miles away. All the fish will be replaced once the reservoir has been refilled.

Temporary works

A bed survey of the reservoir was carried out in May 2001 and showed that the reservoir is gradually silting up. Once it had been decided that the reservoir would be emptied for the works, a means of preventing sediment from being discharged to the downstream watercourse was required.

In January and February 1995 the reservoir was emptied due to problems with the closure of the right hand sluice gate. The cause of the problem was found to be a chisel, which had become lodged into the gate recess. To prevent sediment from being discharged temporary silt retention barriers were installed into the upstream forebay. The barriers consisted of steel plates resting on angled fixed to the sidewalls and floor of the forebay. The barriers were removed from the reservoir after removal of the chisel and were stored for future use.

It was decided to reuse these barriers for the lining works in order to save time. During the works the sluice gate at the upstream end of the culvert which was being lined would be isolated in the closed position; the other gate would be left open to allow water to pass through the reservoir. The space between the barrier and the closed gate would be dewatered during the works so as to provide a buffer zone upstream of the gate. This barrier would be raised such that its crest would be higher than the barrier upstream of the open gate in order to ensure that the flow would be directed towards the open gate.

CURRENT SITUATION

Gate repair

Construction of the lining works and culvert extension concrete is underway; it is anticipated that the works will be completed by the end of March 2002. The works have progressed relatively smoothly but problems due to the weather and sedimentation have hampered progress.

Brent Feeder supply

Supply to the Brent Feeder, a channel that provides a small amount of water to the Grand Union Canal, is by a draw-off approximately 20 m to the left of the spillway structure. The downstream end of the draw-off comprises two 225 mm pipes that discharge to a brick lined culvert approximately 3 m high by 2 m wide. There has been a history of leakage into the culvert through the upstream headwall and sidewalls and works to identify the cause of leakage have been undertaken in the past (Tedd et al, 1998).

To take advantage of the reservoir being empty, further investigations will be carried out during the construction of the lining works in order to identify and stem the leakage.

## CONCLUSION

The leakage around the sluice gates, whilst not a matter that would compromise the integrity of the dam, affected the normal operational regime for the reservoir. Several technically feasible solutions exist but cost and speed of construction resulted in most being discounted.

Every effort was made in the study to avoid the need for emptying the reservoir. The safety issues that arose during the study involving the upstream gates resulted in the draining of the reservoir being required. Close liaison and discussions with interested parties resulted in this being possible provided environmental issues were addressed prior and during to construction.

The selected solution will allow the reservoir to be refilled even if the sluice gate has to be removed from the dam for repair. In addition to this, the remedial works had to be of a relatively straightforward nature to allow it to be constructed quickly so minimising the duration that the reservoir is empty.

## REFERENCE

Tedd P, Dutton D P M & Holton I R (1998). The prospect for reservoirs in the 21st century. Proc. 10th Conf. British Dam Society. Bangor, pp70 – 78. Thomas Telford, London

# Refurbishment of outlet tunnel and associated pipework at Piethorne Reservoir

A BRISCOE, Montgomery Watson Harza, UK
A A GEORGE, Montgomery Watson Harza, UK
I C CARTER, Montgomery Watson Harza, UK
P GRUNDY, United Utilities plc

---

SYNOPSIS. Piethorne Reservoir was constructed near Milnrow in Lancashire between 1858 and 1863. The dam displays many features typical of a "Pennine" embankment of the early–Victorian period. However it is unusual in that the outlet tunnel provides access for some 20 m beyond the central core. The masonry lining to the tunnel was in a very poor structural condition despite several attempts to stabilise it using brick pillars and bulkheads. The 1998 Statutory Inspection drew attention to its inadequacy and safety measures were recommended. The recent works to refurbish the outlet works are described. These included plastic liner installation within the 140-year old pipework and filling of the tunnel with foam concrete.

## INTRODUCTION

Piethorne Reservoir is owned by United Utilities (UU). It is the third reservoir in the Ogden Valley cascade, as shown on Figure 1, and supplies Piethorne Treatment Works immediately downstream. The reservoir has a storage capacity of 1.53 $Mm^3$ and is impounded behind a 22 m high earth embankment.

The downstream slope of the embankment slipped during construction over 140 years ago. Additional earthworks and drainage measures were undertaken to repair the dam. The 6 ft diameter tunnel settled during the incident and its invert was raised with concrete and the masonry lining was made good. However the structure has remained a problem over the years. Open cracks up to 15 mm wide were present in numerous locations prior to the start of the recent improvements. The original circular shape had become squeezed and flattened beneath the central part of the dam. Clearance to the crown had reduced by about 150 mm. Measurements over the last decade had shown that movement was ongoing despite the attempts to strengthen the structure. Original drawings suggest that part of the tunnel was founded upon embankment fill rather than on rock and this is probably one of the main reasons for the structural distress.

*Reservoirs in a Changing World*, Thomas Telford, London, 2002, 236–246

A statutory inspection was carried out in March 1998 and the Inspecting Engineer recommended the following measures in the interests of safety:

- Tunnel to be made structurally sound by filling with concrete or similar.
- Joints to allow the articulation with any further settlement.
- Scour and draw-off pipes to be fully lined throughout the embankment.
- Existing valves on the draw-off pipe to be refurbished and upgraded.
- Guard valves to provide upstream control at the end of the lined pipe.

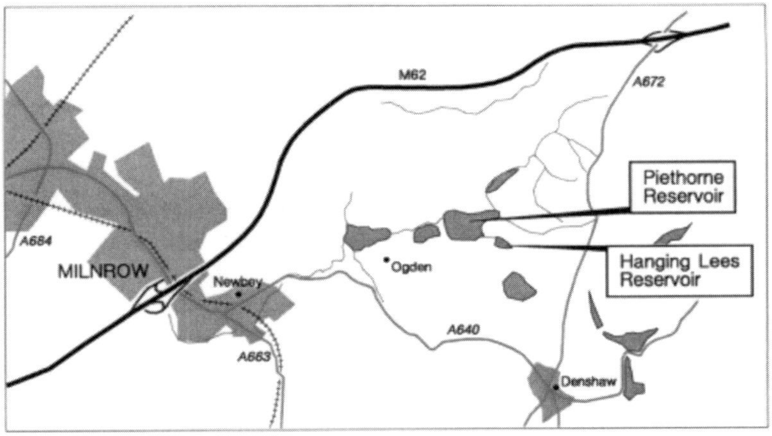


Figure 1 Location Plan

ORIGINAL SCOUR AND DRAW-OFF ARRANGEMENT

The original outlet works consisted of two separate draw-off systems:

- the lower system, which comprises two cast iron pipes that are taken through the tunnel (a 20-inch main and a 12-inch scour), and,
- the higher system, which comprises a single delivery pipe leading from a wet well through the left abutment.

The general arrangement of the tunnel and lower pipework is shown in Figure 2. A bulkhead was positioned upstream from the centre line axis of the embankment. The presence of the tunnel upstream of the bulkhead was not proven during the remedial works and the pipework upstream of it may have been embedded within embankment fill. Masonry blocks supported the larger of the draw-off mains but the lower scour pipe was partially embedded in mass concrete cast into the tunnel invert, as shown in Figure 3.

The draw-off main connected to a cast-iron stack at the upstream toe of the dam that was supported on the pitched face of the embankment. Water was drawn off at three elevations. Each was controlled by a sliding batter valve and protected with a bar screen. The valves were offset in plan to allow mechanical operation by control rods from the dam crest. Modifications were made to the uppermost valve in 1996, when the control rod was replaced by a hydraulically actuated system.

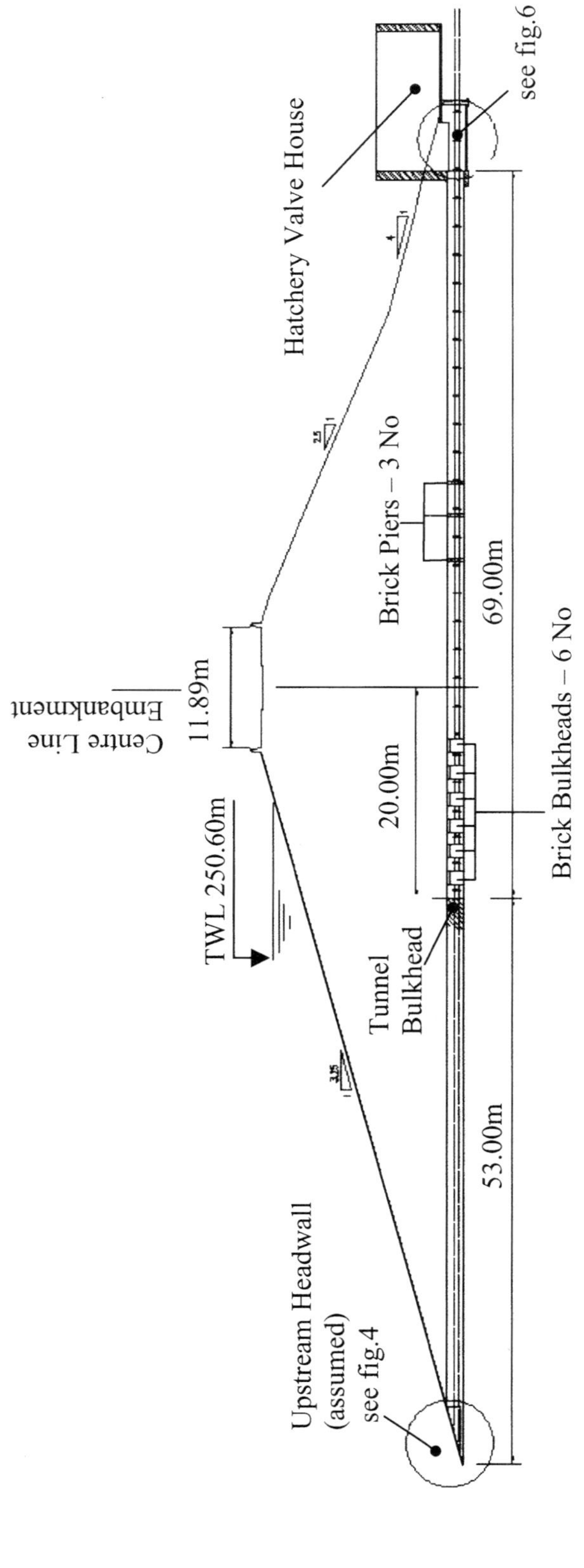


Figure 2 : Longitudinal Section Through Embankment and Masonry Tunnel

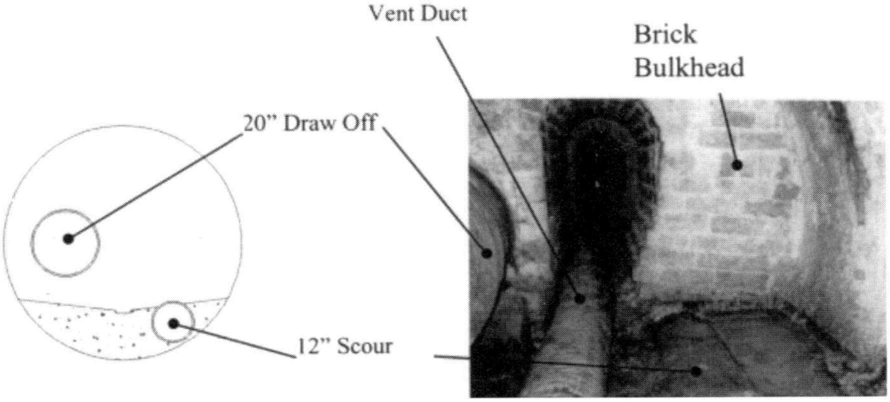


Figure 3 : Cross Section Through Downstream Masonry Tunnel

The scour pipe projected directly into the reservoir and was fitted with a flow control batter valve at its inlet. The lowermost leg of the draw-off stack and the upstream end of the draw-off main were aligned in the vertical plane and the pipes connected with a tee fitting and a radial branch, as illustrated in Figure 4. A headwall is also indicated but Operations staff and divers were unable to confirm its presence.

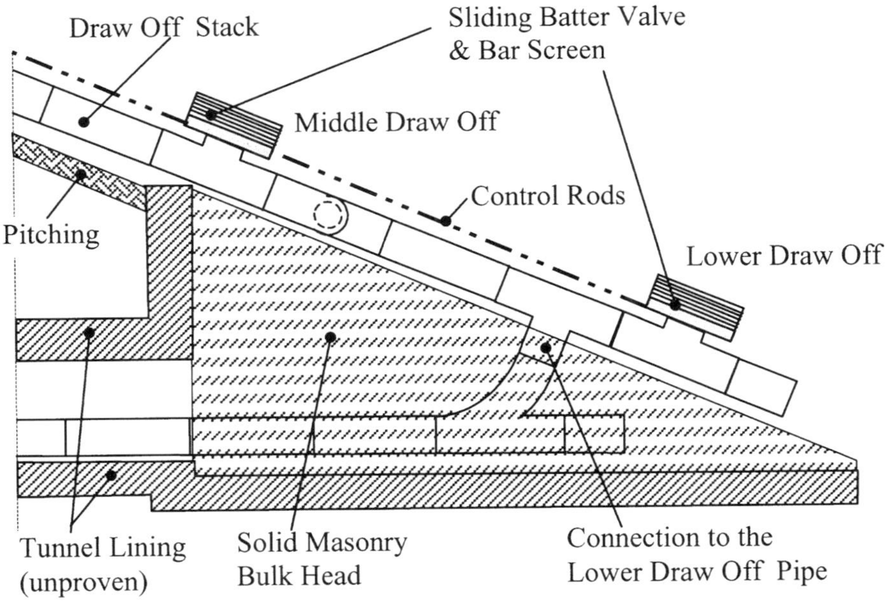


Figure 4 : Original Connection to Lower Draw Off Pipe

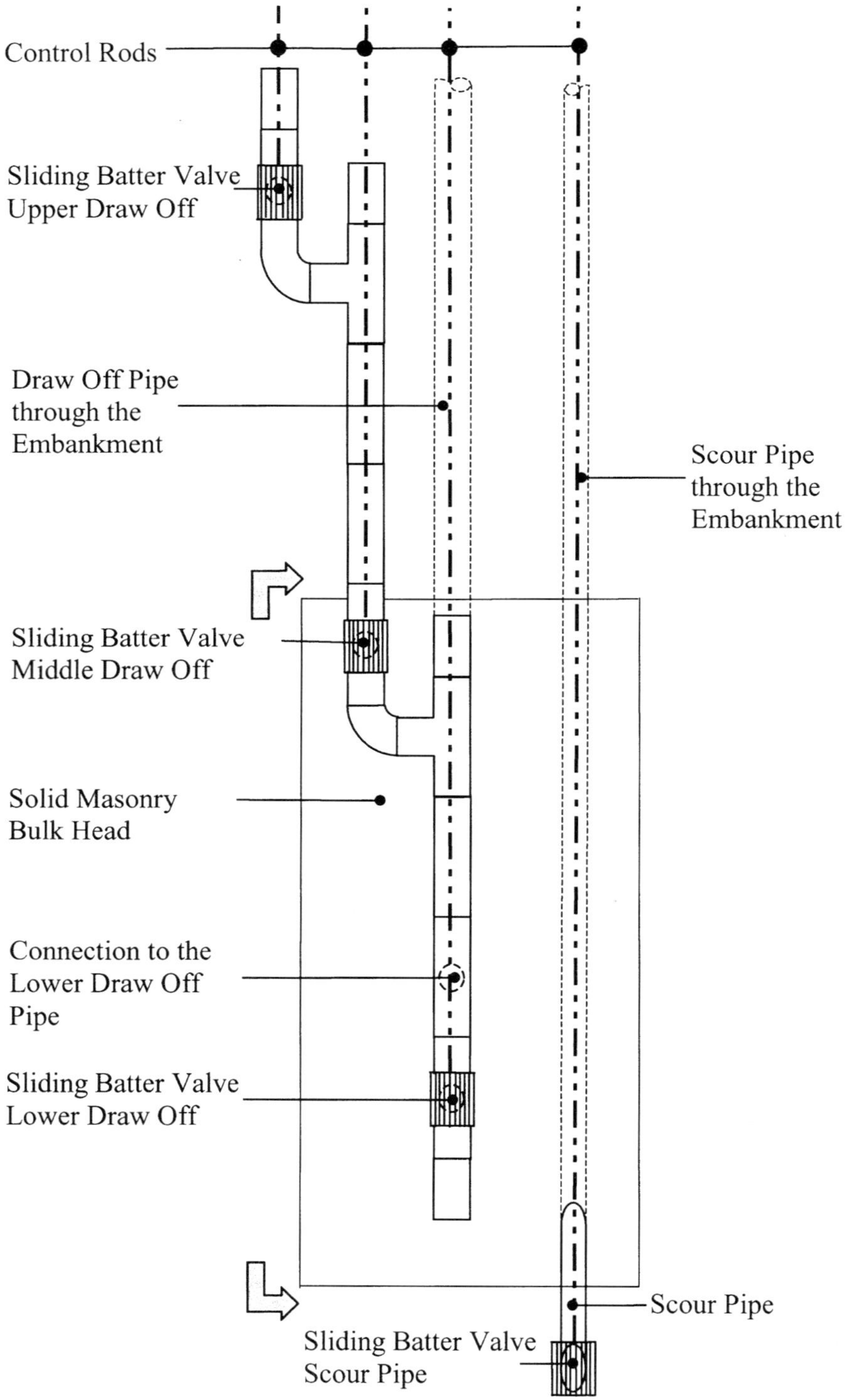


Figure 5 : Original Draw Off - Plan View

The draw-off main and scour emerge directly into a valve house, otherwise known as the 'Hatchery', at the downstream end of the tunnel. Gate valves are located on each pipe and a cross connection exists between the two (Figure 6). A branch from the scour pipe discharges directly to Kitcliffe reservoir downstream. Outside the Hatchery the drawoff and scour pipes combine into a single 28-inch cast iron delivery main.

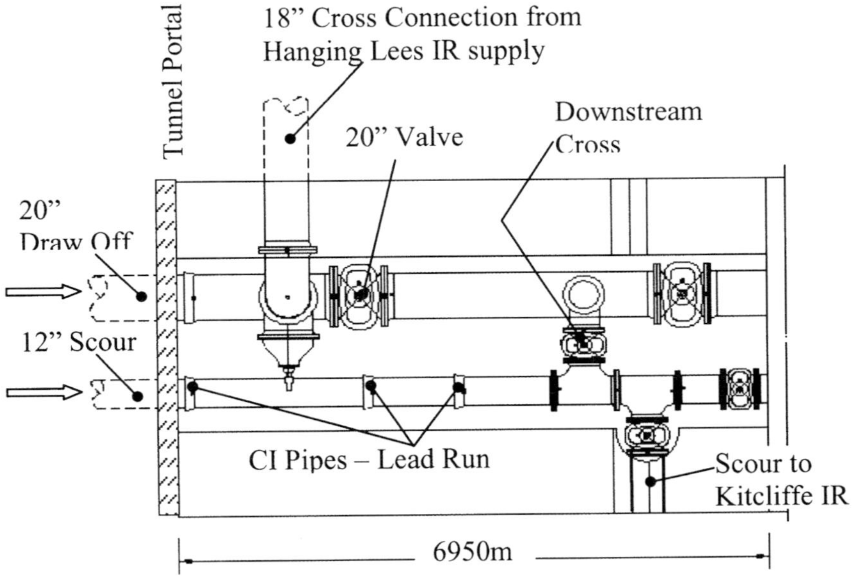


Figure 6 : Hatchery pipework layout - Plan

## DRAW-OFF TUNNEL

The tunnel is of masonry construction, approximately 69 m long. Archive drawings indicate a nominal diameter of 6 ft (1829 mm). Structural surveys and routine monitoring of the tunnel commenced in 1978. Significant internal strengthening was undertaken in the 1960s that involved the installation of six bulkheads and three brickwork pillars, as shown in Figure 2. However cracking of the lining around the supporting pillars indicated that the remedial works had not solved the problem. Survey measurements confirmed that the tunnel was flattening. By 2001 the maximum recorded width was 1.90 m and the minimum height was 1.32 m, although the latter measurement was taken between the soffit and the concrete infill in the invert. Water stood over much of the invert and deposits of silt were also recorded. The severely distorted tunnel was extensively cracked and many of the masonry joints had been re-pointed. Evidence of ongoing structural movement was apparent with open cracks and crack displacements up to 12 mm wide present in the crown of the tunnel. Structural movement was also apparent at the tunnel portal and the Hatchery building.

## CONCERNS OF THE OWNER

Piethorne reservoir is a strategic resource and any extended period without it would have threatened continuity of water supplies in the district. Consideration was given to carrying out the safety works without emptying the reservoir but no practical or economic alternative was available. UU accepted the need to empty the reservoir in order to implement the measures but had three main concerns regarding the remedial works, which were:

- the hydraulic efficiency of the proposed system,
- the utilisation of stored water in the lower part of the reservoir, and,
- completion of the works such that it might be refilled over the winter.

## ENGINEERING THE SOLUTION

Field observations and numerical analysis were used to assess the probable reduction in hydraulic capacity. System head losses were determined by measuring total head loss at a pressure tapping on the cast-iron draw-off in the Hatchery against a measured discharge rate of 16.8 Ml/day. The equivalent roughness ($K_s$) for cast iron pipe over 100 years old was assumed to lie between 6mm and 30 mm and the length of the draw-off pipe was known. The total head loss recorded was 2.32 m, of which 0.5 m to 1.0 m could be attributed to frictional head loss and 1.8 m to 1.3 m attributed to the 'minor' head losses.

The frictional head loss for a 450 mm OD polyethylene-lined main, which has a smaller cross-sectional area, was in the order of 1.0 m for the equivalent discharge rate. Connecting the polyethylene pipe to the existing draw-off stack would therefore reduce the hydraulic capacity of the system. It was decided that the impact could be mitigated by cross-connecting the scour and draw-off mains downstream of the isolating valves such that the system could be run in parallel, if necessary, to reduce the head loss.

Piethorne Treatment Works stands above the toe of the dam but only the upper part of the reservoir can gravitate through the system. A temporary pumping arrangement was provided downstream of the Hatchery in order to make maximum use of the stored water. Additional valves and tee-connections were installed on the 28-inch supply main to allow installation of a portable booster pump. This provided the client with the facility to transfer water from the lower section of the reservoir to the treatment works.

The need to minimise outage time necessitated a tight contract period. Scheduling of activities within the construction programme was further complicated by unknown site factors and constraints, which included:

- How much silt would there be in the reservoir basin?
- How would inflows and silt-laden water be handled?
- How would access be gained to the upstream end of the tunnel?

- What diameter liner could be installed within the existing pipes?
- How was the pipework arranged at the inlet end of the tunnel?
- How could isolating valves be included in the new draw off system?
- Would it be possible to empty the reservoir and hand the site over?
- Would the bypass provide a sufficiently long weather window?

These uncertainties were addressed by formal reviews at key stages during the reservoir emptying and construction phases of the project. The reviews included UU water resources management, operations and reservoir safety staff; the Principal Contractor and key Subcontractors; the Engineering Services Provider (MWH) and an AR Panel Engineer.

Each review took place after site inspection and involved brainstorming sessions in order to resolve difficulties arising out of new information and/or departures from anticipated conditions. A consensus would be reached at the end of each review that allowed the contractor to progress procurement and construction so as to meet programme requirements.

## SCOUR, DRAW-OFF & TUNNEL REFURBISHMENT WORKS

It was decided that the pipes should be lined consecutively such that a permanent outlet from the reservoir could be maintained for the duration of the contract. Luckily, the tunnel inlet was not at the deepest part of the reservoir and a small lagoon formed in a natural basin some distance away. Not only did this allow the works to be undertaken in the dry but it also facilitated control of the silt-laden inflow.

A floating submersible pump decanted cleaner water from the lagoon and pumped it through the draw off pipework to discharge into Kitcliffe reservoir, immediately downstream of the embankment. The procedure provided a simple but effective method of dealing with a potentially difficult problem. Access into the basin was provided by a temporary stone track running down the left abutment.

A structural liner was selected to sleeve both pipes. Steel and polyethylene pipe materials were considered. Appropriately sized stainless steel was available for immediate delivery but would have been prohibitively expensive. Polyethylene pipe had to be ordered but the delivery period would have delayed the programme. An early decision was taken to place an advance order for the polyethylene pipe, not only for the maximum standard size that might be accommodated inside each main but also for the next standard size down. UU accepted that the total cost for the purchase of the polyethylene pipe was less than the cost for providing one stainless steel pipe for the respective draw-off and scour main. PE 100 polyethylene was selected. 450mm and 400mm outside pipe diameters were purchased for the draw-off main and 250mm and 200mm outside diameters for the scour.

The existing draw-off arrangement became clear once the reservoir had been emptied. No headwall was apparent at the upstream end of the tunnel; in its place was masonry blockwork that may have been part of the upstream bulkhead. The masonry was partly broken out in order to provide access to the mains and ease pipe connection works. The cast iron pipework was scraped, 'pigged' and surveyed, using closed circuit television (CCTV), which established the larger sized pipes could be accommodated.

A decision was taken to replace the original cast iron stack with a stainless steel arrangement, as installation of the new isolating valve, refurbishing the existing batter valves and reconnection to the existing pipe stack would have taken too long to resolve. The axis of the new system was offset from the original alignment. Three new draw-off points were installed with hydraulically actuated knife valves and protected by 'top hat' type roses. A gate valve was also installed in the upstream cross connection between the draw-off and scour. Tied flexible joints were provided in the system to accommodate differential settlement. The masonry broken out to install the works was recast in mass concrete.

The polyethylene pipework was delivered in 12 m lengths and fusion butt-welded at the upstream end and pulled through from the downstream end. The annulus between the liner and the cast iron was left un-grouted in order to permit relative movement between liner and host. Restraint had therefore to be provided in the system to cater for the axial thrust from the pressurised polyethylene pipes. This was effected with the use of a flanged pipe, split longitudinally into two for installation purposes, and used as struts to transfer the axial load back from the line valves to bulkheads at either end of each liner pipe.

The pipe system was subsequently tested to 6 bar pressure equating to 1.5 times the head from Hanging Lees reservoir immediately upstream (TWL 269.52m AOD).

To improve the structural integrity of the tunnel and arrest the potential for further movement the tunnel was to be infilled with lightweight foamed concrete. The density of the foamed concrete was 1.1 T/m$^3$ in order to minimise changes to the loading conditions around the tunnel. The Inspecting Engineer recommended that a length, 40m from the upstream bulkhead, should be infilled. Lack of space inside the tunnel, particularly at the brick bulkheads, severely restricted movement and gas detectors had been known to alarm with low oxygen readings. In order to eliminate the need for operational personnel to enter this potentially dangerous confined space it was decided that the concrete infill should be extended over the entire length of the tunnel. The concrete infill was placed following the successful installation and testing of the sliplined pipes.

Figure 7 Upper Draw Off

Figure 8 New Draw Off Pipework

Figure 9 Lower Draw Off and Cross Connection

Final Draw Off Arrangement

The foamed concrete provided structural support along the entire length of the tunnel. Full-face movement joints were installed at approximately 15 m centres to coincided with the lead run joints in the cast iron pipework. The movement joints were made up of laminated polyethylene board filler backed with a rigid capsulated drainage sheet. Each joint drained into a common collector drain running along the tunnel invert, which is to be subject to future monitoring.

## CONCLUSION

The project was successfully completed to time and budget. The reservoir has refilled and is now operational. The refurbished works are performing satisfactorily, the head loss in the system now measuring 1.53m for a discharge rate of 20Ml/day. To date there has been no change in the measured leakage or evidence of further movement. Useful experience was gained with respect to the control of reservoirs. Regular reviews, involving all parties, proved to be an effective way of progressing and completing difficult projects within tight delivery programmes.

## ACKNOWLEDGEMENTS

The authors are grateful to United Utilities plc, for giving their permission to publish this paper. The authors gratefully acknowledge the contributions made by colleagues, in particular, Keith Gardiner, Bill Aighton at United Utilities, Peter Isherwood of Askam Ltd, Jack Taylor of PMP Ltd, and, Alan Brookes of MWH PM.

# Langsett Reservoir: A combined analytical and CFD study of a reservoir side-spillway

D R S WOOLF, Arup, UK
J N HACKER, Arup, UK

SYNOPSIS. The three primary methods for assessing open channel flow in the field of hydraulic engineering are analytical calculations, physical models and, more recently, numerical simulation. This paper examines the relationship between the methods with particular focus on the added insight and benefit gained from a numerical simulation and how analytical calculations can be used to complement the numerical analysis. It also discusses some of the limitations associated with the various methods, such as those relating to surface roughness and scale.

## INTRODUCTION

Analytical (theoretical or semi-empirical) calculations, carried out by hand or through simple numerical methods (e.g. spread-sheet modelling), are the foundation of the work of the hydraulic engineer. Such calculations, however, often involve considerable uncertainty with regard to assumptions and empirical parameters. This is particularly evident in situations that have non-standard features or complex geometry. In such cases, therefore, it is generally necessary to carry out some form of model testing.

The traditional method used is to produce a scale model in the laboratory – so called physical modelling. This approach is widely used and has much to commend it. Principle benefits are the ability to clearly observe the behaviour of the flow in question, to carry out a large number of repeat runs, and to implement and assess the effectiveness of design modifications.

More recently, a third approach – Computational Fluid Dynamics (CFD) – has been used. CFD is a numerical technique which solves the full 3-D equations of fluid flow, in as much detail as is practical (typically, the primary approximations relate to the choice of turbulence sub-model and the constraints of numerical resolution). Use of a CFD model offers several benefits. There can be reduced timescales and costs involved in the model construction and solution, and it may be more cost-effective to test design possibilities using a CFD model than to commission physical testing (this approach, for example, is used widely in aircraft and automotive design). The CFD model can also be stored at minimal cost and re-run should it be needed in the future. However, a principle benefit of the CFD model is that it can provide data for flow quantities at all points within the flow domain.

*Reservoirs in a Changing World*, Thomas Telford, London, 2002, 247–261

This capability makes it far easier to obtain quantitative diagnostics regarding the behaviour of the system (e.g. Froude number and hydraulic head) than is possible from laboratory measurements. Such diagnostics can be of high-value, since they provide insights into how the system is behaving which can be used in the formulation of design modifications.

Although physical and CFD modelling approaches have individual strengths, both also have limitations. Physical models are in general unable to reproduce values of all the non-dimensional groups (e.g. Froude and Reynolds numbers) pertaining to the full-scale prototype. With a CFD model it is possible to model the flow fields at full-scale although there may be uncertainties in the accuracy of the turbulence model. There may also be inaccuracies due to other numerical effects, and it is necessary to carefully validate a particular CFD code against physical data when applying the code to a new situation. Another possible drawback of the CFD approach is that visualisation of the flow may be less immediate than when observed in three dimensions in the laboratory. However, it is possible to animate the computer-generated flow, and to do this for flow variables that are difficult to measure in the lab.

The purpose of this paper is to illustrate the application of the three techniques (analytical, physical and numerical) and to point to some of their strengths and weaknesses (these are also listed at the end of the paper). The case we have chosen to consider is a side-spillway channel at a major UK reservoir for which physical modelling testing had been previously carried out. More recently CFD modelling and an analytical analysis was completed.

## TEST CASE: LANGSETT RESERVOIR

Langsett Reservoir is a category A reservoir situated approximately 3km southwest of Penistone, South Yorkshire. The reservoir was formed by the construction of an embankment dam across the Little Don Valley. The crest of the dam is approximately 300m long by 8.7m wide and carries a public road. Figs. 1–2 show the existing spillway configuration which is of 'side-spillway' design. A weir of 61m length discharges flood water from the reservoir into a 'tumble bay' channel running parallel to the weir crest. The floor of the reservoir rises sharply to the weir crest and the downstream profile is stepped at a gradient of approximately 45°. The tumble bay has a shallow gradient (1 in 100), increases in width from 6m to 16m, and is traversed by a double-arched masonry bridge carrying the public road on the reservoir dam. Downstream of the bridge, the channel slopes more strongly and curves slightly before joining the byewash. The byewash has a steep, stepped gradient (1 in 9) and is 13.7m wide and 168m long. It terminates in a stilling basin at the toe of the dam (Fig. 1).

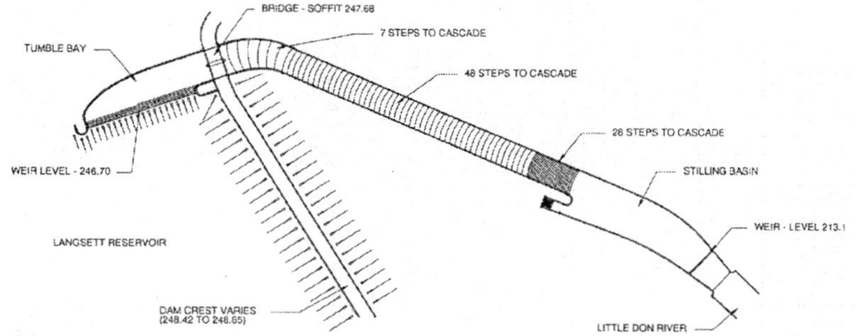


Fig 1. Schematic plan view of the Langsett reservoir spillway.

Fig 2. The tumble bay and byewash.

A feature of this type of spillway is the capacity for the channel to choke the flow from the weir resulting in decreased discharge through the spillway (Ellis 1989). This issue was the primary motivation for the work described in this paper. In response to increased flood requirements, modification works were proposed in 1999 and a 1/35$^{th}$ scale physical model study was carried out. A second study was later required to determine if the spillway performance could be increased further. Rather than using the physical model (which had been dismantled), a CFD model was commissioned for the investigation. Fig. 3 shows the physical model for the original configuration and one of the modified geometries considered in the CFD study. The CFD solution was obtained using the proprietary software, FLOW-3D (FLOW-3D 2000), with a RNG turbulence model. Further details are given in Woolf & Lavedrine (2001) and Woolf (2002).

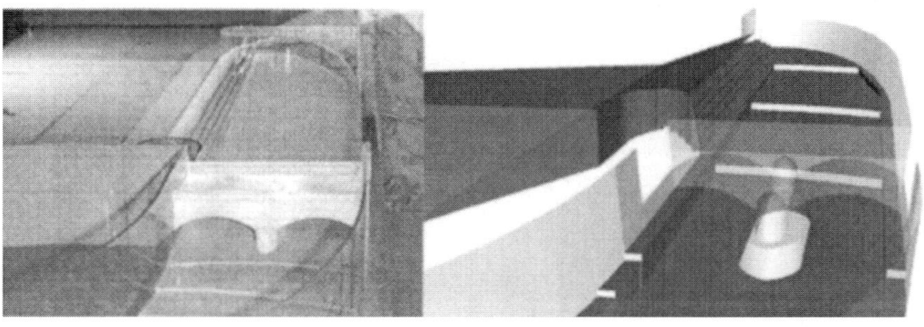

Fig 3. (a) The physical model (original design) (b) and numerical model (modified tumble bay).

The first stage of the study was to validate the numerical predictions against the previous physical model results. It was shown that, with the correct application of surface roughness and turbulence models, the CFD model predicted discharge flow rates that closely match those in the physical model. It was also shown that the predicted primary flow characteristics also closely matched physical observations.

The aim of the present paper is to make more detailed analysis of the CFD results and compare these and the physical model data with some more traditional analytical calculations. The paper focuses on two of the configurations tested in the CFD model. The first of these corresponds to the tumble bay geometry shown in Fig. 4 and the second to the same tumble bay geometry but with the bridge removed. This latter case is representative of the existing bridge replaced with one of single span construction and exerting zero resistance to the flow.

## ANALYTICAL CALCULATIONS

The upper section of the Langsett reservoir spillway from the weir to the top of the stepped byewash channel is shown in Fig. 4. The byewash channel is unlikely to exert any form of control on the discharge due to its steep gradient. However, there are a number of possible control points in the upper section of the spillway system, which are discussed below.

a. Weir. For critical flow conditions at the weir crest, the discharge of the weir is given by

$$Q = C_d L H^{3/2} \quad , \tag{1}$$

where $Q$ is the discharge volume flow rate, $C_d$ is the discharge coefficient, $L$ is the width of the weir and $H$ is the hydraulic head measured to the weir crest (e.g. Chow 1959).

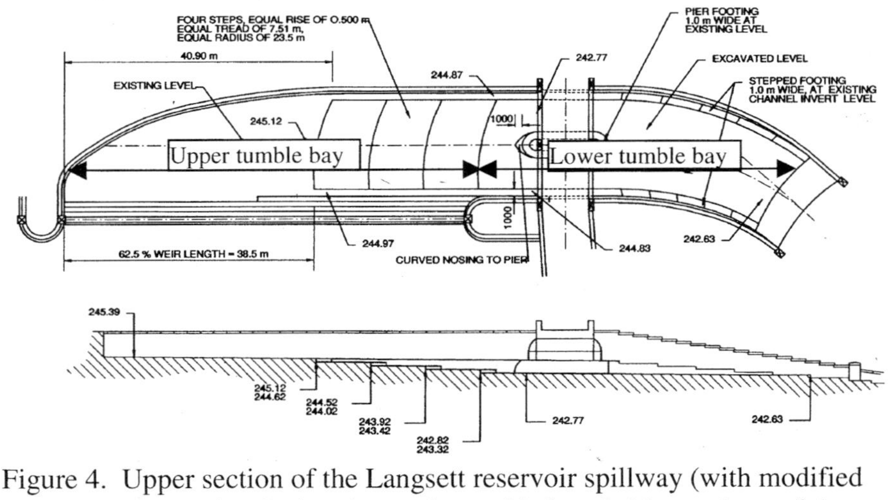


Figure 4. Upper section of the Langsett reservoir spillway (with modified tumble bay invert level) showing weir, tumble bay, bridge and top of the stepped byewash channel.

For the Langsett weir, $L$=61.0m and for the purpose of illustration take $C_d$=1.70, the theoretical value for a broad-crested weir. It is noted that the weir is of more triangular than broad-crested cross-section and so $C_d$ may be slightly larger in reality.

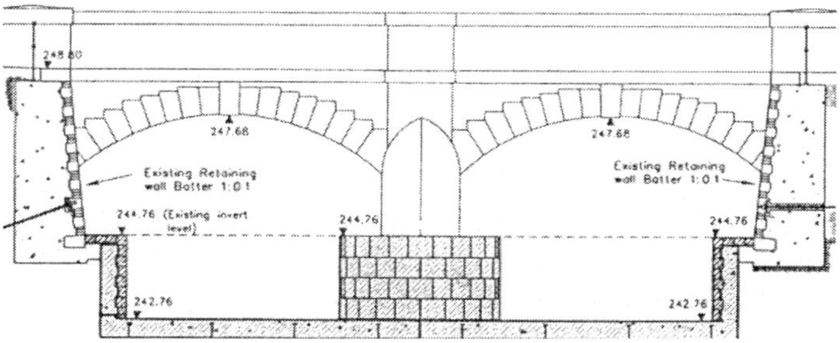


Fig. 5. Bridge elevation showing one modified configuration.

<u>*b. Bridge section.*</u> The bridge represents the narrowest section of the channel and is therefore a potential control point. An elevation of the bridge arches is shown in Fig. 5. The floor of the channel is 5.04m wide beneath each bridge arch and the side and pier footings are 1m wide. The invert levels below the bridge arches and at the top of the footings are respectively 242.76 and 244.76 mOD. The top of the bridge arches and weir crest are respectively at levels 246.68 and 246.70 mOD. For the modified configuration shown in Fig. 4, the invert level beneath the bridge is 3.94m below the weir crest ($z$ = –3.94m in equation (4) below).

Discharge formulae for channels of non-rectangular cross-section are given in standard texts on hydraulics (e.g. Chow 1959, Henderson 1966). The maximum possible discharge through the bridge arches, $Q_c$, which occurs when the flow is critical is given by

$$Q_c^2 = \frac{gA^3(h_c)}{T(h_c)}, \tag{2}$$

where $g$ is the acceleration due to gravity, $A$ is the cross-sectional area of the channel and $T$ is the top width. $A$ and $T$ are both functions of water depth $h$. The value of $h$ at critical conditions, $h_c$, is given by

$$h_c + \frac{A(h_c)}{2T(h_c)} = E, \tag{3}$$

where $E$ is the specific energy. The relationship between maximum discharge $Q_c$ and specific energy $E$ may be obtained by evaluating values of these parameters for each $h_c$ from equations (2) and (3). The final stage is to relate the specific energy at the flow section, $E$, to the hydraulic head in the reservoir, $H$. These quantities are related by

$$E = H - z - \Delta H, \tag{4}$$

where $z$ is the level of the bed (invert) at the bridge relative to datum and $\Delta H$ is the loss of total-energy that has occurred between the reservoir and the bridge. At this stage $\Delta H$ is an unknown parameter which will be determined from the model data.

c. Lower tumble bay. Fig. 4 indicates the line separating the upper to lower tumble bay regions. The lower tumble bay runs from this point to the top of the byewash steps. This region forms a contraction in flow area and so is a possible choke to the flow in the case with the bridge removed. In fact, two potential control sections are identified. The first is the entry point to this region from the upper tumble bay, the lower tumble bay 'channel entry section'. There is a possible control section here because of a combination of reduced channel flow area and 0.5m step. The discharge curve for this section can be calculated as in (*b*) above with appropriate expressions for $A(h)$ and $T(h)$. The invert level at the top of the step is at $z = -3.4$m relative to the weir crest. The second possible control point is at the downstream end of the lower tumble bay where the channel narrows towards the top of the byewash. Although the channel is 2m narrower here (14m wide), the invert level is 0.7m lower than at the lower tumble bay channel entry section, i.e. a 0.2m drop in the slope, and so this location is unlikely to form a primary control unless significant energy dissipation occurs in the lower tumble bay.

d. Upper tumble bay ('side-channel control'). The fourth possibility is that side-channel control occurs within the upper tumble bay. This form of control is due to the increasing discharge of the flow, in the direction parallel to the weir crest, through the side-spillway channel (Chow 1959, Henderson 1966). If the effect of friction on the wall and floor of the channel is neglected, and if it is supposed that (*i*) the direction of inflow into the side channel is perpendicular to the weir, (*ii*) that the rate of inflow per unit width of weir is constant, and (*iii*) that the channel floor is smooth (i.e. no abrupt changes in invert level), then the classical expression (based on the momentum equation rather than energy conservation) for the location of the control section from the upstream end of the tumble bay, $x_c$, is:

$$x_c = \frac{2}{S_0}\left(\frac{Q^2}{gT^2}\right)^{1/3}, \qquad (5)$$

where $S_0$ is the gradient of the channel floor and $Q$ the flow rate through that section. If the water surface depth $h$ is decreasing, the flow at position $x$ from the upstream end of the tumble bay will be subcritical if $x_c > x$ and supercritical if $x_c < x$.

The upper tumble bay is now divided into two sections, an upper section with $S_0 \approx 0.01$ and a lower stepped section (assuming a continuous slope) with $S_0 = 0.08$. The upper section ends at $x = 38.5$m and the lower section at the downstream end of the weir ($x = 61.0$m). Except at very low flow rates ($Q < 4.2\text{m}^3/\text{s}$), equation (5) indicates that the flow will be subcritical in the upper section. For the lower, more steeply sloped section, equation (5) indicates the following flow regimes:

$$
\begin{aligned}
&Q < 95.8\ \text{m}^3/\text{s} && \text{supercritical flow} \\
&95.8\ \text{m}^3/\text{s} < Q < 191.0\ \text{m}^3/\text{s} && \text{transition occurs} \\
&Q > 191.0\ \text{m}^3/\text{s} && \text{subcritical flow}
\end{aligned} \qquad (6)
$$

The occurrence of a subcritical to supercritical transition for the central range of low rates in equation (6) indicates that there is a local control point on the lower slope, possibly (although not necessarily) leading to submergence of the weir upstream of this point.

It would be possible to calculate a full discharge curve for the side channel control. However, to do this requires a calculation of the discharge over the weir for the region $x > x_c$ (i.e. into the part of the channel downstream of the control). This requires each value of $x_c$ to be related to a value of hydraulic head, $H$, at the weir. It is implicit in equation (5) that energy is not conserved in the side-channel. To include this, it would be necessary to calculate the backwater curve through the side-channel with an appropriate expression for the rate of energy dissipation, such as that derived from the Manning's $n$ formula. This calculation is not included here but we note, from equation (6), that side weir control is not predicted by equation (5) to occur for $Q > 191.0$ m$^3$/s.

## COMPARISON OF ANALYTICAL CALCULATIONS WITH PHYSICAL MODEL AND CFD RESULTS

Fig. 6 shows the results from the physical model with the bridge in place (there was no physical model for the modified case without the bridge) and the CFD runs (with and without bridge).

Analytical discharge curves, calculated in the manner described above, are also shown for the weir, bridge and lower tumble bay (LTB) entry control sections. For the latter two sections, discharge curves are calculated assuming no head loss from the value in the reservoir ($\Delta H = 0$) and with constant values of head loss of 1.05 and 1.00m, respectively (as indicated on Fig. 6). Each set of analytical curves has a different value of invert level for each location, i.e. $z$ in equation (4). There are also different gradients due to the different forms of $A(h)$ and $T(h)$ in equations (2) and (3). For each value of head at the weir, $H$, the discharge of the spillway is read from the graph by the curve giving the smallest flow rate, $Q$.

The physical model points approximately follow the broad crested weir discharge curve for lower flow rates indicating that the weir is free flowing in the given range. At around 150 m$^3$/s, however, there is an increase in the gradient of the physical model points, indicating that the weir is discharging freely. At around 190 $m^3/s$ there is a second increase in the slope of the points indicating that another change of control location has occurred.

From the arguments in (*d*) above, it seems reasonable to suppose that the first change of control is due to the onset of side-channel control in the tumble bay. The second change of slope occurs at around 190 $m^3/s$ corresponding to the upper limit in (6). At this point, it is expected that the control has shifted to the bridge section (if present) or the lower tumble bay entry section. The discharge curves for these sections with no energy loss lie some distance from the physical model and CFD results. However, when a head loss $\Delta H = 1.05$m is assigned to the bridge control curve, it closely follows the physical model and CFD results at the higher flow rates (see Fig. 6). This suggests that the magnitude of the head loss is almost constant over this flow range. Similarly, when a head loss $\Delta H = 1.00$m is assigned to the lower tumble bay entry control curve, it passes through the single CFD point without the bridge.

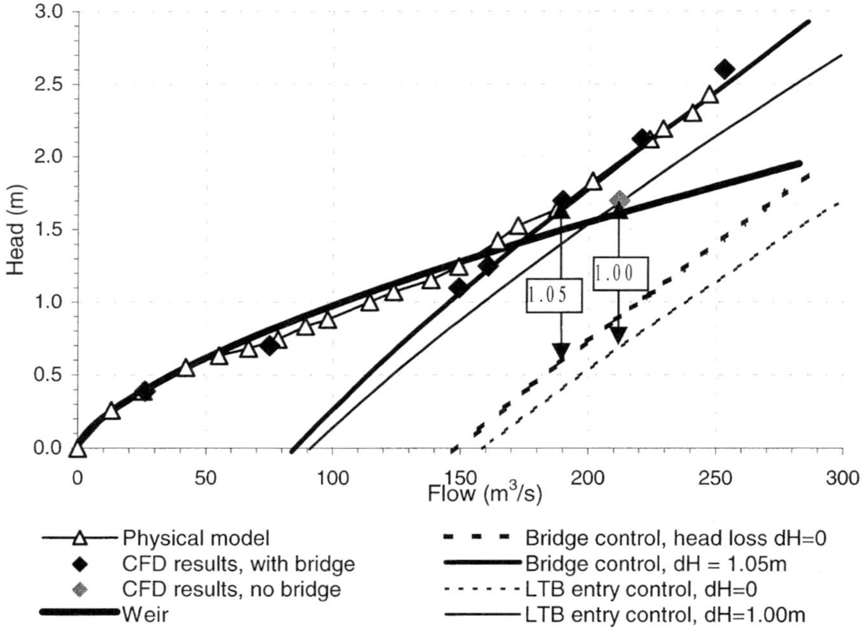


Fig. 6. Physical model measurements and CFD predictions and theoretical discharge curves for bridge and lower tumble bay entry control.

In summary, by comparing analytical curves with physical model measurements and CFD predictions we can draw the following conclusions.

1. For the chosen spillway geometry with bridge in place, the primary flow control moves to the bridge section (with the outflow over the weir being choked) for $Q \approx 190$ $m^3/s$. In this case there is a hydraulic head loss

between the reservoir and bridge section of about $\Delta H$ = 1.05m as inferred from the difference between analytical calculations and model data.

2. When the bridge is removed, the primary control moves to the lower tumble bay entry section. In this case, a hydraulic head loss, $\Delta H$ , between the reservoir and bridge section of approximately 1.0m is inferred.

In order to further understand the validity of these conclusions, it is necessary to analyse the CFD results in greater detail.

FLOW DIAGNOSTICS OBTAINED FROM THE CFD SIMULATION

Two important flow diagnostics are readily obtained from the CFD results. The first of these is the Froude Number

$$Fr = \frac{\bar{u}}{\sqrt{gh}}$$

where $\bar{u}$ is the rms depth-averaged flow velocity. Critical flow occurs at those sections for which $Fr$ = 1.0 across the width of the channel.

The second diagnostic is the total hydraulic head

$$H = \frac{\bar{u}^2}{2g} + h + z \ .$$

For zero energy loss in the flow $H=H_r$, the value in the reservoir. If energy losses do occur, as we expect, then

$$\Delta H = H_r - H \ .$$

Analysis of the results

A number of spillway geometries and discharge flow rates were examined in the CFD study but, as discussed above, only two are presented here. These correspond to the tumble bay geometry shown in Fig. 4 with and without the bridge in place.

Case study without bridge (Figs 7-10). When the bridge is not included in the model, the predicted discharge is about 212m$^3$/s. The plots below show the predicted water depth and derived RMS velocity, Froude number and hydraulic head for the upper section of the spillway between the weir crest and the byewash. The areas where the Froude number is between 0.9 to 1.1 (taken from Fig. 9) are highlighted on each of the plots by two contours. We expect locally a change in the flow regime (e.g. subcritical to supercritical) to occur between these contours.

As the water passes over the weir, the flow regime is subcritical on the left hand side (LHS) where it is flooded but on the RHS the water depth is less than 1.0m with supercritical flow (Fig. 9). This is a common characteristic of a side-spillway of this configuration. The primary control is at the entry point to the lower tumble bay where the Froude number is between 0.9 and 1.1 across the full width of the channel. It is observed, however, that the control line doubles back in a 'V' shape to the free part of the weir crest. This suggests that there is some circumventing of the control section on the RHS close to the wall, although this region will have relatively small volumes of water passing through it. This case illustrates partial control by the side-channel and lower tumble bay entry section.

The second main feature is on the bend on the LHS further downstream where a hydraulic jump was observed. This does not cross the full width of the channel and is limited to the region of superelevation on the outside of the bend, where there is a reduction in velocity and increase in depth.

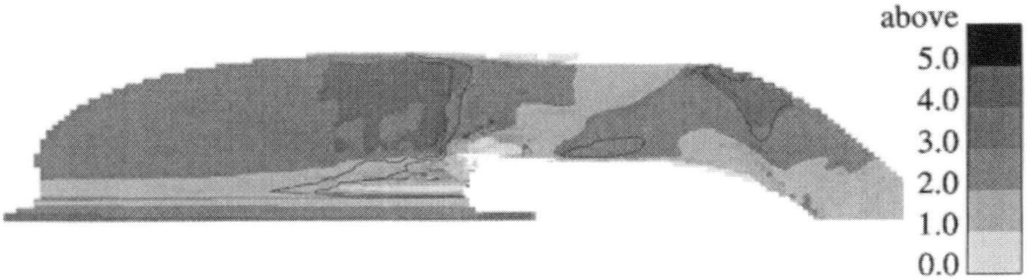


Fig. 7: Water depth for case without bridge (in m)

Fig. 8: RMS velocity for case without bridge (in m/s)

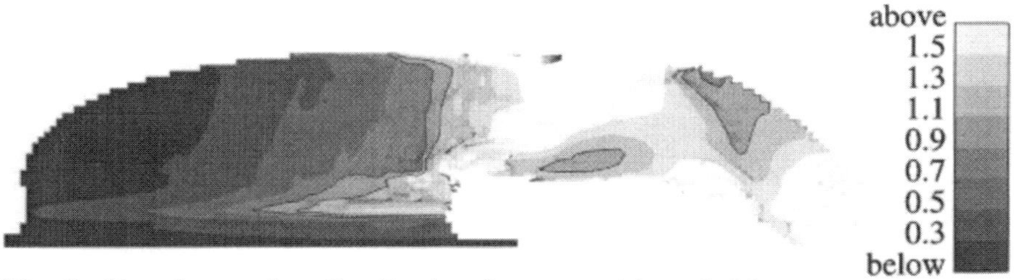


Fig. 9: Froude number distribution for case without bridge

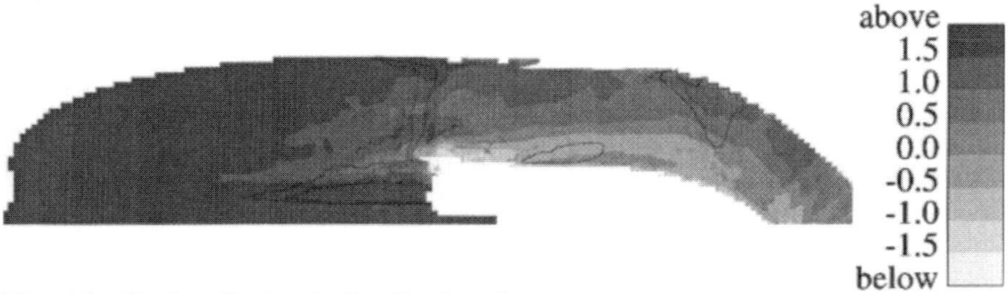


Fig. 10: Hydraulic head distribution for case without bridge (in m)

The hydraulic head distribution can identify the regions of greatest energy loss by the proximity of the shown contour lines. Large losses are shown on the RHS of the tumble bay close to the weir in the previously mentioned supercritical region. This is also where there are highly turbulent regions (turbulent kinetic energy and turbulent kinetic energy dissipation rate plots are not shown here but support this observation). Although it is expected that hydraulic head would reduce down the channel, i.e. the loss in potential energy is not matched by kinetic energy gains because of energy losses in the system, the losses are distorted due to the bends in the channel. It should be noted that cross-channel energy losses can be identified here.

Values of $H$ near the control section are around 1.2m (an approximate mean value) corresponding to a head loss $\Delta H$ of about 0.5m from the weir crest. This value of $\Delta H$ is only about 50% of the value of 1.0m inferred from the lower tumble bay entry discharge curve shown in Fig. 6. This suggests that that the conclusion of primary control at this point is not correct (a point reinforced by Fig. 9) and that the theoretical arguments leading to this conclusion, discussed above, are not applicable in this case. There may be several reasons for this, for example relating to the assumptions on which equation (5) is based. Therefore, it would appear that a form of side-channel control is continuing to operate in this case.

Case study with bridge (Figs 11-14). When the bridge is included in the solution, the predicted discharge reduces to about 190m$^3$/s. The reduced area of the bridge arches causes water to back up in the tumble bay resulting in increased water depth and flow splitting by the bridge pier.

A similar type of Froude number and flow regime as in the case without a bridge is shown close to the RHS of the weir. Apart from this location, the flow first becomes supercritical across the whole span of the channel in the second half of each bridge arch. This indicates that the primary control is at this location. The flow then quickly becomes subcritical again through a weak hydraulic jump. At the top of the byewash, the flow again becomes supercritical as the water accelerates down the slope.

The hydraulic head distribution (Fig. 14) shows less distortion due to the bends. The plot again shows large energy loss close to the RHS of the weir but also losses in the wake of the bridge pier. These losses are also shown by turbulent energy plots showing increased levels of turbulence in each location.

Values of $H$ near the bridge control section are around 0.8m in the LHS arch and 0.6m in the RHS arch which corresponds closely with the value of 0.65m (loss of 1.05m from the reservoir) inferred from the discharge curve shown in Fig. 6.

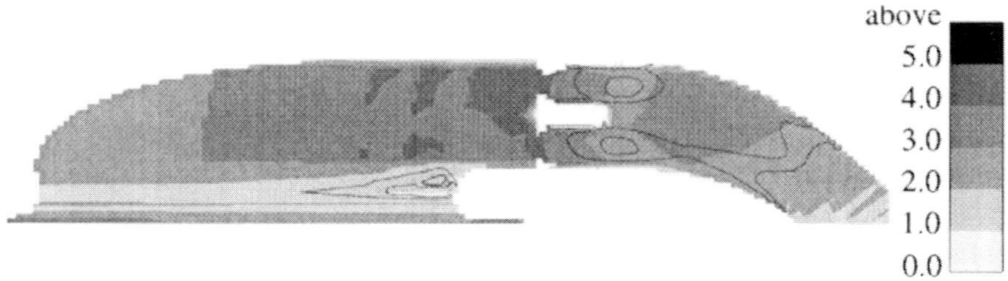


Fig. 11: Water depth for case with bridge (in m)

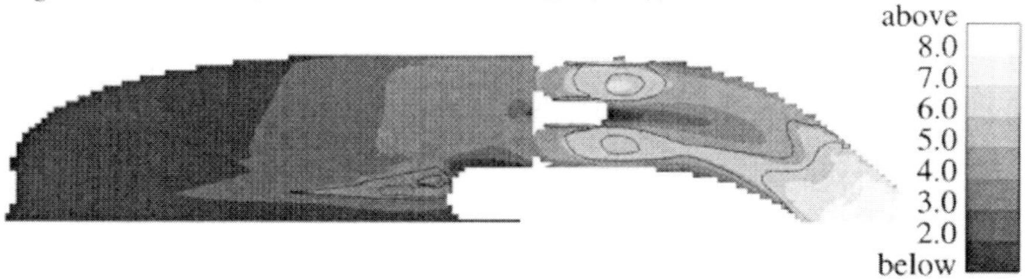


Fig. 12: RMS velocity for case with bridge (in m/s)

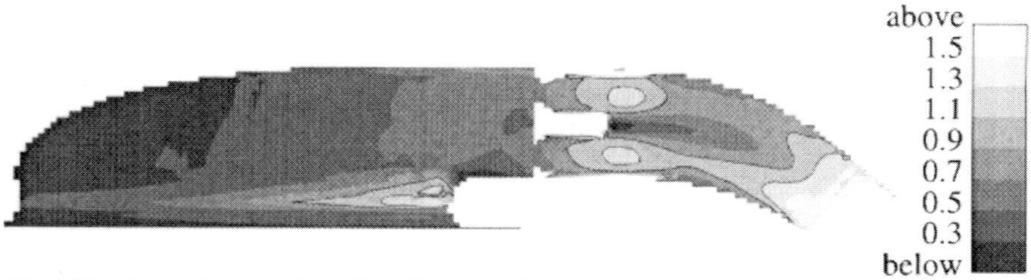


Fig. 13: Froude number distribution for case with bridge

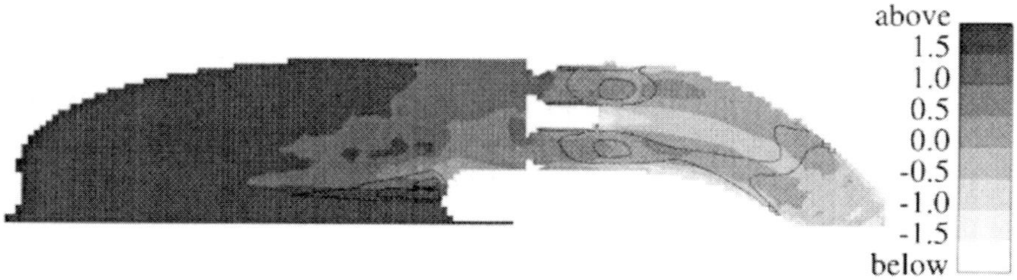


Fig. 14: Hydraulic head distribution for case with bridge (in m)

## CONCLUSIONS

We have shown how theoretical considerations can be used together with CFD modelling to gain enhanced understanding of the flow regimes, features and control points. In particular, it was shown that there is vigorous energy dissipation and associated head loss in the tumble bay. This insight into the spillway dynamics has potentially significant implications with regard to design measures taken in order to optimise hydraulic performance.

In addition, the agreement between the CFD predictions and inferences from the analytic calculations in respect to control point locations and head loss can provide additional confidence in the techniques beyond simple comparison of flow rates. It should be noted, though, that assumptions behind the analytical calculations may lead to inaccurate conclusions as to the location of the primary control point.

The value of physical modelling in establishing discharge curves and providing a benchmark for the CFD results has also been demonstrated. In the validation of the CFD model against the physical model, one of the primary uncertainties was the choice of surface roughness length $k$ to be used in the CFD model. A number of $k$ values were applied to understand the sensitivities of the system to surface roughness and represent the physical model Perspex at full scale once converted. A value of $k$=21.0mm was used in the runs shown in Fig. 6. which lies at the upper end of the range of 3.8mm to 21.0mm given by Ellis (1989) for dressed stone in mortar (the material of the Langsett spillway). Over this roughness length range, the CFD predicted discharges varied by less than 5%.

It has been shown in the cases described above that, where a high level of optimisation is required in the design of a hydraulic structure, that a detailed CFD analysis can provide an attractive methodology to test and develop different design options. In some applications, a physical model may still be required to provide additional confidence in CFD predictions. However, it has been demonstrated that CFD can provide an effective method for model testing, both in terms of time and quality of information provided.

Some of the strengths and weaknesses of the three modelling techniques, numerical simulations, analytical calculations and physical models have also been discussed. These strengths and weaknesses are summarised below.

## STRENGTHS AND WEAKNESSES OF THE MODELLING TECHNIQUES

### Numerical simulations

*Strengths*

- Good accuracy can be obtained using modern CFD codes without need for case-by-case calibration against physical data.
- Models can be constructed and run with a fast turnaround time (depending on hardware resources available) and stored for future use at minimal cost.
- Geometry can be easily modified to test design options and for optimisation of hydraulic performance.
- Roughness lengths of boundaries may be easily varied, to represent different types of materials and to provide sensitivity analyses.
- Full flow diagnostics can be obtained from the simulation and high-quality visualisations and animations of variables produced.
- Transient analyses are easy to carry out.
- A greater understanding of the flow mechanisms may lead to a smaller number of design iterations.

*Weaknesses*

- Exact level of numerical accuracy is not known until benchmarked against physical data.
- Restriction of model size and accuracy due to computer limitations.
- Air entrainment is not simulated.
- Qualitative visualisation from CFD cannot at present reproduce immediacy of observations of laboratory flow.
- Lack of trust and understanding from practitioners and the industry alike.

### Analytical calculations

*Strengths*

- Give insights into hydraulics of flow in order to inform initial design decisions.
- Provide initial estimates of design performance before model testing is carried out.
- Can be used to infer hydraulic parameters, as in the study above.

*Weaknesses*

- Can be time-consuming for non-standard situations and where detailed information is required.
- Only approximate solutions can be obtained for complex geometries and 3D flow.
- For complex flows, too many approximations may give the wrong results.

Physical models

*Strengths*

- Once the model has been set up, steady-state discharge curves can be easily obtained through repeat testing.
- Real fluid is used, so it is not subject to uncertainties regarding numerical errors and sub-grid scale turbulence models.
- Observation of 3D flow in laboratory makes it easy for technical and non-technical witnesses to appreciate flow behaviour and design implications.

*Weakness*

- It is not generally possible to obtain full dynamic similarity, e.g. equivalent Froude and Reynolds numbers.
- Geometry modification can be expensive to implement.
- Detailed flow measurements using non-obtrusive methods can be difficult and expensive to obtain.
- There are uncertainties regarding correspondence between hydraulic roughness of materials at model and full scale.
- Air entrainment is not well reproduced due to scale effects.
- It can be expensive to carry out and obtain data for transient studies.
- Physical models cannot be maintained for a long time. There are problems associated with available space and cost.

## REFERENCES

Chow V T (1959). Open Channel Hydraulics, McGraw-Hill.

Ellis J (1989). Guide to the analysis of open-channel spillway flow, CIRIA Technical Note 134, CIRIA, London.

Henderson F M (1966). Open Channel Flow, Macmillan.

FLOW-3D User Manual Version 7.7 (2000). Flow Science Inc., Los Alamos, NM, USA.

Woolf D R S (2002). Numerical simulation of hydraulic structures, Dams and Reservoirs, February edition.

Woolf D R S & Lavedrine I (2001). Application of CFD modelling to hydraulic structures, Sixth International Conference on Computing and Control for the Water Industry, CCWI 2001, De Montfort University, Leicester.

# Langsett Reservoir: Numerical simulation of hydraulic structures

D R S WOOLF, Arup, UK
I SCHOLEFIELD, Yorkshire Water Services Ltd, UK

SYNOPSIS. As the design flood flow at Langsett Reservoir was increased following the launch of the Flood Estimation Handbook (FEH) in March 2000, a computational fluid dynamics (CFD) model was used to investigate options to increase the spillway capacity.

The study:
- provided confidence in the numerical (CFD) technique by comparing the numerical predictions to observations and measurements taken from a physical model,
- determined the effect of modifications to the spillway on its hydraulic performance,
- quantified the effect of surface roughness and turbulence levels on the flow,
- and assessed the spillway performance under possible future larger flood events.

## INTRODUCTION

Most hydrodynamic models used to model open channels of simple or compound cross-section are one-dimensional. In these models, the flow is assumed to be fully mixed by turbulence over any cross-section so that the effect of turbulence is parameterised solely through the use of wall friction coefficients, such as Manning's n or Chezy's C. This generally represents the turbulence effects adequately when the channel cross-section is simple or if no structures are involved but has a number of practical limitations.

To overcome some of these limitations, two-dimensional depth-averaged models are used which again have some limitations as they do not take into account the vertical variation of flow quantities. Both these techniques are unable to accurately predict the flow around structures, such as weirs or sluice gates as the flows are fully three-dimensional and therefore needing three-dimensional representation either using a physical or numerical model.

In 1999, a 1/35$^{th}$ scale physical model was used to investigate potential improvements to the performance of the existing spillway at Langsett Reservoir such that the flood rise due to the PMF was limited to the level of the dam crest. The magnitude of the design flood was based on the Flood

*Reservoirs in a Changing World*, Thomas Telford, London, 2002, 262–273

Studies Report (FSR) but the effect of the increase in the flood flow from the FEH meant that, even with the proposed spillway modifications, the flood rise would exceed the level of the dam crest by about 200mm. Consequently, the client (Yorkshire Water Services Ltd) commissioned a further investigation into whether the spillway performance could be increased even more than that detailed from the physical model. The physical model had been dismantled, so the decision was made to adopt an innovative approach by using a CFD model which could, at first, be calibrated and validated against the previous physical model study results and then extended to further understand and potentially improve the performance of the spillway.

## BACKGROUND TO THE STUDY

Langsett Reservoir is a Category A reservoir completed in 1905 under powers granted in the Sheffield Corporation Water Act of 1896. The dam is situated approximately 3km south-west of Penistone, South Yorkshire. The reservoir was formed by the construction of an embankment dam, with a central puddle clay core with a maximum height of 32m, across the Little Don Valley. The crest of the dam is approximately 300m long by 8.7m wide and carries a public road. The existing spillway at Langsett Reservoir comprises a 61m long broad crested masonry side weir, the downstream profile of which is stepped. Water flowing over the weir discharges into a tumbling bay that increases in width from 6m to 16m along the length of the weir. A masonry bridge carrying the public road over the spillway is located at the downstream end of the bay. Once through the road bridge, the flood waters pass down a stepped byewash channel 13.7m wide and 168m long to a stilling pool at the toe of the dam. For most of its length the channel has a slope of about 1 in 9 (see Fig. 1).

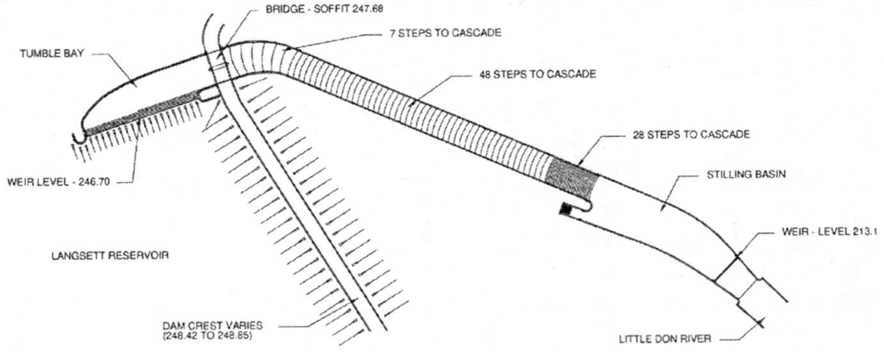


Fig. 1. Schematic plan view of the Langsett reservoir spillway

The recommended option from the physical model study was chosen to form the initial geometry in the numerical model, named Configuration 1, in which four "steps" are created in the tumbling bay before the bridge (see Fig. 2). Configuration 2 included six steps, and Configuration 3 three large steps. The greatest cut volume is in Configuration 3 and the least in Configuration 1. The last geometric variable tested was the removal of the bridge. The geometry between the bridge to the start of the furthest downstream step in all configurations, about 10m upstream of the bridge, is the same.

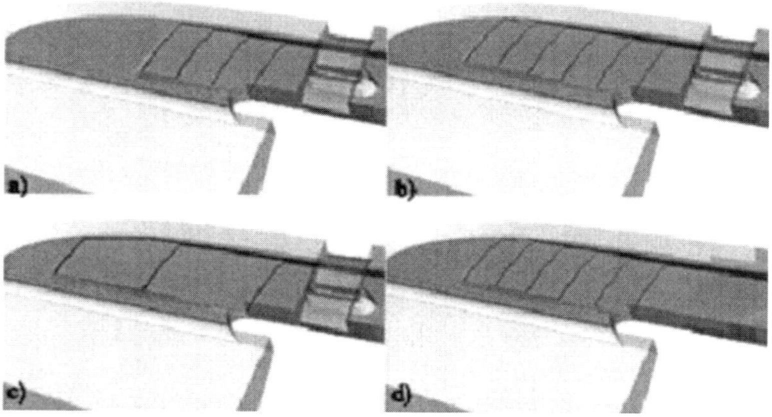


Fig. 2: Tumbling bay geometry showing a) Configuration 1, b) Configuration 2, c) Configuration 3, d) Configuration 2 with bridge removed

The physical model was constructed out of perspex. In order to calibrate the CFD model against the physical one, a uniform surface roughness height was applied to all surfaces. This roughness height was representative of perspex but at full-scale – the CFD is carried out at full-scale. The CFD model enabled a cost-effective sensitivity study to be carried out on the effect of roughness on the flow regime and ultimately on the discharge by changing just one value in the set-up of the model boundary conditions. Comparisons were made with the physical model after scaling results using the appropriate conversion factors.

The software used for the analysis is called FLOW-3D (FLOW-3D 2000). Flow-3D is a general purpose CFD program for simulating fluid flows including those with a free surface. The flow region is subdivided into a cartesian mesh and for each subdivision or "cell", values are retained for the basic flow quantities (e.g. velocity, pressure and density). Finite-volume approximations of the equations of motion are then used to compute the spatial and temporal evolution of these values. An advanced Volume-Of-Fluid (VOF) method is used to track the free surface even in the case of sharp deformation (Hirt & Nichols 1981).

Equilibrium is reached when the volume of water entering the flow domain equals the volume of water leaving it. In fact at no time is the flow in true equilibrium due to the oscillatory nature of the flow regime. To overcome this, in calculating the predicted discharge, each solution was run for 115 seconds with the flow rates averaged over the last 15 seconds of the solution for the calculation of discharge.

Sixteen cases were run which covered the effect of geometric variations, the effect of turbulence and surface roughness, and the effect of change in overflow head. These results were then compared with the physical model.

## RESULTS AND DISCUSSION

Agreement between the physical model and CFD model was good. On a qualitative level, characteristic features picked up by the CFD included (refer to Fig. 3):

- the outward movement of the flow in the tumbling bay,
- the water level as the flow passes through the bridge arches,
- the draw down through the arches,
- the wake zone behind the bridge pier,
- the skewed wave front downstream of the bridge,
- the super-elevation around the bend of the byewash channel,
- and the draw down as the flow accelerates down the byewash.

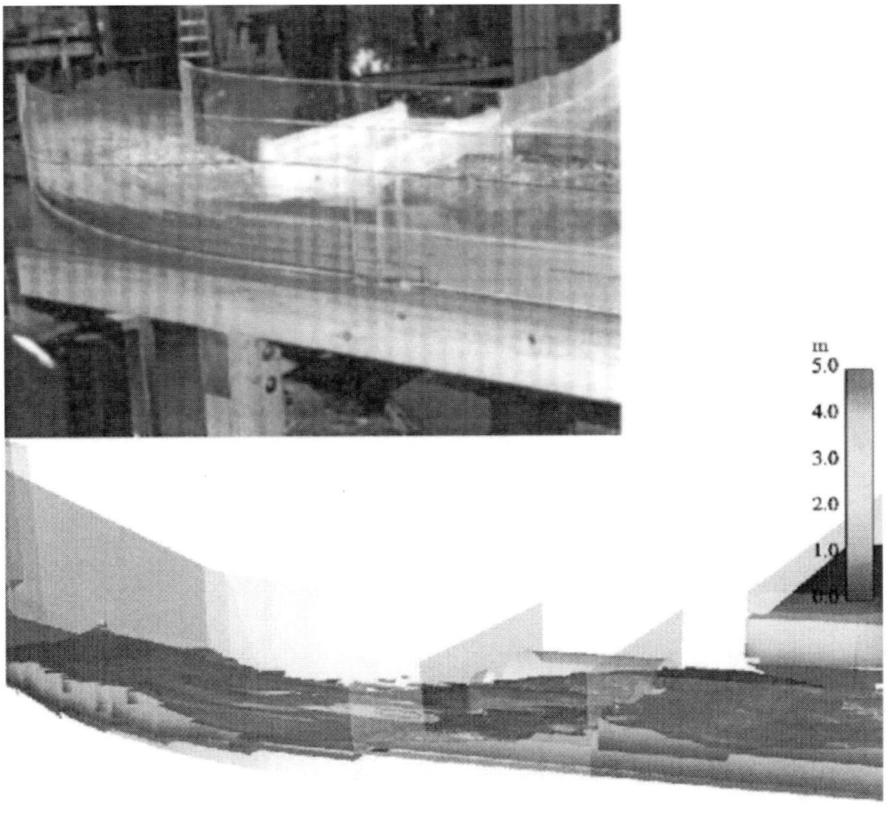


Fig. 3. Comparison between the physical model (upper image) and CFD model (lower image) at a similar discharge

The effect of geometric variations
When comparing Configuration 1 to Configuration 2 (or Case 1 to Case 2 - for case definitions see right-hand indexes in Fig. 5), the discharge increased by 12m³/s to 219m³/s for the same head resulting in an increase in the water level in the tumbling bay of about 0.5m. When compared to Configuration 3 (Case 3), with the greatest cut volume, a further increase in water depth of about 0.5m was predicted in the tumbling bay, but this time there was only a slight increase in depth close to the bridge and no increase in discharge. In fact, a slight reduction to $217m^3/s$ was predicted, indicating the limited benefits of the volume cuts.

When the bridge is removed for Configuration 1 (Case 4), other features besides those relating to the change in surface profile due to the bridge arches become evident. The water depth is generally more even with fluctuations primarily generated by the increased flow on the outer edge, the bend downstream and waves at the surface. By removing the bridge the discharge only increased by $5m^3/s$ to $217m^3/s$ for this configuration, i.e. this was considered not a very significant improvement to spillway performance.

Increased understanding of the flow, for example its control points, can be gained from values derived from the data set. Froude Number distribution was plotted on the free surface (see Fig. 4). This enabled the control points to be highlighted when changing from sub- to super-critical flow (less than 1.0 to greater than 1.0). In addition, the location of the hydraulic jump is highlighted when the flow changes back to sub-critical (shown on the LHS of the channel after the bridge in the middle of the bend). This feature is also shown in the physical model and by the free surface profile. Additional plots of hydraulic head enabled a better understanding of the regions of greatest energy loss.

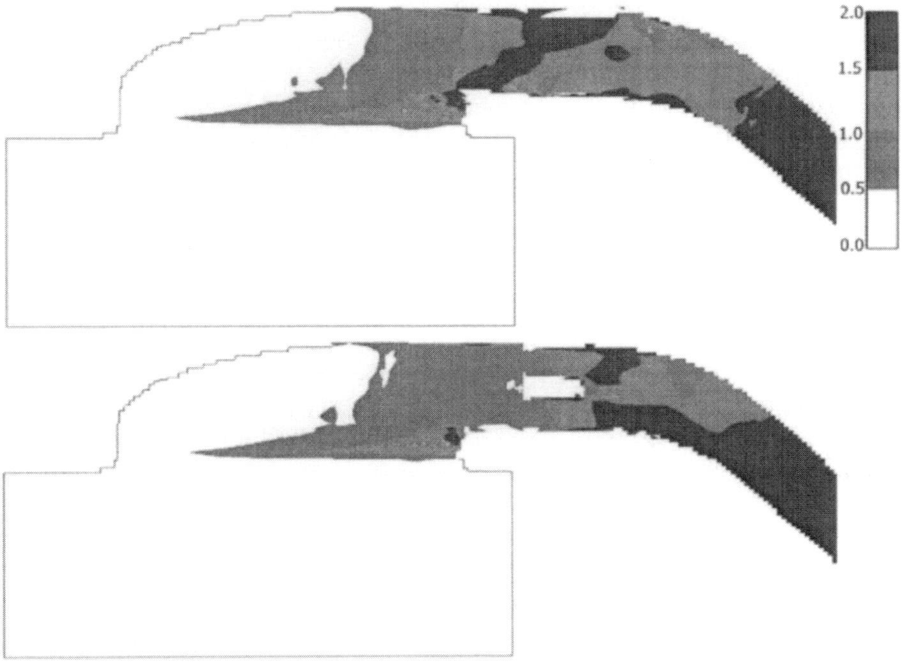


Fig. 4. Froude Number distribution without the bridge (upper image – Case 14) and with the bridge (lower image – Case 2)

<u>The effect of turbulence and surface roughness</u>

In order to improve the hydraulic performance of a spillway it is important to understand the type and nature of the flow at all locations within the channel as well as the likely discharge. The CFD data, having applied a turbulence model, contains the turbulent energy levels and the rate of its dissipation throughout the flow domain. This provides clues to the interaction of the flow with the surfaces and also within the fluid itself.

When the roughness height was increased from 0.0 to 3.2mm (Case 5), the discharge reduces by $3m^3/s$ to $204m^3/s$. A further increase in the roughness height to 21.0mm (Case 6) resulted in a further reduction in the discharge to $190m^3/s$ (a reduction of over 8% from the smooth wall case). The effect of surface roughness was shown in detail by the turbulent energy dissipation rate plotted on the walls of the channel.

Not only did the surface roughness model change the flow rate and the turbulent flow regime but it also had the effect of bringing the downstream primary wave front closer to the bridge. The surface roughness model can therefore have a calculable effect on predicted discharges as well as the general turbulent flow field and hydraulic features within the domain.

The effect of change in overflow head

The head above the weir crest was increased from 1.70 to 2.12 and 2.60m as a check for if the design flood criteria were to increase in the future. With a head of 1.70m (Case 6), the discharge was predicted to be 190m$^3$/s but at 2.12m (Case 7) the discharge increased to 221m$^3$/s. In Case 7, the arches became just about fully submerged but downstream of the bridge a similar surface profile to Case 6 was shown with only a slight increase in water depth around the bend. For a head of 2.60m (Case 8), the flow increased to 253m$^3$/s resulting in the arches becoming fully submerged.

COMPARISON WITH THE PHYSICAL MODEL

In Fig. 5, the predicted flow rates are shown graphically together with the values obtained from the physical model. The rating curve for the flow over a simple broad-crested weir with a free discharge was used as a benchmark against which the relative performance of the various configurations can be judged.

By increasing the roughness height in the model and therefore reducing the discharge, the CFD predictions were able to match more closely the measurements. In fact, with a roughness height of 21.0mm, the results were very similar. The best predicted performance for the spillway was with the bridge removed but this benefit in increased flow rates was only marginal. The cases with the bridge removed were solved with a roughness height of 0.0mm, so the true performance could be about 10% less than the quoted values if it displayed the same flow reduction due to surface roughness as the cases “with bridge”.

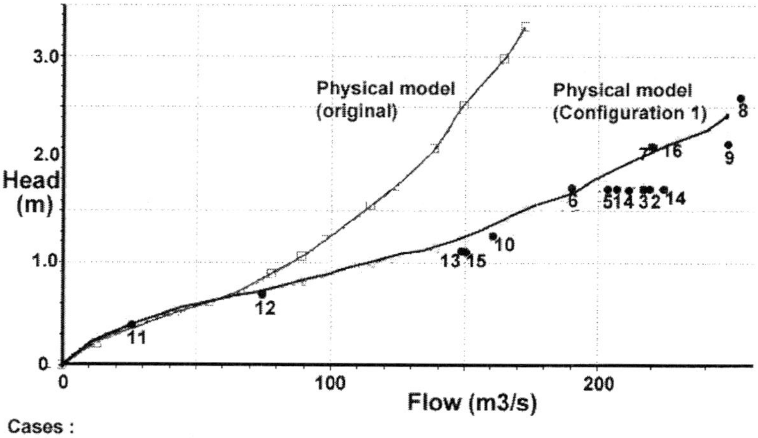


Cases :
(C = configuration, RH = roughness height, BR = bridge removed)
C1, RH = 0.0mm ; 1,9,10,11
C1, RH = 3.2mm ; 5
C1, RH = 21.0mm ; 6,7,8
C1, RH = 0.0mm, BR ; 4
C2, RH = 0.0mm ; 2,12,13
C2, RH = 21.0mm ; 16
C2, RH = 0.0mm, BR ; 14
C3, RH = 0.0mm ; 15

Fig. 5. Graph comparing CFD predictions and physical model measurements.

## OTHER FEATURES

The full three-dimensional data set can be interrogated to provide additional understanding of the various flow features. For example, clipping planes can be used to understand the flow distribution in vertical cross-sections along the flow direction or particle traces can be used to understand the water movement across the width of the channel (see Fig. 6). Sequences of images can then be easily combined into an animation that can provide even greater understanding of the water movement and flow features as well as being used for planning and public relations purposes (see Fig. 7).

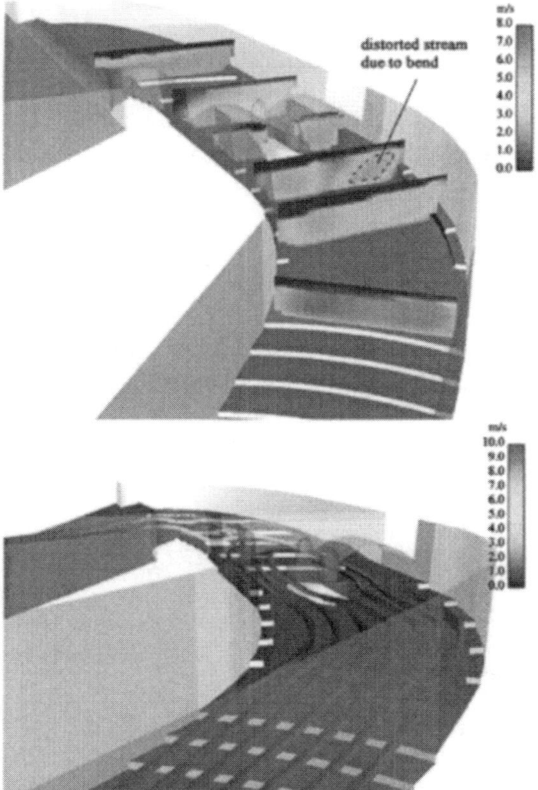


Fig. 6. Clipping planes and particle traces used for interrogation of the data set

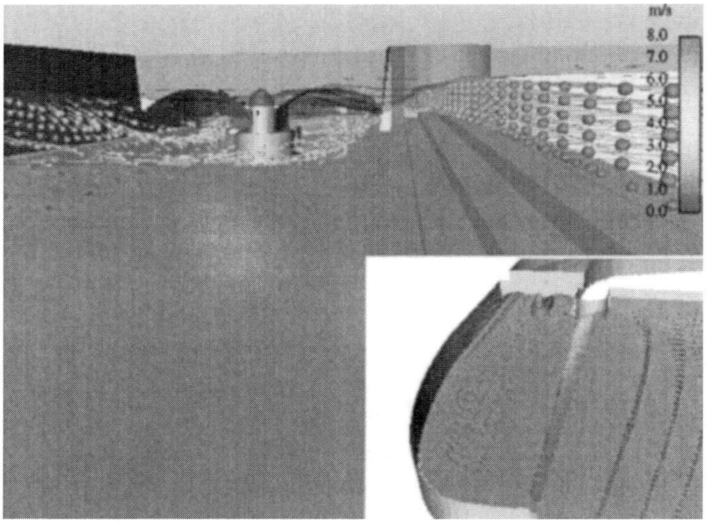


Fig. 7. Image taken from part of an animated sequence

## CONSTRUCTION

The CFD analysis confirmed and extended the predicted discharge curves from the physical hydraulic model and it was considered that the rating curves produced from the analysis could be used to determine the extent of the works required to safely pass the Probable Maximum Flood.

Original Tumble Bay

Deepened Tumble Bay

Concreting Works

Modified Bridge Pier

A scheme was developed with the main work being the excavation of the tumble-bay and deepening of the spillway channel under the road bridge (see photo's above) to improve the characteristics of the overflow weir and upper reaches of the channel. This work has been designed to resist uplift, avoid disturbance to the walls of the channel and to leave the bridge intact and operable throughout the construction phase.

One of the main risks associated with this project was with respect to whether planning permission was required for both the spillway and bridge works. The local planning authority were consulted very early in the design process and it was agreed that the work could be carried out as permitted development, with a few minor restrictions around road closures. Minor raising of the spillway walls was required to keep the peak flood outflow within the confines of the channel and away from the reservoir embankment. Other works at the site included providing restraint to the drawoff pipework, surge pressure control at the water treatment works, extension of the scour pipework within the draw-off tunnel and improving access to the draw-off tunnel.

The main construction works have now been completed. The reservoir is now full and the overflow channel is in operation. The CFD analysis proved to be a valuable tool assisting in the detail design and planning processes and has provided additional confidence in the results of the hydraulic model.

## SUMMARY AND CONCLUSIONS

- CFD can be used to undertake a cost effective sensitivity study on design features, such as the effect of a bridge on channel flow.
- The analysis is carried out at full scale and with time-dependent flows.
- Transient flow features can be examined at relatively little extra cost.
- The results can provide confidence in a design as it is developed and optimised.
- Additional cases can be easily examined at later stages in the design process.
- The data can be examined in great detail providing an insight into the mechanics behind the flow behaviour in three-dimensions and over time.
- CFD can be used either as an alternative or as a complementary technique to a physical model. A calibration exercise against a physical model provides additional confidence in the results.
- High quality images and animated sequences can be used for comparative studies as well as planning and public relation purposes.

It was shown that the upstream section of the byewash, from the bridge, effectively forms the control section at high flows. The layout in Configuration 1, therefore, represents the most cost-effective solution as it increases the spillway capacity to the maximum that can be achieved with the site constraints. To increase spillway capacity further would require either widening or deepening the byewash channel by more than 2m, neither of which are practical solutions.

Through modelling the spillway at Langsett Reservoir, together with the agreement of the results to those obtained from the physical model, this study has helped to establish the credibility and advantages of using CFD to model complex hydraulic structures. The results from the study have been used to assist with the final aspects of detail design and the solution is nearing completion on site.

As a footnote it is necessary to recognise certain limitations within the technique. Scale effect in the physical model was not discussed here but, although these effects were not present in the CFD model, other potential shortcomings should be noted. For instance, the CFD mesh length scale was of the order of about 0.8m in the flow direction. In addition, the applied turbulence model is time-averaged. Transient flow features at smaller length and time scales are therefore not predicted correctly in the CFD. Other considerations, such as entrainment of air at the free surface and spray, are again not taken into account in the CFD solution. This is not considered a significant factor in the prediction of the discharge (the primary objective of the study). Many simplifications and limitations relate to available computing resources and algorithms. However, these are constantly increasing and improving with time and new software releases.

It was shown that, with the correct application of surface roughness and turbulence models, the CFD predicts discharges that match the physical measurements and also generates surface profiles and flow features similar to those observed in the physical model. The CFD can then be used as a cost-effective way to examine the effect of alternative geometric configurations and variations in flow rates. The results can be interrogated in great detail and, if required, the analysis can be transient.

The analysis indicated that there is limited benefit in lowering the invert level of the tumbling bay and/or removing the bridge on increasing spillway capacity. The reason for this is that the upstream section of byewash, from the bridge, effectively forms the control section at high flows. Consequently, to increase spillway capacity further would require either widening or deepening the byewash, which is not a practical solution.

REFERENCES

FLOW-3D User's Manual Version 7.7 (2000). Flow Science Inc., Los Alamos, NM, USA.

HIRT C W & NICHOLS B D (1981). Volume-Of-Fluid (VOF) Method for the Dynamics of Free Boundaries, Journal of Computational Physics, Vol39, pp201-225.

# Rehabilitation of the Upper and Lower Bohernabreena Spillways

D E MacDONALD, Binnie Black & Veatch, UK
J D MOLYNEUX, Binnie Black & Veatch, UK

SYNOPSIS. The Bohernabreena Reservoirs are located south west of Dublin on the Dodder River. The reservoirs are in cascade; the upper reservoir used for supply of raw water to the Ballyboden treatment works; and the lower reservoir for flood relief. Both dams are puddle clay core embankments of the 19th century. The dams are owned and operated by Dublin Corporation. The reservoirs are situated upstream of Dublin and therefore require an 'A' categorisation with a PMF discharge requirement. Due to constrictions formed by the crest road bridges across the chutes, the actual capacities are only of the order of 100 year return period. The spillways are to be replaced with new reinforced concrete structures while the structures remain operational. Careful planning of temporary works and understanding of the hydrological risks are necessary to ensure that the embankments are protected.

## INTRODUCTION

The Upper and Lower Bohernabreena reservoirs are located in the headwaters of the River Dodder, about 13km south west of Dublin. The two impounding reservoirs, together with a treatment works and service works at Ballyboden, form part of the Dodder Water Supply Scheme (Ref. 1). The reservoirs are in cascade; the upper reservoir used for supply of raw water to the Ballyboden treatment works; and the lower reservoir for flood relief. Water from the hilltop catchment is routed around the upper reservoir through a by-wash channel because, at the time of construction, the water was considered unsuitable for treatment. Both dams are puddle clay core embankments of the 19th century, and bear all the hallmarks of Pennine type dams of the same era.

The dams are owned and operated by Dublin Corporation. In the absence of official Irish legislation the Republic generally adopt the UK's Reservoirs Act (1975) on a voluntary basis and Dublin Corporation employs UK All Reservoirs Panel Engineers to inspect their assets. The 10 yearly inspection carried out by Bill Carlyle of Binnie and Partners in 1975, revealed that the spillways were below the required capacity. The situation of the reservoirs upstream of Dublin requires an 'A' categorisation with a PMF discharge requirement.

*Reservoirs in a Changing World*, Thomas Telford, London, 2002, 274–285

There has been a shift in our understanding of hydrology and hydraulics over the years. The original report on the proposed formation of reservoirs on the River Dodder (Ref.2), written in 1844, claimed that the dams would be virtually impossible to overtop. Today we know that, due to constrictions formed by the crest road bridges across the chutes, the actual capacities are only of the order of a 100 year return period. Interestingly, the implied level of protection in 1844 was effectively the same as today's PMF, it is only the understanding of the hydrology and hydraulics that has changed.

The spillways are to be replaced with new reinforced concrete structures over a two and a half-year construction contract while the structures are still operational. A description of the existing and new works is given in a paper entitled "Bohernabreena Reservoirs, Dublin: the impact of Hurricane Charlie" (Ref.3).

The following sections provide details of the:

- design flood studies carried out to determine appropriate freeboard and spillway arrangements for each reservoir;
- selection of the standard of protection to be provided during construction; and
- flood frequency studies undertaken to help select an appropriate construction flood for the Lower reservoir.

## DESCRIPTION OF THE RESERVOIR CATCHMENTS

The total area that can drain by gravity to Lower Bohernabreena reservoir is just less than $28km^2$. It is convenient, however, to divide this total area into three subcatchments, namely:

- The upper Mountain Catchment ($19.7km^2$);
- The direct catchment to Upper Bohernabreena ($6.94km^2$); and
- The direct catchment to Lower Bohernabreena ($1.34km^2$).

The upper Mountain Catchment comprises largely of granite overlain by peat, whereas the direct catchments to the two reservoirs comprise mostly bare and uncultivated mountain hillside. Runoff from the upper Mountain catchment is acidic, and sometimes highly coloured. At the time of construction this water was considered untreatable and a bypass channel, with a maximum capacity of $50m^3/s$, was constructed to carry the highly coloured water around Upper Bohernabreena reservoir and directly into the Lower reservoir. Flows to the bypass channel in excess of $50m^3/s$ spill over a side weir at Old Castlekelly directly into the head of the Upper Reservoir.

Although the direct reservoir catchments contain some granitic formations, these areas are underlain mainly by metamorphosed rocks of Silurian age that are covered by clays and gravels rather than by peat. The runoff from these areas is relatively simple to treat for potable water.

## DESIGN FLOOD STUDIES

The earthfill dams which impound Upper and Lower Bohernabreena reservoirs both fall into Category A (general standard) as defined in Table 1 of Floods and Reservoir Safety (Ref. 4). It follows, therefore, that the freeboard and spillway arrangements of each dam would normally be expected to pass safely a Probable Maximum Flood (PMF) from the total area draining to that dam, with sufficient margin at the height of the flood to provide an adequate wave surcharge allowance.

Table 1 below lists the key levels along the crests of Upper and Lower Bohernabreena dams and shows the freeboard available to contain the flood rise and wave surcharge.

Table 1 - Catchment and embankment data

| **Reservoir** | Height of wave wall (m) | Nominal dam crest level (mAOD) | Spillweir crest level (mAOD) | Res surface area at spillweir crest level ($km^2$) |
|---|---|---|---|---|
| Upper Dam | | 178.00 | 176.17 | 0.23 |
| Lower Dam | 0.91 | 152.94 | 151.11 | 0.12 |

| **Catchment** | Area ($km^2$) | Tp(0) Tyear event (hrs) | Tp(0) PMF event (hrs) | SPR (%) |
|---|---|---|---|---|
| Upper Mountain catchment to bypass channel | 19.70 | 2.09 | 1.39 | 53 |
| Upper Bohernabreena direct | 6.94 | 1.11 | 0.74 | 53 |
| Lower Bohernabreena direct | 1.34 | 0.51 | 0.34 | 53 |
| Total catchment to Lower Bohernabreena | 27.98 | 2.44 | 1.63 | 53 |

PMF inflows to Upper and Lower Bohernabreena reservoirs were computed using the Flood Studies Report (Ref. 5) unit hydrograph rainfall-runoff and losses model. For both reservoirs the version of the model that is described in Flood Studies Supplementary Report (FSSR) No.16 (Ref. 6) was adopted.

In order to derive an appropriate total inflow hydrograph to the lower reservoir from the three subcatchments described in the previous section, the recommended approach for a cascade of reservoirs detailed in FSSR No.10 (Ref. 7) was used.

The key rainfall-runoff model parameters (i.e. unit hydrograph time-to-peak (Tp) and standard percentage runoff (SPR)) were derived from catchment characteristics using the parameter estimation equations recommended in FSSR No.16. The estimated values for each catchment of interest are listed in Table 1.

The rainfall and water level information available for some of the larger historic flood events in the Upper Dodder catchment were used to check the reasonableness of the time-to-peak values listed in Table 1. These efforts were not as successful as initially hoped. For example, analysis of the autographic rainfall data from the raingauge at Glenasmole together with the water level data for Lower Bohernabreena reservoir for the severe flood event of 25/26 August 1986, provided a catchment LAG of 4.5 hours for the total $28km^2$ area draining to the Lower reservoir. This value of LAG equates to a Tp(0) value of 3.3 hours when used in conjunction with the recommended relationship:

$$Tp(0) = 0.604 * LAG^{1.144} \text{ hours (From FSSR No.16)}$$

A separate study (Ref.8) using data for the flood of 18/19 December 1958 also derived a Tp(0) of 3.0 hours for the total catchment to the Lower reservoir.

Both of these values are slightly larger that the corresponding Tp(0) estimate of 2.44 hours derived from catchment characteristics.

During the flood of 25/26 August 1986, however, the peak water level registered by the recorder on the bypass channel around Upper Bohernabreena reservoir was 2 hours earlier than the peak level recorded in Lower Bohernabreena reservoir. Thus LAG for the upper Mountain catchment was only 2.5 hours, which equates to a Tp(0) value of only 1.72 hours. This value is slightly smaller than the corresponding Tp(0) estimate of 2.09 hours derived from catchment characteristics.

Ultimately it was decided to compute the design flood runoff from each subcatchment using the Tp(0) values derived from catchment characteristics. These values were judged to represent a consistent set of estimates that were not significantly different from the corresponding values obtained from local data. As is standard practice, the time-to-peak values for each subcatchment of interest were reduced by one third for use in the PMF computations.

The results of the initial design flood calculations revealed that for both reservoirs:

- A Summer PMF is the critical design event; and
- The existing spillway and freeboard arrangements were too small to pass the critical PMF inflows without the maximum stillwater level in the reservoirs rising to well above dam crest level.

Further sets of flood calculations were carried out for less extreme events to assess the level of protection provided by the existing freeboard and spillway arrangements. The calculations for a 10 000year event revealed that:

- The existing freeboard of 1.83m at Upper Bohernabreena would be sufficient to contain the maximum flood rise of 1.43m likely to occur, provided the bypass channel remained fully operative throughout the flood event;
- A freeboard of 2.41m would be required to contain the maximum flood rise in Upper Bohernabreena should the bypass channel become completely blocked, such that the whole runoff from the upper Mountain catchment flowed into the Upper reservoir;
- The existing freeboard of 1.87m at Lower Bohernabreena would be insufficient to contain the maximum flood rise of 2.41m likely to occur, and the maximum stillwater level in the reservoir would rise to within 0.3m of the top of the wave wall;
- The maximum water level reached in the Lower reservoir would not be significantly affected by any blockage of the bypass channel.

After a full consideration of the available options it was decided that the best means of providing the necessary level of protection was by upgrading the existing spillways and freeboard arrangements at both reservoirs. Provision of additional capacity by other means, for instance supplementary spillways on the opposite abutments, was not feasible.

The proposed modifications were model tested by University College in Dublin. They consist of replacing the crest road bridges to remove the constrictions, and widening and deepening the chutes to increase capacity. The arrangements are described in detail in a paper entitled "Bohernabreena Reservoirs, Dublin: the impact of Hurricane Charlie" (Ref. 3). A further set of PMF calculations and reservoir flood routings were carried out with these new arrangements. The results of the key design flood calculations are summarised in Table 2.

Table 2 - Summary of design flood calculations

Upper Reservoir

| Event | By-pass channel capacity ($m^3/s$) | Peak inflow ($m^3/s$) | Peak outflow ($m^3/s$) | Maximum reservoir stillwater level (m above spillweir crest) |
|---|---|---|---|---|
| With existing spillway | | | | |
| PMF | 50 | 328 | | Dam overtops |
| 10 000 year | 50 | 129 | 113 | 1.36 |
| 10 000 year | 0 | 179 | | Dam overtops |
| With modified spillway | | | | |
| PMF | 50 | 328 | 318 | 1.85 |
| PMF | 0 | 378 | 367 | 2.05 |

Lower Reservoir

| Event | By-pass channel capacity ($m^3/s$) | Peak inflow ($m^3/s$) | Peak outflow ($m^3/s$) | Maximum reservoir stillwater level (m above spillweir crest) |
|---|---|---|---|---|
| With existing spillway | | | | |
| 10 000 year | 50 | 165 | | Dam overtops |
| 1 000 year | 50 | 110 | 106 | 1.23 |
| With modified spillway | | | | |
| PMF | 50 | 379 | 377 | 2.09 |

## DISCUSSION

The design flood calculations summarised above were undertaken prior to the publication of the UK Flood Estimation Handbook (Ref. 9).

Volume 2 of the Flood Estimation Handbook (FEH) provides revised rainfall depth-duration-frequency estimates for mainland Britain and Northern Ireland. For most regions of the United Kingdom, the rainfall depth-duration-frequency estimates provided by the FEH are greater than the corresponding estimates obtained from the FSR. Unlike the maps contained in Volume V of the 1975 UK Flood Studies Report, the corresponding information provided by the FEH does not extend across the border of Northern Ireland.

Volume 4 of the FEH provides a comprehensive technical rewrite of the FSR unit hydrograph rainfall-runoff method incorporating the numerous enhancements contained in various Flood Studies Supplementary Reports, and in a variety of other more recent technical publications. Volume 4 also presents new equations to enable the key model parameters, unit hydrograph time-to-peak (Tp) and catchment standard percentage runoff (SPR), to be estimated from the digitally derived catchment descriptors provided by the FEH CD-ROM. Unfortunately the FEH and the FEH CD-ROM cover only mainland Britain and Northern Ireland.

It should also be recognised that:

- although the FEH updated the key parameter estimation equations associated with the unit hydrograph rainfall-runoff model, it did not significantly alter the basic methodology;

- the FEH team did not attempt to recalibrate the FSR rainfall-runoff model to reflect changes in the rainfall depth-duration-frequency estimates, or the changes in the ways that estimates are derived for the model parameters Tp(0) and SPR.

- the FEH did not attempt to revise the PMP estimates contained in the FSR;

- the FEH rainfall depth-duration-frequency model has not been calibrated for events greater than about 2000 years and, for the reasons set out in the recent paper by MacDonald and Scott (Ref. 10), may well be overestimating extreme rainfalls.

Our judgement is that the design flood calculations described in the previous section, which were completed prior to the publication of the FEH, still provide realistic design flood inflow hydrographs suitable to help to determine the improvements required to the freeboard and spillway arrangements at Upper and Lower Bohernabreena reservoirs.

## CONSTRUCTION FLOOD STUDIES

### Introduction

Careful phasing of temporary works is necessary to ensure that the embankments are protected. Essential to the success of the construction is the ability to pass floods that could occur during the construction period. This process is in three stages:

1. defining a suitable level of protection,
2. hydrological studies to arrive at the magnitude of an appropriate construction flood event, and
3. design of a temporary works scheme.

We have confirmed that a construction temporary works scheme is possible but responsibility for the details of that scheme has been left with the contractor.

### Standard of protection during construction

Dublin Corporation's Inspecting Engineer for the reservoir, Mr Andrew Rowland, defined the necessary standard of protection using guidance from Floods and Reservoir Safety.

Due to the different consequences of exceedance, two levels of protection were defined. If the contractor's working area should be flooded, the consequences are considerably less than the consequences of the embankment being overtopped. On the one hand some works and plant will be damaged, and on the other, the residents of Dublin will be threatened. It was decided that a 10% risk of overtopping would be acceptable for the working area, provided that the embankment remained secure and was exposed to no more than a 1% risk of overtopping during construction. Each reservoir has been considered separately and the contract limits the period for which the reservoirs are at any increased risk.

Hydrological studies

Table 3 lists the annual maximum peak outflows from Lower Bohernabreena reservoir for the majority of water years since 1949/50. These peak outflows are based on:

- records of maximum reservoir levels provided by Dublin Corporation;
- the reservoir level/outflow relationship for the existing spillway shown in Figure 1; and
- an allowance for outflows via the reservoir drawoff pipes.

Figure 1 - Lower reservoir spillway rating curve

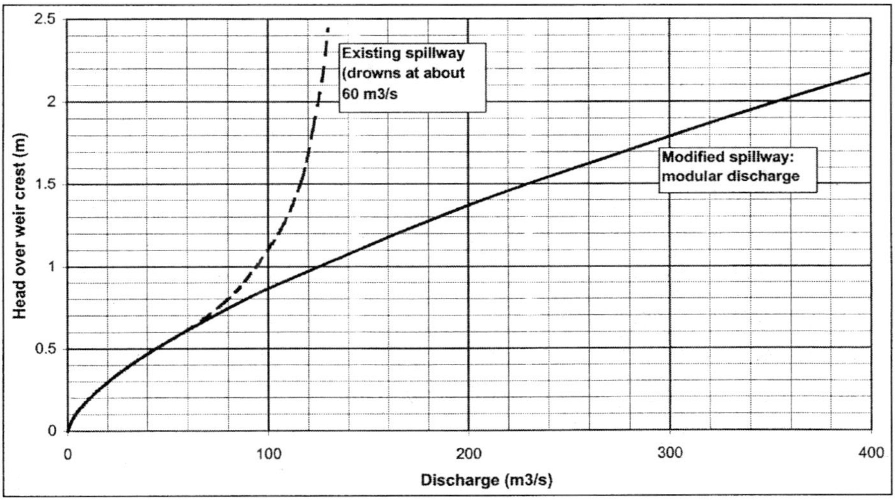


The combined surface areas of Upper and Lower Bohernabreena reservoirs at spillway crest level comprise less than 1% of the total area draining to the lower reservoir. Thus the routing effects of the two reservoirs are unlikely to have significantly reduced the peak flows of any but the smallest annual floods. In our judgement, therefore, a magnitude/frequency analysis of the records of annual maximum outflows from Lower Bohernabreena reservoir provides the best means of selecting an appropriate construction flood for the spillway remedial works.

The annual maximum peak outflows from Lower Bohernabreena reservoir for the period 1949/50 to 2000/01 were analysed using the software package WINFAP, originally marketed by the Institute of Hydrology (Wallingford). A General Extreme Value (GEV) was fitted to the annual maximum events by the method of probability weighted moments. The resultant flood frequency curve based on these data, which is shown in Figure 2, provides a value of 102 $m^3/s$ for peak flow in a 100year event. Unfortunately the standard error of the estimate attached to this value is relatively high such that we can be only 68% sure that the true instantaneous flood peak during a 100 year event lies between 79 and 125 $m^3/s$.

Figure 2 - Lower Bohernabreena flood frequency curve (1949-2000)

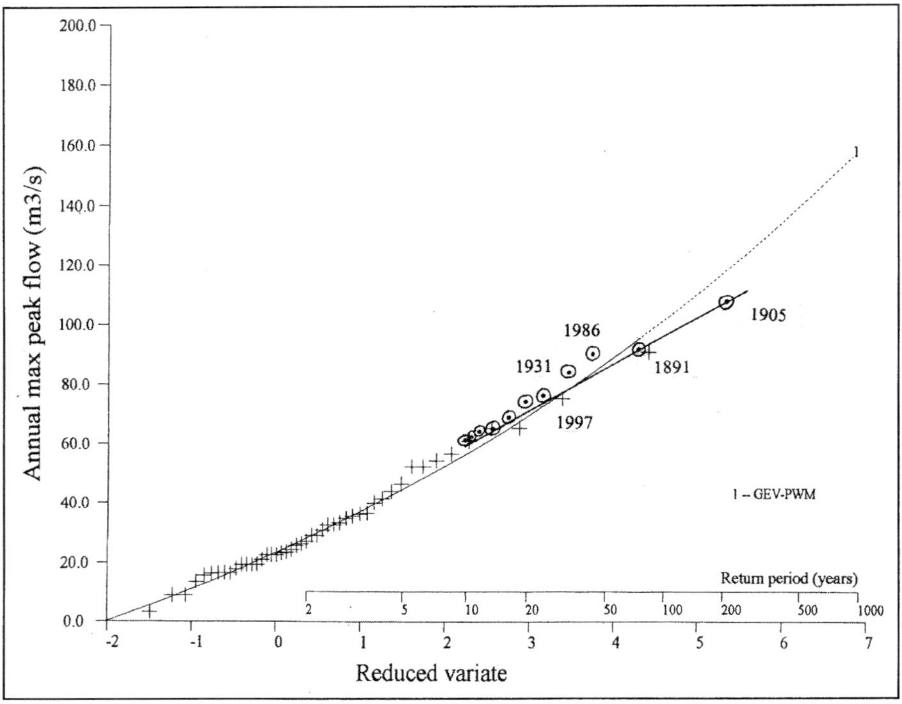


In an effort to reduce the uncertainty attached to the 100 year flood estimate we have collated the information available about major flood events in the Dodder catchment prior to 1949 from various sources (Refs.8, 11 and 12). The flood peaks attributed to these major events, which are also listed in Table 3, are thought to include all the very severe flood events that have occurred in the Upper Dodder catchment since 1886.

Each of the 11 flood events listed in Table 3, with a peak flow above 60 $m^3/s$, were:

- ranked in descending order of magnitude;
- assigned a plotting position appropriate to the 11 top ranking events in an 115 year series according to the Gringorten method (Ref.13); and
- superimposed on the flood frequency graph derived from the series of annual maximum peak flows for the period of records 1949 to date.

A "best fit" line was then drawn by eye through these 11 flood events (see Figure 2). This best fit line provides general support for the flood frequency curve obtained from an analysis of the recorded annual maximum outflows since 1949, but suggests the slightly lower value of 96 $m^3/s$ for the peak outflow in a 100 year event.

Table 3 - Lower reservoir - Recorded annual maximum outflows

| Water year Starting 1 October | Date | Max level above Spillweir crest (m) | Max. flow Over Spillway ($m^3/s$) | Outflow via drawoff pipes ($m^3/s$) | Max. outflow From Reservoir ($m^3/s$) | Rank for events since 1949 | Plotting Position (years) | Rank for events since 1886 | Plotting Position (years) |
|---|---|---|---|---|---|---|---|---|---|
| 1886/87 | 16 Oct | | | | 69.7 | | | 7 | 17.5 |
| 1891/92 | 13 Oct | | 90.6 | 1.56 | 92.2 | | | 2 | 73.8 |
| 1904/05 | 28 Aug | 1.219 | 106.2 | 1.56 | 107.8 | | | 1 | 205.6 |
| 1912/13 | 12 Nov | | 62.2 | 1.56 | 63.8 | | | 10 | 12.0 |
| 1930/31 | 04 Sep | | | | 85.0 | | | 4 | 32.30 |
| 1945/46 | 12 Aug | | 73.6 | 1.56 | 75.2 | | | 6 | 20.7 |
| 1949/50 | 25 Oct | 0.366 | 27.7 | 1.56 | 29.3 | 21 | 2.24 | | |
| 1950/51 | 27 May | 0.311 | 21.7 | 1.56 | 23.3 | 28 | 1.67 | | |
| 1951/52 | 12 Aug | 0.415 | 33.4 | 1.56 | 35.0 | 16 | 2.96 | | |
| 1952/53 | 19 Sep | 0.210 | 12.0 | 1.56 | 13.6 | 43 | 1.08 | | |
| 1953/54 | 02 May | 0.305 | 21.1 | 1.56 | 22.7 | 29 | 1.61 | | |
| 1954/55 | 13 Dec | 0.290 | 19.5 | 1.56 | 21.1 | 32 | 1.46 | | |
| 1955/56 | 06 Sep | 0.347 | 25.6 | 1.56 | 27.2 | 23 | 2.04 | | |
| 1956/57 | 25 Sep | 0.457 | 38.6 | 1.56 | 40.2 | 13 | 3.67 | | |
| 1957/58 | 28 Jul | 0.430 | 35.2 | 1.56 | 36.8 | 11 | 4.37 | | |
| 1958/59 | 19 Dec | 0.564 | 53.0 | 1.56 | 54.6 | 7 | 7.03 | | |
| 1959/60 | 08 Dec | 0.314 | 22.0 | 1.56 | 23.6 | 27 | 1.74 | | |
| 1960/61 | 04 Dec | 0.421 | 34.2 | 1.56 | 35.8 | 15 | 3.17 | | |
| 1961/62 | | | 32.0 | 1.56 | 33.6 | 17 | 2.79 | | |
| 1962/63 | 16 Mar | 0.259 | 16.5 | 1.56 | 18.1 | 37 | 1.26 | | |
| 1963/64 | 14 Mar | 0.235 | 14.2 | 1.56 | 15.8 | 41 | 1.14 | | |
| 1964/65 | | | 23.0 | 1.56 | 24.6 | 26 | 1.80 | | |
| 1965/66 | 17 Nov | 0.640 | 64.0 | 1.56 | 65.6 | 3 | 18.02 | 8 | 15.2 |
| 1966/67 | 12 Dec | 0.460 | 40.0 | 1.56 | 41.6 | 12 | 3.99 | | |
| 1967/68 | | | 15.0 | 1.56 | 16.6 | 41 | 1.14 | | |
| 1968/69 | 02 Nov | 0.506 | 45.0 | 1.56 | 46.6 | 10 | 4.82 | | |
| 1969/70 | 25 Apr | 0.381 | 29.4 | 1.56 | 31.0 | 20 | 2.36 | | |
| 1970/71 | 18 Mar | 0.396 | 31.2 | 1.56 | 32.8 | 18 | 2.63 | | |
| 1971/72 | 02 Feb | 0.341 | 24.9 | 1.56 | 26.5 | 24 | 1.96 | | |
| 1972/73 | 15 Jul | 0.335 | 24.2 | 1.56 | 25.8 | 25 | 1.88 | | |
| 1973/74 | 05 Jan | 0.244 | 15.1 | 1.56 | 16.7 | 38 | 1.23 | | |
| 1974/75 | | | | | | | | | |
| 1975/76 | | | | | | | | | |
| 1976/77 | | | | | | | | | |
| 1977/78 | | | | | | | | | |
| 1978/79 | Oct | 0.244 | 15.1 | 1.56 | 16.7 | 39 | 1.20 | | |
| 1979/80 | Jan | 0.244 | 15.1 | 1.56 | 16.7 | 40 | 1.17 | | |
| 1980/81 | May | 0.274 | 17.9 | 1.56 | 19.5 | 33 | 1.42 | | |
| 1981/82 | Jun | 0.396 | 31.2 | 1.56 | 32.8 | 19 | 2.48 | | |
| 1982/83 | Nov | 0.274 | 17.9 | 1.56 | 19.5 | 34 | 1.37 | | |
| 1983/84 | Dec | 0.305 | 21.1 | 1.56 | 22.7 | 30 | 1.56 | | |
| 1984/85 | May | 0.274 | 17.9 | 1.56 | 19.5 | 35 | 1.33 | | |
| 1985/86 | 25 Aug | 0.923 | 89.5 | 1.56 | 91.1 | 1 | 82.36 | 3 | 45.0 |
| 1986/87 | Apr | 0.274 | 17.9 | 1.56 | 19.5 | 36 | 1.30 | | |
| 1987/88 | Oct | 0.152 | 7.4 | 1.56 | 9.0 | 44 | 1.06 | | |
| 1988/89 | Jan | 0.152 | 7.4 | 1.56 | 9.0 | 45 | 1.04 | | |
| 1989/90 | | | | | | | | | |
| 1990/91 | Nov | 0.061 | 1.7 | 1.56 | 3.3 | 46 | 1.01 | | |
| 1991/92 | Nov | 0.305 | 21.1 | 1.56 | 22.7 | 31 | 1.51 | | |
| 1992/93 | 15 Sep | 0.549 | 50.9 | 1.56 | 52.5 | 9 | 5.39 | | |
| 1993/94 | Oct | 0.427 | 34.9 | 1.56 | 36.5 | 14 | 3.40 | | |
| 1994/95 | Mar | 0.640 | 64.0 | 1.56 | 65.6 | 4 | 12.96 | 8 | (13.4) |
| 1995/96 | May | 0.366 | 27.7 | 1.56 | 29.3 | 22 | 2.14 | | |
| 1996/97 | Jun | 0.579 | 55.3 | 1.56 | 56.9 | 6 | 8.29 | | |
| 1997/98 | Oct | 0.732 | 74.0 | 1.56 | 75.6 | 2 | 29.56 | 5 | 25.2 |
| 1998/99 | Sep | 0.610 | 59.6 | 1.56 | 61.2 | 5 | 10.11 | 11 | 10.9 |
| 1999/00 | Apr | 0.488 | 42.6 | 1.56 | 44.2 | 11 | 4.37 | | |
| 2000/01 | 05 Nov | 0.550 | 51.0 | 1.56 | 52.6 | 8 | 6.10 | | |
| | | | Mean (1949-2000) | | 32.6 | | | | |

In our judgement the value of 96 $m^3/s$, based on the maximum flood events during the 115year period since 1886, represents the best available estimate for the peak flow from the 28$km^2$ catchment draining to Bohernabreena reservoir in a 100year event.

Design of temporary works

It is envisaged that discharge capacity will be maintained during construction by splitting the chutes in two using some form of temporary barrier. Calculations carried out during the design phase demonstrated the feasibility of such a scheme but the contractor has not been constrained in selecting a scheme to suit his own capabilities and methods. The contractor's proposals must be reviewed and approved by his own All Reservoirs Panel Engineer prior to submission to supervisory staff and construction.

REFERENCES

1. Bourke, E.J. and O'Connor, B.E. (1961) "The Dodder Water Supply Scheme" City Engineers Department, Corporation of Dublin.

2. Mallet, R. (1844) "Proposed formation of Reservoirs on the River Dodder, County Dublin" Report to the Commissioners for Drainage of Ireland.

3. Rowland, A. (2000) "Bohernabreena Reservoirs, Dublin: impact of Hurricane Charlie". Published in "Dams 2000", Proc. of the biennial conf. of the BDS, University of Bath. Thomas Telford.

4. Institution of Civil Engineers, (1996) "Floods and Reservoir safety". Third Edition. Thomas Telford, London

5. Natural Environment Research Council (1975) "Flood Studies Report" 5 Volumes, Natural Environment Research Council, London.

6. Institute of Hydrology, (1985) "The FSR rainfall-runoff model parameter estimation equations updated" Flood Studies Supplementary Report No.16. Institute of Hydrology, Wallingford.

7. MacDonald, D.E and others. (1983) "A guide to spillway flood calculation for a cascade of reservoirs" Flood Studies Supplementary Report No.10. Institute of Hydrology, Wallingford.

8. Corporation of Dublin. (1959) "Flooding of the River Dodder – December 1958: Interim Report" City Engineers Department.

9. Institute of Hydrology. (1999) "Flood Estimation Handbook" 5 Volumes. Institute of Hydrology, Wallingford.

10. MacDonald, D.E and Scott, C.W. (2001) "FEH vs FSR rainfall estimates: An explanation for the discrepancies identified for very rare events" Dams and Reservoirs Volume 11 No.2.

11. Tyrell,A.W. (1888) "Description of the Rathmines and Rathgar Township Waterworks" Institution of Mechanical Engineers, October.

12. Keyes,J.M. (1986) "Flooding in Dublin City Rivers – 25th/26th August 1986". Overview paper presented to the Water Engineering Section, Institution of Engineers of Ireland, November 1986.

13. Gringorton, I.I. (1963) "A plotting rule for extreme probability paper" J.Geophysical Res.68.

# Embankment dams performance and remedial works

# Rehabilitation of Irrigation Dams in Albania

J L HINKS, Halcrow Group Ltd, UK
Y DEDJA, Director, Second Irrigation and Drainage Rehabilitation Project, Albania

SYNOPSIS. The Republic of Albania has a large number of irrigation dams of which 33 were inspected in 2000 – 2001 as part of the Second Irrigation and Drainage Rehabilitation Project. A manual was prepared for the calculation of flood peaks and volumes and detailed proposals were drawn up for remedial works estimated to cost $ 6.0 million. A simple prioritisation system was developed to allow the total cost of the works to be kept within the available budget of $ 4.5 million.

Throughout the exercise there was a need to maintain an appropriate balance between the demands of safety at individual dams and the need to spread scarce financial resources as widely as possible.

## INTRODUCTION

Albania is a small country with an area of 28,750 km$^2$ on the east side of the Adriatic Sea. It is bordered by Montenegro and Kosovo to the north, Macedonia to the northeast and Greece to the south. Much of the country is mountainous with the highest peak rising to 2,694 m. There is, however, a significant coastal plain which is farmed quite intensively – more than 50 % of Albania's GDP derives from agriculture which, in 1999, accounted for 60 % of employment.

Annual rainfall varies from 700 mm in the southeast of the country to 3,500 mm in the north. In the main agricultural areas it averages 1,000 to 1,600 mm/year. However much of this rain falls in the winter months with long, hot and dry summers.

Table 1 – Average Monthly Rainfall on the Coastal Plain (mm)

| | Jan | Feb | Mar | Apr | May | Jun | Jul | Aug | Sep | Oct | Nov | Dec | TOTAL |
|---|---|---|---|---|---|---|---|---|---|---|---|---|---|
| Shkoder | 183 | 129 | 142 | 126 | 109 | 71 | 42 | 73 | 128 | 169 | 208 | 223 | 1603 |
| Lezhe | 174 | 148 | 138 | 121 | 98 | 78 | 38 | 54 | 94 | 152 | 190 | 175 | 1460 |
| Durres | 124 | 101 | 88 | 62 | 51 | 32 | 14 | 29 | 56 | 107 | 207 | 210 | 1081 |
| Fier | 128 | 104 | 88 | 77 | 53 | 29 | 26 | 32 | 61 | 101 | 159 | 128 | 986 |
| Average | 152 | 121 | 114 | 97 | 78 | 53 | 30 | 47 | 85 | 132 | 191 | 184 | 1283 |

*Reservoirs in a Changing World*, Thomas Telford, London, 2002, 289–301

The country was under communist control from the end of the Second World War until 1991. The large scale construction of irrigation schemes commenced in 1930 and was enthusiastically continued by the communist administration so that by 1990 there were more than 400,000 hectares covered by such schemes. This represent 60 % of all arable land and almost all that is suitable for irrigation. However only 35 - 40 % of the irrigation schemes have been rehabilitated to date so that some 60 % of the nominal area within these schemes receives irrigation water. The remaining schemes await rehabilitation with only about 10 % of the land in these schemes presently receiving irrigation water.

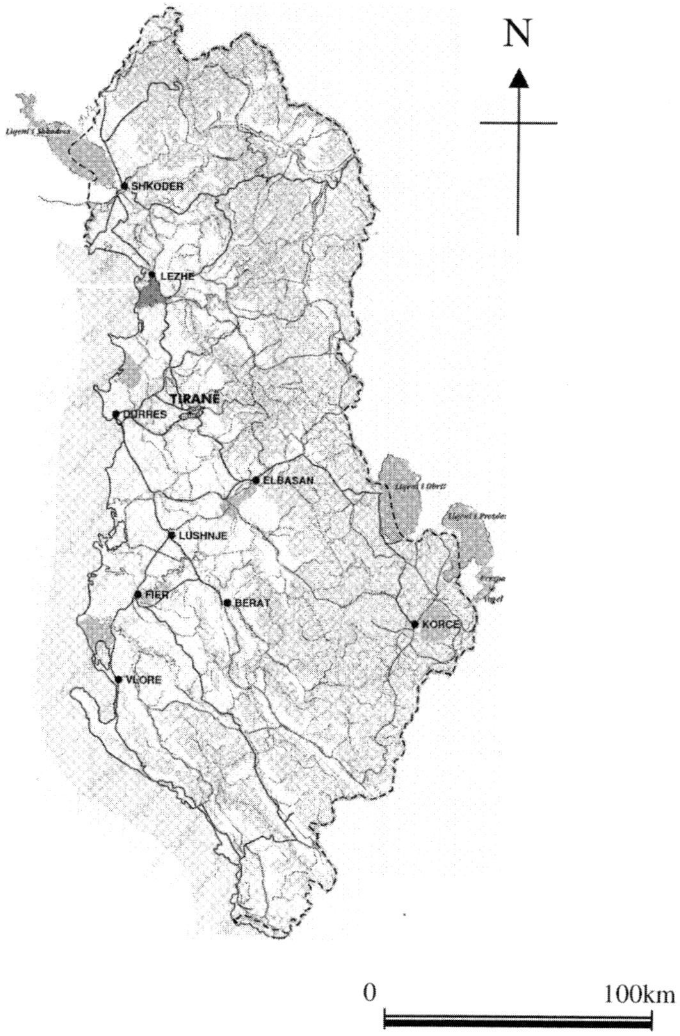


Figure 1 – Map of Albania

To supply the irrigation schemes there are 629 irrigation dams under the responsibility of the Ministry of Agriculture and Food. According to the World Commission on Dams this puts Albania in tenth place in the world for numbers of irrigation dams (WCD, 2000). The typical irrigation dam is between 15 and 30 m high and of homogeneous earthfill construction, possibly with a drainage layer beneath the downstream shoulder. In most cases there are houses downstream.

The Second Irrigation and Drainage Rehabilitation Project aims to restore some 50,000 ha of irrigation schemes. A World Bank requirement is that the associated dams should be inspected, and if necessary brought up to international standards, as a prerequisite for investment in the schemes. Thirty three dams were therefore inspected in 2000 and 2001. Virtually all were found to require remedial works of some kind to ensure their safety.

## CLASSIFICATION OF DAMS

When the dams were constructed they were divided into three categories:

| | |
|---|---|
| Category I | Height >25 m and storage > 1 $Mm^3$ |
| Category II | Height < 25 m and storage > 1 $Mm^3$ |
| Category III | Height < 25 m and storage < 1 $Mm^3$ |

For irrigation dams spillway chutes are understood to have been designed for a return period of 100 years. The overflow sills are understood to have been designed for the following return periods:

| | |
|---|---|
| Category I | 500 years |
| Category II | 200 years |
| Category III | 100 years. |

The above standards were considered too low to meet modern safety requirements. There was considerable debate as to the standards to be used for the rehabilitation works and the following were eventually agreed. It was considered that these standards should apply to the spillway chutes as well as to the sills although lower standards are acceptable for stilling basins.

Table 2. Proposed flood standards for Albanian dams

| **Number of households downstream** | **Return Period of Design Flood** |
|---|---|
| >10,000 | PMF |
| 1,000 to10,1000 | 10,000 years to PMF |
| 25 to 1,000 | 5,000 to 10,000 years |
| 1 to 25 | 1,000 |
| None | 500 to 1,000 |

The above is considered to be at the low end of what is acceptable internationally.

HYDROLOGY

At the start of the project there were very little hydrological data available apart from catchment areas and a start was therefore made using Creager Regional Coefficients (Creager & Justin, 1950 ):

$$Q = C \times 1.303 \times [\ A/2.588]^{B}$$

Where Q is peak discharge in $m^3$/sec

C is the applicable Creager coefficient

A is the catchment area in $km^2$

$$B = 0.936/(A^{0.048})$$

Monthly winter rainfall in the irrigation areas in Albania is approximately similar to that in the English Lake District or Lancashire Pennines where the RSMD is around 75. In these areas it is found that the rapid method in the second edition of "Floods and Reservoir Safety" ( Institution of Civil Engineers, 1989) gives approximately the same peak flow as the Creager formula for a return period of 10,000 years if the Creager coefficient is set at 25. The following Creager coefficients were therefore chosen:

Table 3 – Proposed Creager Regional Coefficients ( for checking purposes )

| Return period Years | Creager Coefficient |
|---|---|
| PMF | 50 |
| 10,000 | 25 |
| 5,000 | 22.5 |
| 1,000 | 15 |
| 500 | 13.5 |

Confidence in the above method was not, however, sufficient to make it the basis for the design of remedial works and it was therefore decided to prepare a Floods Design Manual for Albania incorporating the extensive rainfall records gathered during nearly fifty years of communist rule ( Adamson, 2000 ). This Manual distinguishes six different types of catchment and incorporates rainfall records from 24 different towns.

In the event the average difference between the two methods was only 6% (the flows derived from the Creager coefficients being, on average, the larger) with a standard deviation of 22 %. As would be expected the largest discrepancies were in areas with particularly high or low rainfall.

An interesting footnote is that when twelve Albanian engineers visited five major dams in the UK on a study tour Creager coefficients were calculated for each dam visited and found to range from 16 to 45 with an average value of 36. Creager coefficients thus have value in permitting instant comparisons between spillway sizes in different countries with similar climates. Obviously more sophisticated methods are needed for most purposes but it is suggested that Creager (or Francou Rodier) values are useful as a check or where only minimal hydrological data are available.

The new Floods Manual included a methodology for the calculation of flood volumes and this was extremely useful in a number of cases. For example one dam had a catchment area of 47.5 $km^2$ and no spillway at all. The reservoir was normally empty and had a fairly large outlet although this tended to get blocked with sediment. The capacity of the reservoir was about 2.5 $Mm^3$. A puzzling feature was that there was reported to have been 344.5 mm of rainfall in 24 hours at the nearby town in October 1981 without the dam being overtopped (although the water came within 1 metre of the crest ). About half of the catchment is underlain by limestone formations. After some debate it was decided to treat the catchment as permeable with a runoff coefficient of 21 % (backcalculated from the event in 1981) but with the runoff coefficient increasing to the value of 53 % suggested in the Flood Estimation Handbook ( Institute of Hydrology, 1999 ) for a saturated or frozen catchment. This led to a flood volume of 7.3 $Mm^3$ for a 24 hour storm with a return period of 10,000 years and clearly demonstrated the need for a spillway to protect some 100 houses downstream.

Another interesting case was a dam holding 9.6 $Mm^3$ just upstream of a sizeable town. This had been completed in 1987 just before the end of the communist period and had two spillways. One took the form of a shaft into the outlet culvert with its sill 6.15 m below the crest of the dam and the other was a side entry spillway 3.1 m below the crest of the dam. The catchment area was 30.9 $km^2$. The chief value of the lower spillway, which had a capacity of only about 12 $m^3$/sec, appeared to be to ensure a suitably low water level at the start of the design storm. Calculations showed that the spillway provisions were sufficient for a flood with a return period of 2,900 years ignoring the absorption of 3.8 $Mm^3$ between the levels of the two spillways. When the absorption of the 3.8 $Mm^3$ was taken into account it was considered that the 10,000 year flood could be safely routed through the reservoir. The only safety recommendation was that five smaller reservoirs upstream should be investigated since their failure might threaten the main dam.

At 20 out of 30 dams with direct catchments the spillway capacity was found to be inadequate, when measured against the new standards, and in need of being increased.

## EARTHQUAKES

Albania is in a seismic area; there have recently been significant earthquakes in the northeast of the country ( 1967 ) and in the north ( 1979 ). The World Map of natural hazards published by the Munchener Ruckversicherung shows the country partly in Zone 3 and partly in Zone 4. Zone 3 represents a Modified Mercalli Intensity of VIII ( ie: a Peak Ground Acceleration in the range 0.1 to 0.2g ) and Zone 4 represents a Modified Mercalli Intensity of IX ( ie a Peak Ground Acceleration of 0.2g to 0.5g ).

The above suggests a Peak Ground Acceleration ( PGA ) for a 250 year return period of about 0.2g. However at a seminar attended by specialists from Tirana University it was suggested that the PGA for a return period of 100 years should be set at about 0.4g. Calculations have therefore been performed for both of these values.

## GEOTECHNICAL DESIGN

Ideally it would have been desirable to have carried out site investigations, laboratory testing and stability analyses for each dam. This would however have significantly extended the overall programme and the results would have been meaningless unless there was full confidence in the quality of the laboratory test results.

It was therefore decided to carry out various parametric studies with shear strength parameters established by back analysis from observed slips. Analyses were also carried out for several typical dams assuming them to be provided with free draining buttresses on the downstream face 3 m wide at the top increasing to 5 m at the toe. Factors of safety for a typical 25 m high dam were as given in Table 4 below:

Table 4 – Factors of Safety for Static Load Cases

| **Water Level** | **Drainage** | **No Buttress** | **Small Buttress** |
|---|---|---|---|
| Reservoir full | Poor drainage<br>Good drainage | 1.08<br>1.20 | 1.27<br>1.36 |
| 3m below TWL | Poor drainage<br>Good drainage | 1.20<br>1.30 | 1.36<br>1.45 |

The parametric studies confirmed the intuitive belief that buttresses would provide worthwhile increases to the factors of safety for both static and seismic conditions.

Table 5 - Predicted Seismic Settlements (mm) for Reservoir Full

| | PGA | No buttress | Small Buttress |
|---|---|---|---|
| Poor Drainage | 0.2g | 220 | 19 |
| | 0.4g | N/A[1] | 117 |
| Good Drainage | 0.2g | 52 | 8 |
| | 0.4g | 196 | 76 |

However even with the buttresses, factors of safety for seismic load cases were generally less than unity; seismically induced settlements calculated with the Ambraseys formula (Ambraseys, 1972 ) were however generally small enough to be acceptable. Table 5 gives the calculated settlements for a 25 m high dam under "reservoir full" conditions.

Unless the earthquake were to occur at the start of the irrigation season in about May it is likely that water levels would be some way down. Table 6 gives the predicted settlements on the assumption that water level is 3 m below Top Water Level (TWL) at the time of the earthquake.

Table 6 - Predicted Seismic Settlements (mm) for Reservoir 3 m below TWL

| Buttress | PGA | No buttress | Small |
|---|---|---|---|
| Poor Drainage | 0.2g | 35 | 8 |
| | 0.4g | 162 | 76 |
| Good Drainage | 0.2g | 12 | 4 |
| | 0.4g | 92 | 52 |

[1] The settlement for Poor drainage, no buttress and a PGA of 0.4g lies outside the limits of applicability of the Ambraseys formula.

It will be seen that predicted settlements with water level 3 m below TWL are only 45 to 70 % of those for the reservoir full condition.

Despite the improvement in factors of safety for static load cases they were, in some cases, still less than the minimum value of 1.5 recommended in the Guide to the Safety of Embankment Dams in the United Kingdom ( Building Research Establishment, 1999 ). Consequently it was decided to increase the dimensions of the buttresses so as to be 5 m wide at the top and 10 m wide at the bottom with a 5 m berm at mid-height.

A single "one size fits all" free draining buttress was then specified to go on the downstream faces of all the dams exhibiting serious signs of instability, leakage or collapse features.

The specification required the buttress fill to be free-draining, angular, quarried rockfill graded from 20 to 100 mm with a $D_{50}$ size of 50 mm. It was to be separated from the existing downstream face of the dam by a 500 mm thickness of sand/gravel filter graded from 1.5 to 7 mm with a $D_{50}$ size of 3 mm. The grading of the existing dam fill was not accurately known but is probably finer than shown on Figure 3. Traditional filter rules are not, therefore, fully complied with.

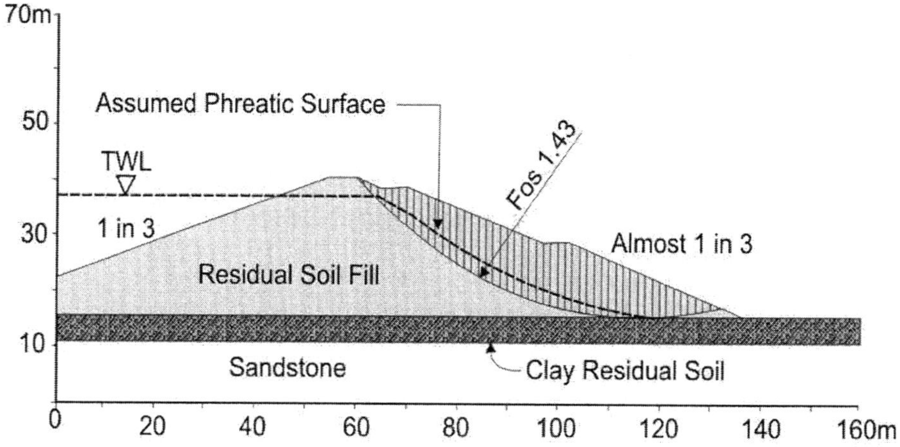


Figure 2 – Typical Dam with Buttress of Final Design

Buttressing was recommended for 7 out of the 33 reservoirs. A more modest toe weight was specified for three others.

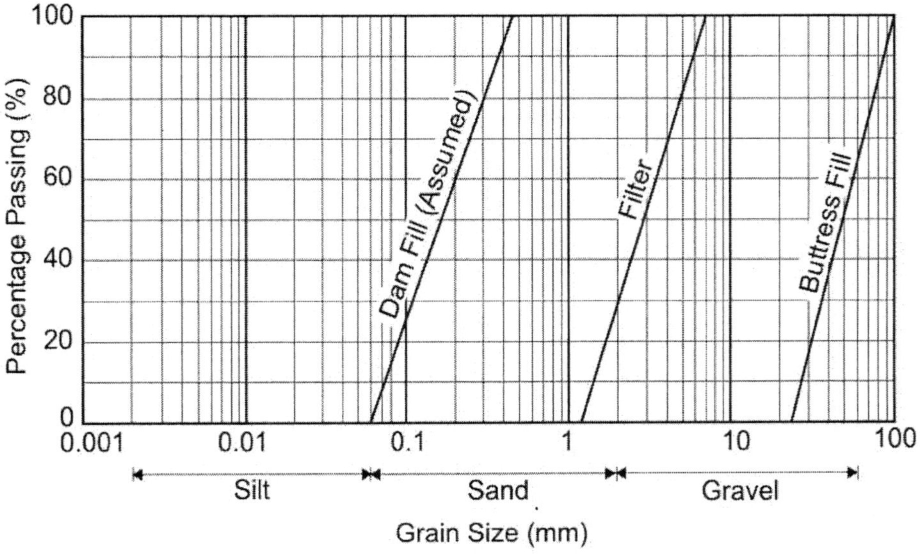


Figure 3 - Grading curves for buttress material

## OUTLET PIPES

Outlet pipes through and under the embankments were a source of considerable concern since, with only five exceptions, these did not have upstream gates or valves. Where, at the larger reservoirs, there were upstream gates these were scheduled for refurbishment since they had, in most cases, been damaged in the riots following the fall of communism. It was not, however, felt to be an efficient use of scarce funds to embark on a programme for the installation of new outlet towers, gates and access bridges where these did not already exist.

## UPSTREAM SLOPE PROTECTION

It is, of course, good practice to provide stone protection to the upstream slopes of earth dams and this would certainly be proposed for any new dams. However there were a number of dams where there was no such protection and where there was little evidence of serious wave erosion having taken place. In these cases it was not proposed to provide wave protection under the project. If wave erosion is observed in the future appropriate measures, such as the tipping of stone, can be taken.

## WAVE FREEBOARD

Necessary wave freeboard was calculated using the method in the third edition of “Floods and Reservoir Safety” ( Institution of Civil Engineers, 1996 ) with mean annual maximum hourly windspeeds provided by the Albanian Instituti Hidrometeorologjik.

## INTERIM SAFETY MEASURES

At reservoirs which were considered dangerous on the grounds of inadequate spillway capacity and freeboard, recommendations were made that water levels should be held down temporarily so as to accommodate 50 mm of runoff from the catchment below Top Water Level. Very roughly this represents 24 hour rainfall with a return period of five years.

## PRIORITISATION

Whilst it is hoped that all the dams will be remediated it is considered prudent to undertake works at the most dangerous first. A simple priority ranking system was therefore developed.

The following steps are needed to obtain the ranking:

(i) the unacceptable portion of the risk of overtopping of the dam during the next 100 years is first calculated and is entered into a spreadsheet after first
deducting the acceptable element of risk ( eg. if the Design Flood has a return period of 1,000 years then 9.5 % is deducted from the probability of overtopping since this 9.5 % is deemed an acceptable risk )

(ii) the Geotechnical risk of failure in a 10 year period is estimated subjectively from observations made at the dam and is multiplied by 10. For calibration purposes the average geotechnical risk at all the dams is assumed approximately equal to the average hydrological risk ( there is some justification for this assumption in ICOLD Bulletins 99 and 109 )

(iii) the Hydrological and Geotechnical risks are added together and multiplied by the number of households downstream to give a Risk Index. The higher the Risk Index the higher the priority for remediation.

Table 7 – Extract from Priority Ranking Table

| Name of Reservoir | Households Downstream No. | Risk in 100 Year period Hydrological % | Geotechnical % | Overall % | Risk Index Houses x Overall |
|---|---|---|---|---|---|
| Liz #1 | 5 | 77 | 60 | 137 | 685 |
| Goricani | 30 | 10 | 10 | 20 | 600 |
| Fishte | 12 | 0 | 35 | 35 | 420 |
| Kurjani | 80 | 0 | 5 | 5 | 400 |
| Malas Grope | 5 | 53 | 15 | 68 | 340 |
| Sadovice | 20 | 0 | 15 | 15 | 300 |
| Shtodri | 50 | 0 | 5 | 5 | 250 |
| Kashta | 20 | 2 | 10 | 12 | 240 |
| Vashaj | 5 | 0 | 40 | 40 | 200 |
| Grizhe | 15 | 0 | 10 | 10 | 150 |
| Troshan | 3 | 20 | 15 | 35 | 105 |
| Shkalle | 5 | 0 | 10 | 10 | 50 |
| Helmes | 0 | 0 | 40 | 40 | 0 |
| Selce | 0 | 45 | 20 | 65 | 0 |

The true probabilities of failure may only be about one tenth of those given in the above table ( ICOLD, 1997 ) because of uncertainties such as the ability of the dams to withstand limited overtopping, the probabilities of high winds concurrent with peak flows and the probabilities of the reservoirs being full at the start of the storms. However the table is thought to be suitable for ranking the dams in a rational order for the implementation of remedial works.

The above table is an extract from the database of 33 dams inspected to date. There are, of course, a number of dams with a Risk Index of zero because there are no houses downstream. Remediation may, nevertheless be important because there may be roads downstream and fields in which people might be working. Failure of a dam would also do long term damage to local agriculture by washing away the topsoil.

## CAPACITY BUILDING

There are thought to be almost 600 further irrigation dams in addition to the 33 inspected to date. In consultation with the Albanian National Dams Committee there are therefore plans to extend the project database to cover as many as possible of these dams. This will then allow proposals to be formulated for their inspection in a logical order.

Wherever possible it is desirable that such inspections should be carried out by Albanian nationals and a training programme is therefore in hand for dam inspectors. Prospective inspectors are being encouraged to produce reports on various dams which are then discussed in seminars at which the Albanian versions of reports prepared by the UK Dams Specialist are also available. It is hoped that such training exercises, together with Study Tours to the UK , will produce a pool of inspectors able to produce inspection reports covering all the principal disciplines such as hydrology, geotechnics, hydraulics etc.

## CONCLUSIONS

The works are designed to make good the deterioration that has occurred since the dams were built and to bring them up to standards appropriate for modern Albania. The chief beneficiaries will be about 5,000 people living downstream of the dams. The estimated cost of the full programme is $ 6.0 million. This could be reduced to fit within the available budget of $ 4.5 million if it is decided to defer remediation of some of the dams with a low risk index.

It is hoped that the project will also pave the way for the systematic inspection and remediation of other significant dams in the country.

## ACKNOWLEDGEMENTS

The authors would like to thank the Ministry of Agriculture and Food for their kind permission to publish this paper and the Albanian Committee on Large Dams for their helpful suggestions made at meetings in Tirana. They would also like to acknowledge the valuable work done by Mr.A.Rowland, who carried out preliminary inspections of some of the dams in 1998, and also the work of geotechnical engineer Ian Moore. The help received from UK dam owners, who acted as hosts for dams visited on the study tours, is also gratefully acknowledged.

## REFERENCES

Adamson P T (2000). A Manual for the Estimation of Design Floods in Albania, Halcrow.

Ambraseys N N (1972). Behaviour of foundation materials during strong earthquakes, Proceedings 4th European Symposium on Earthquake Engineering, London , September 1972, Vol 7, pp 11-12. Sofia, Bulgarian Academy of Sciences.

Bodgani & Selenica (1996). Catastrophic Floods and their risk in the rivers of Albania, Proceedings of IAHS conference held in Anaheim, California, June.

Building Research Establishment (1991). An Engineering Guide to Seismic Risk to Dams in the United Kingdom.

Building Research Establishment (1999). An Engineering Guide to the Safety of Embankment Dams in the United Kingdom.

CIRIA (2000). Risk Management for UK Reservoirs, CIRIA Report C 542.

Creager W P & Justin J D (1950). Hydroelectric Handbook, John Wiley & Sons

Institute of Hydrology (1999). Flood Estimation Handbook.

Institution of Civil Engineers (1989 ), Floods and Reservoir Safety, Second Edition.

Institution of Civil Engineers (1996 ), Floods and Reservoir Safety, Third Edition.

International Commission on Large Dams (1989). Selecting Seismic Parameters for Large Dams, Bulletin 72.

International Commission on Large Dams (1992). Selection of Design Flood, Bulletin 82.

International Commission on Large Dams (1995). Dam Failures – Statistical Analysis, Bulletin 99.

International Commission on Large Dams (1997). Dams Less than Thirty Metres High, Bulletin 109.

World Commission on Dams (2000). Dams and Development: A New Framework for Decision Making. Earthscan Publications Ltd., London.

# River Shannon Hydro-Electric Scheme: Failure of upstream slope of Fort Henry Embankment: Analysis

M LONG, University College Dublin, Ireland
I LYDON, Ove Arup & Partners, UK
E CONATY, University College, Dublin

SYNOPSIS. An 80 m long section of the Fort Henry embankment failed due to rapid drawdown of the impounding reservoir in 1979. Back analyses using the limit state stability approach with total stress soil parameters and the technique favoured by US Corps of Engineers, yielded factors of safety greater than the true value of 1.0. A simple stability chart based effective stress analysis was able to model the slide accurately. Finite element analyses showed that both magnitude and rate of drawdown control the embankment stability. The safety level of the embankment under the current operating regime has been found to be adequate but with little scope for relaxation.

## INTRODUCTION

This paper describes a detailed ground investigation and analysis which was undertaken following a significant failure of the upstream face of the Fort Henry embankment, in the River Shannon hydro-electric scheme in Ireland. The owner and operator of the scheme, ESB, supported two research projects at University College Dublin (UCD), which have examined different approaches to analysing the failure. The ultimate objective was to permit a scientifically based review of the current operating procedures, in particular the allowable magnitude and rate of drawdown.

Details of the scheme, the operating procedures, the failure itself and the subsequent remedial work are given in the accompanying paper to this conference by Casey et al. (2002).

## FORT HENRY EMBANKMENT

Fort Henry embankment is some 3.5 km long, with a maximum height of 8.5 m and an average height of 7.0 m above original ground. The top of the bank is at +35 mOD and it was designed for a range of water levels from +32 mOD to +33.5 mOD. A typical section is shown on Figure 1. Each cross section represents a length of about 25 m. The failure occurred between cross sections (C/S) 55 to 57.

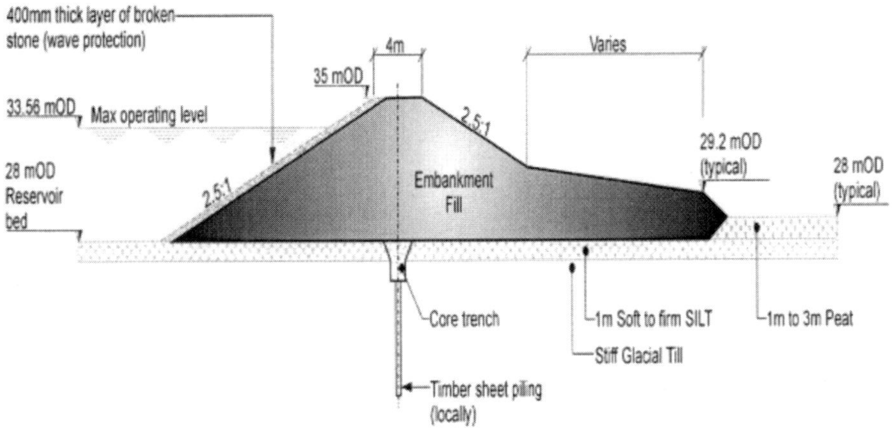


Fig. 1. Typical cross section of Fort Henry embankment

Before construction began the surface soil was stripped from the foundation. A longitudinal trench was cut to intercept land drains and to form a core trench. For Fort Henry fill material was obtained from a borrow pit adjacent to the north end of the embankment. After blasting the material was excavated by face shovel, transported by rail and side tipped from wagons. Compaction was achieved by moving the track as construction proceeded. Some other stability problems have been encountered at Fort Henry. Harty (1953), for example, describes a slide in the downstream side of the embankment at C/S 100.

## CONTROL WATER LEVELS AND RATES OF DRAWDOWN

Details of the allowable rate and magnitude of drawdown and are given by Casey et al. (2002). At the time of failure the water level dropped from approximately +33.2 mOD on 20/6/79 to +32.1 mOD on 30/6/79. This level was the lowest in at least 35 years. Normal maximum operating level is approximately +33.5 mOD and the minimum allowable level permitted after the 1979 slide is +32.6 mOD.

## SITE INVESTIGATIONS

At the time of the slide, minimal detail of the ground conditions at C/S 55 – 57 of Fort Henry embankment were available. Following the slide, a comprehensive investigation was carried out by SWECO Consultants Ltd. The comprehensive investigation carried out by SWECO included static cone penetration tests, Swedish weight soundings, pore pressure soundings, hand auger holes and piston sampling points together with routine and specialist laboratory tests.

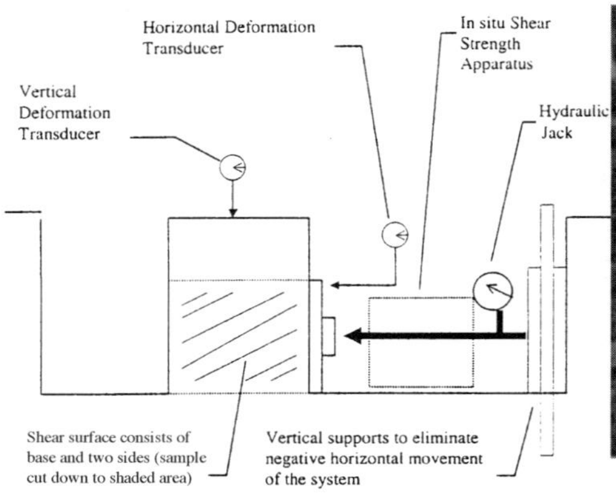

Fig. 2. In situ shear box testing

Further investigations were carried out by ESB International (ESBI) in 1992 and 1994. The purpose of these investigations was to address various specific points raised by the ESB's External Dam Safety Committee and the Chief Civil Engineer. These investigations comprised trial pitting, shell and auger boreholes and Swedish auger holes. Laboratory testing was also carried out.

The above investigations covered a wide area and did not specifically deal with Fort Henry embankment C/S 55 – 57. Therefore as part of the UCD / ESB research project a further detailed investigation was made of this area (Lydon, 1999). This investigation comprised a comprehensive field and laboratory study and included:

- Shell and auger boreholes with U100 sampling.
- Swedish auger boreholes.
- Trial pitting.
- In situ strength testing using a specifically designed device (see Figure 2).
- Laboratory tests for basic index parameters.
- Triaxial (CIU and CAUC) testing
- Small and large shear box testing.

## GROUND CONDITIONS / FAILURE SURFACE

Ground conditions at the location of the failure are summarized on Figure 3. The embankment is founded on a 1 m thick layer of soft to firm silt, which is thought to be alluvial in origin. This layer overlies a loose alluvial sandy layer, also 1 m thick. More competent glacial clay till underlies these layers.

The profile of the failure surface is shown on Figure 3 (Casey et al, 2002). It day lighted on the embankment crest and was located relatively close to the surface of the embankment. It extended mostly through the embankment fill. However, as slip debris was found 25 m into the reservoir, the possibility of the slip surface encroaching into the soft silt layer cannot be ruled out.

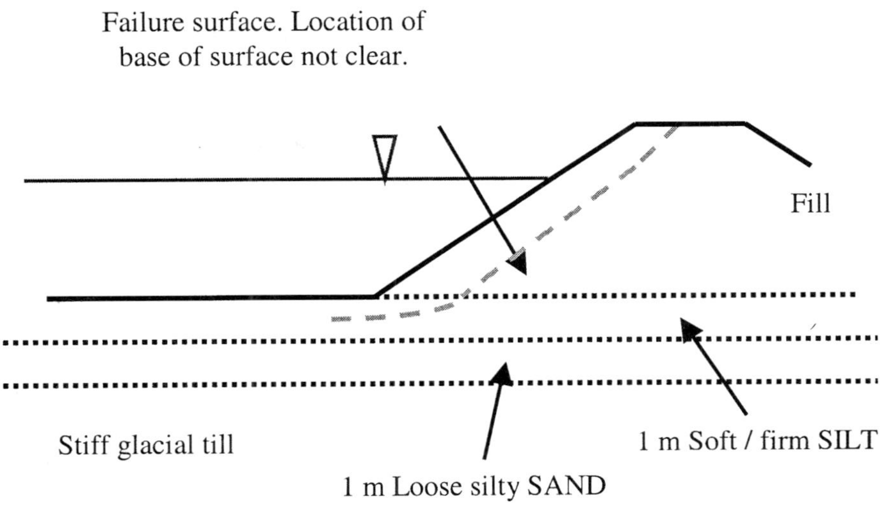


Fig. 3. Fort Henry C/S 55 - 57: Ground conditions / failure surface

SOIL PARAMETERS

The focus of the in situ and laboratory testing was to obtain soil parameters so that the failure could be analysed. Emphasis was therefore given to the embankment fill and the underlying silt.

Embankment fill

Grading curves for this material, shown on Figure 4a, confirm it to be a typical Irish lodgement till, comprising a mixture of clay, silt and sand with a relatively uniform grading. Other basic material data shown on Figure 5, confirm that it has a moisture content of about 16%, a liquid limit of 21% and a plasticity index of 8%. Thus the material can be classified as a sandy silt of low plasticity.

The permeability of the material was measured in triaxial tests and was found to be in the range of $1 \times 10^{-9}$ m/s to $1 \times 10^{-10}$ m/s, i.e. it is of very low permeability. It is therefore reasonable to express the strength of the material in terms of undrained shear strength ($s_u$). $s_u$ values were measured by a variety of means and the results are summarised on Figure 5. It can be seen that the values are in reasonable agreement and suggest that $s_u$ increases from about 30 kPa at the crest of the embankment to 70 kPa towards its base. Thus the material is of a soft to firm consistency.

Effective stress parameters can be determined from the triaxial test (CIU) stress paths as shown on Figure 6a. From these it can be said that the material has effective friction angle ($\phi^1$) and cohesion ($c^1$) values of 28° and 25 kPa respectively. The material can be seen to be highly dilatant, which is typical for silty or sandy deposits.

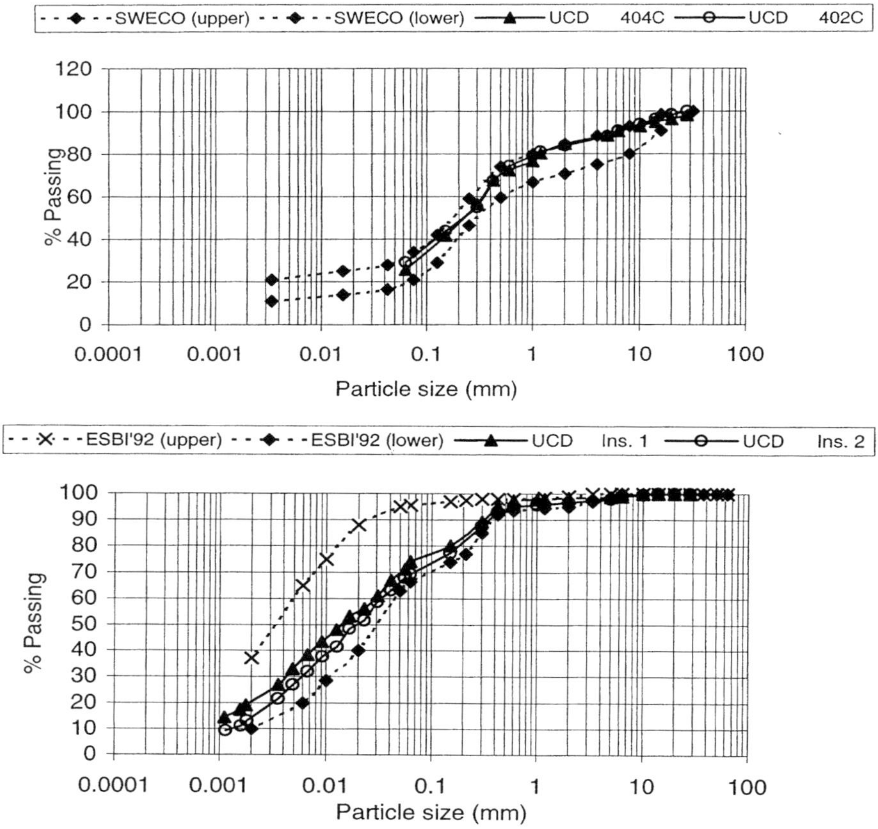


Fig. 4. Particle size distribution charts for embankment fill (a) and silt (b)

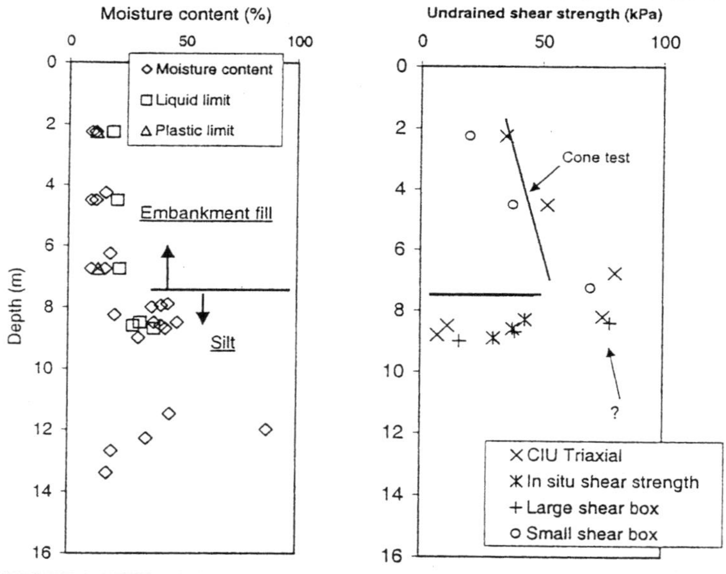


Fig. 5. Basic material parameters

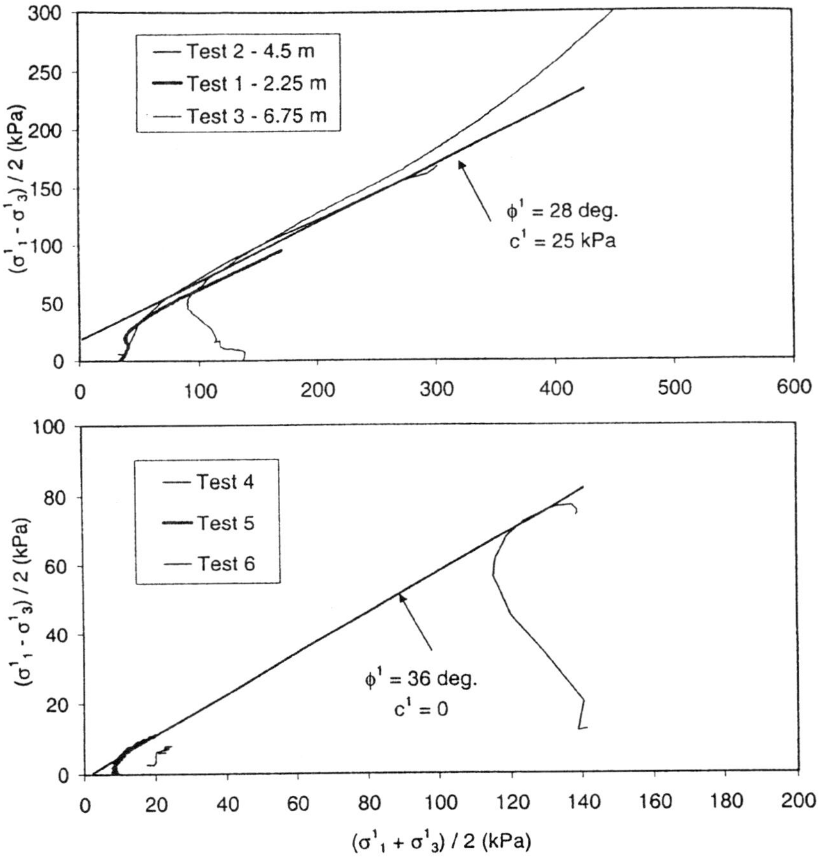


Fig. 6. Triaxial test (CIU) stress paths for embankment fill (a) and silt (b)

It is interesting to note that, as pointed out by Lydon (1999), the cone penetration test (CPT) cone resistance values were lowest around C/S 52 – 57 (≈ 600 kPa), i.e. where the failure occurred. This suggests that this tool has valuable application in demarking vulnerable areas of the embankments.

Soft silt

Grading curves for this material, shown on Figure 4b, confirm it to be predominantly a silt. This is confirmed by the plasticity data (Figure 5), which indicates that it is non-plastic. Its natural moisture is of the order of 40%. Like the embankment fill the material was found to have low permeability with a permeability coefficient of about $1 \times 10^{-10}$ m/s.

A significant scatter was found in the measured $s_u$ values for this material, as can be seen on Figure 5. An average value of about 40 kPa was determined suggesting that this material is also of a soft to firm consistency. The in situ shear tests seem to give the most reasonable values.

Effective stress parameters, obtained from CIU triaxial tests (Figure 6b) are $\phi^1 = 36°$ and $c^1 = 0$. Again the material can be seen to be highly dilatant, which is typical for loose / soft silt.

## ANALYSIS OF FAILURE

### Total stress analysis using undrained parameters

Conventional analyses of failure due to short term loading conditions in low permeability materials are carried out using undrained (or total stress) parameters. This has the major advantage that there is no need to calculate pore water pressures.

Such analyses were carried out by Lydon (1999) using $s_u$ values as described above. He found that for a circular surface, of shape similar to that shown on Figure 3, the calculated factor of safety (FOS) was of the order of 4.0. Similarly if calculations were performed using a non-linear surface, with a shape as close as possible to the actual failure surface, then the FOS obtained was of the same order of magnitude. Lydon (1999) found that the $s_u$ values for the embankment fill and silt layers needed to be reduced to unrealistically low values of 10 kPa and 5 kPa respectively in order to theoretically obtain a FOS of 1.0.

Perhaps it is not surprising that the total stress analysis yields unreliable results. Strength values obtained in undrained tests are very sensitive to test type, specimen size, techniques used for consolidation, rate of application of load etc. $s_u$ values are perhaps of most use as an index parameter rather than a true expression of soil strength. Also the behaviour of the two materials involved is likely to be very different, reaching peak strength at various strain levels and with different development of pore pressure etc.

Similar findings have been found by other researchers. Janbu (1989, 1996), for example, reviewed eight failures in real slopes for which comprehensive ground investigation data was available. He found that if conventional total stress analyses were used then the back-analysed FOS values varied between 0.5 and 3.0 (compared to the true value of 1.0).

### US Corps of Engineers Method

The Duncan et al. (1990) method has recently been adopted by the US Army Corps of Engineers as the standard method of rapid drawdown analysis. It utilises undrained shear strength to avoid the complications / inaccuracies involved in estimating undrained pore water pressure during drawdown. Envelopes are developed, which relate undrained shear strength to isotropic and anisotropic consolidation pressure. An example of such a relationship is shown on Figure 7, for the embankment fill material. The horizontal axis represent the effective stress on the failure plane during consolidation ($\sigma^1{}_{fc}$) and the vertical axis is the shear stress on the failure plane at failure ($\tau_{ff}$). Duncan et al. (1990) showed that the undrained shear strength of soils when plotted as $\tau_{ff}$ versus $\sigma^1{}_{fc}$ varies with the effective principal stress ratio during consolidation ($K_c$) where $K_c = \sigma^1{}_{1c}/ \sigma^1{}_{3c}$.

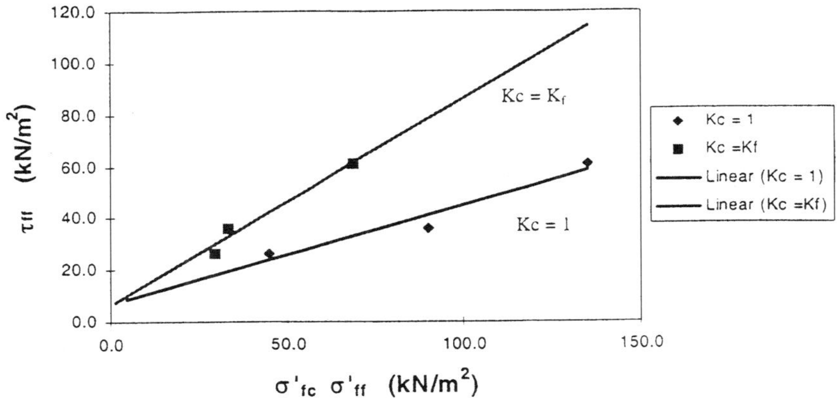


Fig. 7. $K_c$ curves for Fort Henry embankment fill material

As the value of $K_c$ increases the strength increases. For any material there is a family of strength envelopes corresponding to different values of $K_c$ varying from 1.0 to $K_f$. $K_f$ corresponds to the failure line when $\tau_{ff}$ is plotted against the effective stress on the failure plane during failure ($\sigma^1{}_{ff}$).

The first stage of the process is to identify the materials in the section that can drain rapidly and those with a low coefficient of permeability (k) during rapid drawdown. As described above both materials under question here have very low permeability and thus the undrained strength should be considered (Duncan et al, 1990). The next stage of the analysis uses the results of isotropically / anisotropically consolidated undrained (ICU / CAUC) triaxial tests to determine the soil strength envelopes. The final stage of the analysis is to select a slip surface and perform the analysis using the method outlined in Duncan et al. (1990) to obtain a factor of safety after drawdown. The above steps are repeated for other slip surfaces to locate the slip surface with the lowest factor of safety following drawdown.

These calculations were performed by Lydon (1999) for C/S 57 of Fort Henry embankment and a FOS value of 1.7 was determined. Although this value is obviously too high it is much closer to the true value than that produced by the conventional undrained analysis. This is likely to be due to the strength values being closer to reality, because the consolidation process adopted mimics roughly the true stress state in the embankment. Lydon (1999) made a similar finding for his analysis of some rapid drawdown failures in the Headrace embankment on the same hydro-electric scheme.

Effective stress analysis

It has been long recognized that soil behaviour is governed by effective stresses rather than total stresses. In the context of slope stability Janbu (1989) states for example "from years of experience it is strongly advised that effective stress analyses are a must for cuts and excavations in clays and clay silts".

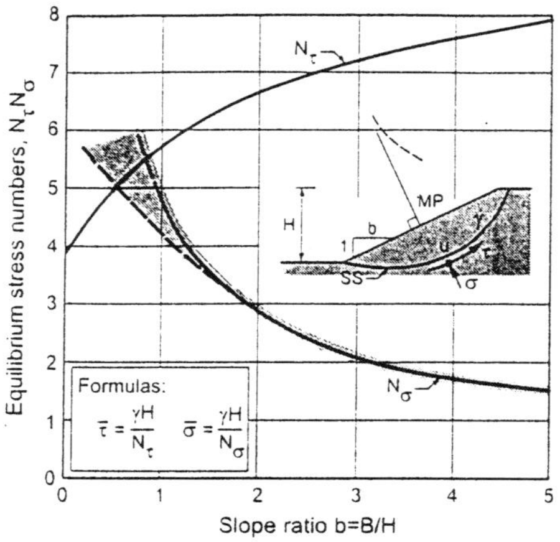


Fig. 8. Janbu's stability charts (Janbu, 1954)

Janbu furthermore goes on to say, "it is not the details of the numerical approach to an overall stability analysis that counts in practice, but the pore pressure regime and the strength parameters". He advocates the use of a relatively simple approach based on effective strength and stability charts. This approach will be applied here.

The factor of safety (FOS) can be expressed as the ratio between average shear strength ($\tau^1_{avg}$) and the average equilibrium shear stress ($\tau^1_e$):

$$FOS = \frac{\tau^1_{avg}}{\tau^1_e} \qquad (1)$$

The average shear strength ($\tau^1_{avg}$) refers to the average in situ effective stress ($\sigma^1_e = \sigma_e - u$) at equilibrium and hence:

$$\tau^1_{avg} = (\sigma_e - u + a^1)\tan\phi^1 \qquad (2)$$

where: u is the average pore pressure along the sliding surface, $a^1$ = attraction = $c^1\cos\phi^1$. Janbu (1954) produced stability charts, which allowed the determination of ($\tau^1_e$) and $\sigma_e$. These are shown on Figure 8.

$$\tau_e = \frac{\gamma H}{N_\gamma} \qquad (3) \qquad\qquad \sigma_E = \frac{\gamma H}{N_\sigma} \qquad (4)$$

For slopes with an external water level (such as at Fort Henry), then the reference stress $\gamma H$ is replaced by $\gamma H - \gamma_w H_w$, where $H_w$ is the height of water. For simplicity u is calculated from a $r_u$ value, where $r_u$ is the ratio of pore pressure to total overburden stress, i.e.,

$$u = r_u \sigma_e \qquad (5)$$

Combining equations, one can obtain the expression:

$$FOS = \left[\frac{(1 - r_u)\sigma_e + a^1}{\tau_e}\right]\tan\phi^1 \qquad (6)$$

For Fort Henry, H = 7 m, b = 2.5, thus $N_\tau = 7.0$ and $N_\sigma = 2.3$. $\tau_e = 12.1$ kPa and $\sigma_e = 37$ kPa. Taking average strength values for the embankment fill and silt from Figure 6 gives $\phi^1 = 32°$ and $a^1 = 10$ kPa. Thus:

$$FOS = 1.911x(1 - r_u) + 0.517 \quad (7)$$

This of course illustrates the difficulty of this approach in that some assumptions must be made about the pore water pressures. It is assumed simply that the reduction in total stress is due to the water level lowering and that the reduction of pore pressure, due to undrained loading, is given by the decrease in the mean total stress (Svanø and Nordal, 1987). Then the $r_u$ value corresponding to the water level of +32 mOD, which occurred at the time of failure, equals 0.78, thus giving FOS = 0.94. This is an encouraging result given that the true factor of safety on this occasion was of course 1.0.

No attempts have been made to refine this value by altering the average strength values or carrying out a more rigorous method of analysis, using for example the method of slices (Janbu, 1973). Janbu (1989) suggests that more refined analyses will only alter the result by "a few percent".

For the situation which exists prior to drawdown (water level of +33.5 mOD) and the current minimum level permitted after drawdown (+32.6 mOD, the relevant $r_u$ values are 0.57 and 0.69 respectively. These values give FOS of 1.33 and 1.11 and are shown graphically on Figure 9. Thus it can be seen that under the current reservoir operating regime, the available level of safety is probably acceptable but there is little scope for altering the regulations as far as level of drawdown is concerned.

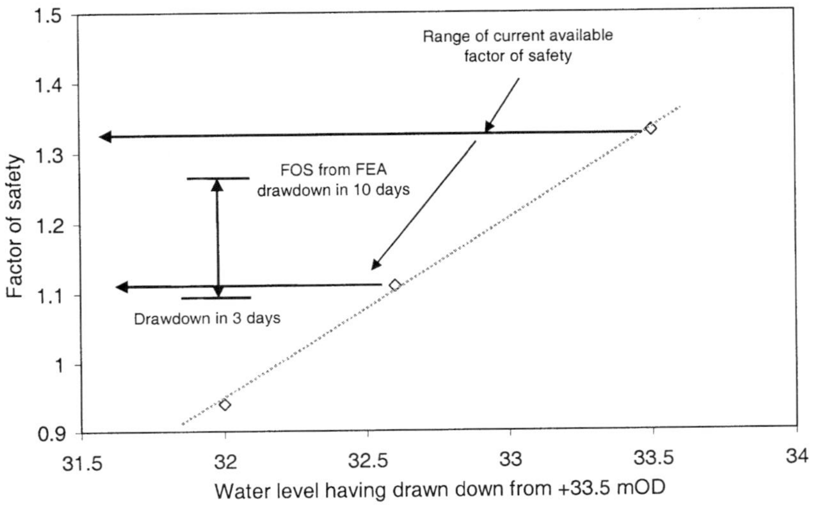


Fig. 9. Relationship between stability and magnitude of drawdown

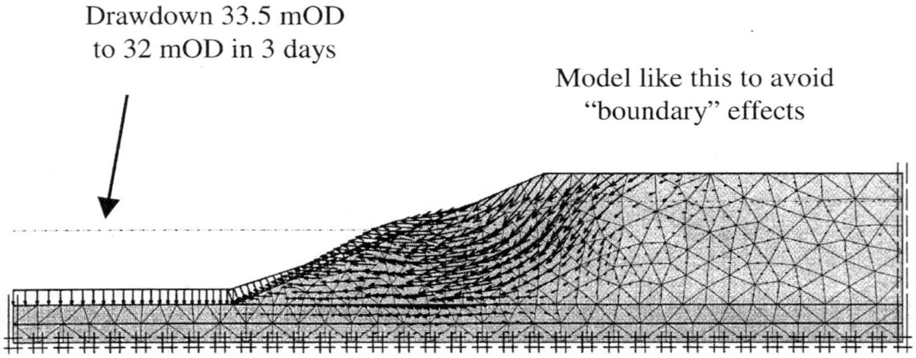


Fig. 10. Finite element mesh for Fort Henry embankment

INFLUENCE OF RATE OF DRAWDOWN

The discussion above has concerned itself on the relationship between stability and the magnitude of drawdown. A question remains as to the influence of the rate of drawdown. Such a stress – stain – time analysis is beyond the capacity of simple hand calculations and resort must be made to more sophisticated techniques.

Such coupled stress - strain / consolidation analysis using the geotechnical finite element software SAGE – CRISP (Conaty, 2001) and also with the code Plaxis. It was assumed that the embankment fill and the underlying silt had effective stress strength parameters and permeability values as described above.

In the analyses the time over which drawdown from +33.2 mOD to +32.1 mOD occurred was varied between 10 days and 3 days. (Currently the regulations give 8 days for drawdown from 33.5 mOD to 32.6 mOD.) Output from Plaxis for the situation over which drawdown took place in 3 days is shown on Figure 10. The plot shows the finite element mesh and boundary conditions used with the arrows used to denote vectors of total displacement. It can be seen that the mode of failure predicted by the anlaysis is similar to that observed on site.

Factor of safety against failure can also be determined by the finite element analysis. Values of 1.1 and 1.25 were determined for the drawdown period of 3 days and 10 days respectively. These are shown, together with those values obtained from the simple effective stress calculations on Figure 9. The results of both sets of analyses are encouragingly close.

It can be concluded that the stability of the embankment is controlled both by the magnitude and rate of the drawdown. The rate of drawdown has a lesser influence. This is probably because of the low permeability of the material involved and dissipation of excess pore pressures are so slow that there are only small changes in periods under consideration here.

## CONCLUSIONS

1. Back analyses of the 1979 Fort Henry slide using total stress (undrained strength) limit state stability analyses yield unrealistically high factors of safety.
2. The technique favoured by the US Corps of Engineers yields a somewhat lower but still unrealistic value.
3. Both simple (stability chart based) effective stress stability analyses and more sophisticated finite element based analyses are able to model the actual slide realistically.
4. These analyses suggest that under the current operating regime the stability of the embankment is probably just adequate.
5. Stability is governed both by magnitude and rate of drawdown, with rate having a lesser influence.
6. There seems little scope for modifying the current operating regime of the reservoir.
7. After the failure remedial works and a review of the operating regime took place. No subsequent problems have been encountered.

## ACKNOWLEDGEMENS

The authors are very grateful to ESB, for their financial and practical support during this work. The help of Jarlath O'Looney, Tony Fitzgibbon and Senan Mc Evoy of the civil works team at Ardnacrusha and the Chief Civil Engineer Jack O'Keeffe is particularly acknowledged.

## REFERENCES

Casey, B. Long, M. & Fitzgibbon, T. (2002). River Shannon hydro-electric scheme: Fort Henry embankment: upstream slope failure and remedial work. Proc. British Dam Society Conference, Dublin. Thomas Telford.

Conaty, E. (2001). A study of the stability of the Shannon embankments. MEngSc thesis, University College Dublin.

Duncan, J.M, Wright, S.G & Wong, K.S. (1990). Slope stability during rapid drawdown. Proc. Seed Memorial, pp253 – 272, Berkeley, CA.

Harty V.D. (1953). Slide in Fort Henry embankment River Shannon, Ireland. Proc. 3rd. ICSMFE, Switzerland, Vol. 2, pp225 – 258.

Janbu, N. (1954). Discussion of different definitions of factor of safety. Chapter 4, Harvard, SM Series No. 46, Cambridge, Mass.

Janbu, N. (1979). Slope stability computations. Embankment dam eng., Casagrande Volume, pp47 - 86. Eds. Hirschfied and Poulos. John Wiley.

Janbu, N. (1989). Slope failures due to rainfall and loss of suction. Proc. De Mello Tribute Volume, pp205 – 1, E. Blucher, LTDA, Sao Paulo, Brazil.

Janbu, N. (1996). Slope stability evaluations in engineering practice. Proc. 7th. Int. Conf. Landslides, Trondheim, Vol. 1, pp17 – 34, Balkema.

Lydon, I. (1999). The stability of the Shannon embankments during rapid drawdown conditions. MEngSc thesis, University College Dublin.

Svanø, G. & Nordal, S. (1987). Undrained effective stress stability analysis. Proc. IXth ECSMFE, Dublin, Vol. 2, pp871 – 875.

# River Shannon Hydro-Electric Scheme: Fort Henry Embankment: Upstream Slope Failure and Remedial Work

B CASEY, ESB International, Ireland
M LONG, University College Dublin, Ireland
T FITZGIBBON, ESB, Ireland

SYNOPSIS. An 80 m long section of the Fort Henry embankment, on the River Shannon hydro-electric scheme failed due to rapid drawdown of the impounded reservoir in June 1979. The water level in the reservoir, at the time of the failure was the lowest for some 35 years. Investigations revealed that the failure plane was non-circular in shape and appeared to be confined to the embankment fill. Remedial work was implemented immediately and was successfully completed within 28 days of the failure. In the aftermath of the failure a review was made of the reservoir operating procedures and changes were implemented to the regulations for the control of the River Shannon. No subsequent difficulties have been encountered. The background to the failure, the remedial work and subsequent review of the reservoir operating procedures are all described in this paper. A description of the ground investigations at the site and the analysis of the slide are given in the accompanying paper to this conference by Long et al. (2002).

## INTRODUCTION

The river Shannon, at 240km long, is Ireland's longest river. The Shannon rises in County Cavan, north of Lough Allen and falls only 13m over the 150km length to Lough Derg. Below the outlet from Lough Derg the river drops 30m in a distance of about 23km. It was this fall that was developed for hydroelectric purposes with construction of the Shannon scheme in the 1920's.

The layout of the scheme is shown on Figure 1. A weir at Parteen diverts water from the Shannon via the 12.5km Headrace canal to the hydroelectric generating station at Ardnacrusha. The water leaves the generating station via the Tailrace canal to Limerick City. Embankments, up to 8.5m in height, were constructed upstream of Parteen at Ardclooney and Fort Henry to expand the storage capacity of Lough Derg. Embankments were also required, up to 20m in height, to contain the Headrace canal for much of its length.

*Reservoirs in a Changing World*, Thomas Telford, London, 2002, 314–323

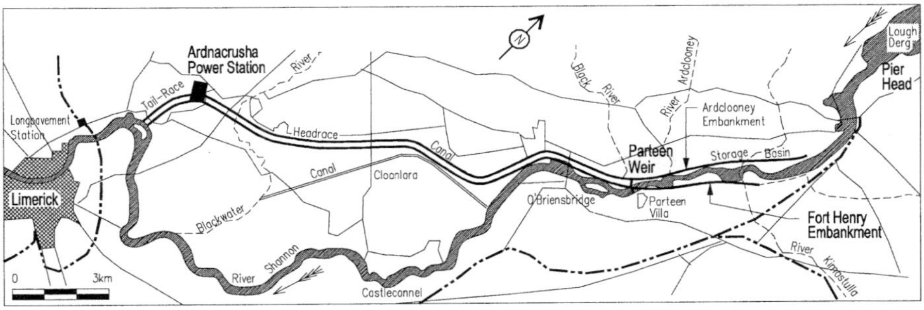


Figure 1. Layout of River Shannon hydro-electric scheme

The embankments were generally formed using locally excavated glacial till deposits and constructed directly on existing ground after removal of very soft deposits where present. Since the time of construction some instances of instability have been recorded on the upstream slopes. The types of problems that occurred prior to 1979 are summarised on Table 1.

Table 1 Upstream embankment failures before 1979

| Year | Location | Description |
|---|---|---|
| Headrace | | |
| 1931 | 3 RB | 20m long slip (Drawdown) |
| 1931 | 175 LB | 6m long crack (Drawdown) |
| 1935 | 105 – 106 RB | Slump at top of wave protection plating |
| 1944 | 458 – 460 LB | 51m long slip in plating |
| 1952 | 495 – 500 LB | Crack over plating |
| 1968 | 459 LB | Slip in plating |
| Ardclooney | | |
| 1928 | 23 – 24 | Slide during first filling |
| 1931 | 330 – 350 | 60m slip (Drawdown) |
| 1949 | 0 | Slump into bed of reservoir |
| Fort Henry | | |
| 1929 | 23 – 31 | 200m long slip |
| 1931 | 23 | Crack due to Drawdown |

Note: LB = left bank, RB = right bank, numbers relate to section markers each section unit equating to 25 m length of embankment

It is noted that six of the incidents occurred during embankment construction and reservoir filling and that there was only one event in the 25 years before 1979. The upstream failure described in this paper took place between C/S 55 and C/S 57 on Fort Henry embankment.

## EMBANKMENT CONSTRUCTION AT FORT HENRY

Fort Henry embankment is some 3.5 km long, with a maximum height of 8.5 m and an average height of 7.0 m above original ground. The crest level of the embankment was designed at 35 mOD to provide adequate freeboard for a range of water levels from 32 mOD to 33.5 mOD. A typical cross section through the embankment is shown on Figure 2.

Before construction began Topsoil and Peat was stripped from the foundation, and a longitudinal channel cut to intercept land drains and to form a core trench. At some locations, there was evidence of high permeability in the underlying glacial till and timber sheet piles were sunk through the ground along the centre of the embankment. Fill material was obtained from a borrow pit adjacent to the north end of the embankment. The Fill is primarily sandy silt of low permeability and is of glacial origin. Excavation from the borrow pit was by face shovel with the material transported by rail to the point of deposition and side tipped from wagons. Compaction was achieved by slewing the track as construction proceeded.

After the embankments had been given an appropriate amount of time to settle, the crest and slopes were dressed by hand and a top layer of surface soil 20cm thick was placed on the crest and on the downstream slope. The upstream slopes and the reservoir bed were protected from the eroding effect of the water by means of a 40cm thick layer of broken stone.

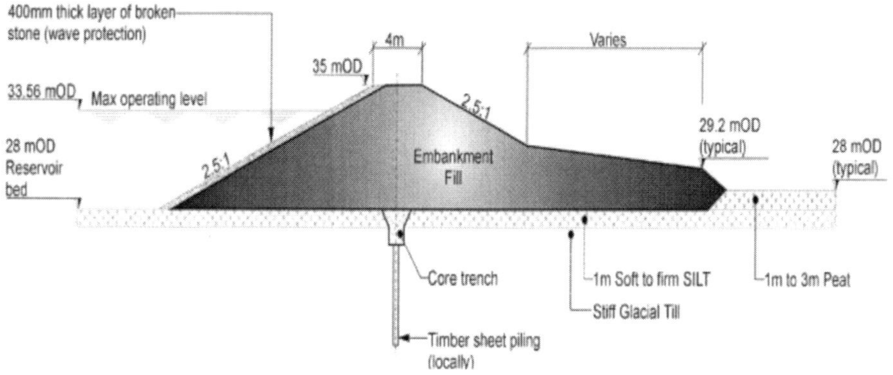


Figure 2. Typical cross section through Fort Henry embankment

## REGULATIONS RELATING TO RATES OF DRAWDOWN AND ALLOWABLE WATER LEVELS BEFORE THE FAILURE

Water levels to control operation of the generating units at Ardnacrusha are measured at 'Pier Head', which is near the Lake outlet. A minimum level of 33.1 mOD is required to sustain the necessary flow and gradient for operating the four turbines on maximum output. At the time of construction, the limiting allowable rate and level of drawdown was determined empirically by Professor Peter Meyer of the Technology University of Zurich, a consultant on the project.

The regulation levels and controls on drawdown, as pertained prior to 1979, were those originally established by Professor Meyer and are outlined in Table 2. It should be noted that the minimum permissible level at the Pier Head Gauge was set at 32.0 mOD.

Table 2. Regulation Levels/Controls at Pier Head Gauge Prior to 1979

| **Description** | **Level (mOD)** |
|---|---|
| Max. normal operating level | 33.56 |
| Min. normal operating level | 33.10 |
| Min. statutory low water level | 32.00 |
| Max. avg. rate of drawdown | 20cm/day |

Water levels are also measured at Parteen Weir, which is approximately 5km downgradient of Pier Head. The water level at Parteen Weir is typically 20-30cm below that at Pier Head and is representative of the average water level in the basin at Fort Henry embankment. Prior to 1979, there was no explicit regulation regarding allowable levels at Parteen Weir.

## GROUND CONDITIONS AND EMBANKMENT PROFILE AT THE FAILURE LOCATION

Little detail of the ground conditions at cross sections 55 – 57 of Fort Henry embankment was available at the time of the slip. However it was known that the embankment was founded on a layer of soft to firm silty clay, approximately 1m thick, which was thought to be alluvial in origin. Investigations in the aftermath of the failure showed that this layer was in turn underlain by 1m of loose silty sand above more competent glacial clay.

## DESCRIPTION OF SLIP

### Background

In spring and early summer 1979, ESB had severe difficulty in meeting peak power demands and the hydropower capacity at Ardnacrusha was utilised to the full. Heavy running of the power station saw water levels in the basin at Fort Henry drop from approximately 33.2 mOD on 20th June to 32.1 mOD on 30th June, refer to Figure 3.

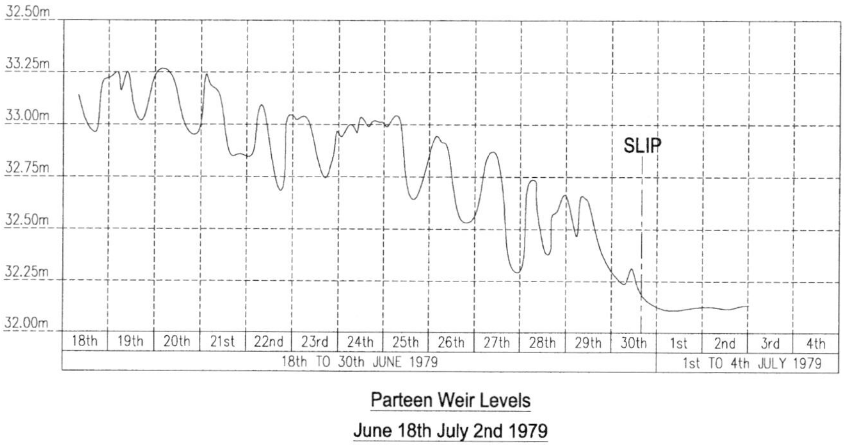


Figure 3. Water levels at Parteen Weir 18/6/79 to 2/7/79

The first report of the slip came from an ESB tug skipper at 6pm on 30th June. In the 10 hours between 8am and 6pm, the drawdown had been 0.35 m and 0.55 m in 24 hours.

## Nature of slip

The slip, with an overall length of 80m, was non-circular and resulted in an irregular slip scarp between 3m and 4m in height (Figure 4). The head of the slip daylighted on the embankment crest. The toe of the slip appeared to emerge towards the base of the slope below water but, owing to obscurity created by the disposition of the slip mass, the possibility of slip involving the underlying silt cannot be discounted. After the slip, divers estimated that the slip mass extended up to 25m into the reservoir. The slip did not result in a breach of the embankment.

The inferred shape of the slip surface is shown on Figure 5.

Figure 4 Photograph of Slip Scarp on day after failure, showing displaced stone protection in foreground

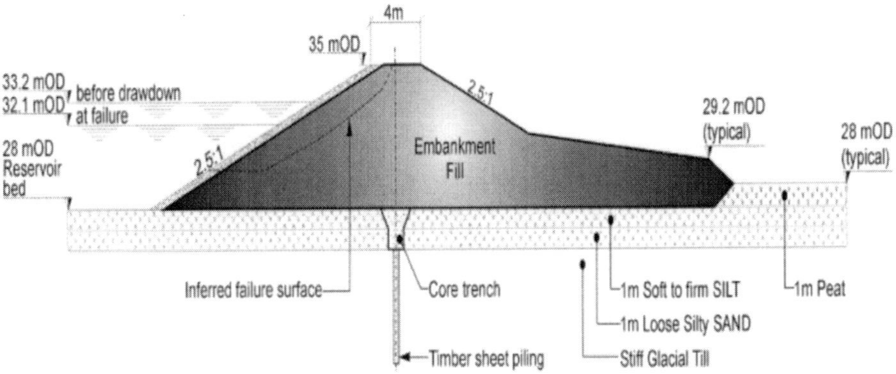


Figure 5 Inferred shape of slip surface and ground conditions at C/S 55-57 Fort Henry

## REMEDIAL WORK

Following a preliminary inspection of the slip, the slip area was covered with visqueen sheeting to prevent ingress of rainwater. The control room at Ardnacrusha was alerted and instructed to hold a steady load on the sets so as to maintain a constant water level at Parteen. A judgement was made that the potential for progressive collapse leading to a breach of the embankment was remote, and that a public alert was not appropriate.

A topographical survey and borehole investigation was commenced on the day following the failure, and the shape of the slip surface estimated by careful excavation from the crest. Following analysis of the failure, a programme of remedial measures were designed with a target that repairs up to water line be completed before the end of July.

The primary concern was to remove the possibility of the occurrence of a deep seated slip that would pass through the silt layer below the embankment. To this end, it was decided to construct an underwater berm within the reservoir to provide a surcharge against a potential slip. The full programme of remedial measures is indicated on Table 3 and on Figure 6 below:

Table 3. Programme for Remedial Measures

| Remedial Measures Stages | Scope of Work |
|---|---|
| Stage 1 | 150mm layer of pit run gravel placed through water on to slip mass in reservoir to act as filter |
| Stage 2 | Rock berm constructed to a level of 30mOD by deposition through water over pit run gravel |
| Stage 3 | Slip scarp above water trimmed and overlaid with 150mm layer of pit run gravel |
| Stage 4 | Slope reinstated to original profile using well graded gravel and cobble mixture in the range 20mm to 125mm |
| Stage 5 | Balancing berm constructed on the downstream side of the embankment |

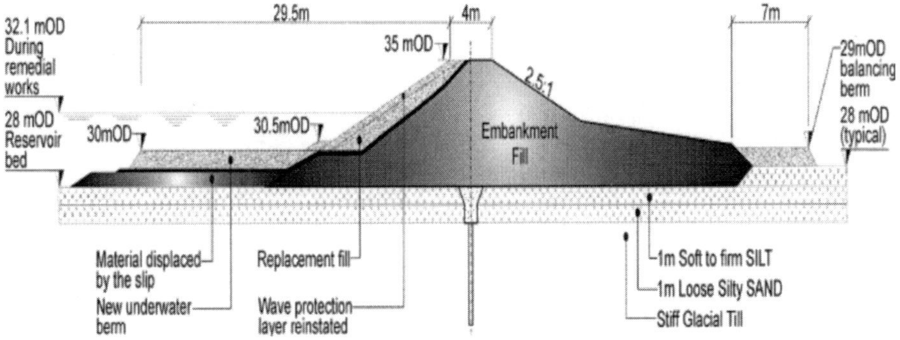


Figure 6 Remedial Works Details

At first it was uncertain whether or not the slip material within the reservoir could support the filter blanket and rock berm. As a precautionary measure rolls of geotextile were brought to site. In the event, although a considerable amount of the pit run gravel did infiltrate the mud, it was possible to form a sound base that carried the rock berm without settlement. The entire operation was continuously monitored by divers.

A 200t hopper barge, with a laden draft of approximately 2m was used for deposition of material within the reservoir. The barge was towed to site by tug. A smaller 20t hopper barge was used for accurate placement of material close to the water edge. Leveling of the berm stone below water was achieved by drawing a heavy steel beam behind the tug.

A 12 hour one shift day was operated for seven days a week and the repairs were substantially completed on 28th July 1979.

## REVIEW OF OPERATING PROCEDURES

Following the slide and the remedial work, a review was made of the operating procedures pertinent to the Fort Henry embankment and the reservoir in general. The review involved an analysis of the maximum previous rates of drawdown in the Parteen basin together with the historic lowest levels that had taken place, without any reported cases of failure. These rates formed the basis of *Extreme Limits* of Drawdown in new regulations that are set out in Table 4. The new regulations also define *Normal Maximum Limits*, set at 5cm below the Extreme Limits, which should be adhered to in normal operation. Significantly, the new regulations also established a minimum allowable level at Parteen Weir of 32.6m OD.

Table 4. Revised Drawdown regulations for the reservoir

| **Duration of Drawdown** | | **Drawdown (cm)** | |
|---|---|---|---|
| **Hours** | **Days** | **Normal Maximum Limits** | **Extreme Limits** |
| 1 | | 10 | 12.5 |
| 2 | | 20 | 25 |
| 5 | | 40 | 45 |
| 7 | | 45 | 50 |
| 24 | (1) | 50 | 55 |
| 48 | (2) | 57 | 62 |
| 72 | (3) | 64 | 69 |
| 96 | (4) | 71 | 76 |
| 120 | (5) | 78 | 83 |
| 144 | (6) | 85 | 90 |
| 192 | (8) | 90 | 95 |

## CONCLUSIONS

In the aftermath of the failure, a major review of stability of all of the embankments on the Shannon scheme was undertaken. A substantial programme of phased improvements was initiated including piezometer installations, filter drain construction and berm construction.

Since the failure described and the implementation of the new regulations, there have been no further instability instances on the upstream face of Fort Henry (or Ardclooney) embankments.

## ACKNOWLEDGEMENTS

The authors acknowledge the help of their colleagues in the preparation of this paper. The authors would also like to thank ESB for permission to publish this paper.

## REFERENCES

Harty V D, (1953). Slide in Fort Henry Embankment River Shannon, Ireland. Proc. 3rd International Conference on Soil Mechanics and Foundation Engineering. Vol II, Session 8.

Long, M., Lydon, I. and Conaty, E. (2002). River Shannon hydro-electric scheme: failure of upstream slope of Fort Henry embankment: analysis. Proc. British Dam Society Conference, Dublin. London. Thomas Telford.

# Long term behaviour of Portumna embankments

E CONATY, University College Dublin, Ireland
M LONG, University College Dublin, Ireland

SYNOPSIS. Monitoring of the settlement of Portumna flood protection banks has been carried out since construction in 1929, producing unique long-term data. Analyses show that it is possible to make accurate predictions of such long term settlements, provided issues such as lateral squeezing of the soils and creep are accounted for. It is also shown that sample disturbance effects play a significant role in obtaining parameters for calculations. The samplers commonly used in Ireland and the UK may not be adequate for important works on soft soil. Finite element analyses suggest that the state of stress in the ground beneath the embankments is such that future loading is feasible.

## INTRODUCTION

This paper describes a detailed ground investigation and analysis of the behaviour of some flood protection embankments, which were constructed for the River Shannon hydro-electric scheme in Ireland. The embankments have settled continuously since their construction and have required regular maintenance including raising to their original design level. To facilitate this repair work unique settlement records have been kept since these embankments were constructed in 1928. There are very few examples of such long-term data sets available in the geotechnical literature.

The main objective of the research was to compare the results of simple conventional hand calculations and more sophisticated computer based methods of analysis to the actual field measurements. A secondary objective was to examine sample disturbance effects for the two different soil samplers used. This was achieved by comparing laboratory determined parameters to those back-calculated from field data. As the embankments have been raised on several occasions, this work also presented an opportunity to examine the state of stress in the ground with future raising cycles in mind.

The owner and operator of the scheme, ESB, supported this work, which was carried out at University College Dublin (UCD). A full account is given by Conaty (2001).

Fig. 1. River Shannon

PORTUMNA EMBANKMENTS

The River Shannon is 240 km long from its source in County Cavan to Ardnacrusha in County Limerick. It drains 10,000 km$^2$ in this distance, which is 1/8 of the area of Ireland (see Figure 1). Between 1925 and 1929 a hydro-electric power scheme was constructed on the river with a power station at Ardnacrusha, County Clare. Several embankment stability problems have been encountered since construction. For example, the 1979 slide in the Fort Henry embankment is described in two other papers presented to this conference (Casey et al., 2002 and Long et al., 2002).

Embankments were also constructed further upstream at Portumna, Co. Galway. Their purpose is to permit higher operating levels in Lough Derg without flooding low lying land along the right bank of the river. They also serve as flood protection against storm floods. The embankments extend mostly upstream of Portumna Bridge to Keelogue Weir. There is also a short section of embankment downstream of Portumna Bridge. In all, the total length of the embankments is about 21 km. The top of the embankments varies between approximately 34.55 mOD and 35.85 mOD. Their height is on average about 1.25 m (maximum 3 m) with a crest width of about 3 m. The side slopes are 1 on 2. Generally they are situated about 25 m from the edge of the river and therefore are operative only when the river is in flood.

For this project a typical cross section of the embankments was chosen for study: on Pump System VI, C/S 14, chainage 620 m, which is located just upstream of Portumna bridge.

Fig. 2. Shell and auger borehole and sampling at Portumna

## SITE INVESTIGATIONS

Prior to this project ESB had carried out several ground investigations mostly by means of Swedish auger borings. It was known that the embankments are underlain by compressible organic, alluvial and post-glacial soils. A further investigation was carried out specifically for this project in November 1999 at the study site. It comprised:

- Two shell and auger boreholes, with samples being retrieved using the 101.4 mm ELE fixed piston sampler, which is the standard sampler used for high quality soft ground sampling in Ireland and the UK. Samples were obtained up to a depth of 8 m.
- One borehole, with sampling using the 95 mm stainless steel thin walled Norwegian Geotechnical Institute (NGI) sampler. This sampler is known to give reasonably high quality specimens.

A photograph of this work is shown in Figure 2. Due to no undisturbed samples of the upper organic layers being obtained at Portumna a further sampling exercise took place at Meelick, approximately 12.5 km upstream of Portumna in March 2000, where the stratigraphy was known to be very similar to that at Portumna. Sampling was carried out using ELE tubes, NGI tubes and U104 tubes which were pushed into the ground using the bucket of a caterpillar excavator.

## GROUND CONDITIONS

The ground conditions revealed at Portumna, together with a cross section of the embankment are summarized on Figure 3. The embankment is founded on a 1 m thick peat layer, overlying about 1.5 m of calcareous silt (known locally as marl). Both of these layers are of recent origin and are highly compressible.

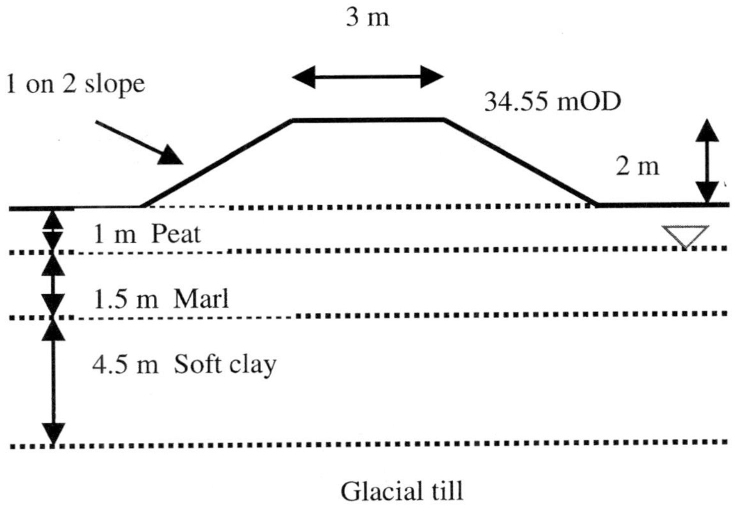


Fig. 3. Cross section

They are underlain by 4.5 m of soft clay, which is thought to be a post glacial lake clay, deposited during the retreat of the glaciers some 18,000 year BP. Competent glacial clay till underlies these layers. Groundwater level is typically at the base of the peat but varies as influenced by the river level. The embankment itself comprises a peaty material, which was obtained from shallow excavations immediately adjacent to the works.

## SOIL PARAMETERS

Conaty (2001) undertook an extensive program of laboratory testing in order to characterize the material and provide input data for analysis. Some data from an earlier ESB investigation carried out by Soil Mechanic Ltd. are also included in the following plots.

### Basic parameters

A plot of moisture content with depth is shown on Figure 4. Moisture contents in the upper two layers are very variable but are typically 60% in the peat layer and close to 200% in the marl. For the underlying soft clay the values are more uniform and fall from about 60% to 45% through the stratum. Bulk density values reflect the above and increase from just over 1 Mg/m$^3$ in the peat to about 1.25 Mg/m$^3$ in the marl. For the soft clay layer two zones are apparent, the upper layer having a density of about 1.65 Mg/m$^3$ and the lower about 1.8 Mg/m$^3$.

Grading curves are available for the soft clay only and confirm that it is made up entirely of clay ($\approx$ 45%) and silt sized particles. It has a liquid limit of about 55% and a plasticity index of 30%, which means it is classified as "clay of high to very high plasticity".

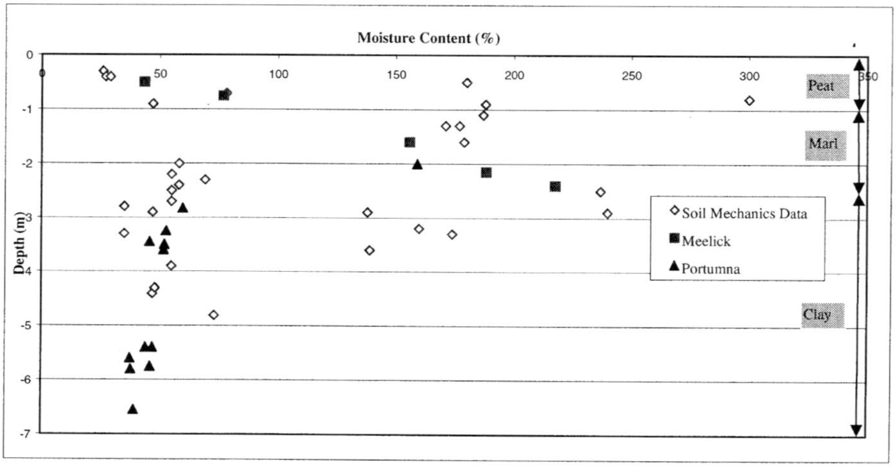


Fig. 4. Moisture content

<u>Compressibility and consolidation parameters</u>

As the nature of the study concerned long-term settlement of the embankments, most emphasis was given to measuring the compressibility and consolidation parameters of the materials using standard oedometer (1D consolidation) tests with 24 hour load increments. Typical test results are shown on Figure 5. It can be seen that the tests are not of particularly high quality, given the rounded nature of the curves. This is especially the case for the NGI sample. Preconsolidation pressure was on average slightly larger than the in situ effective stress, resulting in an overconsolidation ratio between 1.0 and 1.2. These values are consistent with the geological history of the material. A summary of the average parameters obtained from the tests is given in Table 1. Many authors have produced correlations between consolidation properties and basic parameters such as moisture content (e.g. Mesri et al, 1994).

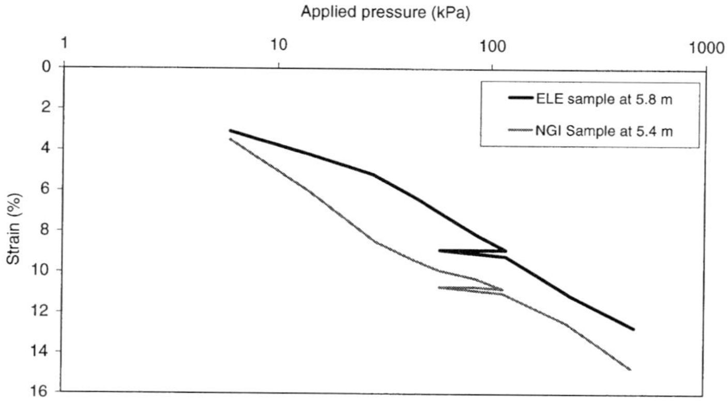


Fig. 5. Typical oedometer test results

Table 1. Summary of average soil parameters

| Geo Profile | Compression index $C_c$ | Swelling Index $C_s$ | Secondary compression $C_{sec}$ | Permeability $k_v$, m/s | Secondary compression $C_{\alpha\square}/C_c$ |
|---|---|---|---|---|---|
| Peat | 0.171 (0.49) | 0.143 (0.095) | 0.1406 | $2.3 \times 10^{-7}$ | 0.022 (0.06) |
| Marl | 1.978 (2.65) | 0.145 (0.76) | 0.064 | $2.5 \times 10^{-7}$ | 0.005 (0.04) |
| Clay top | 0.272 (0.86) | 0.01 (0.1) | 0.017 | $3.5 \times 10^{-9}$ | 0.017 (0.04) |
| Clay Bottom | 0.168 (0.69) | 0.01 (0.1) | 0.008 | $3.5 \times 10^{-9}$ | 0. 024 (0.04) |
| Till * | 0.075 | 0.01 | - | $1 \times 10^{-10}$ | - |

* Typical values assumed for glacial deposits
(Values in brackets were derived from Mesri et al., 1994 using correlations)

Conaty (2001) has described using such correlations for these soils in detail and he found that the stiffness values measured in the oedometer tests were larger than would be expected from the correlations with basic parameters. It is well known that disturbance increases stiffness by compacting or densifying the samples. Corresponding values, which were obtained from this correlation, are shown in brackets in Table 1.

<u>Strength parameters</u>
Some results of anisotropically consolidated undrained triaxial tests (CAUC) are shown on Figure 6 in the form of a deviator stress – strain curves and effective stress paths. The clay material behaves as a conventionally normally consolidated deposit as was expected.

The clay material can be seen to be very soft. Undrained shear strength ($s_u$) values are in the range 13 kPa to 16 kPa and the effective stress friction angle ($\phi'$) and cohesion ($c'$) are of 25° and 5 kPa respectively. It can also be said that the test results are reasonable as far as sample quality is concerned, with perhaps the ELE specimen being superior as found for the consolidation tests.

PERFORMANCE OF EMBANKMENTS
Since their construction the embankments have continuously settled. They have been raised regularly, usually on a 5 year cycle, see Figure 7. A settlement history of three sections of the embankments, from 1929 to 1994, is shown in Figure 8. For example, the raising of the embankments at Pump System V Chainage 610m by 450mm, 170mm and 400mm in 1952, 1961 and 1973 resulted in settlement of 170mm, 125mm and 180mm respectively.

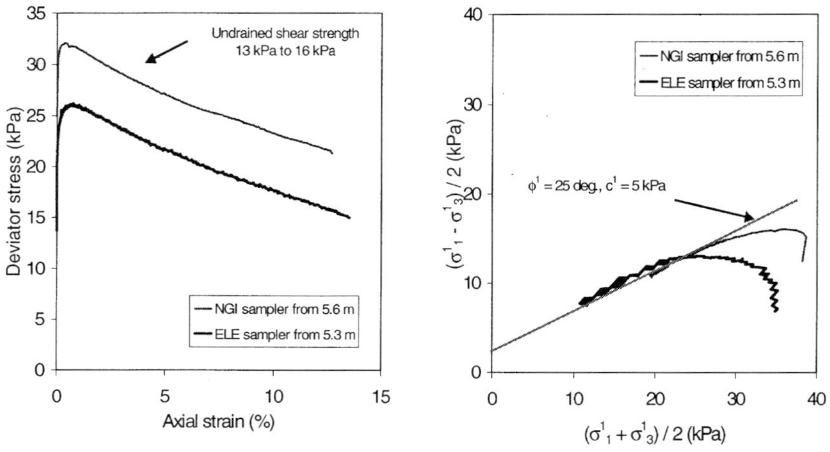


Fig. 6. Typical CAUC triaxial test results

It is however evident that the rate of settlement has decreased with time. This is most likely due to the progressive consolidation and corresponding gain in shear strength and stiffness of the foundation material.

The embankments were breached by high river levels in 1954, which resulted in extensive flooding of the adjoining land. In 1991 a repeat of the 1954 river level occurred. However due to the embankment being raised since the previous flood no flooding occurred. Locally derived topsoil has been used for the maintenance work (see Figure 7). No attempt has been made to compact this soil due to its peaty nature.

Fig. 7. Maintenance work / leveling on Portumna embankments

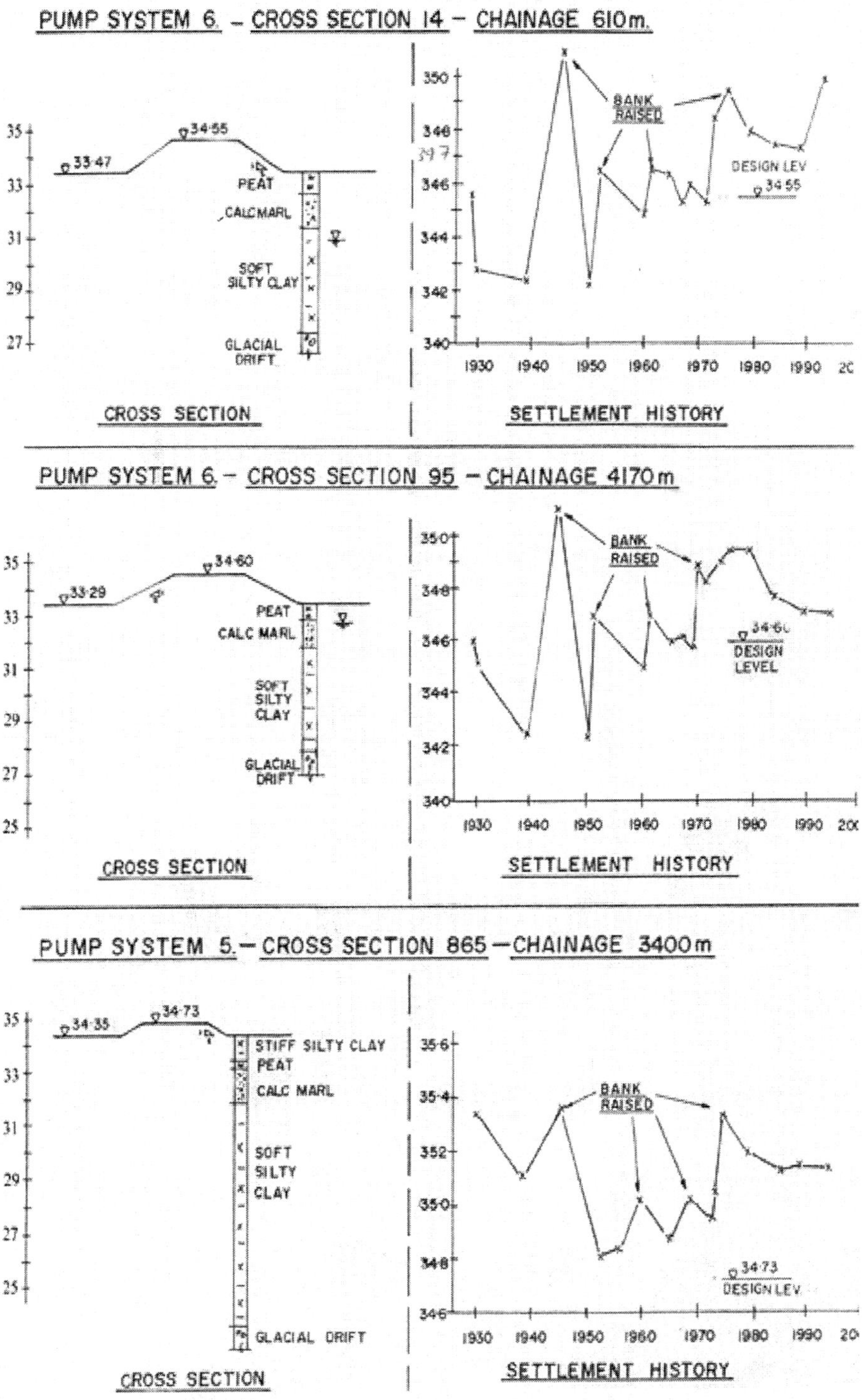


Fig. 8. Portumna embankments – typical settlement history

SETTLEMENT ANALYSIS

Simple conventional approach using directly measured parameters

Total settlement, $\rho_{total}$ is defined as follows:

$$\rho_{total} = \rho_{immediate} + \rho_{primary} + \rho_{secondary}$$

Immediate settlement is mostly due to the expulsion of small quantities of air during construction and has been therefore ignored in the calculations. Primary or consolidation settlement occurs as a result of the gradual reduction in volume of a soil due to dissipation of excess pore water pressure. Secondary settlement or creep is thought to be due to the gradual readjustment of the soil particles into a more stable configuration after the excess pore water pressure has dissipated to zero. Both $\rho_{primary}$ and $\rho_{secondary}$ are clearly of importance for the Portumna embankments and can be calculated from the following equations:

$$\rho_{primary} = \frac{C_c}{1+e_0} * H_o * Log_{10} \frac{\sigma'_{vo} + \Delta\sigma'_{vo}}{\sigma'_{vo}}$$

$$\rho_{secondary} = C_{sec} * H_o * Log_{10}\Delta t$$

where : $e_o$= initial void ratio, $H_o$ = layer thickness, $\sigma'_{vo}$ = vertical effective stress, $\Delta\sigma'_{vo}$ = change in vertical effective stress and $\Delta t$= change in time.

The results of such a calculation for Pump System VI, C/S 14 are shown on Figure 9. The parameters which were directly measured in the laboratory (Table 1) were used. Clearly actual settlement far exceeds calculated values. Conaty (2001) made a similar finding for Pump System V, chainage 86.5.

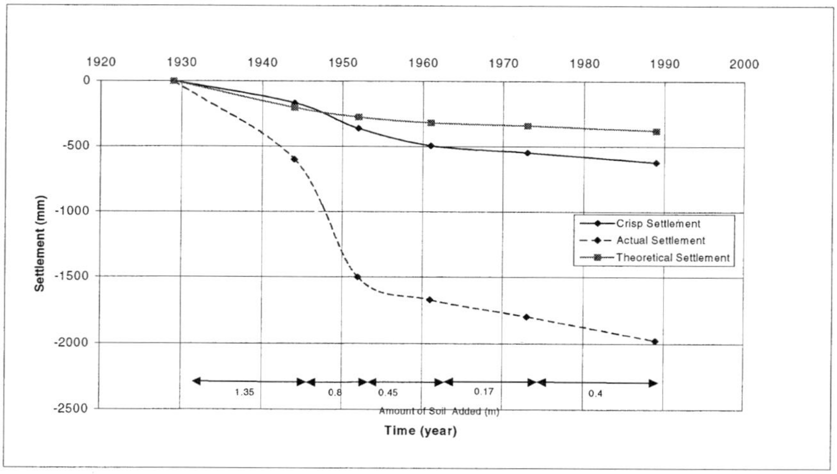


Fig. 9. Actual / calculated settlements – Pump System VI, C/S 14.

Conaty (2001) concluded that some reasons for the under–estimation were that, in the simple calculations, no account is taken of lateral soil movement or of compression of the embankment itself. He therefore decided to attempt a more sophisicted finite element based method of analysis.

Finite element analysis

Use was made of the geotechnical finite element software SAGE – CRISP. It was assumed that the behaviour of the foundation soil material followed the Modified Cam clay constitutive law. This model is commonly applied to analyses of soft normally consolidated clays. The classical Mohr – Coulomb model was used for the embankment fill. The same soil parameters (i.e. laboratory measured) as discussed above were used. In addition it was necessary to assume a Young's modulus (= 8 MPa) for the embankment and a Poison ratio of 0.25 for all the soils.

The CRISP results are also shown on Figure 9. After the first 15 years the theoretical hand settlement is some 40mm greater than that predicted by CRISP. However after this the settlement predicted by CRISP is marginally closer to the real behaviour. This is what might be expected as the CRISP calculations include for lateral displacement which would not be accounted for in the theoretical calculations. However the settlement obtained using CRISP (622mm) is only 31.5% of the measured settlement (1975mm) whilst the theoretical hand settlement is a mere 17% (337mm). Note that CRISP computes primary compression only, ignoring creep.

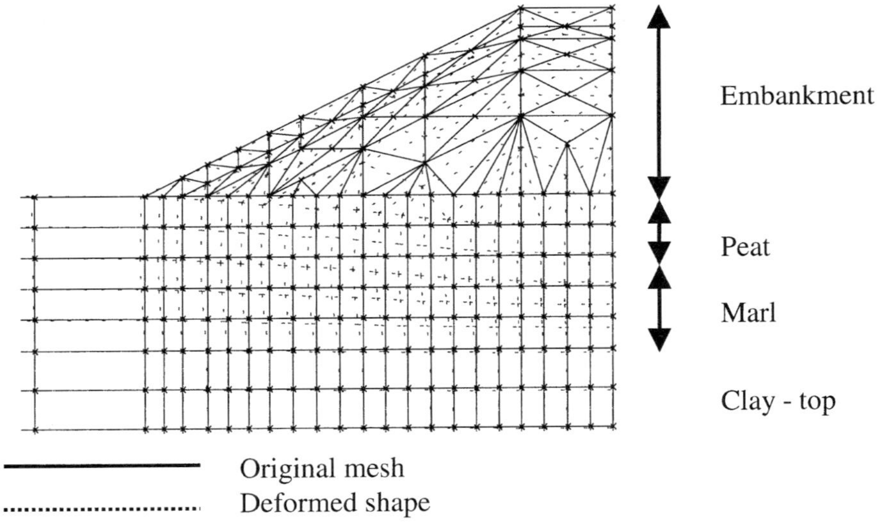


Fig. 10. Deformed finite element mesh (part of)

Figure 10 is a representation of the distorted shape of the mesh after all layers have been added and allowed to consolidate. It can be seen that majority of the settlement has occurred in the marl layer (63% of the total predicted settlement). This is to be expected given that the consolidation properties calculated from the laboratory tests were much higher for the marl than those of the peat or clay. However one might expect the settlement to be larger in soils such as the peat and soft clay.

Influence of sample disturbance

As it was known that the samples were not of the highest quality, these results prompted a more detailed examination of the degree of disturbance. Andresen & Kolstad (1979) and others proposed using the strain required to re-consolidate a specimen back to its in situ effective stress in order to quantify disturbance. Theoretically for a perfect sample it should be possible to re-consolidate with no strain.

Their assessment criteria, together with the Portumna data are shown on Figure 11. It is evident that the results are less than satisfactory. No test is in the quality range described as "excellent" and only one test could be said to be of "very good to excellent" quality. The majority of the samples would be described as of "poor" quality and some even fall into the "very poor" category. The samples recovered using the ELE sampler were of better quality than those from the NGI sampler.

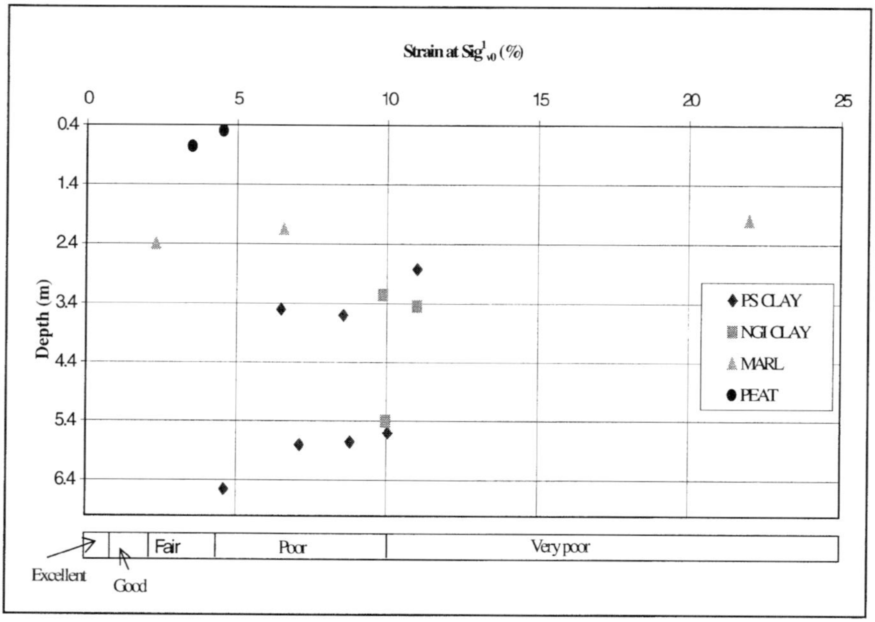


Fig. 11. Assessment of specimen quality (Andresen & Kolstad, 1979)
PS = ELE sampler, NGI = NGI sampler

Revised analysis

Therefore a new set of analyses were undertaken using a revised set of parameters, based on correlations with simple index parameters as described above. The parameters used are summarised on Table 1. The results are shown on Figure 12. In this case the hand calculations underestimate the actual settlement but the finite element analysis yields a reasonable prediction of the recorded data. It is clear that lateral squeezing of the soil (allowed for in the finite element analysis) plays an important role. If the finite element model had been extended to include creep, then the actual and predicted values would be very close.

## STATE OF STRESS

In his finite element analyses, Conaty (2001) found that, following the successive stages of filling, the excess pore pressures which initially develop in the foundation soils have had time to dissipate before the next stage of filling. Therefore the soils have had time to consolidate and gain strength. The analyses showed that some strain hardening of the soil has taken place but the critical state has not been reached except in some small localized areas. Therefore some future loading of the embankments is feasible.

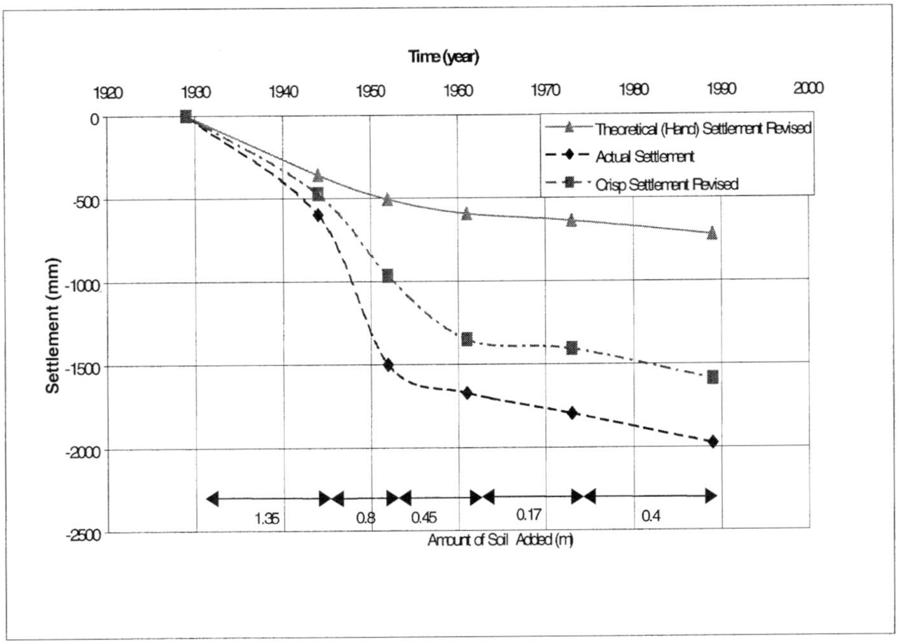


Fig. 12. Comparison of revised settlements

## CONCLUSIONS

1. It is possible to make accurate predictions of the long-term behaviour of embankments on very soft ground. However this study shows that:
2. It is necessary to allow for lateral squeezing of the soil beneath the embankment. Simple 1D calculations may underpredict settlement, especially if the foundation is thick compared to the embankment width.
3. Sample disturbance effects play a major role in the determination of parameters for the calculations.
4. Conventional best practice in Ireland and the UK may not yield satisfactory specimens and it may be necessary to obtain higher quality samples such as block samples if the project is important.
5. Parameters obtained in the laboratory should be compared with those which can be obtained from simple correlations.
6. The ELE 100 mm sampler (standard for Ireland and the UK) yields better quality specimens than the NGI 95 mm sampler.
7. The finite element analyses suggest that the soils beneath the embankment have strengthened due to consolidation and will be able to accept some future loading.

## ACKNOWLEDGEMENTS

The authors are grateful to ESB, for their financial and practical support during this work. The help of Jarlath O'Looney, Tony Fitzgibbon and Senan Mc Evoy of the civil works team at Ardnacrusha and the Chief Civil Engineer Jack O'Keeffe is particularly acknowledged.

## REFERENCES

Andresen, A. & Kolstad, P. (1979). The NGI 54 mm sampler for undisturbed sampling of clays and representative sampling of coarser materials. Proc. Int. Sym. SOA Current Practice of Soil Sampling, Singapore, pp 13 - 21.

Casey, B. Long, M. & Fitzgibbon, T. (2002). River Shannon hydro-electric scheme: Fort Henry embankment: upstream slope failure and remedial work. Proc. British Dam Society Conference, Dublin. Thomas Telford.

Conaty, E. (2001). A study of the stability of the Shannon embankments. MEngSc thesis, University College Dublin, November.

Long, M, Lydon, I & Conaty, E. (2002). River Shannon hydro-electric scheme: Failure of upstream slope of Fort Henry embankment: analysis. Proc. British Dam Society Conference, Dublin. Thomas Telford.

Mesri, G., Lo, D.O.K. & Feng, T.W. (1994). Settlement of embankments on soft clays. Geo. Sp. Pub. No. 40, ASCE.

# The influence of climate and climate change on the stability of abutment and reservoir slopes.

P R VAUGHAN, Imperial College, London, UK
N KOVACEVIC, Geotechnical Consulting Group, UK
A M RIDLEY, Geotechnical Observations Ltd., UK

SYNOPSIS. The stability of abutment slopes, if they are steep enough to slide, depends on trends in both long term and short-term rainfall. The risk of sliding can be related to these trends in a manner similar to the determination of design floods. Slides may have moved to a more stable position when climate was worse, giving a reserve of stability. Speed of movement strongly influences danger. Reactivated slides should not move rapidly but first-time slides may do so. Risk of sliding could be estimated by limited monitoring. Prophylactic remedial drainage works could then be used to control risk.

## INTRODUCTION

Along with other slopes, the abutments of dams and reservoirs have been affected by climate since the last glaciation. Many are too flat to slide. Some involve old slides or have the potential to develop new ones. At present, inland Britain is virtually free from tectonic uplift. There is little downward erosion and steepening of slopes[1]. The climate is relatively benign and has been since we built our dams. Sliding events are rare. However, massive landsliding has occurred since the last glaciation, probably when the climate was more extreme. This should have made critical slopes flatter and more stable. However, we do not know by how much, nor is it always true.

One of the main causes of landsliding is extreme rainfall. We require dam spillways to be big enough to deal with extreme rainstorms. Such storms might subject slopes to increases in seepage pressure more severe than they have experienced before. The risk from slope failure may be comparable to risk from inadequate spillway capacity. A wetter and more extreme climate would increase the risk from both.

Typical seepage regimes in slopes are described in a companion paper (Vaughan

[1] There are exceptions. The Severn in the Ironbridge Gorge is downcutting actively (Hutchinson, 2000), as is the small Cod Beck near Osmotherly on the North York Moors.

*Reservoirs in a Changing World*, Thomas Telford, London, 2002, 337–352

*et al*, 2002). In Britain the water available for infiltration exceeds evapotranspiration and, on average, the surface boundary is wet. Two profiles can be identified. The nomenclature of Vaughan *et al* (2002) is followed. In profiles A.1 the permeability of the soil is lower than the water available for infiltration and high pore pressures develop with saturated flow and a superficial zone of climate-induced fluctuation. (A.1) profiles generally exist where permeability is low, in clays. The stability of such slopes has been discussed fully in the accompanying paper for embankment slopes. The same arguments apply to clay abutment and reservoir slopes and they will not be repeated here. In profiles A.2 the water available for net infiltration is less than the saturated permeability of the soil. Superficially, water percolates vertically downward under a low suction through a zone of partly saturated flow. It then reaches a zone of saturated flow at depth, from which it flows laterally out of the slope. The relationship between infiltration and permeability which determines which type of flow develops is given by Fig. 2 of Vaughan *et al* (2002). Profile A.2 is typical of the fissured rock slopes of many British upland valleys.

SEEPAGE PRESSURES AND SLIPS IN ABUTMENTS SLOPES.

The British inland landscape has a good cover of vegetation which largely prevents surface erosion. Rivers cut down very slowly. Many valleys have filled with sediment since they were at their steepest. There is no significant tectonic uplift. Thus there is little current landslip activity due to slope steepening[2]. Slips are usually caused by high or increasing pore pressure.

Many slopes are quite flat due to the much more active climate and the sparse vegetation during the immediately post-glacial period. Thus new 'first time' inland slides are rare, other than those caused by the works of man. The risk of such slides can only be considered for a particular site. A more common risk in land is associated with the reactivation of old landslides which may date from immediately post-glacial times.

Both types of movement can be triggered by increased seepage pressures due to increased rainfall[3]. Possible increases in seepage pressure due to climate change can be estimated if the relationship between current climate and seepage pressure is established by observation. The effect of an estimated climate change on future seepage pressures can then be predicted and the effect of this on slope stability estimated.

---

[2] This occurs on coasts where marine toe erosion is taking place

[3] Climate and rainfall has varied over the last few thousand years. It is likely that current slopes have been subject to more severe rainfall conditions in the past. If a slope then adjusted to these more severe conditions, it may now have a reserve of stability to resist future increases in seepage pressure.

Variation of seepage pressure with rainfall in slopes with the A.2 profile. Sometimes the abutment slopes (or some strata in them) are very permeable. Then the water table in the abutments can be low and changes in infiltration rate will have negligible influence on seepage pressures. More usually the infiltration generates a relatively high hydraulic gradient in the saturated zone of the abutment and the water table is quite high, matching the geometry of the slope.

The raising of a water table by infiltration in the UK usually involves the filling of cracks and fissures and any drained and open pores above the existing water table. The increase, $\Delta h_w$, due to an inflow, $\Delta q$ is apparently given by :-

$$\Delta h_w = \Delta q / n_e \qquad (1)$$

where $n_e$ is the volume of fissures, etc. filled by water per unit volume of soil or rock. In the field $\Delta q$ is unknown, and it is convenient to substitute $\Delta q_r$, the rainfall per unit area over the time in which the rise $h_w$ occurs. This gives a magnification factor, $C_1$, where

$$C_1 = \Delta h_w / \Delta q_r \qquad (2)$$

Fig. 1 (after Knill, 1966) shows a plot of water table level measured by a continuous float recorder in a borehole in the abutment of the Cow Green Reservoir in Upper Teesdale prior to its construction. There is a small seasonal change in water level (Upper Teesdale has a wet climate in both winter and summer) and short term increases due to periods of heavy rain. A similar situation was deduced by Skempton *et al* (1989) in the Mam Torr landslide in Derbyshire. Using data from elsewhere, they deduced that the multiplication factor could be as high as eight, although most of the data they used was for (A.1) profiles. The Teesdale data shows a magnification factor of the order of 50. The rises correlate with heavy rainfalls (Fig.2). However, the meteorological station used was some 4km away and 100m lower. Local variation in rainfall may account for the scatter of the data points.

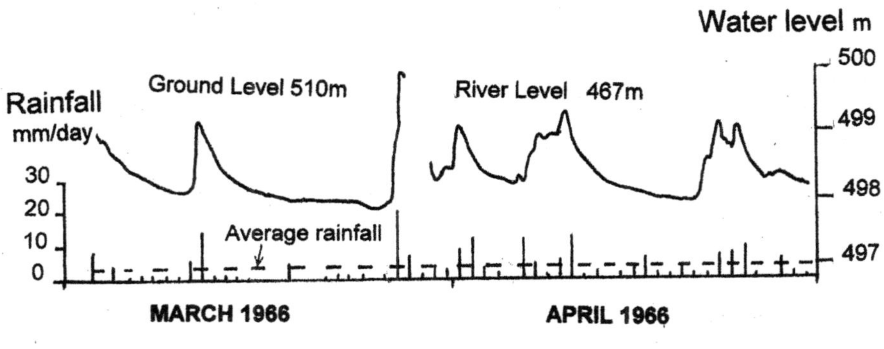


Fig. 1. Fluctuation of water table in the abutment of Cow Green reservoir due to rainfall (Knill, 1966).

Re-activation of existing landslides.

The occasional movements of the Mam Torr landslide over the last 150 years are known from the history of the road which crosses it (Skempton *et al,* 1989; Waltham & Dixon, 2000) . They occur in winter conditions when the general water table is high, but only after periods of unusually heavy rain, when significant short term increases in seepage pressures occur. Movements are small. The residual strength operating on the shear surface is rate dependent and increases with displacement rate as shown on Fig. 3 (Skempton *et al,* 1989). Wedage (1997)gives similar results.

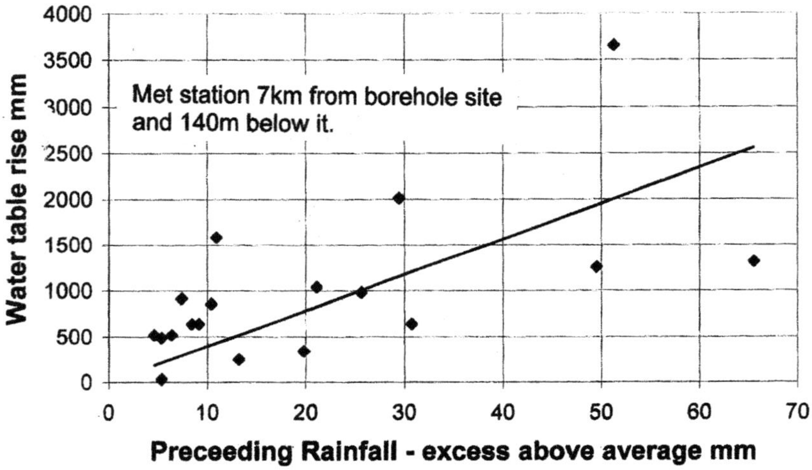


Fig. 2. Ratio of increase in water table level to preceding rainfall in the abutment of Cow Green dam.

As the water table rises the 'static' factor of safety drops below unity and the slide starts to move. The strength then increases, restricting the slide velocity to that which gives the strength necessary to prevent significant acceleration. The relationship between the short term water level pulse, the drop of safety factor below unity and the velocity can be calculated, giving a displacement/ time relationship as shown on Fig. 4. This is derived for a semi-infinite slide by assuming a water level increase and the shape of the pulse taking one day to increase and ether 10 or 50 days to decay in the manner of Fig. 1. The strength velocity function is the average from Figure 3. The new strength required for stability is calculated from the water level. and the matching velocity is taken from Fig. 3. The displacement is obtained by integrating velocity with time. The period of visible movement is much shorter than the period of the water table pulse. Skempton et al (1989) and Vaughan (1993a) present similar analyses. Wedage *et al* (1997) incorporate similar rate effects in a finite element analysis.

An oddity in the rainfall induced pulses in water level is that the post-peak decay is quick. Once the pulse is over, the decay rate becomes much slower. This is not predicted by conventional seepage theory . The same rapid decay is seen in similar pulses in, for instance, residual soil slopes subject to severe cyclonic rainstorms and elsewhere (Shimizu, 1999). In principle, the rapid recovery

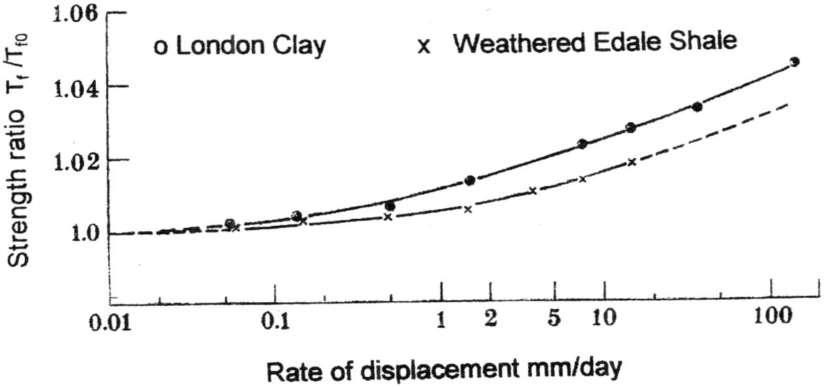


Fig. 3. Relationship between residual strength and rate of displacement (after Skempton et al, 1989).

might be due to a more permeable layer above the starting level, but it happens at various starting levels in the same slope. Thus the explanation of varying strata seems unlikely.

Cabarkapa (personal communication) has suggested that the effect could be due to trapped air. Both the increase and the decay due to a rainfall pulse will be controlled by the value of $n_e$ (Equation 2). If a sudden and substantial seepage downflow due to rain traps occluded air bubbles below the rising water table, the value of $n_e$ would be reduced, giving a higher magnification factor and a higher rate of increase and decrease of water level in the pulse.

The rate of decay of the pulse is important. Fig. 4 shows that movement is about four times higher for the slower recovery time. The sensitivity of the landslide movement to the height of the rainfall pulse is shown in the last part of Fig. 4. A 1m rise produces 0.1m movement; 2m produces 1m and 4m produces 20m movement. Extreme rainfall will produce large movements. Occasional rises of 1m would produce movements which look like creep. Bracegirdle *et al* (1992) show movement of a wharf destabilised occasionally by low spring tides. It looked like creep but consisted of small pulses.

If the link between maximum water level and climate is established, so can the risk of movement and the change in this risk due to a change in climate. Continuous recording of water level and of climate at the site of the observation[4] is required. This can now be done automatically, with remote reading. Existing standpipe piezometers can be used.

Present stability of existing landslides

Most inland landslides are several thousand years old. They have been subject to a history of climatic change. Probably, conditions more critical than those at present have existed in the past. For instance, it seems that there was more landslide activity than usual in the 'little ice age' of the 17th and 18th century. Climate during the Loch Lomond Stadial, some 10,000 years ago was much more severe and lead to many more slope movements and landslides (Hutchinson, J. N., personal communication).

Periodic rainfall/water table/movement events such as those identified at Mam Torr (Skempton *et al*, 1989) cause downslope movement of the whole landslide. With the same seepage pressures, this may make the slide more stable. A simple analysis of the effect is given on Fig. 5. In principle, if the slide angle and the ground water level remain constant when it moves, the safety factor remains unity and the slide remains equally vulnerable to rainfall events. If the slide moves out onto ground of lower slope angle, then it becomes more stable and the vulnerability to rainfall decreases. Each movement increases the rainfall required to cause a further movement. Fig. 5 presumes a landslide typical of Mam Tor. Four cases are presented; where the length of the slide, $L_1 + L_2$ remains constant and where the back of the landslide is recharged by falls from the back scarp ($L_1$ constant), each with an upper slope with $\beta_1 = 14°$ and with alternate lower slopes, $\beta_2 = 7°$ and $\beta_2 = 0$.

The movement required to raise the critical water level by 1m in the 20m deep slide is approximately 5m with the lower slope half the upper one and 3m when run-out is on level ground. It can be seen that modest historical movements will substantially improve stability and increase the size of the rainfall event required to cause further movement. The Mam Tor landslide does not change angle significantly with movement and it is still periodically active. On the other side of the Mam Tor ridge, the Mam Nic landslide, in the same geology and of similar size, has run out onto near level ground at the toe and is almost stable (Waltham & Dixon, 2000).

There is good evidence that old landslides may have a useful reserve of stability under current conditions. On the south side of Longdendale, east of Manchester, the Laurence Edge landslide, 500m high, has been destabilised by the unloading of the toe for the construction of the Lancashire and Yorkshire Railway and by the flooding of the toe by the Woodhead Reservoir. No movement has been reported. However, on the northen side of the dale, the Didsbury Intake landslide was reactivated by heavy rain (120mm in 6 days) in February 1852. The movement was probably exacerbated by excavation for

---

4 In the short term, upland climate varies locally to a significant extent.

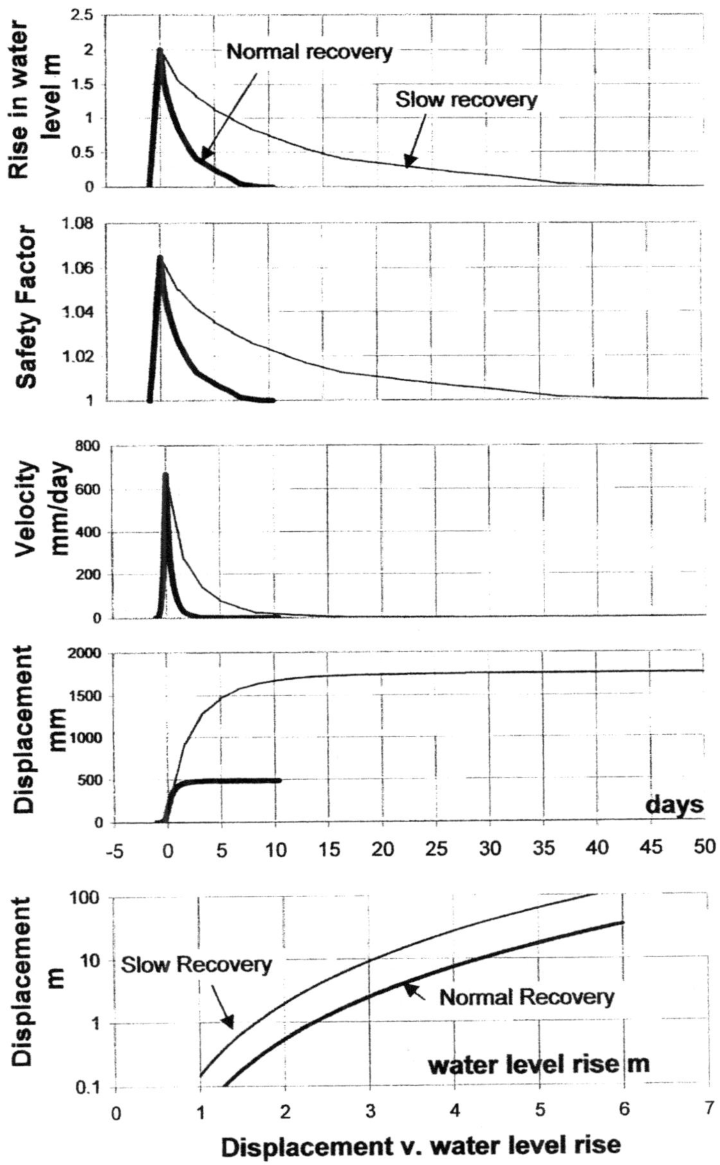


Fig. 4. Movement of an existing landslide due to a rainstorm (Vaughan, 1993a)

Rhodeswood dam, then under construction. The landslide was stabilised by a drainage adit and a buttress of fill. Further up the slope, the Millstone Rocks landslide is still active (Skempton *et al*, 1989; Johnston & Walthal, 1979).

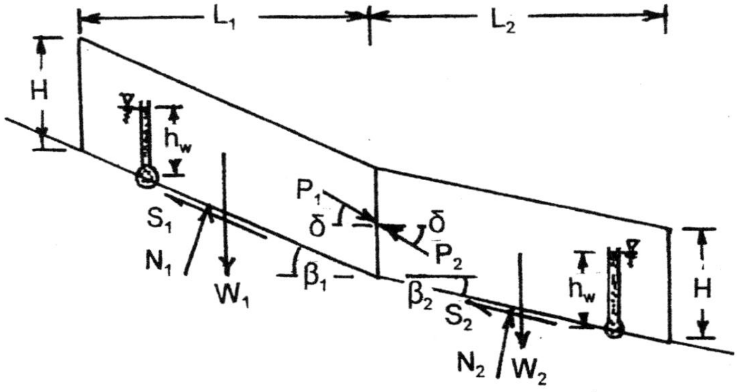


$$\frac{h_w}{H} = \frac{\gamma}{\gamma_w} \cdot \frac{C_1 - C_2 . \dfrac{D_1 . L_2}{D_2 . L_1}}{\sec\beta_1 - \sec\beta_2 . \dfrac{D_1 . L_2}{D_2 . L_1}}$$

where $C_1 = \cos\beta_1 - \sin\beta_1 . \cot\varphi'$, $C_2 = \cos\beta_2 - \sin\beta_2 . \cot\varphi'$

$D_1 = \cos(\delta - \beta_1) . \cot\varphi' - \sin(\delta - \beta_1)$, $D_2 = \sin(\delta - \beta_2) - \cos(\delta - \beta_2) . \cot\varphi'$

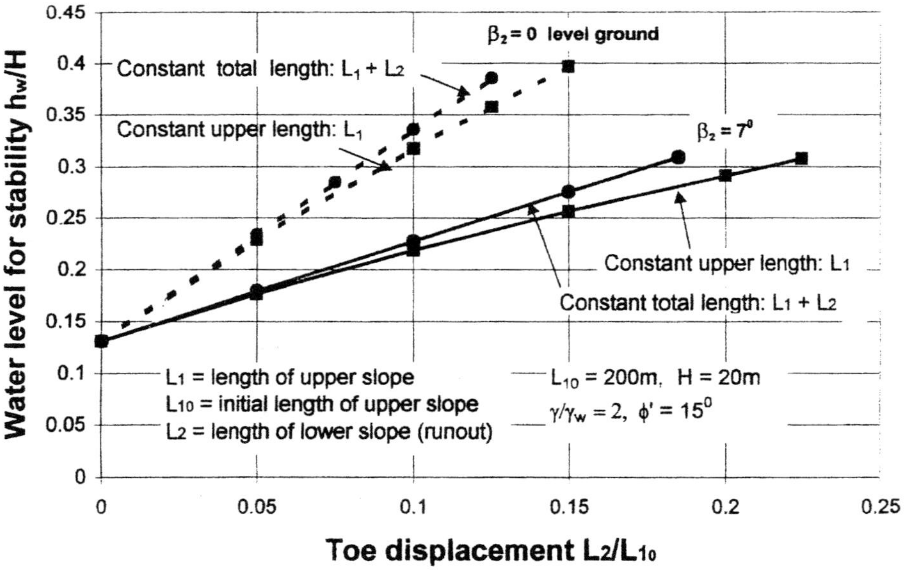


Fig. 5. Stabilisation of a landslide due to movement - the increase in the threshold water table, $h_w$.

Velocity of movement

Since the very fast landslide in the slope above the Vajont reservoir caused a wave which overtopped the dam, the potential velocity of a slide into a reservoir has been of great concern, yet the problem is poorly understood. If a rockfall occurs and generates debris at the top of a slope steeper than its natural angle of repose, the debris turns into an avalanche and accelerates to high speed. High speed involves three dangers; loss of life because there is no time to escape from the landslide path, a much larger travel distance as the momentum reached allows the slide to travel well past the point at which it should be in static equilibrium at its reduced strength, and the possibility of generating dangerous waves if it slides into water.

The principal cause of acceleration is brittle failure; the material loses strength once it has failed. Disturbing forces are not decreased until large movements occur. The slide then has the potential to accelerate to high speed. However, it cannot do so if strength increases with velocity to maintain equilibrium[5]. The general behaviour of a first time slide where this is true is illustrated on Fig 6.

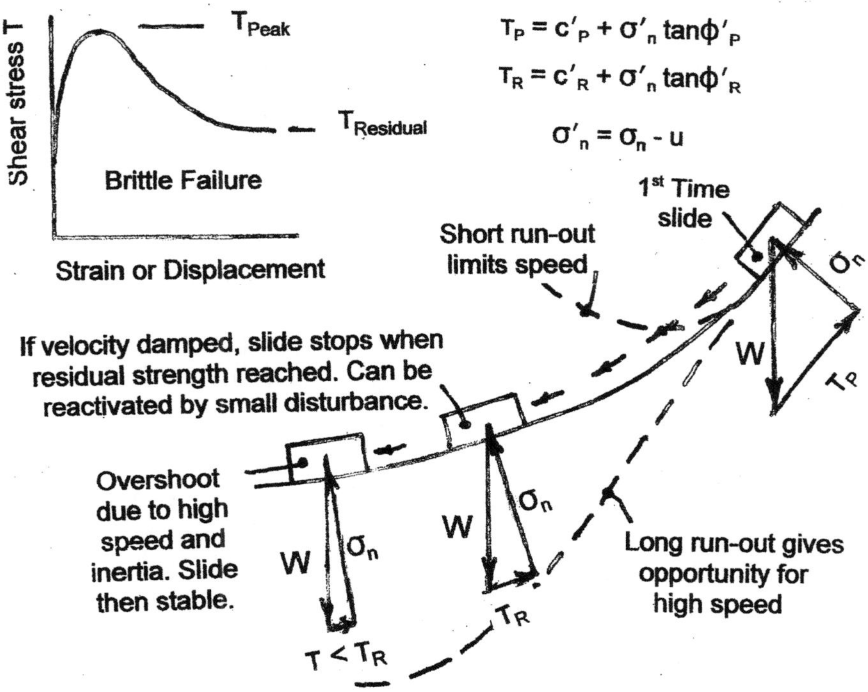


Fig.6. Behaviour of a first time slide and its position when it comes to rest. Slide shown as a block and pore pressure assumed to be zero for clarity (Vaughan, 1993).

[5] In a brittle first-time slide a part of the eventual rupture surface forms after the collapse starts. This fails undrained and may develop reduced pore pressures. There is also load transfer as the slide moves which changes pore pressure. The slide may be braked by these undrained pore pressure effects and its velocity may be controlled by the rate at which they dissipate.

Some saturated soils (quick clays and loose sands) are very brittle due to an unstable and highly contractive micro-structure and to the development of high excess pore pressures when they are sheared to large strains. Other soils and weak rocks have weak bonds which are destroyed by shear, leading to loss of strength. In all these cases the strength drops post peak due to properties inherent to the pre-failure material.

Reactivation of a fully formed existing slide cannot be brittle as it is already operating at its minimum strength[6]. Reactivation by a rising water table produces a small destabilising effect. This may be balanced by a strength increase as the slide accelerates. There is then no risk of rapid movement.

However, there is a further potential cause of rapid movement. A drop in strength on a residual surface has been observed when the rate of shearing of cohesive soils in the Imperial College ring shear apparatus (Bishop *et al*, 1971) exceeds a certain velocity (Tika *et al,* 1996). A typical result is shown on Fig. 7. The authors show that the drop in strength is an apparatus effect. The apparatus was designed for accurate, slow, fully drained tests and has a gap between the two halves of the box to prevent metal to metal sliding. When the base is rotated quickly, it seems that the top half of the box bounces a little and water from the water bath usually enters the shear surface. The loss of strength only occurs when this happens. However, no excess pore pressure develops until the fast shearing stops.

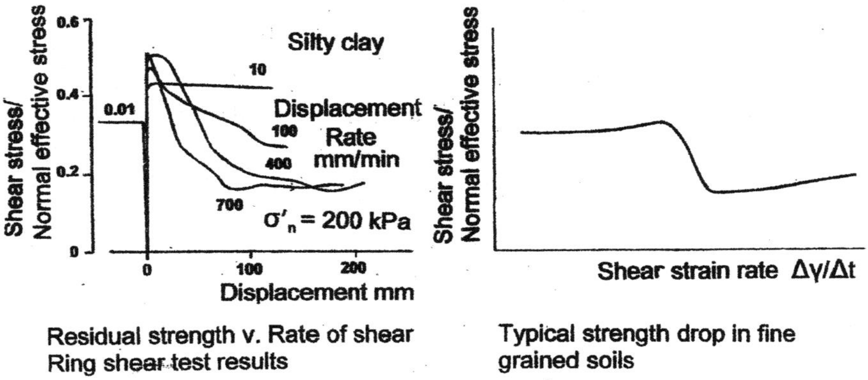


Fig. 7. The influence of shear rate on residual strength measured in the ring shear apparatus - loss of strength at a critical rate (Tika *et al*, 1996)..

The mechanism operating seems to be the same as that identified by Bagnold (1954 &1956). It involves the collision of particles, which generate a stress which has an effect similar to that of an effective stress but which is of different

[6]Second order effects may develop due to aging and slight changes in geometry.

origin. Bagnold performed experiments in which a mixture of wax spheres and water was sheared in a vertical drum at different rates and different concentrations. The spheres had the same density as water, so low concentrations could be tested without them segregating due to gravity. The lateral normal stress and the shear stress was measured; the concentration and the rate of shear was varied. The equivalent result for a ring shear test (constant normal stress and varied rate with shear stress measured) can be interpolated from the results. Such an interpolation is shown on Fig. 8. It is of the same form as the ring shear result shown on Fig. 7. However, the concentration of the particles (here expressed as a void ratio) is known for the Bagnold tests. Fig. 8 shows that when the drop in strength occurs the void ratio increases to a value which indicates that the particles are no longer in contact.

The Bagnold approach and equations have been used quite extensively in the study of avalanche movement (see, for instance, Irgens & Noram, 1996 and Melosh, 1987). They do not seem to have been fully recognised within Soil Mechanics and the study of slope stability. The evidence that they can also operate on the micro scale in the shear zone of a ring shear apparatus is strong. The inference from this is that they can operate in shear zones in the field. Such a loss in strength could cause the rapid acceleration of a landslide to high speed. It is a function of the sliding mechanism, not of the intrinsic nature of the slide material before failure. It can develop in dense soils and in rocks.

The ring shear evidence indicates that two conditions are necessary for loss of strength to occur. First, the rate of shear displacement must reach the critical value. The ring shear tests indicate that this is about 100mm/min, but this might be a function of the apparatus. The increase in pre-drop strength up to this speed is considerable and it is likely to prevent a landslide reactivated by an increasing water table from reaching the critical speed (see Fig. 4). First-time slides are inherently more likely to reach critical speed as a substantial loss of strength occurs immediately the landslide starts.

Second, free water (or possibly air[7]) must be available in the shear surface or shear zone. This is perhaps unlikely when shearing is in clay and other materials of low permeability. It seems most likely in shear in, say, mud-rocks along bedding planes, particularly if there are fissures containing free water. The two sides of the shear may not mate perfectly, leaving gaps into which water can flow from the fissures.

An underwater slide will move onto free water. Such slides in sand tend to flow quite often, perhaps more often than is consistent with the presence of very loose and contractive soil as a necessary pre-condition.

---

[7] In principle, no fluid is required - hence the occurence of flow slides on the Moon and Mars, as reported by Casagrande (1971 ) and Melosh (1987).

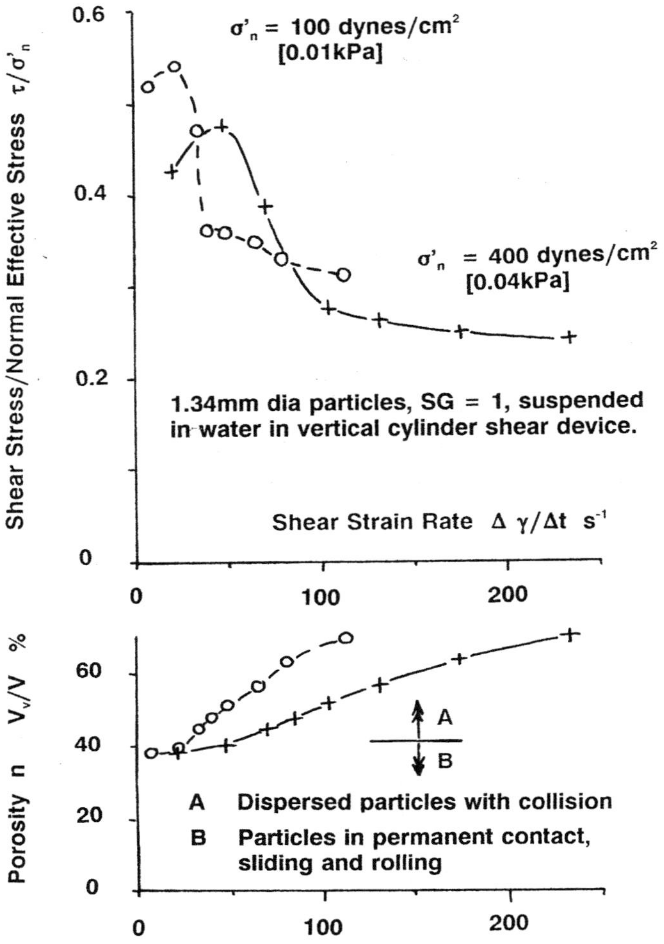


Fig. 8. Gravity free shear of a suspension of spheres interpolated as a simple shear test (Bagnold, 1954; Vaughan, 1993a).

Blight (1997) noted that two out of four slides in tailing dams in South Africa, which occurred during very heavy rain, turned into flow slides. Two which occurred in dry conditions did not. It may be that a wet surface with puddles is sufficient to provide the water required. Vaughan (1993b) noted that the toe of the slide in the cliff at Holbeck Hall, Scarborough flowed when it passed over the rocky foreshore which was wet with rock pools (it was low tide). The sliding material was from Jurassic rock and quite dense glacial till, which could not have liquefied spontaneously. The same phenomenon could account for the development of flow slides in slopes of tropical residual soil when these are triggered by heavy rain and move onto a very wet surface.

The presence of a reservoir at the toe of a slide could promote loss of strength if the slide enters it at sufficient speed for the Bagnold mechanism to come into play. However, the approach illustrated on Fig. 4 indicates that the movement likely to be involved when an old slide is triggered by rainfall is too small to seriously influence stability, even if the shear surface loses strength over this length. It looks a greater risk with first-time brittle slides. More work needs to be done on these potential mechanisms.

From the above arguments, the greatest risk of rapid movement is from first-time slides. It will be exacerbated if the slide occurs during heavy rain and/or the slide moves into a reservoir. Rapid movement leaves easily recognised signs of overshoot, with the landslide debris often climbing the opposite hillside. The movement may then lead to a landslide dam (Schuster, 1986). There is evidence that such movements have occurred in UK in early post-glacial times (Hutchinson, personal communication; Hutchinson & Millar, 2001).

There seems to be much less risk of rapid movement when an old landslide is reactivated by heavy rain. However, even slow movements can be inconvenient, damaging and expensive.

MONITORING.

The stability of abutment and reservoir slopes where these are steep enough to involve the possibility of sliding is complex. It is not the purpose here to argue in favour of extensive and expensive monitoring and remedial work for a problem which has not yet manifested itself and may not do so in the foreseeable future, if ever. However, given the present evidence for climate change, it may well be that the geotechnical effects of this could increase risk to a higher level than, for instance, is due to extreme floods. Therefore it is relevant to consider how this risk might be managed.

If the climate change (precipitation and evapo-transpiration) can be predicted then this can be linked to potential ground water level change and to changed risk of sliding, provided that the present relationship between water level and short term water level change is established. There is no reason why risk from this cause should not be put on the same basis as risk from severe floods - both depend on the same extreme rainfall events. A few relevant slopes where weather and water level (in old or new boreholes) can be monitored continuously should suffice.

The most useful general observation which can be made is of movement. Sliding is likely to be preceded by small movements not apparent in visual inspection. Small pulses of movement of an otherwise stable slope which accompany high rainfall and water table levels indicate that the rainfall is just sufficient to reduce the static factor of safety to one. Survey measurements are required to give advanced warning. Vulnerable locations can then be identified. Modern developments in electronic surveying make this possible without excessive expense. An example is the measurement of movement of the Mam Tor landslide from 1990 to 1998 (Waltham & Dixon, 2000). An accuracy of ±5mm is claimed. These surveys were made by undergraduate students as part of their training. The accuracy obtained from GPS satelite systems is now sufficient for useful surveys, further simplifying this type of measurement. Collaboration between utilities and educational establishments might be fruitful in this area.

Surveys are unlikely to pick up the nature of movement and whether it is due to occasional pulses due to winter storms or due to creep. This can be best done by using inclinometer tubes in boreholes equipped with permanent but recoverable inclinometers connected to automatic recorders. These give a continuous record of movement with depth, to an accuracy of better than 0.1mm, which will show movement during rainstorms. Such an installation might be combined with weather and water level observations where a potential hazard has been identified, as a second stage of observation.

REMEDIAL WORKS.

The most cost effective method of stabilising movement and improving safety factor against movement is likely to be drainage. Vertical holes must be pumped. Naturally discharging drains must be near-horizontal. In the past this has required either deep trench drains or adits, which are expensive. Modern developments in lateral directional drilling allow the construction of drillholes into rock below a slide, or even into the slide itself, which discharge naturally, at much more modest cost. Water levels can then be lowered to stop movement or as a reserve against future movement if water levels rise.

CONCLUSIONS

Seepage conditions in Britain involve a 'wet' surface boundary. The zero pressure seepage line (the water table) will be high if rainfall is higher than permeability (profile A.1). If it is lower (profile A.2) then there will be a (sometimes) partly saturated zone with downward percolation above a deeper water table. Old puddle clay cored dams and abutment slopes are usually in the second (A.2) category. Dam slopes with profile (A.2) are little influenced by climate or climate change.

Abutment slopes often have seepage profile (A.2). They are then subject to seasonal changes in water table, and to large abrupt changes due to rainstorms. The risk of exceptional rainstorms as taken into account in spillway design is not taken into account in assessing the stability of abutment slopes.

Owing to shear rate effects, reactivation of old slides by rainstorms will generally involve small movements of short duration. Slides eventually become flatter, more stable and able to resist larger rainstorms. They are likely to have been subject to a more severe climate in the past. Thus many slides now have a reserve of safety. However, some slide movement is occurring currently, indicating that this is not universal.

Slide movements are much more dangerous if they become fast. There seems little risk of rapid movement in the reactivation of old slides with shear surfaces in clayey material. However, large first-time slides may become fast. There is evidence that such slides have occurred in Britain since the last ice age.

Techniques for monitoring and remediation have greatly improved in recent times, which could allow a more active role to be taken in slope management.

REFERENCES

Bagnold R A (1954) Experiments on gravity free dispersion of large solid spheres in a Newtonian fluid under shear. Proc. Royal Soc. London. A, Vol 225. 49 - 63.

Bagnold R A (1956) The flow of cohesionless grains in fluids. Proc. Royal Soc. London. A. Vol 249, 235 - 297.

Bishop A W,Green G E,Garga V K, Andresen A & Brown J D (1971) A new ring shear apparatus and its application to the measurement of the residual strength. Geotechnique, Vol 21, No 4, 272 - 328.

Blight G E (1997) Destructive mudflows as a consequence of tailings dyke failures. Proc. Inst. Civ. Engnrs., Geotech. Engng., Vol 125, 9 - 18.

Bracegirdle A, Vaughan P R & Hight D W (1992) Displacement prediction using rate effects on residual strength. Proc. 6th Int. Symp. Landslides, Christchurch. Vol. 1, 343 - 347.

Casagrande A (1971) On liquefaction phenomena. Report of Lecture to BGS by P A Green & P A S Ferguson. Geotechnique, Vol 21, No 3, 197 - 202.

Hutchinson J N (2000) Fourth Glossop Lecture : Reading the ground: morphology and geology in site appraisal. Quart. Jour. Enging. Geology & Hydogeology. Geol. Soc. Of London. Vol. 34, 7 - 50. Part 1, Feb. 2001.

Hutchinson J N & Millar D L (2001) The Craig Goch landslide dam, Meirionnydd, Mid Wales. in The Quaternary of West Wales - Field Guide. Quaternary Research Association. 113 -125.

Irgens F & Norem H (1996) A discussion of the physical parameters that control the flow of natural landslides. Proc.7th. Int. Symp. Landslides, Trondheim. Vol 2, 1251 - 1256.

Johnson R H & Walthal S (1979) The Longdendale landslides. Geol. J. Vol 14, 135 - 158.

Knill J L (1966) Cow Green Geological Report. Tees Valley & Cleveland Water Board.

Melosh H J (1987) The mechanics of large rock avalanches. Debris Flows/Avalanches: Process, Risk & Mitigation. Ed J. E. Costa & G. F. Wieczorek. Geol. Soc. America. Reviews in Eng. Geol. Vol. VII. 41 - 49.

Schuster R L (Ed.) (1986) Landslide dams: Processes, risk and mitigation. Am. Soc. Civ. Eng. Geotech. Spec. Publication No 3.

Skempton A W, Leadbetter A D & Chandler R J (1989) The Mam Tor Landslide, North Derbyshire. Phil. Trans. Royal Soc. A 329, 503 - 547.

Shimizu M (1999) Reinforcements effects in the tank-model prediction of slope failure due to rainfalls. Proc. Int. Symp. Slope Stability Engng. Shikoku, Japan. Vol 1,453 - 458.

Tika T E, Vaughan P R & Lemos L J (1996) Fast shearing of pre-existing shear zones in soil. Geotechnique, Vol 46, No 2, 197- 233.

Vaughan P R (1993) Possible actions to help developing countries mitigate hazards due to landslides. Proc. Conf. Landslides Hazard Mitigation. Royal Academy of Engineering. 1995. 114 - 122.

Vaughan P R (1993) Engineering behaviour of Weak Rocks; some answers and some questions. Keynote Address; Int. Symp. Geotech. Engng. Hard Soils - Soft Rocks, Athens. Balkema. Vol 3, 1741 - 65.

Vaughan P R (1994) Assumption, prediction and reality in geotechnical engineering. Geotechnique. Vol. 44, No. 4, 573 - 609.

Vaughan P R, Kovacevic N & Ridley A M (2002)The influence of climate and climate change on the stability of embankment dam slopes. This conference.

Waltham, A. C. & Dixon, N. (2000) Mam Tor landslide, Derbyshire, UK. Quart. Jour. Engng. Geology & Hydrogeology. Vol. 33,105 - 123.

Wedage A M P (1997) Rate effects on residual strength of clays. Proc. 14th Int. Conf. Soil Mech. & Foundation Engng., Hamburg. 225 - 228.

Wedage A M P, Morgenstern N R &Chan D H (1997) Analysis of Mam Tor Landslide considering rate effects on the residual strength of the shear zone. Proc. 14th Int. Conf. Soil Mech. & Foundation Engng., Hamburg. 229 - 232.

# The influence of climate and climate change on the stability of embankment dam slopes.

P R VAUGHAN, Imperial College, London, UK
N KOVACEVIC, Geotechnical Consulting Group, UK
A M RIDLEY, Geotechnical Observations Ltd., UK

SYNOPSIS. Seepage pressures in the downstream slopes of embankment dams are largely controlled by surface climate and by the permeability of their fill. These influences are considered in this paper. The types of seepage regime which can result are described. Shallow movement may develop in clays due to seasonal effective stress changes. Such stress changes are prevented by a layer of free draining gravel below the topsoil. The influence of climate change on the stability of embankment slopes is likely to be small. The stability of abutment slopes is considered in a companion paper in this conference (Vaughan *et al*, 2002).

## INTRODUCTION

This paper considers the influence of climate on the slopes of embankment dams and the types of seepage regime which can result. Changes in rainfall infiltration can change the unit weight of fill and this is considered first. The mechanisms by which climate controls seepage and pore pressures within the slopes of embankment dams are then described. The ways in which climate change could modify these pressures are discussed. The causes of superficial slipping in clay slopes is considered.

## THE EFFECT OF SOIL UNIT WEIGHT ON SLOPE STABILITY

Infiltration from rain can increase the water content and unit weight of soil. This can increase shear stress in a slope and reduce safety factor. It also increases normal effective stress which increases safety factor. The two effects tend to cancel. They do so exactly when $c' = 0$, leaving the safety factor unaffected.

As an example, the factor of safety of a typical slope is shown in Table 1, as calculated by stability coefficients (Chandler & Peiris, 1989) for two differing unit weights and two values of $c'$. The potential decrease in safety factor due to potential uniform change in unit weight is negligible. In the event that the upper, active part of a slope became heavier and the lower passive part did not, there would be a case dependent reduction in stability.

*Reservoirs in a Changing World*, Thomas Telford, London, 2002, 353–366

Table 1. Factors of safety, 2:1 slope 15m high, $r_u == u/\gamma.h = 0.3$ and $\phi' = 25^o$

| Unit weight | 19 kN/m$^3$ | | 21 kN/m$^3$ | |
|---|---|---|---|---|
| Cohesion, c′ | 0 | 10kPa | 0 | 10kPa |
| Factor of safety | 0.700 | 1.242 | 0.700 | 1.207 |

## THE SEEPAGE REGIME IN SLOPES AS CONTROLLED BY SURFACE CLIMATE

Seepage pressures in slopes are mainly controlled by the surface climate (infiltration and evapo-transpiration), run-off and the permeability of the ground. It is convenient to subdivide the range of possible conditions into two pairs, illustrated in Fig. 1 by profiles for level ground. In the first pair of profiles (A) the infiltration available exceeds the evapo-transpiration. On average, the boundary is 'wet'.

Profile (A.1) develops when the inflow available is greater than the permeability of the soil. The water table then rises to the ground surface, except for a zone of seasonal and shorter term fluctuations when the surface condition turns from wetting during rain to drying after it. These fluctuations, which are strongly dependent on surface vegetation, give an average small suction at the surface. In (A.2) the average annual inflow is less than the permeability of the soil. In the top half of the profile the infiltration percolates down through a partly saturated zone which contains continuous air. Below this is a near saturated zone without continuous air from which the water flows, usually laterally down slope. The top of this zone (the water table) will vary with the weather and the season according to the net inflow. Pore pressures in the partly saturated zone will be zero (or a small suction). Occasionally, transient flow in heavy rain may choke the top of the profile and give small positive seepage pressures.

In the second two profiles (B) the average rate of evapo-transpiration exceeds the rainfall infiltration. In (B.1) the soil is fine grained and retains water continuity even at high suction. On average, water is then sucked out of the profile and deep desiccation results. In (B.2) the soil is relatively coarse and permeable. It de-saturates when subject to a suction from surface drying sufficiently for the water to become discontinuous. Water only moves due to vapour transfer and the soil then has a very low permeability. This limits loss of water from surface evaporation. If and when it rains the water in the near surface soil becomes continuous and the permeability becomes high. Some of the water runs in. The surface layer acts like a 'one way valve', only allowing water in.

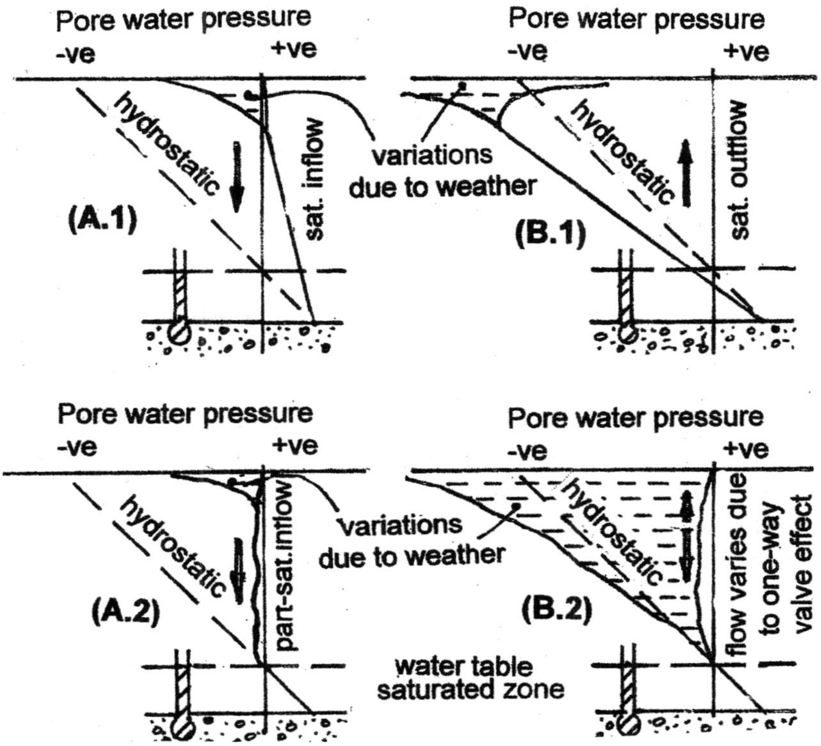


Fig. 1. Ground water profiles resulting from infiltration, evapo-transpiration and high and low permeability; (A) Infiltration >evapo-transpiration: (A.1) Infiltration>permeability: (A.2) Infiltration<permeability; (B) Infiltration <evapo-transpiration: (B.1) Permeability low: (B.2) Permeability high.

Conditions giving profiles (B) do not exist in the British Isles, although the balance between infiltration and evapo-transpiration in Essex is marginal (Hutchinson, 1992). The average zero pressure line is unusually deep. A small change in climate could produce a large change in groundwater conditions to give profiles (B)[1]. This would generally improve slope stability, although there might be problems in clays, with deep cracking and quite large settlements due to desiccation. There are few dams in the area which could be affected and condition (B) will not be considered further in this paper. (A.1) profiles generally exist where fills are granular and permeability higher than annual rainfall; (A.1) profiles exist in clay fills.

[1] Note that only (A) or (B) can exist. There is no 'half-way house'.

## THE INFLUENCE OF CLIMATE AND CLIMATE CHANGE ON EMBANKMENT SLOPES

Upstream slopes

The seepage pressures in upstream slopes are controlled by reservoir operation. Climate change could only affect them if it affects drawdown cycles.

Downstream slopes

Infiltration of rainfall into the downstream slope of an embankment dam is usually more important in determining seepage pore pressures than seepage from the reservoir. The flow from rainfall may be of the order of 0.5m per year, equivalent to a permeability of $1.5 \times 10^{-8}$m/s. Typical dam cores in UK are less permeable than this and the flow through the core will be much smaller than flow from rain, unless the core is leaking.

In fills of higher permeability condition (A.2) will develop. Fig. 2 shows the relationship between the average infiltration flow, the flow out of the base of the fill and the depth of the saturated zone. It can be seen that this zone will be too thin to cause significant pore pressures in the slope unless the permeability of the fill is low. If climate changes, a substantial increase in the infiltration rate will be required to produce a significant increase in downstream seepage pressure. Moreover, the increase will be at depth where it has least effect on stability. Changes in surface climate will have little influence on stability[2]. There

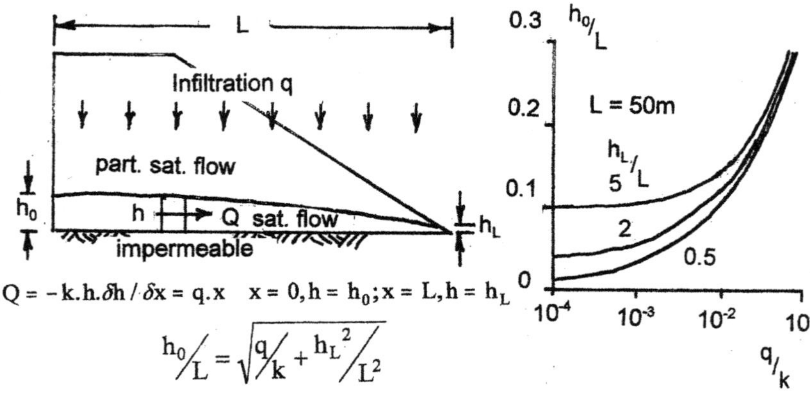


Fig. 2. (A.2) slope with net average inflow less than its average permeability - Ratio of infiltration to permeability for situation to change to (A.1).

---

[2] Downward percolation is sometimes impeded by a relatively impermeable layer. If this occurs then a point or line of seepage may develop on the downstream face where flow runs out laterally. This could develop or become larger if percolation increases due to climate change.

could be some effect on surface erosion by flood water if severe storms increase.

These (A.2) 'low water table' conditions are common in the downstream fills of older British embankment dams with puddle clay cores. Clay fills were seldom adopted even when clay was more readily available than other fill. One reason might well have been that more permeable granular materials (river gravels, weathered rock, etc.) were much easier to dig and tip by hand or with very light machinery than cohesive soils (Skempton, 1995). They could be taken from small local areas. The habit of using granular fill wherever possible persisted up to the use of modern plant. Problems with clay fills in railway construction (Skempton, 1995) may have had an influence.

An early use of loose dumped clay was for the 6m high Aldenham embankment dam built near Watford in 1799 to provide compensation water for local mill owners. Problems were experienced with shrinkage cracks (Jessop, 1802). London clay was not used for a dam again until 150 years later, in the smaller of the Hanningfield dams, 7m high, in 1955. For the fill of the larger Hanningfield dam, 19m high, gravel was brought 13km from off site. The embankment dam at Foxcote, of Gault clay, 9m high, started at the same time.

Some older dams have lower permeability fills, however, and clays from weathered mudrocks, glacial clays and tills were sometimes used, mainly from necessity. This seems to be generally true in Northern Ireland, where there is little to use as fill except glacial till.

Fills of lower permeability.

The second seepage profile (A.1, Fig. 1) develops when the permeability of the fill is lower than the available average water supply for infiltration. On average the slope surface remains 'wet' and water flows into the ground. The surface boundary condition is not constant. It is only at zero pressure when free water is present. During dry periods free water will evaporate. A surface suction will develop due to capillarity and transpiration from vegetation. Thus the surface boundary condition changes continuously, with major trends from winter to summer. These boundary changes cause pore pressure changes within the fill which become less with depth. The pore pressures are near constant at depth, only influenced by long term average boundary changes. Extrapolated to the surface, measurements at depth give the mean annual surface boundary condition, which is about -10kPa in the UK[3] (less in wet upland areas). In the zone of fluctuation there are significant seasonal cyclic pore pressure and

[3] Hence, in clay soils a piezometer at shallow depth but below the zone of fluctuation tends to show a 'ground water table' about 1m below ground level.

effective stress changes.

Evapo-transpiration is strongly influenced by surface vegetation. Fig. 3 (Black et al, 1958) shows suctions measured at the same site, depth and time, near Heathrow in Harmondsworth brickearth, under a bare soil surface and under grass. The suctions below grass were higher and varied more.

The Soil Moisture Deficit (SMD) as calculated from meteorological data for use in hydrology and agriculture (Hough & Jones, 1997) has been found valuable in indicating changes in superficial soil suctions and in correlating landslide activity with weather (Hutchinson & Gostelow, 1976; Hutchinson, 1995). The SMD is the amount of water which can be added to the soil profile before absorption stops. It varies with soil type and vegetation cover. It is computed country wide by the Meteorological Office as a commercial service. Fig. 4 shows the soil suction measured near a tree in dumped London clay fill in an embankment of London Underground Limited, plotted against time (Ridley, 2002). Also shown is the SMD calculated for clay soil and deciduous trees. The soil suction measured varies closely with SMD, with little lag.

Fig. 5 shows maximum and minimum seasonal pore pressures in grassed clay embankment and cutting slopes in eastern and central England. Negative pore pressures were measured by high air-entry value piezometers (Walbancke, 1976). Pore pressures are highest and stability most critical at end of winter.

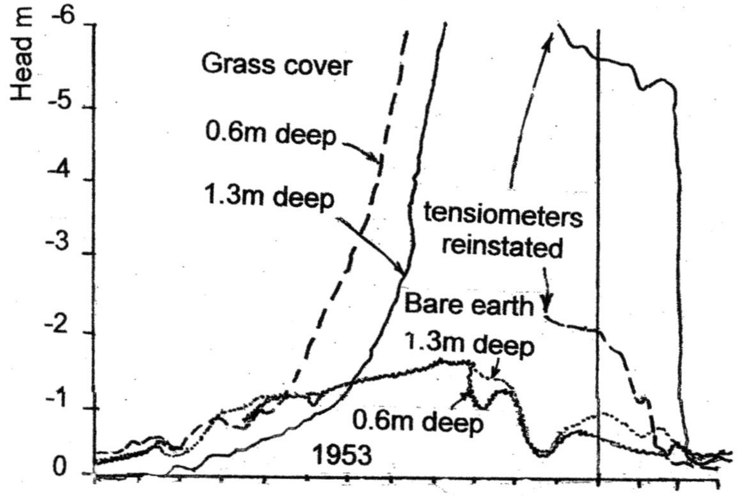


Fig. 3. Measurements of pore pressure change in Harmondsworth Brickearth, Heathrow, in response to climate, below bare earth and below grass cover (Black *et al*, 1958)

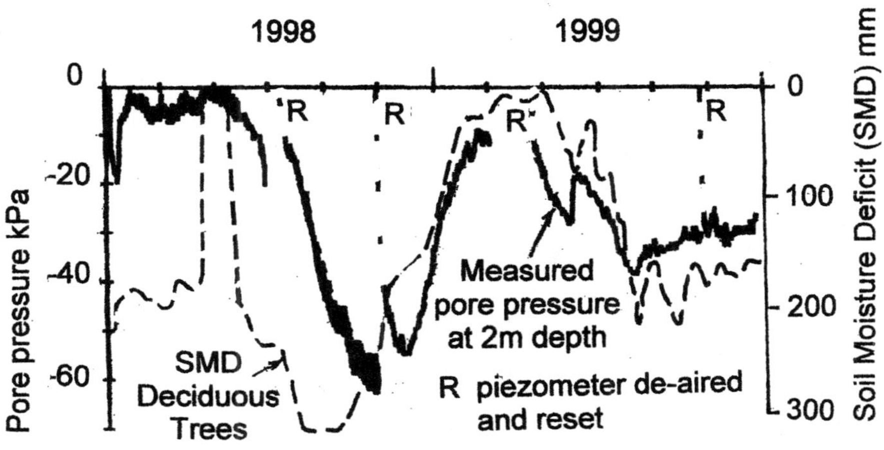


Fig. 4. Measurement of suction in dumped London clay fill with trees (London Underground Ltd. railway embankment) and Soil Moisture Deficit (SMD).

Plant roots require oxygen as well as water. Fresh capillary roots grow each year. In a near saturated clay the roots must crack the soil to obtain an oxygen supply. When the slope wets up the pore pressure distribution reflects this cracking. The upper bound pore pressure of Fig. 5 is equivalent to flow parallel to the slope surface in the cracked zone, with a water table at the slope surface The cracked zone probably controls the swelling process and may account for the lack of lag in pore pressure response at depth to surface changes in climate shown on Fig. 3 & 4. The rate of root growth probably controls the shrinkage process.

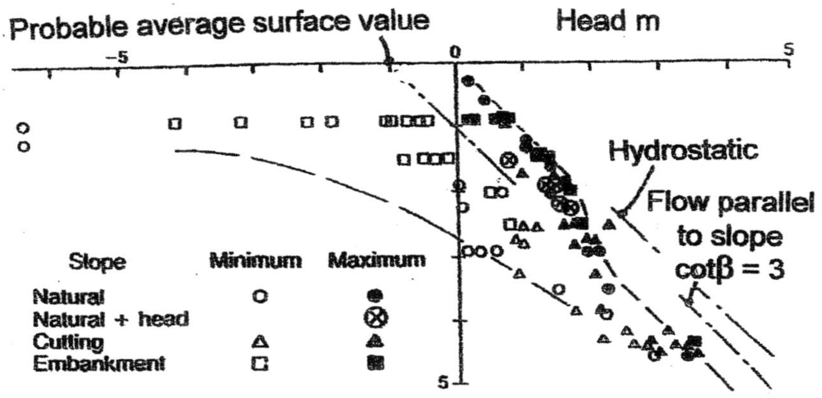


Fig. 5. Measurements of maximum winter and minimum summer pore pressures below grass cover in clay slopes in south east and central England (Walbancke, 1976; Vaughan, 1994).

SMD is also a reasonable indicator of the risk of slope failure. Fig. 6 shows SMD for grass cover for London plotted for a number of years. Also shown are the dates of several small embankment slips on the London Underground network. The coincidence of these slips with periods of low and zero SMD is noticeable. SMD for London dropped to zero in October 2000, unusually early, and did not increase again until April 2001, a wet period of 6 months, compared to the more usual period of 2 - 4 months. Railtrack reported five times the normal number of slips in the south-east over this winter.

The seasonal fluctuation in superficial pore pressure is complex as it is controlled by root growth and cracking, as well as by the consolidation properties of the intact soil. It cannot be predicted theoretically at present. However, piezometers capable of measuring suctions have been developed which provide continuous field records of pore pressure on automatic recorders and which operate for long periods untended (Ridley, 2002). Thus field conditions can be monitored.

The superficial stress cycles cause slope movement and failure. Fig. 7 (Kovacovic *et al*, 2001) shows the finite element analysis of the effect of cycles of pore pressure simulating the effect of deciduous trees[4], applied to an embankment of dumped London clay fill. An elasto-plastic stress-strain soil model was used, with drained strength dropping from peak to residual with

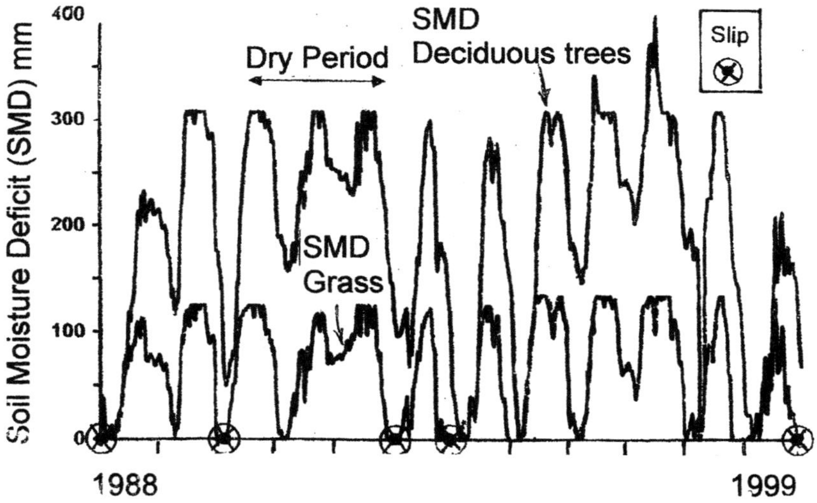


Fig. 6. Changes in SMD in the London area over several years and the occurrence of slips in clay railway embankments.

[4]Field observations show that pore pressure changes induced by tree roots go down to the full height of typical railway embankments.

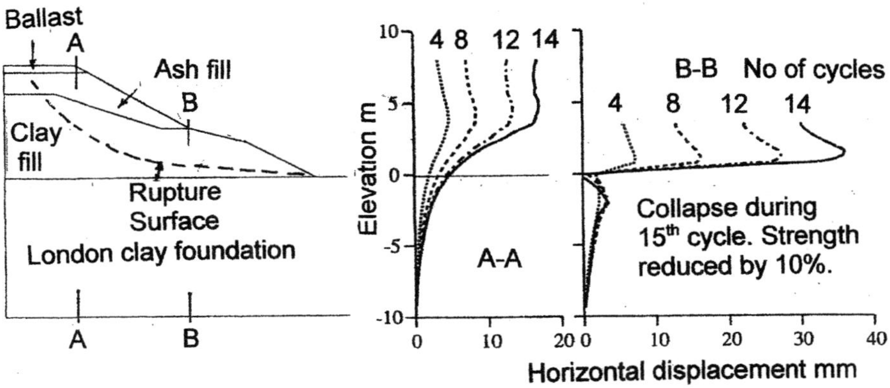


Fig. 7. Prediction by finite element analysis of movement and collapse of a plastic clay embankment by progressive failure when it is subject to seasonal cycles of effective stress (Kovacevic *et al*, 2001).

post-peak strain. An equal unload/reload elastic modulus was adopted, which was stiffer than first loading. During each swelling stage the fill expands; mostly upwards but with a small outward component. This is not recovered during shrinkage, indicating some plastic strain, and there is a small permanent outward movement per pore pressure cycle. As this occurs a shear zone develops in the base of the fill on which strength drops towards residual. Eventually the strength drops enough for a gravity-driven slide to develop. The number of cycles required for this depends on how quickly the strength drops from peak to residual and the magnitude of the pores pressures at the end of winter.

On grassed slopes the zone of pore pressure change is shallower (see Fig. 5). The mechanism is the same but it results in shallow slab slides, at most 1 - 1.5m deep. Such slides are numerous on modern motorway slopes of plastic clay in cut and fill (Perry, 1989). They occur on a rather random basis, with a significant but variable delay between construction and sliding. Only part of a slope at the same angle and in the same soil will fail The flattest clay slope subject to this type of sliding is around 3.5:1 (Perry, 1989). A simple analysis can be made of the potential situation when a slide occurs. This involves an adaptation of the analysis for the semi-infinite slope. Assuming that a shear surface has formed with strength given by $c'_R$ and $\phi'_R$ at depth d and that the pore pressure on the shear surface is given by $h_w$ with flow parallel to the slope, the factor of safety is given by :-

$$F = \frac{\left(1 - \frac{\gamma_w . h_w}{\gamma . d}\right)\tan\varphi'}{\tan\beta} + \frac{c'.\sec\beta}{\gamma . d} \qquad (1)$$

With F = 1 this equation predicts the limiting slope angle, $\beta_f$, for a given strength, slide depth and water table. The residual strength of clays cannot be measured at the very low effective stresses operating in these shallow slips. Application of the equation to shallow field failures using strengths measured at higher stresses suggests that they should occur on flatter slopes than they seem to do. It might be that the residual friction angle developed is higher than that indicated by tests at higher stresses, or it might be that pore pressures when slides are moving are lower than those indicated by Fig. 5.

The stability of an external slope is also controlled by the development of topsoil and grass (Vaughan, 1994). Without grass the slope will flatten by erosion. Grass will increase evapo-transpiration and decrease average pore pressure at depth. However, grass and topsoil may slide at shallower angles than the underlying fill. The ultimate slope will depend on the angle at which topsoil and grass can establish itself. This is a lengthy process. Grass and topsoil on steep slopes is sometimes terraced and displaced by grazing sheep.

Field observation is currently the best guide to behaviour. Table 2 shows some observations of limiting slopes on different in-situ clays. It may be seen that the stronger the soil the steeper the limiting slope and the shallower the failures. The shear strengths calculated from Equation 1 for these slopes show considerable scatter and confirm that a safe and serviceable slope cannot be predicted usefully from current stability theory.

Superficial climate-induced slips have the greatest effect on slopes of plastic clay. Such slopes exist mainly in the south and centre of England, on geology of Jurassic age or later. There are few embankment dams with such slopes.

A historical habit developed of placing a layer of gravel below topsoil over downstream slopes of clay. Such a layer becomes sufficiently dry at a low suction for its pore water to loose continuity. It then has a very low permeability to water (transfer only by water vapour) which prevents loss of water and summer drying of the clay below. Thus the surface boundary of the clay below the gravel is permanently 'wet'. Fig. 8 shows changes in pore pressure measured over a year under a 300mm gravel layer covering plastic Gault clay fill at the Foxcote Dam (Walbancke, 1976; Vaughan, 1994). The changes are of the order of ±0.25m of water, probably about the resolution of the measurements. The pressure changes of Fig. 3 and 4 without a gravel layer are some twenty times greater. Thus the seasonal pressure changes which can cause superficial failure are virtually eliminated by the gravel layer.

Jessop (1802), suggested placing a layer of sand on the downstream slope of Aldenham Dam, to alleviate the cracking problem, as discussed previously. The effect of a capillary break was understood empirically, although it seems unlikely that the early engineers knew why a gravel layer worked!

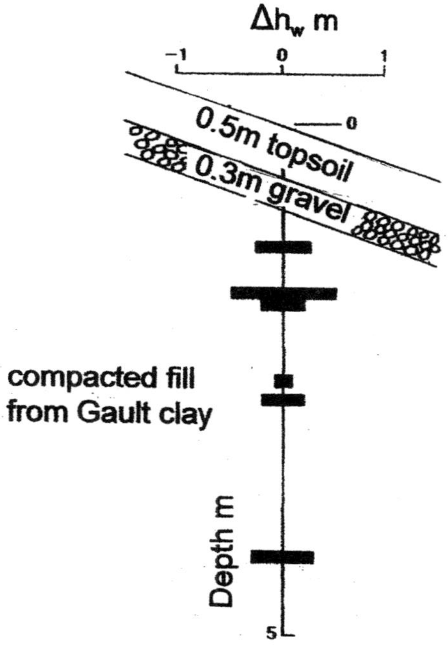


Fig. 8. Seasonal change in pore pressure in Gault Clay fill under topsoil and a gravel layer; Foxcote Dam, Buckingham (Vaughan, 1994).

Clay fill was avoided in dams as far as possible, although there were many such slopes on the railways which gave trouble. Even in 1953, while London clay fill was used in the lower (7m high) of two dams at Hanningfield, Essex under a gravel layer, it was not used in the main (20m high) dam, for which gravel was hauled from 8 miles away. Slopes with gravel layers as steep as 2.5:1 have been used in embankment dams without developing superficial slope failures, compared with slopes without gravel layers on motorway embankments which do so when as flat as 3.5:1 (Perry, 1989). The gravel layer is likely to reduce the average surface pore water suction, which will increase pore pressures at depth, although only by a very small and insignificant amount. These dams were built using modern plant and compaction.

A gravel layer carries its own hazard, however. Such a layer was provided on the 2.5:1 slopes of fill from weathered Lower Lias mudstone at the Draycote dam near Coventry (Vaughan, 1994). The slopes remained stable for some years. However, the gravel was insufficiently permeable (or became so) to shed rainfall infiltration. It developed pore pressures itself and the gravel and topsoil slid off the slope on a shear surface in the top of the clay fill. A gravel layer requires careful specification and site control.

Table 2 Surface stability of clay slopes. Limiting slopes for three different soils

| Soil Type | Properties % < 2μm & $\gamma$ | Nature of sliding | Slope angle $\beta$ | Shear Strength from Eq. .(1) |
|---|---|---|---|---|
| Pennine Boulder clay, Teesdale | 10 - 15 21kN/m³ | Topsoil failure (2) | 37° (1.3:1) | c'=0, $\phi$ =48° |
| | | Slipping of eroded slope | >45° | |
| Chalky Boulder Clay, Whitby | 20 - 30 20kN/m³ | Very shallow sliding in clay | 27° (2:1)(3) | c′=0, $\phi'$=40° or c′=3kPa, $\phi'$=24° |
| London (4) clay, Essex | >45 19kN/m³ | Slab slides 1 - 1.5 m deep | 10 -10.5° (5.5-5.8:1) | c′=0, $\phi'$=21 - 22° |

(1) With $h_w = d. \cos^2\beta$ (flow parallel to slope). (2) The slopes are old, uniform and are now covered with stable topsoil and grass. (3) Slopes probably controlled by first time failures. (4) Skempton & De Lory (1957)

In considering the risk of instability of dam slopes it should be remembered that the shallow slides developed within the zone affected by cyclic surface stress changes do not necessarily imply inadequate stability against more deep seated slides which are potentially dangerous. They may best be considered as expensive nuisances rather than as threats of disaster.

Changing climate can have virtually no adverse influence on superficial or deeper seated stability of clay slopes with an effective gravel layer under the topsoil. If there is no gravel layer then the seasonal cycles of stress change may be modified. The high 'end-of-winter' pore pressures which cause the movement (Fig. 5) cannot get higher[5], but they might penetrate deeper due to more time with a wet boundary and due to deeper cracking. This could then cause slightly deeper instabilities and more rapid generation of superficial slipping. However, there would be no threat to the overall stability of a dam. Reconstruction with a gravel layer should cure the problem.

## SEEPAGE PRESSURES IN FOUNDATIONS

Seepage pressures in foundations are generated largely by rainfall on abutment

---

[5] Provided they are sometimes already at the limit of flow parallel to the slope with a water table at the slope surface.

slopes. Abutment flow will be deflected towards downstream by a dam and reservoir (Bishop et al, 1964) and direct seepage from the reservoir will usually be the smaller part of total flow to downstream. Thus seepage pressures will change if infiltration rates change.

Seepage pressure in the base of a valley under and downstream of a dam are influenced by lateral flow from the abutments and will usually be artesian at depth. This effect can be accentuated if the foundation is stratified. For instance, if the valley is completely blanketed by clayey till, seepage pressure under the till can be strongly artesian. Seepage pressures of this kind are easily monitored and correlations can be established with climate. Then the effect of climate change can be predicted. Artesian pressures are easily reduced by vertical boreholes.

MONITORING.

It has been shown that there is little risk of climate changes causing problems with embankment dam slopes Thus there is no need to introduce monitoring above the level normally employed unless there are signs of instability. Modern survey techniques greatly ease the problems of making accurate surface surveys.

Superficial movement of slopes of clay can best be examined by instrumentation used diagnostically. This can involves modern piezometers which can measure and automatically record the seasonal stress cycles (Ridley *et al*, 2002) together with precise shallow inclinometer installations permanently installed and automatically recorded over at least an annual cycle.

REMEDIAL WORKS.

Some caution is required in stabilising shallow movements, if any develop. If flattening of the slope is difficult, then probably it can be stabilised at its existing angle. This must involve removal of any slip surfaces and could involve placing of a gravel layer to control cyclic stresses. Care is required to maintain adequate stability of the slope during contstruction.

CONCLUSIONS

Changes of water content and unit weight of fill have negligible influence on stability. Climate in UK creates a 'wet' surface boundary. The zero pressure line (the water table) will be high if rainfall is higher than permeability (profile A.1). If it is lower (profile A.2) then there will be a (sometimes) partly saturated zone with downward percolation above a deeper water table. Old puddle clay cored dams and abutment slopes are usually in the second (A.2) category. Dam slopes with profile (A.2) are little influenced by climate or climate change.

Near the surface slopes are subject to cyclic seasonal effective stress changes as climate fluctuates. These cause movement and can cause shallow slipping in clays, which will have the (A.1) profile. Soil Moisture Deficit as used in agriculture is a useful measure of climate and climate change as it affects these slopes. Limiting safe slope angles are best established by field observation. A layer of permeable gravel below topsoil stops the cyclic stress changes and the movements they cause.

REFERENCES

Bishop A W, Kennard M F & Vaughan P R (1964) The development of uplift pressures downstream of the grouted cut-off during the impounding of the Selset Reservoir. Proc. Symp. Grouts & Drilling Muds in Engng. Practice. Butterworths, London. 98 - 104.

Black W P M, Croney D & Jacobs J C (1958) Field studies of the movement of soil moisture. Technical Paper No 41. Road Res. Lab., UK. 78 pages.

Chandler R J & Peiris T A (1989) Further extensions to the Bishop & Morgenstern slope stability charts. Ground Engineering. May, 33 - 38.

Hough M N & Jones R J A (1997) The United Kingdom Meteorological Office rainfall and evaporation calculation system: MORECS version 2.0. Hydrology and Earth System Sciences Vol. 1, No 2, 227 - 239.

Hutchinson J N & Gostelow T P (1976) The development of an abandoned cliff in London Clay at Hadleigh, Essex. Phil. Trans. R. Soc. Lond. Vol. A283, 557 - 604.

Hutchinson J N (1992) Landslide hazard assessment. Proc. 6th Int. Conf. Landslides, Christchurch. Vol. 3, 1805 - 1842.

Jessop W (1802) Report on Aldenham Reservoir. MS Tanscript in Minute Book of General Committee, Grand Junction Canal, 1802 - 1805. British Transport Archives.

Kovasevic N, Potts D M & Vaughan P R (2001). Progressive failure in clay embankments due to seasonal climate changes. Proc. 15th Int. Conf. Soil Mech. & Geotech. Engng., Istanbul, 2127-2130.

Perry J (1989) A survey of slope condition on motorway earthworks in England and Wales. Research Rep. 199, Transport & Road Res. Lab., Crowthorne.

Ridley A M (2002) The measurement of pore water pressure/suction in clay slopes. Report for Transport Research Laboratory. To be published.

Skempton A W (1995) Embankments and cuttings on the early railways. Construction History. Vol 11,33 - 49.

Skempton A W & DeLory F A (1957) Stability of Slopes in London Clay. Proc. 4th. Int. Conf. Soil Mech. & Foundn. Engng., London. Vol.2, 378 - 381.

Vaughan P R (1994) Assumption, prediction and reality in geotechnical engineering. Geotechnique. Vol. 44, No. 4, 573 - 609.

Vaughan P R, Kovacevic N & Ridley A M (2002) The influence of climate and climate change on the stability of abutment and reservoir slopes. This conference.

Walbancke H J (1976) Pore pressures in clay embankments and cuttings. Ph D. Thesis, Univ. London.

# Settlement of old embankment dams and reservoir drawdown

P TEDD, Building Research Establishment Ltd, UK
J A CHARLES, Building Research Establishment Ltd, UK
A C ROBERTSHAW, Yorkshire Water Services Ltd, UK

SYNOPSIS. The monitoring of crest settlement has an important role in assessing long-term performance and safety of embankment dams. Over the last fifteen years field studies have been carried out at five old embankment dams in the Yorkshire Pennines. Published settlement data for these dams are updated and compared with published data for other old embankment dams. The effect of major reservoir drawdowns on the settlement behaviour of dams with central puddle clay cores and upstream blankets has been confirmed.

## INTRODUCTION

Changes in condition or behaviour of an old embankment dam may be detected by visual surveillance or instrumentation readings and a modified form of observational method has been recommended for safety evaluation (Johnston et al, 1999). Monitoring of crest settlement has a particularly important role in assessing long-term performance. The observations need to be related to a likely mechanism of behaviour so that it can be determined whether or not there is an incipient problem.

Over the last fifteen years field studies have been carried out at five old embankment dams which belong to Yorkshire Water. The dams are located in the Pennines and some details of their construction are given in Table 1. The puddle clay for the watertight elements will have come from surface deposits of Recent and Pleistocene boulder clay and the shoulder fill from a mixture of variably weathered mudstone, siltstone and sandstone derived from the Millstone Grit series of the Carboniferous Period. All five dams have a central puddle clay core, but Ogden and Holmestyes also have an upstream clay blanket. For these two dams, a key question is whether the internal core or the upstream blanket is the effective watertight element.

Observations at the five Yorkshire Water dams have shown the major effect of reservoir drawdown on crest settlement for dams with a central puddle clay core (Tedd et al 1994, 1997a, 1997b). This paper summarises the most recent observations at these dams and compares their monitored behaviour with the settlement behaviour at other dams for which data have been published. Basic information about these other dams is included in Table 1.

*Reservoirs in a Changing World*, Thomas Telford, London, 2002, 367–377

Table 1. Central puddle clay core (cpcc) embankment dams with observations of crest settlement

| Dam | Date | $H$ (m) | Embankment watertight element | Foundation cut-off |
|---|---|---|---|---|
| Ramsden | 1892 | 25 | cpcc | Concrete |
| Walshaw Dean Lower | 1907 | 22 | cpcc | Deep puddle clay |
| Yateholme | 1872 | 17 | cpcc | Concrete |
| Ogden | 1858 | 25 | cpcc + upstream clay blanket | Shallow puddle clay |
| Holmestyes | 1840 | 25 | cpcc + upstream clay blanket | Shallow puddle clay |
| Ardsley | 1888 | 22 | cpcc | |
| Brownhill | 1932 | 30 | cpcc | Concrete |
| Challacombe | 1945 | 15 | cpcc | Concrete |
| Cwmwernderi | 1901 | 22 | cpcc | Shallow puddle clay |
| Digley | 1954 | 49 | cpcc | |
| Ladybower | 1944 | 43 | cpcc | Concrete |
| March Haigh | 1838 | 21 | cpcc | |
| Widdop | 1878 | 22 | cpcc | Shallow puddle clay |

SETTLEMENT INDICES

Two settlement indices have been proposed to assist in the interpretation of settlement measurements and as a means of comparing the settlement behaviour of different dams. Charles (1986) proposed the settlement index $S_I$ to quantify settlements associated with long term steady state loading and Vaughan et al (2000) have proposed a drawdown settlement index $D_I$ for the purposes of comparing settlements due to changes in reservoir level. These indices can also be used to estimate the magnitude and rate of movement that is likely to result from ageing related processes associated with steady state loading and reservoir drawdown which do not indicate any threat to reservoir safety, thus facilitating the identification of movements which could indicate the onset of a hazardous situation.

Where settlement is not influenced significantly by fluctuations in reservoir level, it can be expressed by a settlement index $S_I$ such that:

$$S_I = s/(1000H \log[t_2/t_1])$$

where $s$ is the crest settlement measured in mm between times $t_2$ and $t_1$ since the completion of embankment construction at a section of the dam which is $H$ metres high.

The drawdown settlement index $D_I$ is defined as the permanent settlement ($s_p$) measured in mm per metre height of dam ($H$), per cumulative drawdown ($\Delta h_w$) measured in metres:

$$D_I = s_p/(H\, \Delta h_w)$$

The drawdown settlement index provides a means of comparing the settlement of dams taking into account the height of the dams and the magnitude of the reservoir drawdown. It can indicate that settlement resulting from reservoir drawdown represents an acceptable vertical strain and need not cause concern. However, this index does not take into account the duration of the drawdown.

BRE – YORKSHIRE WATER STUDY

The five dams of the study are briefly described with an emphasis on the more recent observations.

Ramsden. The dam embankment has settled 1m since it was constructed 100 years ago, with an average rate of settlement of 8mm per year between 1977 and 1985. During a detailed deformation study between 1987 and 1989, the reservoir was emptied completely (Tedd et al 1990, 1997a, 1997b). This major drawdown caused significant crest settlement which continued while the reservoir remained empty. Reservoir refilling reversed the movements, but the net result of emptying the reservoir and refilling was a large crest settlement. Continuity of crest observations was lost in 1990/91 when the dam was raised and a new wave wall was built. The reservoir level and crest settlement from 1993 onwards are shown in Figure 1. During 1993/94, when the reservoir was not drawn down, there was no measurable crest settlement. Since construction of a new treatment works in 1994, a revised operating regime has resulted in less extensive reservoir drawdowns. The relatively small reservoir drawdowns of 5m in 1994 and 1995, and 3m fluctuations in 1996 resulted in a permanent crest settlement of 25mm at the deepest section, 25 m high, and 15mm where the embankment is 18m high. The 1995 drought did not result in large drawdowns, but the dry weather could have caused some drying out of the upper part of the dam. Since 1996, with relatively small drawdowns, there has been very little settlement.

Walshaw Dean Lower. Between completion of construction in 1907 and final commissioning in 1915, there was a series of leaks and repairs and a major loss of material occurred. By 1943 settlement had reduced freeboard to 0.56m and the crest and puddle clay core were raised to their original level. A settlement of 1m has occurred since the end of construction. The dam is subjected to substantial annual drawdowns as the reservoir is used for compensation. Instrumentation to measure settlements and pore pressures in the upstream fill and stresses in the puddle clay core was installed in 1992 prior to the prolonged drawdown of 1993/94 which was required to facilitate construction of an additional spillway (Holton et al, 1996). Reservoir level and crest settlement are shown in Figure 2. The largest movements have been confined mainly to the crest and upstream fill close to the core. Despite repeated reservoir drawdowns for more than 80 years, only partial recovery of the crest settlement induced by drawdown occurs. The average annual rate of settlement from January 1990 to 1997 including the prolonged drawdown of 1993/94 has been 8mm. Relatively wet summers since 1998 have reduced the rate of settlement to 4mm per year as shown in Figure 2. Drawdown induced settlement has occurred throughout the depth of the core but very little settlement has been recorded in the cut-off trench, probably because of arching across the narrow trench in the relatively incompressible rock foundation.

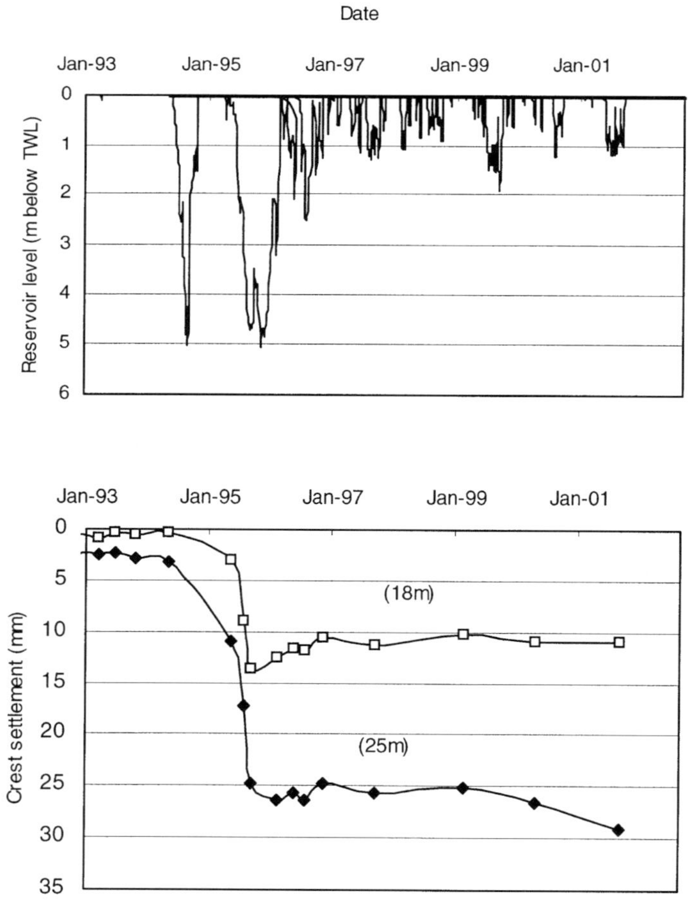


Fig. 1. Reservoir level changes and crest settlement at Ramsden dam

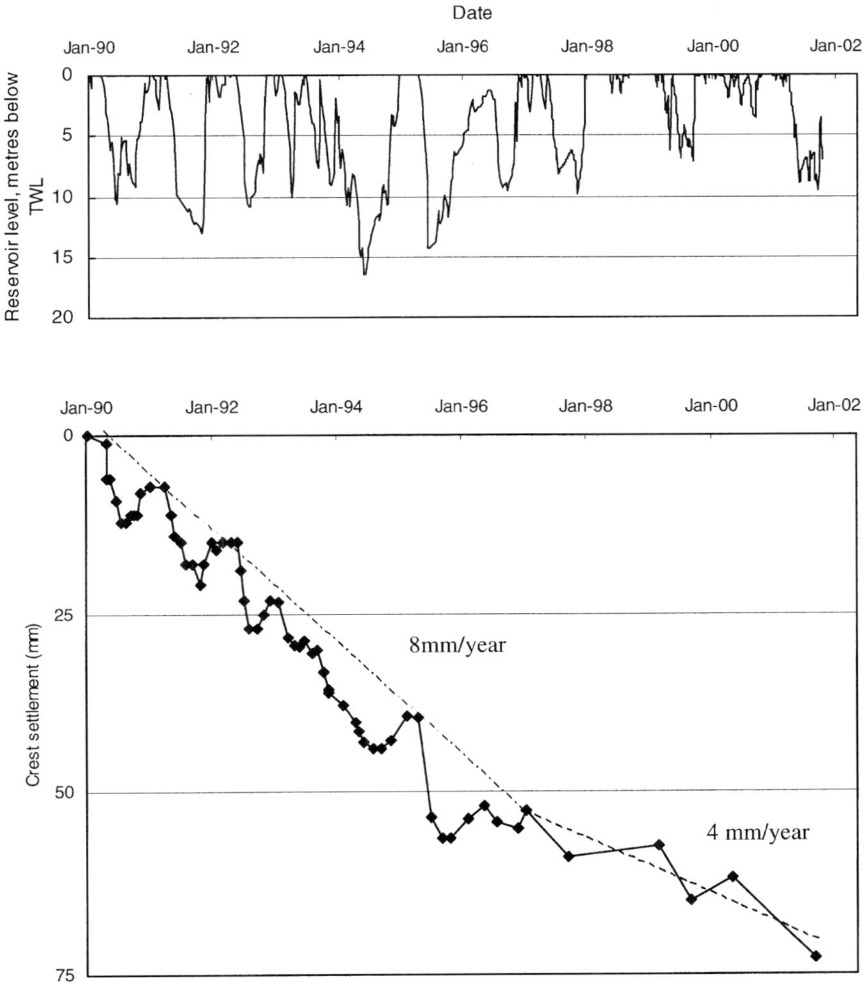


Fig. 2. Crest settlement at Walshaw Dean Lower

Yateholme. Unlike the other dams in the study, which lie in steep-sided valleys, Yateholme was constructed on the side of a hill. Of those dams in the study which have effective central puddle clay cores, only Yateholme does not appear to have settled significantly. With relatively small reservoir drawdowns since 1992, the average annual settlement was less than 2mm until the 1995 drought when the settlement rate doubled during a 10 month reservoir drawdown with a maximum depth of 8m. Only 7mm settlement has occurred between 1996 and 2001.

Ogden. The embankment dam has an upstream clay blanket in addition to a central clay core. Observations of settlement and piezometric pressure have indicated that the central clay core is the effective watertight element and that large settlements have occurred with significant reservoir drawdowns. However, the very large settlement of 140mm that occurred during the emptying of the reservoir in 1990/91 was not repeated in the drought of 1995/96, probably because the drawdown was not as large or prolonged. Figure 3 shows that the reservoir has not been drawn down much since 1996 and there has been no settlement of the crest until the recent 7m drawdown in 2001.

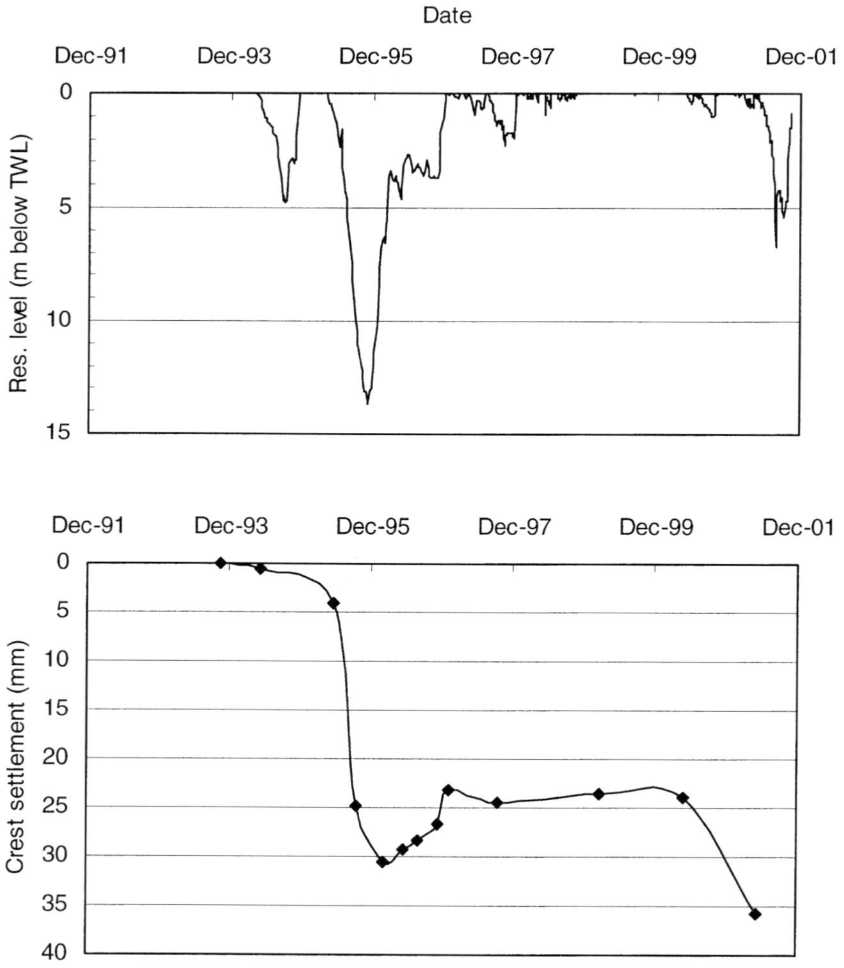


Fig. 3. Crest settlement at Ogden dam

Holmestyes. This embankment dam also has an upstream clay blanket in addition to a central clay core. The water level in the upstream fill is 15m below top water level when the reservoir is full, which indicates that, in contrast to Ogden, the upstream blanket has the major role as the watertight element. Reservoir drawdown has caused very small crest heave and refilling has caused slightly larger settlement resulting in a net settlement (Tedd et al, 1993). The settlement observations at Holmestyes dam are plotted in Figure 4. A crest settlement of 19mm has been monitored between 1991 and 2001. This corresponds to $S_I$ = 0.027, a somewhat larger value than would be expected at a dam where crest settlement has not been influenced significantly by fluctuations in reservoir level. For example, at two central puddle clay core dams, Cwmwernderi and Challacombe, where the reservoir has not been drawn down for many years, $S_I$ has been of the order of 0.01. The settlement rate at Holmestyes from 1992-96 was 2mm/year and involved two drawdowns of more than 8m, as shown in Figure 4. When the reservoir was emptied in 1997, an 18m drawdown for major remedial works, the crest settled only 3mm. This contrasts with the large settlements of 50 mm and more measured at similar height dams with central puddle clay cores.

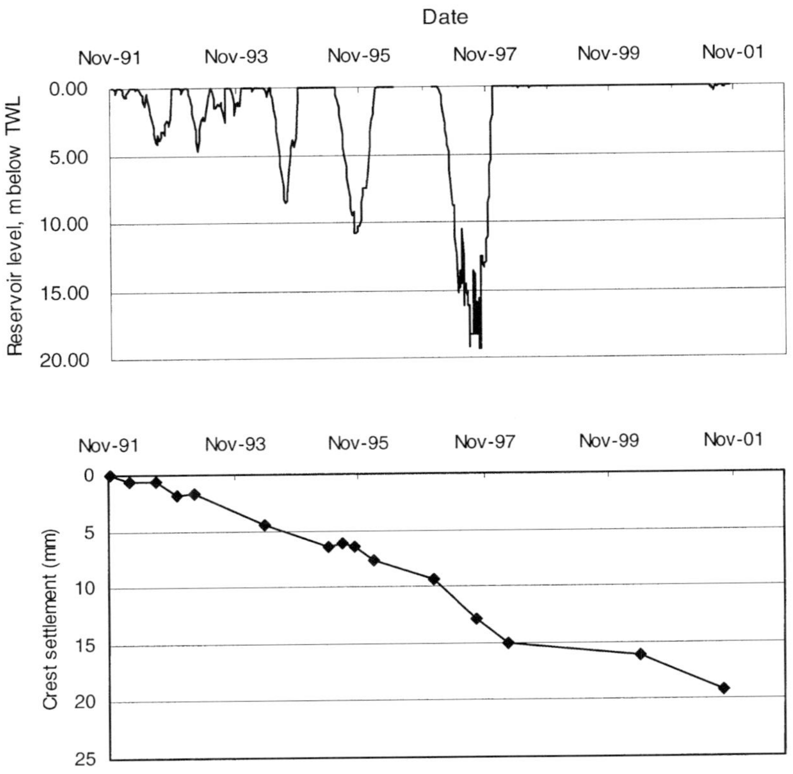


Fig. 4. Crest settlement at Holmestyes dam with effective upstream blanket

## COMPARISON WITH BEHAHAVIOUR OF OTHER DAMS

Similar patterns of deformation behaviour on reservoir drawdown have been measured at the dams in the study with effective central clay cores, although there are differences in the magnitude of movements. Figure 5 shows the permanent settlement of the crest of the embankments ($s_p/H$) as a function of reservoir drawdown. Since the BRE - YW study commenced, data have become available for a number of other old embankment dams.

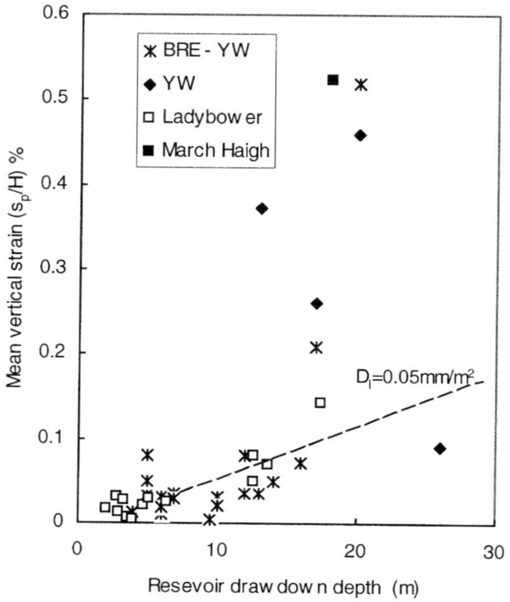


Fig. 5. Crest settlement plotted as mean vertical strain as a function of depth of reservoir drawdown

Ladybower. Since the completion of construction in 1944, the crest of the dam is reported to have settled 1.5 m at the deepest section, which is 3.75% of the height. This is comparable with the dams in the BRE-Yorkshire Water study. The continuing settlement of the dam has necessitated raising the crest on a number of occasions. Vaughan et al (2000) have correlated the crest settlement with the amount by which the reservoir is drawndown. The data for Ladybower have been plotted on Figure 5 and are similar to data obtained in the BRE-Yorkshire Water study.

March Haigh. The dam has a central, very sandy puddle clay core with granular fill in the shoulders. Problems occurred in 1997 when water was discovered to be leaking into the downstream valve tunnel. With some difficulty the reservoir was completely emptied. It is believed that the crest settled between 90 and 130mm as a result of the prolonged drawdown (Dutton, 2001).

Yorkshire Water reservoirs. The drought of 1995 resulted in major drawdowns at Yorkshire Water reservoirs. Brownhill settled 138mm in response to a reservoir drawdown of 20m (Robertshaw, 1998). High values of $D_I$ were also measured at Ardsley dam where the reservoir was drawn down by 13m during the drought. Robertshaw and Dyke (1990) have described the settlement at Widdop when the reservoir was emptied in 1988/90.

Although Figure 5 shows a general correlation between the maximum mean vertical strain in the core and the depth of the drawdown which has caused it, the relationship does not correspond to a particular value of $D_I$. The value of $D_I$ can be much larger with major drawdowns. This may be because such drawdowns are very infrequent at most reservoirs.

The pattern and magnitude of deformations due to reservoir drawdown depend upon the position of the watertight element and on the properties of the fill. The majority of the larger old embankment dams in Britain have a narrow central puddle clay core with relatively free draining fill in the upstream shoulder. Lowering the reservoir level causes an increase in vertical effective stress in the upstream fill with resulting settlement. The maximum settlement usually will occur at the crest of the dam at the deepest section. Re-submergence of the upstream fill causes a decrease in the effective stress and heave of the upstream slope. The effect of reservoir level changes on the central clay core itself is complex (Tedd et al, 1997b).

Reservoir drawdown at a dam with an upstream watertight element causes a decrease in water pressure acting on the upstream slope with consequent heave of the upstream slope. Subsequent refilling causes settlement of the upstream slope but virtually no settlement at the crest. This pattern of behaviour assumes that the upstream watertight element is fully effective and reservoir level changes have no effect on pore water pressures in the upstream fill.

## CONCLUSIONS

The uncertainties surrounding the composition of most old embankment dams are so great that no purely analytical approach to dam safety can be adequate, and an observational approach to safety evaluation is essential. Deformation measurements, particularly crest settlement, are commonly undertaken and simple models of behaviour are required so that such measurements can be interpreted reliably.

The major difference between the deformation behaviour on reservoir drawdown of a dam with a central watertight element and a dam with an upstream watertight element has been confirmed by field measurements. With a central core, significant crest settlement can be expected on reservoir drawdown, but this is not the case where there is an upstream watertight element. Because the effective watertight element of Holmestyes dam is an upstream clay blanket rather than a central puddle clay core, relatively little crest settlement has taken place despite a large reservoir drawdown in 1997.

Settlement during reservoir drawdown at dams with central puddle clay cores and submerged upstream fill is dominated by the magnitude and duration of the drawdown and does not normally indicate a threat to the stability of the dam, except that after many years of reservoir operation the freeboard may be lowered to an unacceptable level. When relatively small reservoir drawdowns follow a substantial drawdown, crest settlement is negligible, often less than expected from creep or secondary compression.

## ACKNOWLEDGEMENTS

This paper has been prepared as part of the DEFRA Reservoir Safety Research Programme. The collaboration of Yorkshire Water Services in the programme of field studies of embankment deformations is gratefully acknowledged.

## REFERENCES

Charles J A (1986). The significance of problems and remedial works at British earth dams. Proceedings of BNCOLD Conference, Edinburgh, vol 1, pp 123-141.

Dutton D (2001). British Waterways' dams: the last 25 years. Dams & Reservoirs, vol 11, no 2, October, pp 1927.

Holton I R, Tedd P and Charles J A (1996). Walshaw Dean Lower - embankment behaviour during prolonged reservoir drawdown. The Reservoir as an Asset. Proceedings of 9th Conference of British Dam Society, York, pp 94-104. Thomas Telford, London.

Johnston T A, Millmore J P, Charles J A and Tedd P (1999). An engineering guide to the safety of embankment dams in the United Kingdom. (Second edition) Building Research Establishment report BR363. Construction Research Communications, London.

Robertshaw A (1998). Settlement of Brownhill reservoir embankment. Contribution to discussion, Session 2, 1996 BDS Conference, Discussion volume, p50.

Robertshaw A C and Dyke T N (1990). The routine monitoring of embankment dam behaviour. The embankment dam. Proceedings of 6th BDS Conference, Nottingham, pp 177-183.

Tedd P, Charles J A and Claydon J R (1990). The deformation of Ramsden dam during reservoir drawdown and refilling. The Embankment Dam. Proceedings of 6th British Dam Society Conference, Nottingham, pp 171-176. Thomas Telford, London.

Tedd P, Robertshaw A C and Holton I R (1993). Investigation of an old embankment dam with an upstream clay blanket and central clay core. Dams & Reservoirs, vol 3, no 1, February, pp 6-9.

Tedd P, Charles J A, Holton I R and Robertshaw A C (1994). Deformation of embankment dams due to changes in reservoir level. Proceedings of 13th International Conference on Soil Mechanics and Foundation Engineering, New Delhi, vol 3, pp 951-954.

Tedd P, Charles J A and Holton I R (1997a). Settlement of old embankment dams: a guide to measurement and interpretation. Dams & Reservoirs, vol 7, no 1, March, pp 18-23.

Tedd P, Charles J A, Holton I R and Robertshaw A C (1997b). The effect of reservoir drawdown and long term consolidation on the deformation of old embankment dams. Geotechnique, vol 47, no 1, March, pp 33-48.

Vaughan P R, Chalmers R W and Mackay M (2000). Ladybower dam: analysis and prediction of settlement due to long term operation. Dams 2000. Proceedings of 11th British Dam Society Conference, Bath, pp 360-376. Thomas Telford, London.

# Internal erosion in European embankment dams

J A CHARLES, Building Research Establishment Ltd, UK

SYNOPSIS. The safety of an ageing population of embankment dams is a cause for concern in many European countries. A European Working Group on Internal Erosion in Embankment Dams was formed in 1993 and has examined the hazard posed by internal erosion to existing dams. From a study of case histories, the features in an embankment dam that are critical in rendering it vulnerable to internal erosion have been identified. The significance of the risk of failure and the warning signs that might be expected prior to failure have also been assessed.

## INTRODUCTION

The long-term performance and safety of an ageing population of embankment dams is a cause for concern in many European countries. Many of the dams were not built to modern standards and there is likely to be deterioration with time. Most European countries with sizeable populations of embankment dams have experienced cases of internal erosion and this hazard is of crucial significance in assessing long-term safety. The national experience of serious incidents will, to some extent, reflect the number of embankment dams in the country and the age of these dams. The type of incidents and their seriousness will also depend on the regional geology and the prevalent type of embankment dam construction. Modern practice, incorporating specially designed filters, should greatly reduce vulnerability to internal erosion.

A European Working Group on Internal Erosion in Embankment Dams was formed in 1993. The Working Group currently has representatives from Austria, Bulgaria, Finland, France, Germany, Italy, Norway, Romania, Spain, Sweden, Switzerland and the United Kingdom. A progress report was presented at the Barcelona symposium (Charles, 1998). The activities of the Working Group have been confined to a study of the hazard posed by internal erosion to existing dams. Filter design for new dams has not been studied because the extensive research that has been carried out on this subject has been comprehensively reviewed in ICOLD Bulletin 95.

Internal erosion involves the removal of solid material, usually in suspension, from within an embankment or its foundation by the flow of water. The various mechanisms of internal erosion may be associated with construction defects and weaknesses or with the deformations and stress

*Reservoirs in a Changing World*, Thomas Telford, London, 2002, 378–393

conditions within the embankment. Internal erosion may cause increased leakage and some surface settlement, possibly in the form of a sinkhole.

A helpful distinction can be made between localised and mass internal erosion. The former may be related to some local defect or may be associated with piping or hydraulic fracture. The term 'piping' is usually applied to a process that starts at the exit point of seepage and in which a continuous passage or pipe is developed in the soil by backward erosion. Where a central clay core is supported by stiffer granular shoulders and internal stress transfer is caused by differential settlement, cracks formed in the clay core may be associated with hydraulic fracture by the reservoir water pressure. Flow along preferential paths also may result from the inherent heterogeneity of the core fill. In a cohesive soil which is capable of sustaining an open crack, concentrated leaks may occur with erosion of soil particles along the walls of the cracks. In contrast to these localised forms of erosion, the type of mass erosion known as 'suffosion' can occur by seepage flow in soils which are internally unstable.

Soils inadequately compacted at low moisture contents may be susceptible to collapse compression on saturation. Layers of better compacted material may arch over the collapsing soil resulting in the formation of loose, erodible, wet seams. Certain clay soils disperse or deflocculate in the presence of relatively pure water and are therefore highly susceptible to internal erosion. The tendency for dispersive erosion depends on the mineralogy and chemistry of the clay, and dissolved salts in the pore water and the eroding water (ICOLD, 1990).

Internal erosion is often associated with the presence of structures such as outlet conduits and culverts which pass through an embankment. The contact between the embankment fill and the structure can be a potential zone of weakness as the fill may have been inadequately compacted making suffosion and piping more probable. Leakage into a culvert may cause internal erosion and, where an unprotected outlet pipe has been placed in the embankment fill, any leakage from the pipe may result in severe erosion of the embankment fill.

## CASE HISTORIES

A study of internal erosion incidents at European dams has been undertaken by the Working Group. A selected group of 47 case histories has been summarised in Appendix 1 and has been cross-referenced to an extensive bibliography which forms Appendix 3. Examples have been included from most European countries where there is a substantial population of embankment dams.

Internal erosion incidents involving sink holes and turbid leakage have been relatively common in Scandinavian dams with moraine cores, but very rarely do they appear to have caused failure. There has been a high incidence of sinkholes and concentrated leakage at Swedish dams (Bartsch, 1995). Internal erosion usually appears to be self-healing in these moraine soils, but some serious events have occurred. A number of small dams have breached in Sweden due to internal erosion, but no large Swedish dam has failed.

The United Kingdom has a much larger population of embankment dams than other European countries. There are numerous examples of internal erosion at these dams, many of which were built to a traditional design using a central puddle clay core, but there have been very few instances of failure.

Although there has been no failure of a major dam in France since 1970, there have been 70 internal erosion incidents (Comité Français des Grands Barrages, 1997). Ten serious incidents occurred at small dams, including three that breached the embankment. There are not many published cases of internal erosion at German dams. However, there have been major failures of canal embankments. Concrete dams generally predominate in southern Europe and, consequently, there are fewer cases of internal erosion. Nevertheless, there are a number of interesting cases of internal erosion in embankment dams in Spain.

In Romania many hydro-power plants have reservoirs formed by embankments up to 30 m high and 2 km to 3 km in length with a thin upstream concrete facing. Problems are occurring due to deterioration of the bituminous sealing in the joints between concrete slabs and leakages have increased (Hulea, 1997).

## ANALYSIS OF CASE HISTORIES

An analysis of the case histories has identified some critical factors associated with internal erosion. In order to assist in this analysis, each dam listed in Appendix 1 has been given a classification under four headings:
(A) severity of problem or incident
(B) cause of problem or incident
(C) symptoms of problem or incident
(D) remedial works.
Classifications for each of the above headings are defined in Appendix 2.

### (A) Severity of problem or incident

There are fourteen case histories of dams where the severity of the problem or incident has been classified as either A1 (failure) or A2a (serious incident involving emergency action or drawdown where, without emergency action, a breach was likely). In eight of these cases the problem occurred during, or immediately following first filling of the reservoir. Such incidents during the early life of the dam are of considerable significance for new dams, but, from the aspect of long-term safety, the other six dams are of greater relevance. These six dams are now considered. References for each of the dams are listed in Appendix 1.

Peruca. At 10.48 on 28 January 1993 explosions were activated by military action at five locations in the inspection gallery of the 63 m high Peruca rock fill dam in Croatia. The dam, which has a narrow central clay core, suffered major damage and 3000 people in the most endangered area close to the dam were evacuated. The reservoir level had been lowered 5 m below the maximum water level prior to the explosions and was lowered at a rate of 0.9 m per day subsequent to the attack. Nevertheless, between 30 January and 5 February the leakage rate increased from 400 l/s to 570 l/s. Subsequently the flow rate decreased. It was calculated that 1500 $m^3$ of clay was eroded out of the dam by this leakage.

Saint Aignan. The 8 m high, homogeneous earth fill embankment dam failed catastrophically almost 20 years after construction. The embankment was constructed of variable fills and had no internal drainage. It was believed that suffosion had eventually turned into piping. No alarm was given prior to failure, but the downstream slope had shown signs of saturation. Failure was attributed to variable permeability in the fill inducing high seepage rates, the lack of an internal drain, and the absence of monitoring and maintenance.

Sapins. In 1988, ten years after first filling, flows of water and a shallow slip occurred in the lower part of the downstream slope of the 16 m high, homogeneous sand fill embankment. The situation rapidly worsened, and the reservoir was emptied. The homogeneous embankment was composed of a sand fill with a chimney drain which stopped 2 m below the top water level. The problem was attributed to suffosion within the embankment.

Sorpe. The 69 m high embankment was damaged during air raids on the nights of 16/17 May 1943 and 15 October 1944. On the latter occasion craters 12 m deep and 25 m to 30 m in diameter were produced. A breach was avoided largely because the reservoir level had been lowered by 6 m prior to the air raid. The central concrete core wall of the earth and rock fill embankment was severely damaged and subsequently the dam suffered serious leakage problems with remedial works undertaken in the 1950s.

Uljua. On 29 May 1990, twenty years after construction, a major incident occurred in which a sinkhole appeared in the upstream slope near the crest. The upstream slope dropped 3 m over a length of 7 m and the rate of leakage increased to 100 l/s. Rapid remedial measures saved the dam from total collapse. An erosion channel 1 m in diameter across the dam core was found during the repair works.

Warmwithens. During remedial works between 1964 and 1966, a 1.5 m diameter tunnel, formed in concrete segments, was driven through the 10 m high embankment to contain new outlet pipes. On 24 November 1970 the dam failed and it is thought that internal erosion took place along the line of the tunnel. The breach, 20 m wide at crest level, extended down to the tunnel, which was washed out, large sections of the concrete segments being deposited downstream.

At Peruca, Sorpe and Warmwithens internal erosion followed external actions which may be regarded as unusual. At Peruca and Sorpe, lowering of the reservoir water level before military action took place was a critical factor in avoiding breaching of the dams. The other three cases, Saint Agnan, Sapins and Uljua, demonstrate that long-term safety of embankment dams is ultimately contingent on adequate maintenance and surveillance, together with preparedness for emergency lowering of the reservoir.

(B) Cause of problem or incident
Although internal erosion problems are not restricted to one particular type of fill or form of dam construction, the type of problem experienced by the various dams is related to the type of embankment construction. Of the fourteen dams classified in the A1 and A2a severity categories, four are homogeneous earth fill embankments, five have upstream membranes, four have internal earth fill cores and one has a central concrete core wall.

In five cases, erosion at the contact between the embankment and a structure or pipe passing through the embankment (B1) was a major factor in the incident. This type of interface within an embankment clearly represents a significant potential hazard. It is noteworthy that, following incidents at two dams, Italian regulations prohibited this type of construction (Dolcetta, 1997). In three cases, the failure of the upstream membrane (B4) was critical. The hydraulic gradient across such a membrane is very high and flows can be very great at a localised defect.

(C) Symptoms of problem or incident
Of the total 47 case histories, internal erosion manifested itself in the form of a sinkhole (C1) in 20 of the dams. In one case, a vortex (C8) was observed in the reservoir. In the majority of the 47 case histories, seepage and leakage gave some indication of a problem. This could be in the form of excessive or increasing flows (C2), turbid flows (C3), wet areas on the downstream slope (C4) or leakage into or around a culvert (C5).

(D) Remedial works
In about half the 47 case histories, grouting (D2) formed a major part of the remedial works. In ten cases diaphragm walls (D1) were installed. In about one third of the cases, remedial works included earthmoving operations in

the form of slope flattening, berm construction and reconstruction of the embankment.

## CONCLUSIONS

The work of the group has addressed three basic questions:

*Is there a significant risk that an old embankment dam, with a long record of apparently satisfactory behaviour, could fail suddenly and catastrophically due to internal erosion?* In many European countries there are a significant number of embankment dams which could pose a threat to public safety and their long-term satisfactory performance is of considerable importance. Internal erosion is likely to be the major hazard for many of these old dams. While internal erosion problems often occur during first filling of the reservoir, there are many cases where the phenomenon does not manifest itself until a much later stage. A lack of problems during the early life of a dam, therefore, does not guarantee continuing satisfactory performance. The case histories indicate that the risk of an old embankment dam, with a long record of apparently satisfactory behaviour, failing suddenly and catastrophically due to internal erosion is very small unless some unusual external circumstance arises. It would seem that the risk of internal erosion causing embankment breaching can be reduced to almost negligible proportions by appropriate maintenance, surveillance and, where necessary, remedial works, together with preparedness for emergency lowering of the reservoir. Smaller dams where surveillance may not be so effective may pose the greatest risk.

*What features in an embankment dam are critical in rendering it vulnerable to such a development?* Internal erosion is not confined to a particular soil type, although the way the erosion develops will be strongly influenced by the nature of the soil. For example, the broadly graded moraines typically found in Scandinavia are prone to internal erosion but are usually self-healing. Internal erosion may be less likely in clay soils but, if it does occur, it may be more likely to lead to breaching of the dam. In almost half the cases where failure occurred (A1), or where failure almost certainly would have occurred very quickly if the reservoir had not been rapidly drawn down (A2a), the problem was associated with a structure passing through the embankment. The conditions at the interface of the soil and the structure are critical. In several cases the failure of an upstream membrane was a critical factor. There are very high hydraulic gradients and failure of the membrane can, in some circumstances, lead to rapid internal erosion of the fill.

*What warning signs could be expected and how much warning would these give prior to failure?* Internal erosion is difficult to analyse and the continuing safety of embankment dams is dependent on an approach based on the observational method. However, internal erosion is a hidden phenomenon and until some feature such as a sinkhole appears at the surface of the soil, it is difficult to identify and investigate.

## RECOMMENDATIONS FOR FURTHER WORK

Research on internal erosion is being undertaken in many countries around the world and the European Working Group fulfils a useful role in promoting European collaboration and bringing together dam engineers and research workers. It is recommended that it should continue its work.

The study of incidents has provided information on the types of dams affected and the speed with which internal erosion problems develop. This should assist in identifying analytical, laboratory and field studies which are needed to complement current national research programmes. The identification and investigation of internal erosion is not easy as it can be very localised. Many different techniques have been used including electrical resistivity, ground-probing radar, self potential and temperature measurements. It is important to evaluate their effectiveness.

It is recommended that risk assessment methodology should be applied to the extensive information now contained in dam databases. This should assist in the identification of potentially hazardous conditions and, possibly, the quantification of the risk of internal erosion. There is a need to collaborate closely with the European Working Group on Risk Assessment. Detection and investigation are key issues and field investigation methods need to be critically assessed. Fundamental work in the laboratory to develop improved understanding of erosion mechanisms, particularly the erosion resistance of clays, would be valuable.

## ACKNOWLEDGEMENTS

This paper is a progress report of the European Working Group on Internal Erosion in Embankment Dams of which the author is Chairman. The report was presented at the ICOLD European Symposium, *Dams in a European Context*, held at Geiranger, Norway, in June 2001. The contributions of the members of the Working Group are gratefully acknowledged.

## REFERENCES

Bartsch M (1995). Filtration processes and internal erosion in dams of broadly graded material. Technical Report, Royal Institute of Technology, Stockholm

Charles J A (1998). Internal erosion in European embankment dams – progress report of Working Group on internal erosion in embankment dams. Dam safety (ed L Berga). Proceedings of international symposium, Barcelona, vol 2, pp 1567-1576.

Comité Français des Grands Barrages (1997). Internal erosion: typology, detection, repair. Barrages & Reservoirs, no 6, May. 126pp.

Dolcetta M (1997). Discussion on Q73, Transactions of 19th International Congress on Large Dams, Florence, vol 5, p 339.
Hulea D (1997). Private communication.
ICOLD (1990). Dispersive soils in embankment dams: review. Bulletin 77. International Commission on Large Dams, Paris.
ICOLD (1994). Embankment dams: granular filters and drains. Bulletin 95. International Commission on Large Dams, Paris.

APPENDIX 1
SUMMARY OF CASE HISTORIES

| Dam | Refs | Country | Date built | Watertight element | Problem or incident | | | |
|---|---|---|---|---|---|---|---|---|
| | | | | | A | B | C | D |
| Arbon | 50, 51 | Spain | 1967 | Central gravel and silty sand core | 2b | 6 | 1, 3 | 1, 2 |
| Balderhead | 11, 55 | United Kingdom | 1965 | Central rolled clay core | 2b | 6, 7 | 1, 2<br>3 | 1, 2 |
| Buget | 12, 19 | France | 1980 | Homogeneous clayey earth fill embankment | 2a | 2 | 2, 4 | 4 |
| Caspe | 14, 60 | Spain | 1988 | Clay core | 2b | 9 | 1, 2 | 2 |
| Elbe-Seitenkanal | 20, 33<br>45, 46 | Germany | 1976 | Upstream asphaltic membrane | 1 | 1 | 4 | 11 |
| Fonte Longa | 41 | Portugal | 1988 | Homogeneous earth fill | 3 | 1, 3 | 5 | 4 |
| Gostei | 41 | Portugal | 1993 | Homogeneous earth fill | 3 | 1 | 5 | 4 |
| Gourdon | 12, 17<br>19 | France | 1983 | Upstream geomembrane, clay blanket on foundation | 2b | 5 | 4, 6<br>7 | 6 |
| Greenbooth | 10, 11<br>16 | United Kingdom | 1961 | Central puddle clay core | 2b | 6 | 1 | 2 |
| Grossee | 58 | Austria | 1980 | Upstream asphaltic membrane | 3 | 4 | 2 | 5 |
| Grundsjoarna | | Sweden | 1972 | Central moraine core | 2b | 7 | 1, 7 | 11 |
| Hallby | 1, 9 | Sweden | 1970 | Central moraine core | 2b | 10 | 1, 2 | 2 |
| Hyttejuvet | 22, 28<br>59 | Norway | 1965 | Central moraine core | 2b | 6 | 1, 2 | 2 |
| Ibra | 7 | Germany | 1975 | Upstream geomembrane | 1 | 1, 4 | 4, 7 | 5, 8<br>9,10 |
| Jukla | 23, 26 | Norway | 1974 | Central moraine core | 2b | 6, 7 | 2 | 2,10<br>11 |

| Dam | Refs | Country | Date built | Watertight element | Problem or incident | | | |
|---|---|---|---|---|---|---|---|---|
| | | | | | A | B | C | D |
| Juktan | 29 | Sweden | 1978 | Central moraine core | 2b | 6, 7 | 2 | 1,11 |
| La Prade | 12, 17 19 | France | 1982 | Wide central clay core | 2b | 2 | 4 | 2, 4 |
| Lavaud-Gelade | 12, 17 18, 19 | France | 1943 | Homogeneous sand fill embankment | 3 | 8 | 4 | 10 |
| Lluest Wen | 10, 11 | United Kingdom | 1896 | Central puddle clay core | 2b | 3 | 1 | 1, 2 |
| Lovon | 24 | Sweden | 1973 | Upstream sloping moraine core | 2b | 6, 7 | 1 | 2 10 |
| Main-Donau-Kanal | 15, 20 | Germany | 1978 | Upstream asphaltic membrane | 1 | 1 | 4 | 11 |
| Martin Gonzalo | 27 | Spain | 1987 | Upstream geomembrane | 2a | 4 | 1, 2 | 5 |
| Moravka | 8 | Czech Republic | 1965 | Upstream asphaltic membrane | 2a | 4 | 1, 2 | 5 |
| Motru | 42 | Romania | 1984 | Central clayey core | 3 | 10 | 2 | 1 |
| Mysevatn | 22, 23 | Norway | 1973 | Moraine core | 2b | 6 | 1, 2 | 2 10 |
| Nepes | 17, 18 34 | France | 1945 | Reinforced concrete core wall | 2b | 10 | 2, 3 6 | 2 |
| Nyrsko | 49 | Czech Republic | 1970 | Upstream reinforced concrete facing | 2b | 4 | 2 | 5 |
| Peruca | 13, 37 48 | Croatia | 1958 | Narrow central clay core | 2a | 12 | 1, 2 3 | 1 11 |
| Porjus | 25 | Sweden | 1980 | Central moraine core | 2b | 6, 7 | 1 | 2 |
| Rengard | | Sweden | 1970 | Moraine core | 2b | 6, 7 | 1, 2 3 | 2 |
| Saint Aignan | 12, 17 | France | 1965 | Homogeneous earth fill embankment | 1 | 8 | 4 | 11 |
| St Julien des Landes | 12, 17 19 | France | 1969 | Homogeneous earth fill embankment | 2a | 1 | 5 | 4, 11 |
| Saint Pardoux | 6, 12 17, 19 34 | France | 1975 | Homogeneous sand fill embankment | 2b | 8 | 4, 7 | 1, 2 |
| Sapins | 6, 12 17, 18 19, 34 | France | 1978 | Homogeneous sand fill embankment | 2a | 7 | 4, 9 | 1 10 |
| Seitevare | 5 | Sweden | 1967 | Central moraine core | 2b | 10 | 4 | 2 |

| Dam | Refs | Country | Date built | Watertight element | Problem or incident | | | |
|---|---|---|---|---|---|---|---|---|
| | | | | | A | B | C | D |
| Songa | 53 | Norway | 1962 | Central moraine core | 4 | 6, 7 | 2, 3 | 12 |
| Sorpe | 21, 30 57 | Germany | 1935 | Central concrete core wall | 2a | 12 | 1, 2 | 2, 3 |
| Stenkullafors | 38, 39, 40 | Sweden | 1983 | Central moraine core | 2b | 1, 6 7 | 1 | 10 |
| Suorva | 1, 3 9, 39 40, 54 | Sweden | 1972 | Central moraine core | 2b | 6, 7 | 1, 2 3 | 2 |
| Sylvenstein | 4, 35 | Germany | 1958 | Central vertical soil-cement core | 3 | 7 | 2, 7 | 2 |
| Taibilla | 2 | Spain | 1973 | Upstream sloping clay core | 2a | 9 | 8 | 11 |
| Torcy Vieux | 52 | France | 1800 | Homogeneous earth fill embankment | 2b | 11 | 4, 7 | 9 10 |
| Uljua | 31, 32 36, 43 44 | Finland | 1970 | Central moraine core | 2a | 10 | 1, 2 3 | 1, 2 11 |
| Viddalsvatn | 22, 56 | Norway | 1971 | Central moraine core | 2b | 7 | 1, 2 3 | 2 |
| Warmwithens | 10, 11 | United Kingdom | 1870 | Central puddle clay core | 1 | 1 | | 12 |
| Winscar | 11, 47 | United Kingdom | 1975 | Upstream asphaltic concrete membrane | 3 | 4 | 2 | 5 |
| Withens Clough | 11 | United Kingdom | 1894 | Central puddle clay core | 3 | 6 | 2 | 1, 2 |

## APPENDIX 2
## CLASSIFICATIONS

A – severity of problem or incident

A1 Failure

A2 Serious incident involving emergency action or drawdown
- (a) Without emergency action, a breach was likely
- (b) Little danger of immediate breach

A3 Incident causing concern, major investigation and remedial works

A4 Symptoms causing concern

B – cause of problem or incident

B1 Erosion at contact with pipe or structure

B2 Outlet pipe failure

| | |
|---|---|
| B3 | Erosion into pipe or culvert |
| B4 | Upstream membrane failure |
| B5 | Fracture of clay foundation blanket |
| B6 | Fracture of core |
| B7 | Inadequate filter |
| B8 | Absence of internal drain/filter |
| B9 | Foundation solubility |
| B10 | Inadequate foundation treatment |
| B11 | Rotting tree roots |
| B12 | Military action |

C – symptoms of problem or incident

| | |
|---|---|
| C1 | Sinkhole |
| C2 | Excessive or increasing seepage and leakage |
| C3 | Turbid seepage and leakage |
| C4 | Wet areas or seepage and leakage on downstream slope |
| C5 | Leakage into or around culvert |
| C6 | Piping |
| C7 | Excessive or increasing pore pressures |
| C8 | Vortex in reservoir |
| C9 | Slip |

D – remedial works

| | |
|---|---|
| D1 | Diaphragm wall |
| D2 | Grouting |
| D3 | Jet grouting |
| D4 | Pipe or culvert repair |
| D5 | Upstream membrane repair |
| D6 | Clay blanket repair |
| D7 | Drainage gallery |
| D8 | Relief wells |
| D9 | Filters |
| D10 | Slope flattening or addition of berm |
| D11 | Partial or total reconstruction |
| D12 | None |

## APPENDIX 3
## REFERENCES FOR CASE HISTORIES

1. Alemo J, Bronner N & Johansson N (1991). Long term stability of grout curtains. Transactions of 17th International Congress on Large Dams, Vienna, vol 2, pp 1311-1325.
2. Alonso Franco M & Yague Cordova J (1997). Earth materials for cores and homogeneous earth fill dams; singularities. Transactions of 19th International Congress on Large Dams, Florence, vol 2, pp 381-401.

3. Andersson P M (1991). Borehole radar as a tool for detecting dam ageing. Transactions of 17th International Congress on Large Dams, Vienna, vol 5, pp 251-257.
4. Beier H, List F & Lorenz W (1979). Subsequent sealing in the impervious core of the Sylvenstein dam. Transactions of 13th International Congress on Large Dams, New Delhi, vol 2, pp 103-116.
5. Bernell C E (1976). Control of leakage through dams founded on glacial till deposits. Transactions of 12th International Congress on Large Dams, Mexico City, vol 2, pp 937-950.
6. Bister D, Fry J J, Costaz J, Houis J, Dupas J M, Degoutte G, Lino M & Rizzolli J L (1994). Reevaluation de la securite des barrages en remblai exemples d'application. Transactions of 18th International Congress on Large Dams, Durban, vol 1, pp 645-670.
7. Breth H & Romberg W (1984). Failure of an earth dam on Triassic Sandstone: causes of the failure and criteria for reconstruction. Proceedings of International Conference on Safety of Dams, Coimbra, pp 15-20. Balkema, Rotterdam.
8. Brezina P, Satrapa L & Skokan T (1998). The reconstruction of the collapsed bituminous concrete facing of the Moravka dam, Czech Republic. Rehabilitation of Dams. Proceedings of symposium held at ICOLD 66th Annual Meeting, November 1998, New Delhi, pp 223-226. Balkema, Rotterdam, 1999.
9. Bronner N, Fagerstrom H & Stille H (1988). Bedrock cracks as a possible cause of leakage in two Swedish dams. Transactions of 16th International Congress on Large Dams, San Francisco, vol 2, pp 1029-1052.
10. Charles J A, Tedd P & Holton I R (1995). Internal erosion in clay cores of British dams. Research and Development in the Field of Dams. Proceedings of European Symposium, Crans Montana, pp 59-70. Swiss National Committee on Large Dams.
11. Charles J A, Tedd P, Hughes A K & Lovenbury H T (1996). Investigating embankment dams: a guide to the identification and repair of defects. Building Research Establishment Report BR 303. 81pp.
12. Comité Français des Grands Barrages (1997). Internal erosion: typology, detection, repair. 126pp.
13. Croatian Committee on Large Dams (1995). Peruca dam remediation. Proceedings of international conference, Brela, 472pp.
14. De Celix Caballero M R & Sanchez-Albornoz A A (1995). Internal erosion of the Caspe dam. Research and Development in the Field of Dams. Proceedings of European Symposium, Crans Montana, pp 47-58. Swiss National Committee on Large Dams.
15. Ferguson H (1979). Germans tight lipped as investigation starts into canal disaster. New Civil Engineer, 5 April, pp 16-18.
16. Flemming J H & Rossington D T (1985). Repair of Greenbooth dam following localised settlement in the crest. Transactions of 15th International Congress on Large Dams, Lausanne, vol 4, pp 875-897.

17. Fry J-J (1995). Developpements concernant les risques d'erosion interne et des seismes relatifs aux remblais. Research and Development in the Field of Dams. Proceedings of European Symposium, Crans Montana, pp 35-46. Swiss National Committee on Large Dams.
18. Fry J-J, Brun P & Royet P (1994). Internal erosion problems in some old embankment dams in France. Reservoir Safety and the Environment. Proceedings of 8th Conference of British Dam Society, Exeter, pp 23-34. Thomas Telford, London.
19. Fry J-J, Brun P, Lautrin D, Albalat C, Droux J M, Mercier A, Riesterer J M, Thareau L & Degoutte G (1997). Typologie de l'erosion interne et detection. Transactions of 19th International Congress on Large Dams, Florence, vol 2, pp 819-844.
20. Hager M (1982). Stand der Risikobewertung bei durchstromten, gewachsenen und geschutteten nichtbindigen Dammbaustoffen im Kanalbau. Vortrage der Basugrundtagung in Braunschweig, DGEG e V Essen.
21. Heitefuss C, Krug H-J & Ruhnke H (1998). Die Sanierung des Sorpes-Dammes mit einer HDI-Dichtwand (Rehabilitation of Sorpe dam with a jet grout curtain), Wasserwirtschaft, vol 88, p11.
22. Hoeg K, Valstad T, Kjaernsli B & Johansen P M (1993). Norwegian experience with embankment dams. Maintenance of older dams: accidents, safety assessment and repairs. Proceedings of European Technical Symposium, Chambery, vol 1, pp 41-61. Comité Français des Grands Barrages.
23. Hoff T & Nilsen K Y (1985). Erosion and leakage problems on some Norwegian dams. Transactions of 15th International Congress on Large Dams, Lausanne, vol 4, pp 573-586.
24. Johansson S (1996). Seepage monitoring in embankment dams by temperature and resistivity measurements. Repair and Upgrading of Dams, Proceedings of European Symposium, Stockholm, June 1996, pp 288-297.
25. Johansson B, Johansson S & Nilsson R (1996). Investigations and repair of the embankment dam at Porjus power station. Repair and Upgrading of Dams. Proceedings of European Symposium, Stockholm, pp 218-227. Royal Institute of Technology, Stockholm.
26. Johansen P M & Eikevik J H (1997). Internal erosion and rehabilitation of Jukla rockfill dams. Transactions of 19th International Congress on Large Dams, Florence, vol 2, pp 245-253.
27. Justo J L (1988). The failure of the impervious facing of Martin Gonzalo rockfill dam. Transactions of 16th International Congress on Large Dams, San Francisco, vol 5, pp 252-262.
28. Kjaernsli B & Torblaa I (1968). Leakage through horizontal cracks in the core of Hyttejuvet dam. Norwegian Geotechnical Institute publication no 80, pp 39-47.

29. Kjellberg R, Norstedt U & Fagerstrom H (1985). Leakage in and reconstruction of the Juktan earth and rockfill dams. Transactions of 15th International Congress on Large Dams, Lausanne, vol 4, pp 553-572.
30. Koenig H W (1967). Damage of the Sorpe dam by vibrations and repair work. Transactions of 9th International Congress on Large Dams, Istanbul, vol 4, pp 485-498.
31. Kuusiniemi R (1991). Internal erosion of the Uljua earth dam. Transactions of 17th International Congress on Large Dams, Vienna, vol 5, pp 257-262.
32. Kuusiniemi R & Loukola E (1996). Repairing works of earth dams owned by the state in Finland. Repair and Upgrading of Dams. Proceedings of European Symposium, Stockholm, pp 195-206. Royal Institute of Technology, Stockholm.
33. Lackner E & Hager M (1980). Der Schadensfall an der Unterfuhrung Luneburg/Nutzfelde des Elbe-Seitenkanals und seine Folgerungen. Jahrbuch der Hafenbautechnischen Gesellschaft e V 37 Band, Springer-Verlag.
34. Lino M, Degoutte G, Goguel B, Chassard G, Saint-Marcel A, Gouvenot D & Delmas F (1994). Mesures prises pour ameliorer l'etancheite des barrages et de leurs fondations et reduire le risque d'erosion interne. Transactions of 18th International Congress on Large Dams, Durban, vol 1, pp 731-761.
35. List F & Strobl T (1991). Changing of tightness in the grout curtain and core of the Sylvenstein dam by ageing. Transactions of 17th International Congress on Large Dams, Vienna, vol 5, pp 1295-1309.
36. Loukola E & Slunga E (1982). Observations on the seepage control through and under dams made of glacial till. Transactions of 14th International Congress on Large Dams, Rio de Janeiro, vol 1, pp 293-304.
37. Megla T, Rupcic J, Petricec M, Sever Z, Macan J, Vilovic M, Kuk V, Herak D, Herak M, Dadic V, Grbec B, Salkovic A, Dujmic D, Nonveiller E & Grgic P (1993). The Peruca dam, Croatia. Proceedings of International Workshop on Dam Safety Evaluation, Grindelwald, Supplementary volume, 71pp. Dam Engineering, Sutton, Surrey, England.
38. Nilsson A & Norstedt U (1991). Evaluation of ageing processes in two Swedish dams. Transactions of 17th International Congress on Large Dams, Vienna, vol 2, pp 23-47.
39. Norstedt U, Eriksson H & Nilsson A (1993). Research and development programs in Sweden. Maintenance of older dams: accidents, safety assessment and repairs. Proceedings of European Technical Symposium, Chambery, vol 1, pp 359-383. Comité Français des Grands Barrages.
40. Norstedt U & Nilsson A (1997). Internal erosion and ageing in some of the Swedish earth and rockfill dams. Transactions of 19th International Congress on Large Dams, Florence, vol 2, pp 307-319.

41. Oliveira Lemos F & Machado do Vale J (1997). Culverts buried in earthfill dams; some cases and problems. Transactions of 19th International Congress on Large Dams, Florence, vol 2, pp 421-429.
42. Popovici A, Sarghiuta R, Abdulamit A & Teodorescu C (1998). Study concerning the excessive leakage from a large reservoir. Rehabilitation of Dams. Proceedings of symposium held at ICOLD 66th Annual Meeting, November 1998, New Delhi, pp 425-433. Balkema, Rotterdam, 1999.
43. Rathmayer H & Polla J (1998). Finnish dam safety code prohibited collapse of the Uljua earth dam. Dam Safety (ed L Berga). Proceedings of International Symposium on New Trends and Guidelines on Dam Safety, Barcelona, vol 2, pp 903-910. Balkema, Rotterdam.
44. Ravasaka O (1997). Piping susceptibility of glacial till. Transactions of 19th International Congress on Large Dams, May, Florence, vol 2, pp 455-471.
45. Reina P & Marshall T (1976). German canal fails after only 6 months. New Civil Engineer, 22 July, p6.
46. Reina P (1976). Six-man team wades into Luneburg collapse. New Civil Engineer, 29 July, pp 16-18.
47. Routh C D (1988). The investigation, identification and repair of the asphaltic concrete facing of Winscar dam. Transactions of 16th International Congress on Large Dams, San Francisco, June 1988, vol 2, pp 655-677.
48. Rupcic J (1997). Damage and remediations of Peruca dam. Transactions of 19th International Congress on Large Dams, Florence, vol 4, pp 93-112.
49. Simek M (1976). Discussion. Transactions of of 12th International Congress on Large Dams, Mexico City, vol 5, pp 215-221.
50. Telleria J (1995). New works (1994) to improve the fissured core of the Arbon dam in Spain. Research and Development in the Field of Dams. Proceedings of European Symposium, Crans Montana, pp 121-132. Swiss National Committee on Large Dams.
51. Telleria J & Gomez-Laa G (1991). Arbon dam – a didactic experience about problems of a dam built on a deformable foundation. Transactions of 17th International Congress on Large Dams, Vienna, vol 3, pp 135-155.
52. Testemale J P, Parron C, Royet P, Faure Y H, Lambert S & Fagon Y (1999). Torcy Vieux dam (France – 71) sampling, tests and results on geotextile filters. Recontres Geosynthetiques, Bordeaux, vol 2, pp 311-318.
53. Torblaa I & Rikartsen C (1997). Songa, sudden variations of the leakage in a 35 years old rockfill dam. Transactions of 19th International Congress on Large Dams, Florence, vol 2, pp 255-267.
54. Triumf C-A & Thunehed H (1996). Two years of self-potential measurements on a large dam in northern Sweden. Repair and Upgrading of Dams. Proceedings of European Symposium, Stockholm, pp 307-315. Royal Institute of Technology, Stockholm.

55. Vaughan P R, Kluth D J, Leonard M W & Pradoura H H M (1970). Cracking and erosion of the rolled clay core of Balderhead dam and the remedial works adopted for its repair. Transactions of 10th International Congress on Large Dams, Montreal, vol 1, pp 73-93.
56. Vestad H (1976). Viddalsvatn dam, a history of leakages and investigations. Transactions of 12th International Congress on Large Dams, Mexico City, vol 2, pp 369-389.
57. Waelter E (1976). Determination of the origin and quantity of the leakage and its influence on the behaviour of the Sorpe dam. Transactions of 12th International Congress on Large Dams, Mexico City, vol 2, pp 1-16.
58. Wellacher H & Tschernutter P (1985). Repair of a high altitude rockfill dam with an upstream asphalt concrete membrane. Transactions of 15th International Congress on Large Dams, Lausanne, vol 4, pp 633-645.
59. Wood D M, Kjaernsli B & Hoeg K (1976). Thoughts concerning the unusual behaviour of Hyttejuvet dam. Transactions of 12th International Congress on Large Dams, Mexico City, vol 2, pp 391-414.
60. Yague Cordova J & Alonso Franco M (1997). Buried conduits in embankment dams: experience in Spain. Transactions of 19th International Congress on Large Dams, Florence, vol 2, pp 321-336.

# The use of temperature measurements for detection of leakage in embankment dams - British Waterways experience

D P M DUTTON, British Waterways, UK

SYNOPSIS. The investigation of an issue of water from an embankment dam is an essential part of any safety assessment. Seepage or leakage may be an early indicator of slope instability or internal erosion and the knowledge of where the water is percolating through an embankment is fundamental to any stability analysis and the design of remedial works. British Waterways' experience in the use of ground temperature measurements to locate leakage paths within earth embankments is described in this paper.

## INTRODUCTION

During the last two years ground temperature measurements have been used to investigate water issues seen at six British Waterways' dams and nine canal embankments. The investigative work and interpretation has been undertaken by GTC Kappelmeyer, for whom British Waterways Technical Services now acts as sole agent for use of the patented technique in the UK.

The theory underlying the technique has been set out in a number of papers, notably Dornstadter (1997) and Aufleger et al (2000), and the investigative method is described in Andrews et al (2000). Essentially, the technique relies on the differing seasonal temperatures within the ground and surface waters and the effect that water, percolating through the ground, has on its temperature.

In ground where there is no water movement the temperature at any depth is a function of the temporal variation in surface temperature and the thermal conductivity of the soil; this results in a reduction in the seasonal temperature variation amplitude with increasing depth and also in a phase shift. Where there is water movement, heat transport by advection causes a measurable change in ground temperatures provided, of course, there is a difference between the water temperature and the temperature of the ground through which it is percolating

To locate the leakage path the ground temperatures within the embankment are measured by lowering a string of temperature sensors into small diameter hollow metal probes, driven into the ground using hand portable rammers. The distance between probes is generally 10m or 20m, depending

*Reservoirs in a Changing World*, Thomas Telford, London, 2002, 394–402

on the nature of the ground and the length of time that seepage has been occurring, but may be reduced to 5m under adverse conditions.

INVESTIGATION RESULTS

During the last 12 months water temperatures have been sampled at 41 reservoirs in England and Scotland in order to provide a typical seasonal temperature distribution curve for the UK. Figure 1 shows a sinusoidal exponential function fitted to this data; there is, as would be expected given

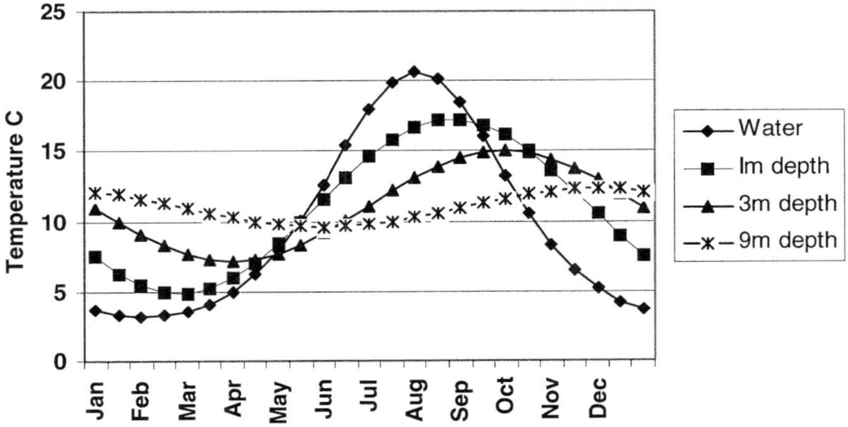


Fig. 1. Seasonal temperature distributions in the reservoir and the ground

the geographic spread of the reservoir sites, considerable variation around this best fit equation (+6°C/-4°C). A sinusoidal function has been fitted to the measured temperatures from the leakage investigations, to provide the non-percolated ground temperatures shown in the figure and given in Table 1; there is considerably less variation in these figures (±2°C).

Table 1 Reservoir water and ground temperatures

| Month | Water °C | 1m depth °C | 3m depth °C | 9m depth °C |
|---|---|---|---|---|
| January | 3 | 6 | 10 | 12 |
| February | 3 | 5 | 8 | 11 |
| March | 4 | 5 | 7 | 11 |
| April | 6 | 7 | 7 | 10 |
| May | 10 | 10 | 8 | 10 |
| June | 15 | 13 | 10 | 10 |
| July | 20 | 16 | 12 | 10 |
| August | 20 | 17 | 14 | 11 |
| September | 16 | 17 | 15 | 11 |
| October | 11 | 15 | 15 | 12 |
| November | 7 | 12 | 14 | 12 |
| December | 4 | 9 | 12 | 12 |

From Figure 1, the reduction in temperature variation amplitude with increasing depth and the increasing phase shift can be seen. The figure also clearly illustrates the basis of the technique. With normal dam freeboards, the scope for using the reservoir water as a tracer during winter and summer months is apparent.

## INVESTIGATION CASE HISTORIES

### 1. Drayton Reservoir

This 324Ml reservoir is impounded by a 335m long, 9m high embankment dam. The A361 road runs on a berm on the downstream face, 4m below the dam crest. Some of the run-off from the road surface discharges into a roadside ditch on the reservoir side of the road, from where a piped drain, at the lowest point of the ditch, takes the water under the road to the toe of the dam.

In 1971 steel sheet piles, up to 10m long, were driven along the centre of the crest, to stop leakage that was occurring through the embankment and for many years this work appeared to have been successful. Recently, following a proposal to widen the dam crest to provide additional space for car parking, a ground investigation was undertaken consisting of 4 boreholes with piezometers. The data from the piezometers indicated a very low phreatic surface within the downstream shoulder, well below the level of the road berm. Shortly after this, the roadside ditch was cleared out and, with the reservoir at Top Water Level, water was reported seeping into the ditch in a number of places over a length of 60m, and one or two small runs of water could also be seen. In view of the anomalous piezometric information,

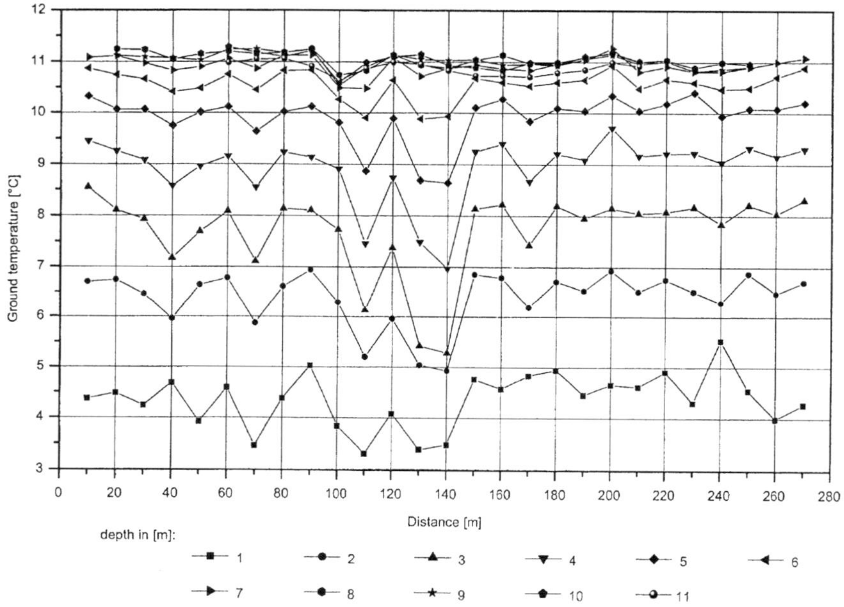


Fig. 2. Measured ground temperatures at all sections

it was decided to undertake a leakage investigation and this was carried out in March 2001. At the time there was a thin covering of ice on the reservoir and the water temperature, just below the surface, was 3.5°C.

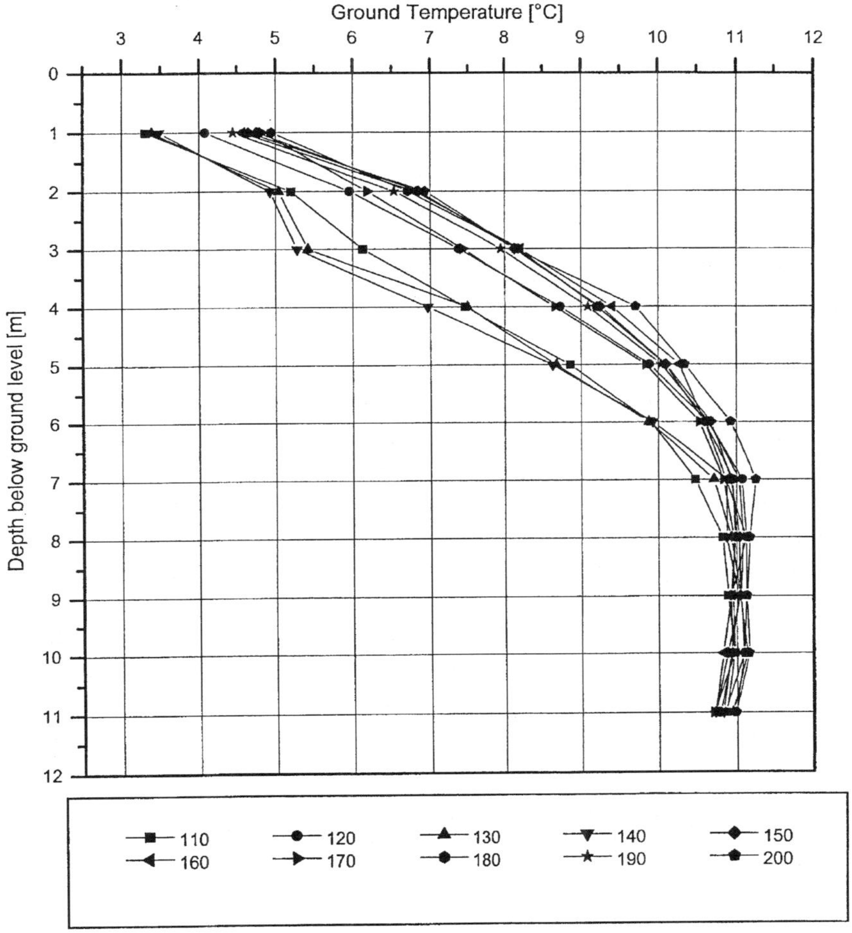


Fig. 3. Measured ground temperatures at Sections 110m to 200m

A 260m length of the dam was investigated with probes installed at 10m intervals along the downstream edge of the crest to 1m below the toe of the piles; ground temperatures were measured at 1m intervals in all the probes. Figure 2 shows a longitudinal section of the dam and the measured ground temperatures. Figure 3 shows the measured ground temperatures at sections 110m to 200m. The departure from the non-percolated section temperature distribution is clearly visible for sections 110m, 130m and 140m, with leakage occurring through, not under, the piles.

Following the investigation six additional piezometers were installed in the dam at locations where leakage was indicated, and these have confirmed that pore pressures within the dam are higher than originally thought. This finding has caused the stability of the downstream shoulder, and the proposal for additional car parking, to be reassessed.

2. Clattercote Reservoir

Clattercote Reservoir is impounded by a 10m high embankment dam of homogeneous construction with no discernible clay core; the fill material is a sandy silty clay. There is a suggestion that the dam was raised 3m early in its life; dam freeboard today is just under 1m. The spillweir is at the righthand end of the dam and the spillway channel runs down the right mitre.

In 1986 leakage appeared on the downstream face close to the righthand end of the dam; this was stopped by reconstructing the spillweir with a sheet pile cut-off extending 20m into the dam, and grouting voids beneath the spillway channel.

By 1992 seepage was being reported at the toe of the dam whenever the reservoir was close to Top Water Level, and this was visually monitored for a number of years. In 1998, following a gradual increase in the flow, an attempt was made to stop the leakage (without prior investigation, other than to note the reservoir level at which the wet area dried up) by excavating and repuddling a 0.6m wide, 2m deep trench along the righthand half of the dam crest. Water was seen entering the open trench in a number of places at about a metre below dam top level. This work appeared, at first, to have been successful but, after only one emptying/refilling cycle, the leakage reappeared.

A shallow ditch runs down the left mitre. Small flows of water are seen in the ditch whenever the reservoir is within 0.3m of Top Water Level. The central section of the dam is built on the Lower Lias Clay. It was originally thought that the flow in the ditch might be coming from the strata's upper horizon, at the dam's mid height, as this is recorded on the geological map as a spring line. However, some trial holes, dug in the downstream shoulder, revealed water flowing down the face just below the ground surface. Ditch flow temperatures as low as 2°C, measured during the 2000/2001 Winter, confirmed that this was not groundwater.

Following a recommendation in the report of the last Section 10 Periodical Inspection, to increase the dam freeboard, it was decided that all the leaks should be stopped before the crest was raised and, in June 2001, with the reservoir full, a leakage investigation using ground temperature sensing was carried out. Temperature probes were installed at 10m intervals along the crest of the dam, about 1m from its downstream edge, over a length of 120m. Alternate probes were rammed to 1m below original ground level to ensure that any low level leakage was detected; the remainder were installed

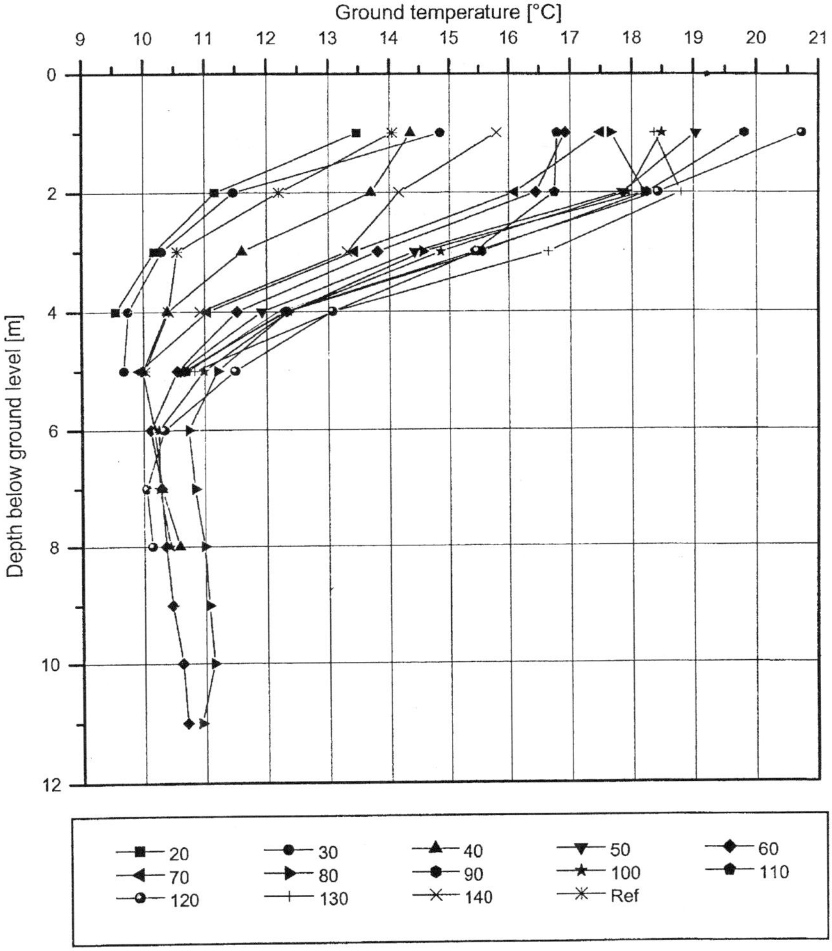


Fig. 4. Measured ground temperatures at all sections

to 5m below crest level. Reservoir water temperature was 21°C at the surface but only 12°C at bottom outlet level. Figure 4 shows the measured temperatures at all investigated sections. Sections 20m and 30m show the expected temperature profile for a non-percolated bank but the remainder indicate considerable ground warming due to leakage through the upper layers of the crest. The shape of the curves indicates that most of the leakage is occurring through the upper 2m but that there is some flow down to about 3m. Remedial work will be undertaken before more material is placed on the crest to increase the freeboard.

3. Wilstone Reservoir

Wilstone Reservoir has a history of problems associated with leakage. Following the first inspection under the 1930 Act the then owners, the Grand Union Canal Company, were advised either to reduce Top Water

Level or to install steel sheet piling along the reservoir's north west bank to reduce leakage flows. The canal company chose the cheaper option and cut a 300mm deep notch in the overflow sill, increasing dam freeboard to 1.6m, and removed the facility to place stoplogs on the sill. Fifty years later a drain had to be constructed along the northern third of the embankment after leakage was again observed at the toe. High groundwater levels in the surrounding land and poor outfall conditions restricted the depth of the pipe to 600mm. The remaining two thirds of the bank has the spillway channel running at its toe. It is known that water leaks into the channel whenever the reservoir is higher than 0.4m below weir level but, as soon as the reservoir is overflowing by any appreciable amount, these leakage flows cannot be monitored. This unsatisfactory situation can occur continuously for a number of months in the early part of the year because the reservoir is spring fed.

In June 2000 a 50m length of the drained portion of the embankment became saturated at the toe and a small leakage flow was visible; water was also found to be seeping and dripping into the spillway channel at locations not seen before. With no knowledge of pore pressures within the bank there was some concern over its stability (the adjoining north east embankment has suffered two downstream face slips in the past, although these may have occurred when a higher water level was being retained).

A leakage investigation was undertaken in July 2000, along 560m of the embankment; temperature probes, up to 10m long and extending 1m below original ground level, were installed at 20m intervals along the downstream edge of the crest. Reservoir water temperature at the time of the investigation had cooled to 16.5°C, from an earlier high of 19°C measured at the end of June. Figure 5 shows the ground temperature profiles between sections 460m and 560m at the northern end of the bank. The 'Ref' profile is for a probe rammed into the ground in the shade of a hedge close to the outfall of the toe drain; the influence of high groundwater levels on its temperature profile is clearly visible. Sections 540m and 560m provide the expected temperature profile for the bank with no leakage and the effect of leakage down to about 3m below the crest can be seen in the 2°C-3°C increase in temperatures in the other sections. The effect of the earlier higher temperatures in the month before the investigation took place is also apparent, with ground temperatures still 1°C or so above the temperature of the percolating water.

The results of the leakage investigation were used to position nineteen window sampler probe holes in the crest and downstream shoulder, at three sections showing the worst leakage. In confirmation of the earlier interpretation of the ground temperature data, groundwater inflows were recorded in three exploratory holes in the crest at depths of 2.2m, 2.7m and 2.9m. Piezometers were installed in the holes to enable stability analyses to

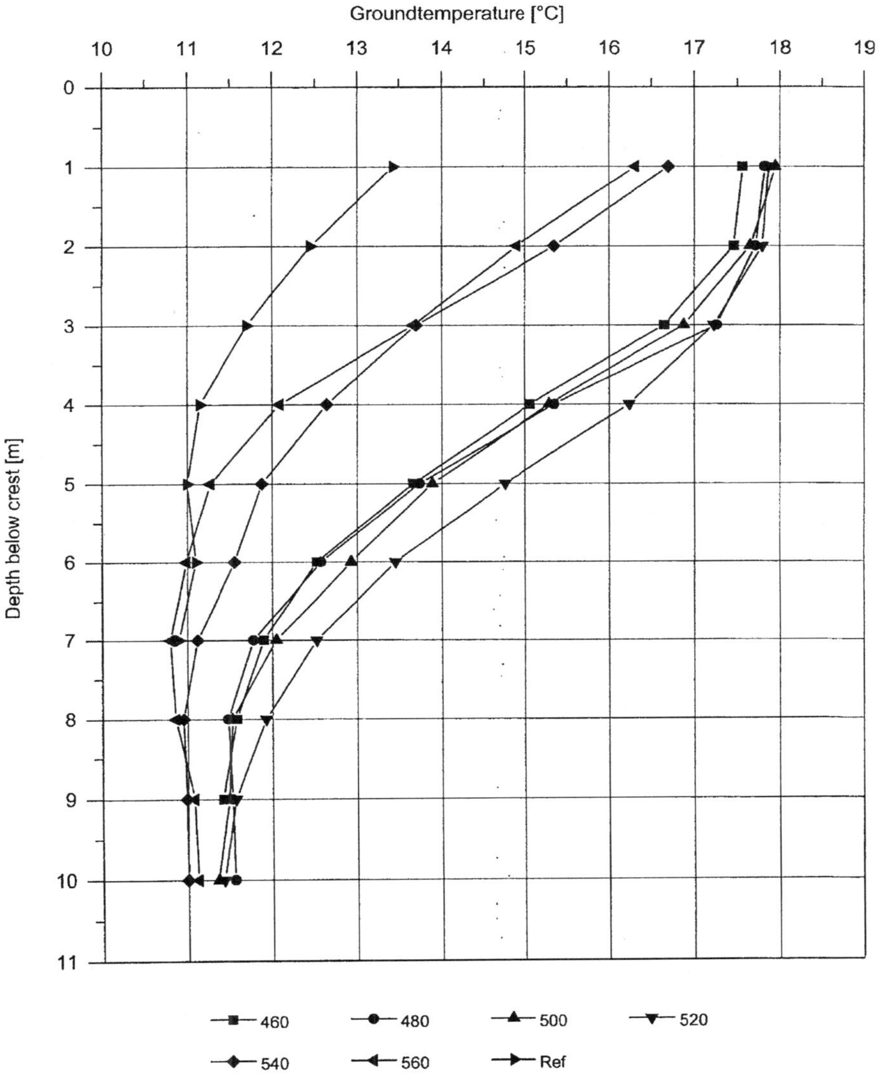


Fig. 5. Measured ground temperatures at Sections 460m to 560m

be carried out. These indicated that factors of safety were a little less than ideal and, with a CCTV survey of the toe drain showing the pipe to be distorted and blocked at one point, it was decided, as a part solution to the problem, to relay the drain and dredge a length of the stream into which it discharges, to provide better outfall conditions.

## CONCLUSIONS

During the last two years ground temperature sensing has been used to investigate issues of water at six dams and nine canal embankments; leakage from the reservoir or canal was positively identified at eleven of the sites.

Ground investigations have been undertaken subsequently at three of the reservoirs and these have confirmed the findings of the leakage investigations.

It is considered that ground temperature sensing provides a valid method for investigating the source of issues from embankment dams, with the ability to distinguish between leakage and groundwater. As all the equipment is hand portable, investigations can be undertaken at sites with limited access and, because results are available almost immediately, any unexpected findings can be further investigated there and then without the need for a return visit to site. The theoretical basis of the technique is easily understood but, because its application is seasonally dependent, knowledge of local temperature conditions is essential to ensure a successful investigation.

REFERENCES

Aufleger M, Strobl T & Dornstadter J (2000). Fibre optic temperature measurements in dam monitoring - four years of experience. Proc. of 20th International Congress on Large Dams, Beijing.

Andrews M E & Dornstadter J (2000). Investigations into seepage at Rotton Park Reservoir using temperature distribution measurement. Proc. of 11th Conference of British Dams Society, Dams 2000, Bath, pp 388-401, Thomas Telford, London.

Dornstadter J (1997). Detection of internal erosion in embankment dams. Transactions of 19th International Congress on Large Dams, Florence. Q 73, R 16

# The successful grouting of Heapey embankment, Anglezarke reservoir

C D PARKS, Bechtel Ltd.
S WALTHALL, Bechtel Ltd.

SYNOPSIS. Since the 1960's there have been a succession of reports of leakage of water from the downstream face of Heapey embankment, close to the eastern shoulder. In November 1997, during the refilling of the reservoir after a dry summer, the issue of water was sufficiently large to require an emergency draw-down of the reservoir. The draw-down procedure has been described by Page (2002). A ground investigation into the structure of the embankment and surrounding strata was undertaken and was followed by a phased grouting operation to control the leakage. On raising the reservoir level after the first phase grouting operation the leak continued in its original location. A second phase of grouting was undertaken during November and December 1999 which was successful.

## CONSTRUCTION AND HISTORY

Heapey Embankment, which forms the northern limit to Anglezarke reservoir, was built in 1870 as part of the Rivington reservoirs for Liverpool Corporation. Figure 1 reproduces the original "tender" drawings showing the expected plan and geological setting of the embankment.

The embankment was constructed with an 18" cast iron drawoff pipe to supply White Brook to the north of the embankment. A search of the historical records in NWW archives showed that details of excavation and backfilling of the core trench were recorded on the Clerk of Works records at the time of construction. These construction records show that the cut off-trench was constructed much deeper than planned and extended into solid rock in the bottom of the valley, through a glacial meltwater channel, Figure 2. It was however not extended into the sides of the valley. The shoulders of the dam were founded on glacial till overlying mudstone and sandstones of Lower Coal Measure Age.

## LEAKAGE FROM THE EMBANKMENT

The dam has suffered from a history of leakage on the downstream east abutment (See Figure 1) and these incidents are recorded in the inspecting engineers reports from the 1960s onward. In 1970 the unused draw-off pipe for White Brook was grouted internally in an attempt to ensure that if this

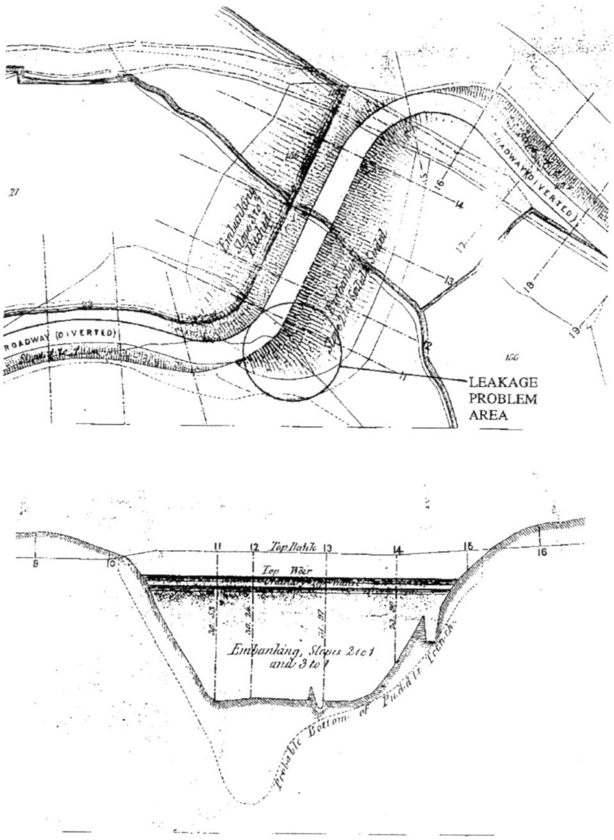


Figure 1. Original "tender" plan and section of Heapey embankment

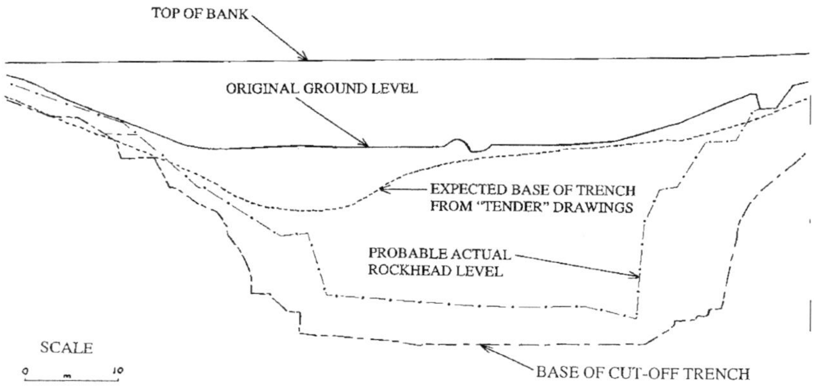


Figure 2. As Constructed Drawing (Adapted from the original of 1870)

were the cause of the leakage, the embankment would be protected. Further wet areas were recorded in the same area in subsequent inspections, however, these were not considered significant. During refilling in November 1997, after a dry summer, a significant issue of clean water was observed in the right downstream mitre and an emergency drawdown was instigated.

## GROUND INVESTIGATION

After the discovery of the leak, a site investigation was carried out to determine the nature of the bank, its construction and the possible source of the leakage. The ground investigation consisted of six cable tool percussion boreholes and eight trial pits designed to investigate:

a) the nature of the downstream fill material,
b) the location of the core,
c) the geology compared to that shown on the core trench inspection drawings,
d) groundwater levels by the installation and monitoring of piezometers.

Laboratory tests were carried out on both the core and shoulder materials sampled in the boreholes to determine its properties.

## ASSESSMENT OF THE GROUND CONDITIONS

The desk study and borehole record confirmed that the embankment consisted of a clay core of the order of 2m wide supported by shoulders of clay and mudstone fill. The depth of the core trench and of the local rock head found in the boreholes was in agreement with the "as constructed" records (See Figure 2). Unlike other embankments in the area the core was easily identified as it consisted of a uniform soft plastic clay whereas the shoulders were of a mixture of clay and mudstone. The identification and location of the core are a fundamental part of any work to be carried out on dams of this age.

It was suggested that the leakage could be caused by:

a) a direct waterpath though the embankment (ie a 'hole' in the bank),
b) the water passing around the core through the abutment where the cut off trench was very shallow
c) a diffused path through the embankment fill materials.

No direct evidence was found for the location of the leakage path during the investigation other than the emission of water from a number of locations on the downstream face very close to the right mitre. It was known that the leakage only occurred when the water level in the reservoir was above 0.8m below top water level.

Since a number of possible alternatives existed a phased grouting operation was planned. Phase one was an attempt to grout any direct water paths in the location of the old draw off and adjacent areas. This was to be tested by raising of the water level in the reservoir and monitoring the leakage. Should this not prove effective then a second phase of grouting would be carried out to grout a larger part of the bank and also the right abutment.

## PHASE 1

### Phase 1 Grouting

The objective of the Phase 1 operation was to grout any possible direct leakage path adjacent to the old draw off pipe. A pattern of grout holes was drilled directly above the old draw-off pipe and the holes were taken down to a depth of 1m below the invert of the pipe. Tube á Manchette (TAM) grouting techniques were used with grout ports at 1m intervals. There was a concern that excessive grout pressures could cause hydraulic fracture within the embankment resulting in further problems in the future.

A trial was carried out to determine the grouting procedures, in particular the optimum grout mix, maximum pressures and quantities to be injected at each pass. The working procedure was to use a maximum grout pressure in each stage equivalent to the overburden pressure, calculated simply as the depth of the port times the unit weight of the soil. A unit weight of 19kN/m$^3$ was used and these pressures were corrected for level and converted to psi (pounds per square inch) for direct reading with the grouting equipment. The maximum quantity injected in any one stage was restricted to 45litres (~10 gallons). The initial breakout pressure for the TAM was observed and recorded.

Grout tubes were installed using rotary percussive air flush drilling at 54mm diameter and then placing the TAM grout tubes in a 1 cement to 1 bentonite mix. The grout for the main grouting operation was a 4 cement to 1 bentonite mix with a water content which varied from 0.4 to 0.6 depending on the nature of the ground and the grout take. A minimum of 24 hours was left between installing the grout tubes and the first phase of grouting.

Grouting was carried out from the bottom port of the holes upwards. An individual port was determined to have been grouted when a grout pressure equivalent to overburden pressure could be maintained during injection.

Where pressure was not maintained, grouting was stopped at that port after the injection of 45 litres of grout. Although the TAM tubes were provided with ports at 0.5m intervals grouting was only carried out at 1m intervals.

Very few of the ports achieved pressure during the first pass and it was found necessary to revisit most of the grout ports for reinjection. A detailed record of the pressures and quantities for grout injections into each of the ports was made and plotted graphically on site. On completion of grouting the grout tubes were completely filled with grout. Ground level monitoring stations were established on the crest of the embankment and were monitored daily during the grouting operations.

Phase 1 Testing -Monitoring of the leakage.

As part of the Phase 1 grouting contract, the leakage points were provided with collection drains, headwalls and flow monitoring facilities. It is essential to have good, easy to-use and reliable monitoring of the outflow of water to evaluate the effect of the grouting. Additional drainage on the downstream toe was constructed to remove the wet spot at the toe of the bank. This was not thought to be part of the problem but the work was carried out to improve the state, of the embankment and to allow proper monitoring of the area.

Results of Phase 1 Grouting

Phase 1 grouting was carried out over a limited area and a trial refilling of the reservoir was carried out on completion. The monitoring of the flows after the Phase 1 grouting is presented in Figure 3. This shows that the flows were related to the water level in the reservoir and that flow started when the level was about 1m below top water level. The rate of flow increased dramatically when the water level reached 0.6m below top water level. The reservoir was not raised above this point in the interests of safety. The plot of flows against precipitation (Figure 4) does not show any correlation. This check was important as it had been suggested that the increased flow might have been due to surface water runoff after periods of heavy rain.

## PHASE 2

Phase 2 Grouting

Following the unsuccessful attempt to stop the leakage in Phase 1 the second phase of grouting was carried out in autumn 1999 using the same techniques as for Phase 1. The objective of this phase of grouting was to prevent diffuse leakage or a water path round the east abutment. For this phase the number and density of the holes was increased and, due to surface space limitations, a pattern of angled holes was required to grout the right abutment, Figure 5. In addition to the requirement for holes to be grouted at

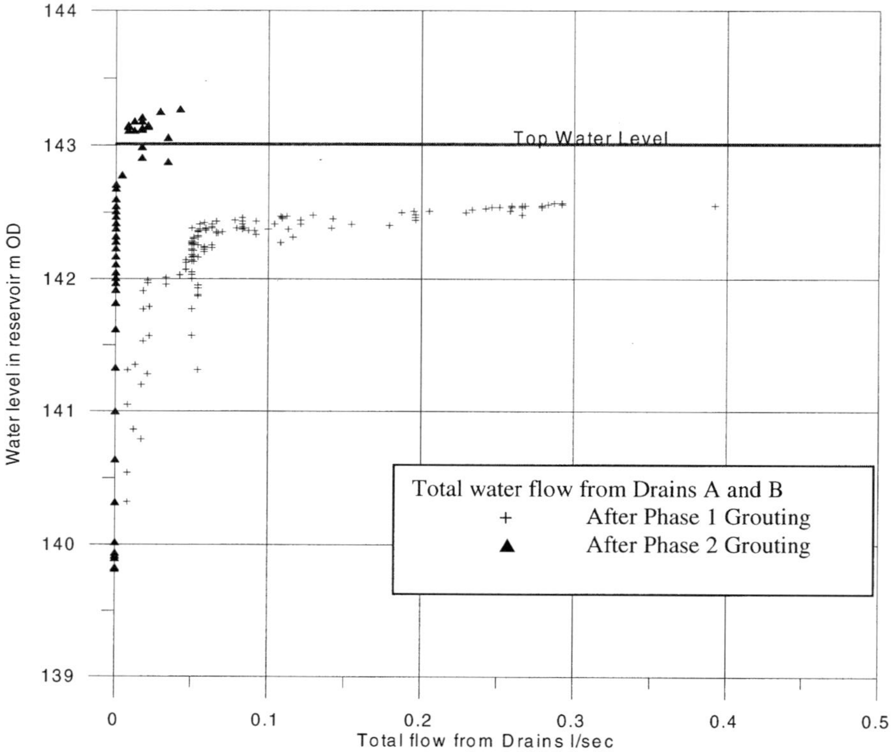


Figure 3. Flow from Drains A and B after Phase 1 and after Phase 2 Grouting vs. reservoir level

one metre centres over their full depth provision was made for some grouting at half metre intervals. The same controls on grouting pressures and quantities to be injected were used as had been developed during Phase 1.

Air flush rotary percussive drilling was used and, during the drilling of the access holes, inter-connectivity between the holes with both flush and grout returns was observed in adjacent holes. This showed that the material within the bank was locally very poor and required great care on behalf of the operatives to establish a safe grouting regime. The order of drilling and grouting was modified as necessary to ensure that there was as little interference between holes. In addition enhanced flows of silty water were observed at flow monitoring point 1 whilst drilling some of the deeper boreholes. This flow stopped as soon as drilling stopped.

The results of the drilling and grouting operations were recorded and stored in the geotechnical database. Plots of the grout take for each location from this program are given in Figure 6. These plots indicate that the maximum grout take is at the base of the embankment fill.

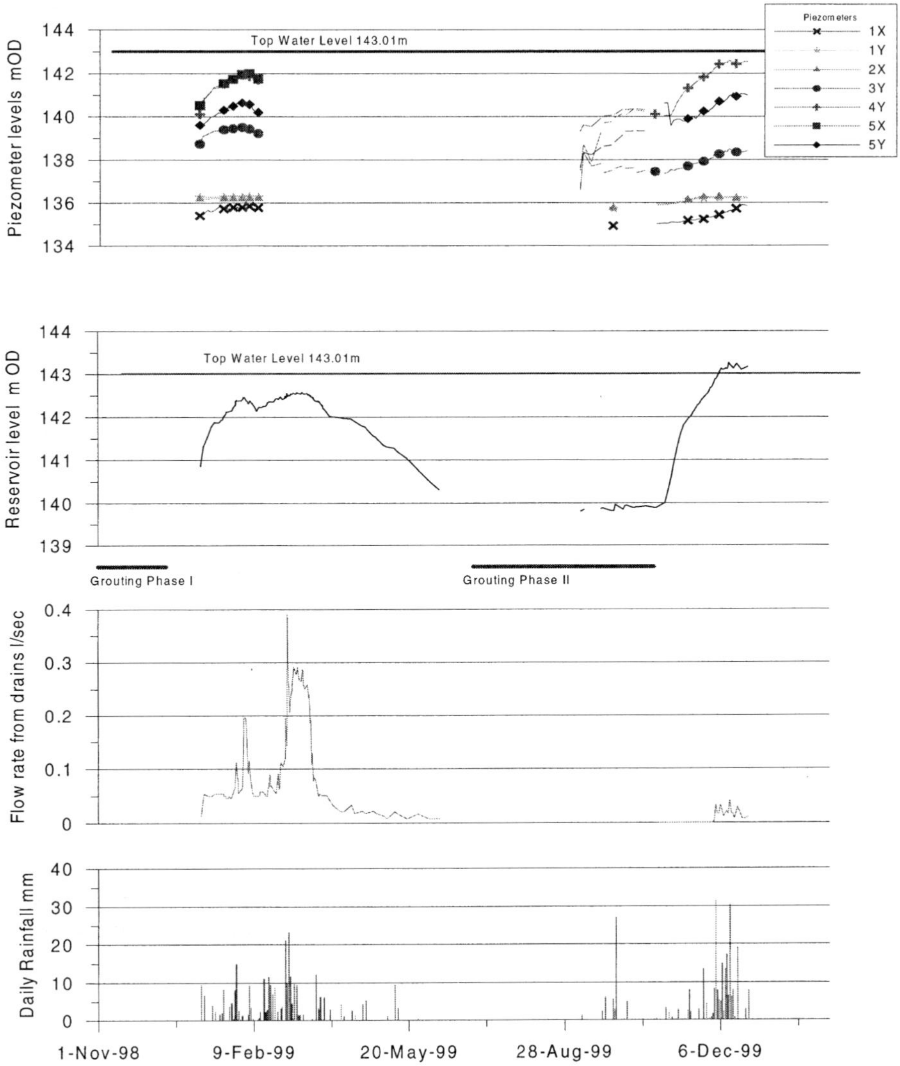


Figure 4. Composite monitoring results – comparison of piezometer readings, reservoir level, flow rate from drains and daily rainfall

During grouting, despite maintenance of the strict flow and pressure regimes grout did emerge on the upstream face on one occasion. This may have indicated a direct flow path through the upstream shoulder of the embankment. No grout was observed on the down stream face at any stage.

In areas of large grout takes the facility to grout at half metre intervals was used to confirm that grouting had been successful. This was achieved by obtaining the required grouting pressure within a few passes. The required grout pressure was achieved in all ports although it did take up to eight

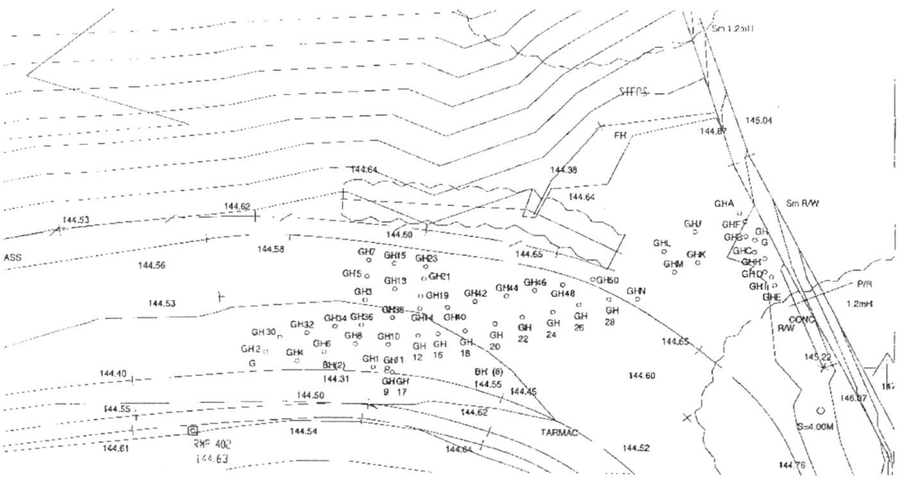


Figure 5. Phase 2 grout hole layout at east abutment

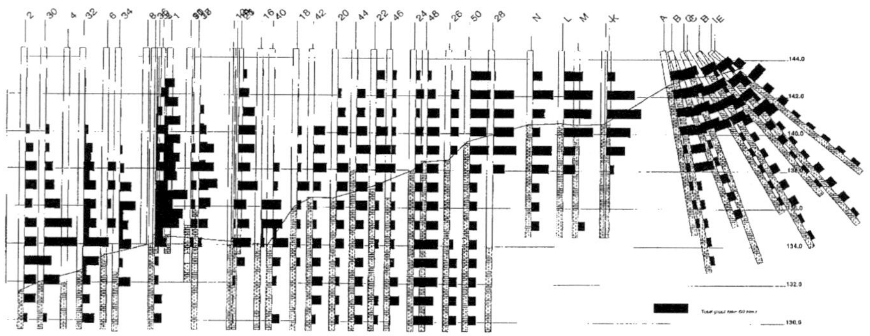

Figure 6. Phase 2 Grout Take

passes, i.e. over 300 litres of grout for certain ports. In those ports where more than four passes were required, the grout mix was thickened for the later injections.

After grouting had been completed a small number of investigation holes were drilled to examine the effectiveness of the grouting. Since rotary flush drilling had been very disruptive of the materials within the bank and it was considered that conventional cable tool percussion may have a similar effect, the investigation were carried out with the 'Archway 150 Explorer' rig which recovered one metre long 75mm diameter continuous cores by percussive means in a continuously cased hole. These proved very effective, providing good quality samples whilst causing minimum disturbance to the ground. Evidence of interstitial grout was found in a number of samples including those from the clay core.

Phase 2 Testing - Raising the Reservoir

The monitoring of the flows after the Phase 2 grouting is presented as Figure 3. This shows that there was no increase in flow as the water level rose even when the water level was above top water level. The effect of water level in the reservoir on the piezometers installed during the investigations can be seen, Figure 4.

## CONCLUSIONS

Both phases of grouting were carried out and the results of the monitoring show the second phase grouting has been successful. A number of lessons were learnt during this project. These were: -

- A carefully thought out monitoring system is required to observe the effects of any construction work.
- Good archive records are essential when working on old embankments.
- A phased approach is necessary using the most appropriate and effective methods and techniques.
- Grouting is a very dependant on skilled operators and cannot be rushed on old embankments that are very delicate structures.
- The expertise and advice of the contractor is acknowledged, particularly on the procedures for the grouting operations.
- Appropriate contract conditions are required to enable an observational method to be used on site. The phases can either be separate contracts as was used at Heapey or they could be built into one contract.

## REFERENCE

Page R (2002). Operational response to leakage and repair at Anglezarke Reservoir. Dams and Reservoirs, vol 12, no1, February 2002, pp 34-36.

# Performance and repair of upstream membranes

# Improving the watertightness of Winscar Reservoir

I C CARTER, Montgomery Watson Harza, UK
J R CLAYDON, Yorkshire Water Services, UK
M J HILL, Montgomery Watson Harza, UK

SYNOPSIS. Leakage has been a problem at Winscar dam since it was filled for the first time in the late 1970s. Supplementary foundation grouting was carried out as well as repairs to the upstream facing of dense asphaltic concrete at that time and the reservoir was operated without major problems for 20 years. A pattern of rising seepage was observed during late 2000, which eventually led to the appearance of a large spring at the downstream toe six months later. This paper describes the investigation of the leak and outlines the actions taken to improve reservoir watertightness. A geocomposite liner was installed over the upstream face and the grout curtain was reinforced. Early results show that these works have been effective in reducing water losses.

## INTRODUCTION

### Description of dam and reservoir

Winscar Reservoir was constructed between 1972 and 1975 for public water supply (Collins & Humphreys, 1974). The reservoir is impounded by an embankment dam, which has a maximum height of 53 m and a crest length of 520 m. The reservoir storage capacity is 8.3 $Mm^3$ and the reliable yield is 22 Ml/day.

The dam is built over a foundation composed of *Namurian* sandstones and shales. The embankment is made of compacted rockfill with an upstream slope of 1V on 1.7H and 1V on 1.4H downstream, as shown in Figure 1. A two-layer membrane of dense asphaltic concrete covers the 25,000 $m^2$ upstream face of the dam. A cement grout curtain continues beneath the upstream toe to depths of up to 70 m. Provision was made for the collection and measurement of seepage, as shown in Figure 2.

### Background history

A pattern of rising seepage was recorded during first filling. Leakage increased significantly once the water reached permeable strata on the left abutment. The reservoir was held down for remedial grouting but this failed to improve the situation. Tracer studies and chemical analyses suggested the presence of a defect in the asphaltic facing. The reservoir was emptied and a hole was found in the membrane (Routh, 1989). Leakage reduced dramatically after repair work and the reservoir was successfully filled.

*Reservoirs in a Changing World*, Thomas Telford, London, 2002, 415–430

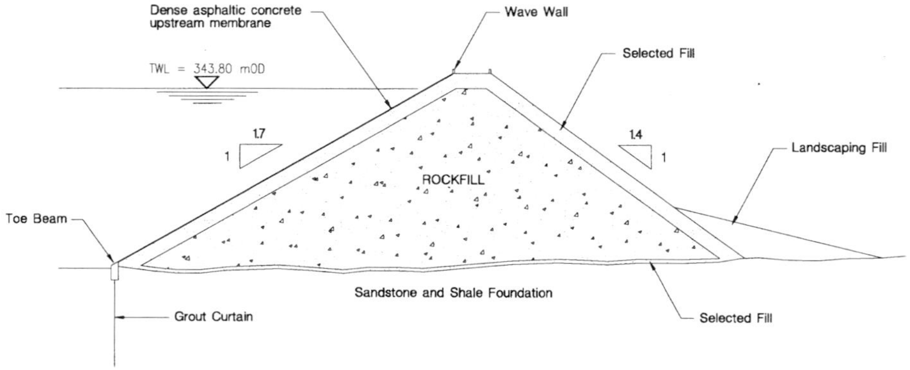


Figure 1 Cross section of Winscar Dam

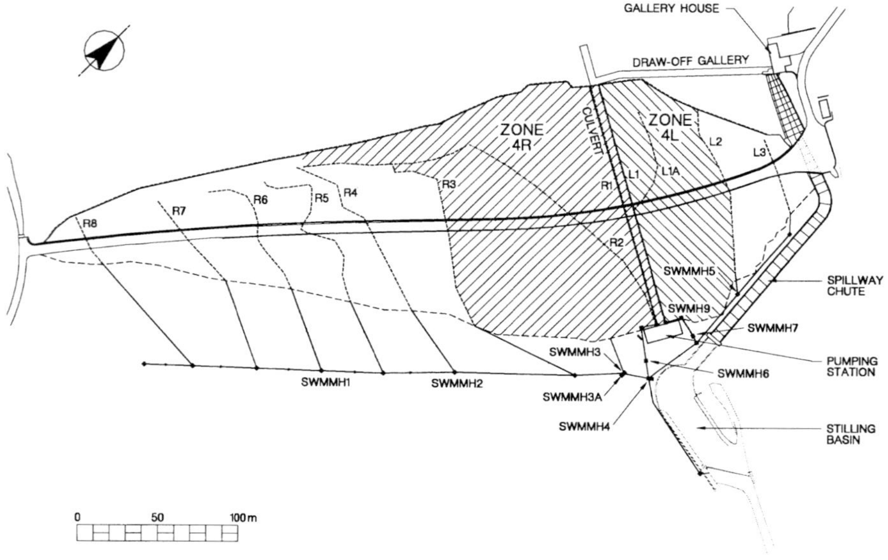


Figure 2 General arrangement of dam and underdrainage system

Leakage remained relatively stable during the 1990s with flow dependent on reservoir level (Carter *et al*, 1999), although a rising trend became evident in the latter part of the decade. Sharp rises were noted in June and October 2000 and a large spring issued from the toe of the dam in January 2001 with an estimated flow of about 15 litres/second. The reservoir was drawn down as a precaution and the seepage flow response was closely monitored. Flow reduced at approximately half-depth of the reservoir, suggesting a major defect at this level.

## INVESTIGATIONS INTO THE CAUSE OF THE NEW LEAKAGE

### Past performance of the reservoir

Seepage flow records since construction indicate a clear relationship between reservoir level and flow. The behaviour patterns have changed over the years, as illustrated on Figure 3, which shows the response during:

- first filling of the reservoir
- reservoir cycles during the 1990s
- relatively static reservoir level during 2000, and,
- emptying of the reservoir in 2001

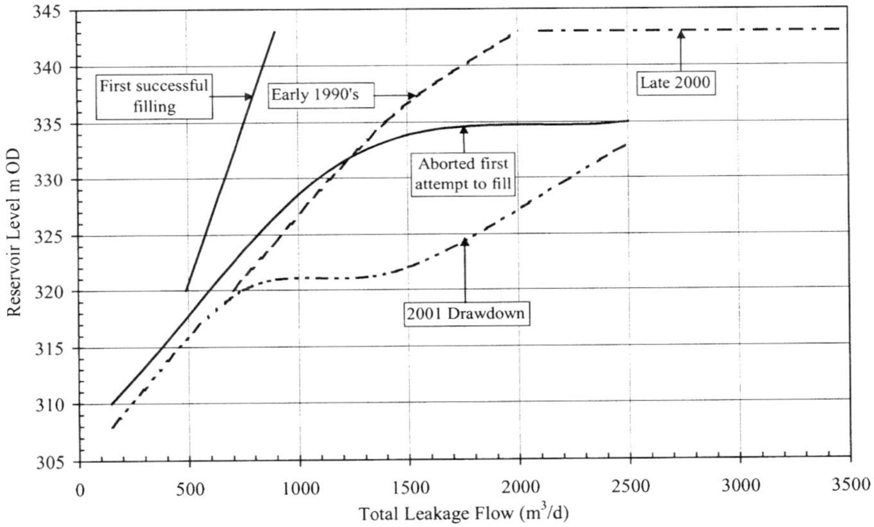


Figure 3 Relationship between reservoir level and seepage/leakage

Seepage through and under the dam collects in a series of underdrains, as shown in Figure 2. Examination of the data indicated that the new leak was originating in the zone between the culvert and the *Huddersfield White Rock* outcrop at the upper left abutment. Shales form the foundation in this area and reconnaissance in the reservoir basin immediately upstream of the dam revealed deposits of re-deposited clay and weathered shale, which might be considered to be reasonably impermeable. No sink-holes were found.

### Inspection of the upstream face

About 60 new defects were observed in the upper half of the asphaltic membrane as the reservoir was drawn down. All had developed since the last inspection in 1996 (Wilson & Robertshaw, 1998). Most were blisters and small cracks but there were also more persistent cracks that coincided with construction joints between panels. This type of defect had not been previously encountered at Winscar and initially seemed to be the most likely explanation for the new leak. However, core drilling at crack locations was unable to confirm full-depth cracking.

The investigation initially concluded that:

- the asphaltic membrane had deteriorated markedly over the past five years despite its submergence for long periods of time,
- while numerous defects were evident, none of them could be cited as the principal cause of the new leak, and,
- there was no evidence to suggest the formation of any concentrated leaks through the foundation or of any leaching of the grout curtain.

The lack of conclusive evidence as to the location of the leakage led to the decision to draw the reservoir down completely. Emptying the reservoir is not straightforward: the mechanical facilities are adequate but there is a problem with the release of silt-laden waters into a trout stream. A range of measures were needed, which included blending of different quality waters; filtration through straw bales, geotextiles and hessian; settling basins; and, timing of discharges. However, the need to avoid environmental damage meant that emptying of the reservoir took over six months.

Taking the reservoir out of use for most of the year was handled by supplying its local treatment works with water from another reservoir using a pumping station built for the drought of 1995/96. Supplies were balanced across the rest of the region using the Yorkshire Grid at increased operating cost. The reservoir is slow to refill (nominally two years), as it is large compared to its catchment area. The uncertain condition of the asphaltic membrane concerned YWS since similar leaks in the future might take the reservoir out of service for long periods and thereby render the source unreliable. This prompted the decision to defer patch repairs and embark instead upon a major refurbishment of the entire waterproofing element. Also, in view of the slow refill it was decided that all works should be completed before the winter rains of 2001/2002.

Thick sediment over the lower half of the face initially obscured new defects in the membrane. However several large cracks became apparent just above the concrete toe beam once the face had been pressure washed. One such crack is shown in Figure 4 : flow had eroded the crack and made a bowl-like depression in the membrane itself. The cause of the crack was unknown but it was probably induced by settlement of deep fill immediately behind the toe beam. Slaking of the shale foundation and the settlement may have exacerbated by the penetration of bouldery rockfill into the formation.

## PROCURING AN ENGINEERING SOLUTION

### Objectives

The objective of the refurbishment works was to secure a reliable water resource for the next 40 years. This was to be achieved by the installation of an appropriate, well-proven and robust waterproofing system over the asphaltic concrete face and by the reinforcement of the grout curtain. The goal was to reduce leakage through the dam and its foundation significantly.

Figure 4 Major defect discovered behind toe beam at chainage 200m

Assessment of alternative engineering solutions

The technical feasibility of various engineering solutions was investigated. The logistics, environmental and planning aspects were considered and the costs were estimated. Two options gained high evaluations, these were:

- Removal of part of the existing facing and its replacement with a dense asphaltic concrete membrane of modern design
- Installation of a geomembrane liner over the existing face of the dam

Suitability of geomembrane

The ICOLD register indicates that synthetic geomembranes have been used since 1959 to repair over 100 dams world-wide. Most have been concrete gravity dams, although its application on embankment dams has grown in recent years. Various geomembrane materials have been used, although PVC is the most common and much is known about its characteristics and performance. The main alternative is polypropylene. The mechanical and physical characteristics of the materials are similar but PVC is much easier to weld, as lower temperatures are required. Also, only PVC can be welded to polyethylene geotextiles during extrusion and this simplifies installation.

The PVC based geomembrane proposed by *Carpi Tech* had never been used previously in a water supply reservoir in the UK and had not been submitted for approval by the Drinking Water Inspectorate under Section 25(1)(a) of the Drinking Water Regulations, 1989. A concern about plasticised PVC relates to loss of material with time by leaching of phthalates, which make up about 30% of the overall mass. Spanish research into the phenomena during the 1980s (Aguiar & Blanco, 1990) indicated that while losses could be substantial, there was considerable variability between different PVC products. Phthalates are in widespread use and while concern has been expressed that it may act as an endocrinal disruptor for some species, there is no evidence that it is harmful to human beings.

The potential loss of plasticiser was also of concern because it might render the liner less flexible, more permeable and more prone to damage with time. However, accelerated age testing indicated insignificant loss of performance during the first 50 years for exposed membrane and an effective life of over 200 years when submerged. Product details and performance data were presented to DWI who stated that this was an issue that should be dealt with by YWS under the provisions of Section 25(1)(b), i.e. the company might well consider that approval is unnecessary because of the small risk posed by the use of unapproved material. This course of action was followed.

Tenders

Priced tenders were sought from contractors and specialist subcontractors for both alternatives. The construction schedule was very challenging, particularly in the case of the asphaltic concrete solution because of the fast-track nature of the project. The mobilisation of the necessary specialist plant was a problem, as was the timing of the works, which might have extended into a second year if inclement weather conditions prevented completion of the asphalt and mastic placing. The geomembrane was not significantly influenced by these factors and on this occasion there was also a clear cost advantage. It was therefore decided that the geomembrane solution should be accepted and a contract was awarded to *Morrison Construction Ltd*, with *Carpi Tech* as specialist Subcontractor.

## DESIGN AND CONSTRUCTION OF THE GEOCOMPOSITE LINER

### Design of the geocomposite liner

The performance criteria for the design of the geocomposite liner was that it should be capable of reducing leakage through the face of the dam below 1 litre/second against full reservoir head.

*Sibelon CNT 3750*, a Flag S.p.a. product manufactured in Italy, was chosen for the liner. It is a flexible PVC geomembrane, 2.5 mm thick, which is heat bonded during manufacturing to a 500 $g/m^2$ non-woven, needle-punched polyethylene geotextile. The geomembrane is impermeable (nominally $k = 10^{-12}$ m/s) while the geotextile backing enhances puncture resistance, dimensional stability, friction characteristics and drainage capability.

The geocomposite liner is mechanically fixed to the dam using watertight anchorages around the periphery and by tensioning devices on the face. Tensioning prevents wrinkle formation that may ultimately lead to cracking. The tensioning assembly, which is patented by *Carpi*, comprises coupled stainless steel profiles, as shown on Figure 5. The lower profile is fastened to the upstream face by anchor rods embedded in epoxy resin. The embedment depth was determined on the basis of 50-year return gust wind speed. The liner covers the lower profile and is clamped and fastened by the upper profile. A PVC cover strip overlies the coupled profile and is welded to the underlying liner.

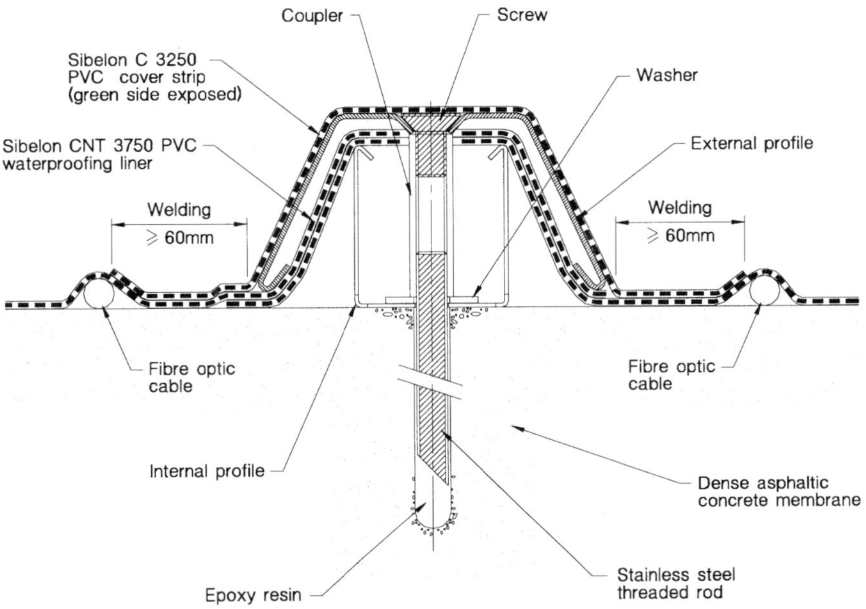


Figure 5 Arrangement at longitudinal joints

The PVC liner is anchored to the concrete toe by two watertight seals, as shown in figure 6. A primary seal was formed by insertion of the liner into an epoxy resin-filled slot, 25 mm wide, cut through the asphalt and into the toe beam underneath. The secondary seal consists of a stainless steel flat mechanically fastened to the face of the toe beam by resin anchors, which compressed the liner against an epoxy bedding mortar.

The secondary seal reduces the head on the primary seal. The twin seals form separate drainage compartments that allow the source of any future leakage to be determined. The top of the liner is fixed to the base of the crest parapet wall by a similar linear anchorage and waterproofed by an additional flap of PVC.

The liner has a dedicated drainage system that is independent from the existing underdrainage system beneath the embankment. It consists of the geotextile underlayer, the steel anchorage profiles (which act as conduits) and transverse geonet grids between the existing asphaltic membrane and the PVC liner that are spaced at regular intervals.

Any leakage through the membrane or past the seals will find its way into one of the two compartments where it will be intercepted. Each compartment has an independent discharge connection into the culvert. The performance of the new system is monitored by measurement of discharge into the culvert.

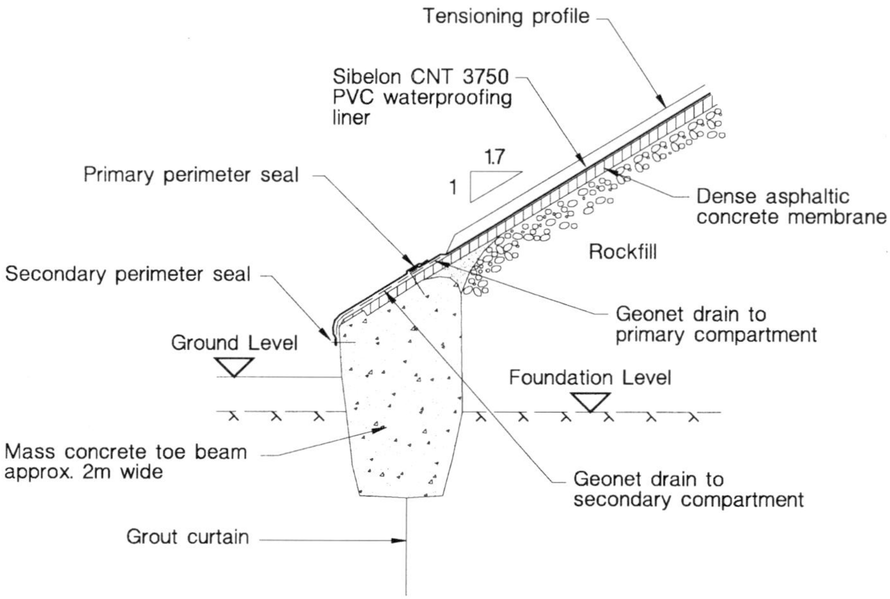


Figure 6 Toe details and watertight seals at perimeter

Leak detection system

A leak detection system has been installed to pinpoint the source of any future leakage through the new liner. A network of fibre-optic cables was fastened to the asphaltic concrete surface and installed under the liner in a series of loops. The system uses a laser source to measure the wavelength of light reflected back down the cable and sense temperature. The temperature at any point along the cable route varies depending on the season, reservoir temperature profile and other factors. However, passing a current through the external sheath of the fibre-optic cable can induce an artificial increase in temperature. In dry conditions the temperature rise would be constant but energy is lost in the presence of water and anomalies can be detected.

Membrane Protection

The physical properties of PVC are considered to be compatible with impact by floating debris, sailing craft, ice, etc, and in-service observations seem to support this view. Vandalism could affect the membrane but its detection and repair is straightforward and simple, hence it was decided that the extra cost of mechanical protection was unjustifiable at this time. Nevertheless, several strategies were identified that could be introduced to manage membrane protection, if this should become necessary. These included the retrofitting of a light cover system, surveillance, and exclusion by barriers.

Installation of the liner

The geocomposite liner was supplied by the factory to length in standard 2.10 m wide rolls. Three rolls were then welded by automatic twin track machines in a pre-fabrication yard to construct wider panels in order to maximise quality and minimise installation time on site.

Installation of the lower of the coupled profiles commenced after clearance of debris from the surface of the dam. Very little surface preparation was necessary, other than the removal of sharp edges, smoothing of the major blemishes and edge preparation of the asphaltic membrane above the toe beam.

The prefabricated panels were lowered by crane on to the slope and unrolled into the correct position. Clamps were used to secure the rolls temporarily in position prior to welding and installation of the upper profiles. Adjoining liner panels overlap each other and were welded together to produce a continuous liner membrane across the full width of the dam. Field welds were made by one-track hot air method with manual welding guns.

The placement of the profiles, the liner and welding were performed by workers either on travelling platforms or using standard roped access procedures. Installation was carried out in three phases during the five-month period prior to Christmas 2001. The work was performed in a variety of weather conditions but only strong wind, heavy rain and very low temperatures prevented welding and placement.

## REINFORCEMENT OF THE GROUT CURTAIN

Original grout curtain

The purpose was to form a barrier of effectively watertight rock across the full width of the valley and most particularly through the permeable *Namurian* sandstones, as indicated on Figure 7.

The curtain was aligned beneath the upstream toe wall and extended some 80 m into the hillside on the left abutment in order to reduce the potential leakage through the *Huddersfield White Rock* formation. The depth of the curtain varied between about 70 m in the valley bottom to about 20 m at the abutment wings. The ratio between the depth of the curtain to the reservoir head was about 1.5:1, which is relatively high compared to most modern dams.

About 800 blanket holes were drilled in two rows of inclined holes on the upstream side of the toe beam prior to curtain formation. These holes were normally 5 m deep but occasionally deepened to 20 m. The mass concrete toe beam acted as a grout cap for the curtain. Some 220 vertical holes were drilled in 3 m stages using the split spacing method. Primary holes were spaced at 6 m.

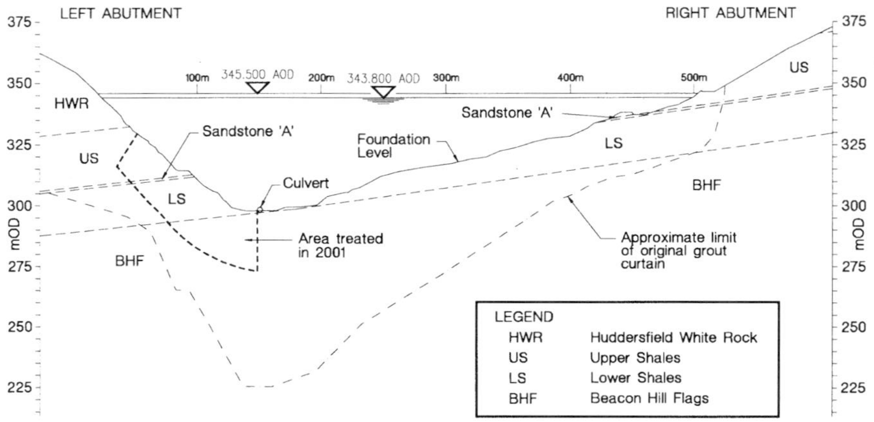


Figure 7 Geology and general arrangement of grout curtain

The designers originally intended to follow the ascending stage method, with grout holes water tested in 3 m stages. Where the result was less than 2 Lugeons then the packer would be raised by a stage and re-tested. If the result was higher than 2 Lugeons or if the stage had reached 9 m in length, then the hole was pressure grouted. However, records show that grout was also introduced into the hole whenever full water returns were lost during the drilling. Most of the grout seems to have been injected in this fashion.

Testing on completion of the initial grouting showed the curtain to be not sufficiently watertight. Raked holes were drilled at 12 m centres to 20 m depth in order to ensure that vertical fissures were effectively filled.

Grout pressures varied between 0.1 to 0.15 bars/m depth of hole. A variety of thin grout mixes were used, mainly with ratios of either 10:1 or 7:1. The records indicate that some 2,325 tonnes of materials were injected, including OPC (39%), Pozament (24%), sand (38%) and bentonite (trace). The total area of the original curtain was estimated to be about 20,000 m$^2$ and the average consumption of cement was recorded as 120 kg/m$^2$. The majority of the grout was injected into the 4,800 m$^2$ section occupied by the *Huddersfield White Rock & Upper Shales*, which took 335kg/m$^2$ compared to the much larger area occupied by the *Beacon Hill Flags & Lower Shales* (14,640 m$^2$), which only took 51 kg/m$^2$.

Supplementary grouting phases

Three phases of additional grouting were undertaken during the late 1970s in response to the excessive water losses during first filling. The main purpose of the first two phases was to treat the *Huddersfield White Rock* and underlying sandy shales on the left flank. The treated area extended down to about 315 m OD and covered an area of about 6,000 m$^2$. The split spacing method was used but with the descending stage approach. Holes were raked and the spacing was reduced during the latter phase.

A thin grout mix was used at the outset with the mix being progressively thickened with cement until a 1:1 ratio was reached. Sand and bentonite were used to thicken and to stabilise the mix. Similar grout pressures were specified for the first phase but increased during the second by about 50%. The total take for the first phase was 390 T cement and 65 T sand with an average consumption of about 50 kg/m$^2$. The quantity injected during the second phase reduced to about 60 T with an average consumption of 28 kg/m$^2$, which suggested that the ground had tightened up satisfactorily.

A further round of grouting was carried out in 1980 but on this occasion the focus was aimed at a series of beds midway between the Huddersfield White Rock and the Beacon Hill Flags. Once again the treated area was located on the left bank above 304 m OD. Primary holes were spaced at 3 m intervals and the majority of holes were vertical.

Grout pressures of 0.2 bars/m depth were used and over 420 T cement was injected. The average grout take in Sandstone "A" and underlying Shales was nearly 100 kg/m. The effectiveness of this treatment was assessed by water testing. The results varied widely along the curtain and with depth but 65% of all tests gave Lugeon values less than 3 and only 3% gave a final value over 10 Lugeons. It was concluded at that time that the treatment had been successful, although some concern was expressed about the permeability of the *Beacon Hill Flags* immediately beneath the toe beam.

Grouting operations during 2001

A possible explanation for the increase in seepage beneath the dam over the life of the reservoir was that the grout curtain had been subject to leaching. In this respect it is pertinent to note that the original cement grout was thin. Thin mixes often tend to bleed and they tend to be less durable. In addition, runoff into the reservoir flows from peat and heather moorland and normally has a low pH. Reservoir water seeping through fractures in the foundation could therefore be expected to react with alkaline grout forming the curtain.

Attention during 2001 was directed at the foundation between the culvert and 327 m OD, since the new leakage was concentrated in compartment 4L of the foundation underdrains. Two holes were drilled for water testing so that the general condition of the curtain could be assessed and the grouting needs determined. The exercise proved difficult; variable results were obtained, six tests gave results ranging from 6 to zero Lugeons while seven others indicated values above 50. No correlation could be found between these Lugeon values and the actual grout consumption. It was therefore decided that pressure grouting of this section would be prudent followed by a review before continuing elsewhere on the curtain.

Two phases of grouting were undertaken with an interim review to allow methods to be evaluated and modified. The first phase of 12 raked groutholes were installed at 2.5m spacing immediately alongside the culvert with the treatment aimed at the *Beacon Hill Flags*, as seen on Figure 8.

Each drillhole was fitted with *tube à manchette* (TAM). External grout filled separator bags divided each grouthole into 3 m stages. Grout was injected into each stage sequentially under the control of a Jean Lutz pump monitoring computer. The grout pressures were higher than those adopted during previous grouting episodes; the initial stage was injected at 3 bars while lower stages were subject to 2.25 x overburden pressure.

The standard grout was a 1:1 mix of water and microfine cement with 2% of dispersant and retarder, although there was provision for the water-cement ratio to be lowered to 0.8 :1 mix, if large amounts of grout were accepted. The grout had a specific gravity of 1.52, a Marsh cone value of 35 seconds and less than 1% bleed after 2 hours. No large takes occurred and the average consumption was 13 kg/m, which suggested that concerns about the integrity of the grout curtain alongside the culvert were unfounded.

The second phase of treatment covered the upper 25 m of the curtain between 304 m and 327 m OD. The TAM approach was replaced by the split spacing method using ascending stages. The grout mix was identical but the pressure factor on vertical overburden depth was reduced first to 0.59 bars/m and later to 0.4 bars/m, due to cross connections between holes.

The second phase of grouting was more successful than the first. Over 20 T of microfine cement were injected and the average consumption was 35 kg/m. Larger takes were recorded in the shales between Sandstone 'A' and the Beacon Hill Flags, as can be seen on Figure 5, while the largest recorded in a single stage was 570 kg/m. Grouting is imprecise in its application but it is unusual that so much could be injected in a section of the curtain that had been previously subjected to two phases of grouting.

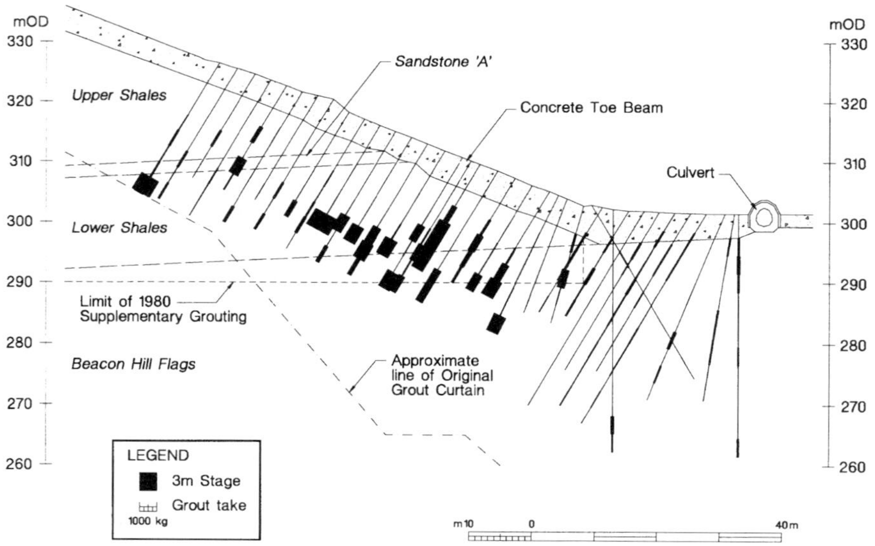


Figure 8 Grout takes recorded during 2001 works

## MONITORING DURING REFILLING OF THE RESERVOIR IN 2002

### Refilling

Refilling of the reservoir commenced at the end of December 2001. The winter of 2001-02 was relatively wet and some 500 mm of rain fell in the first three months after the closure of the control valves. The reservoir rose quickly and at that time the level stood at 335 mOD, which is approximately ¾ of full depth and about 9 m below Full Supply Level.

### Performance of the geocomposite liner

The geocomposite liner has performed well since impounding was resumed. Flow from the primary compartment, which covers the original asphaltic concrete facing, is very small. It has shown a faint increase with reservoir level and seepage was less than 0.02 litres/second at ¾ full reservoir head.

Flow is greater from the secondary compartment, which covers the toe beam between the secondary and primary seals. Leakage can come from various sources, such as defects in the liner seal, the toe beam waterstops, and imperfectly filled groutholes. The non-linear relationship between water level and leakage is shown in Figure 9.

Initially flow increased with rising water level but the trend reversed at about 319 mOD. It is thought that the defects had become clogged by sediment or calcareous deposits, although a number of other explanations can be suggested. However, it is clear that this leakage is insignificant and was less than 0.6 litres/sec at half depth.

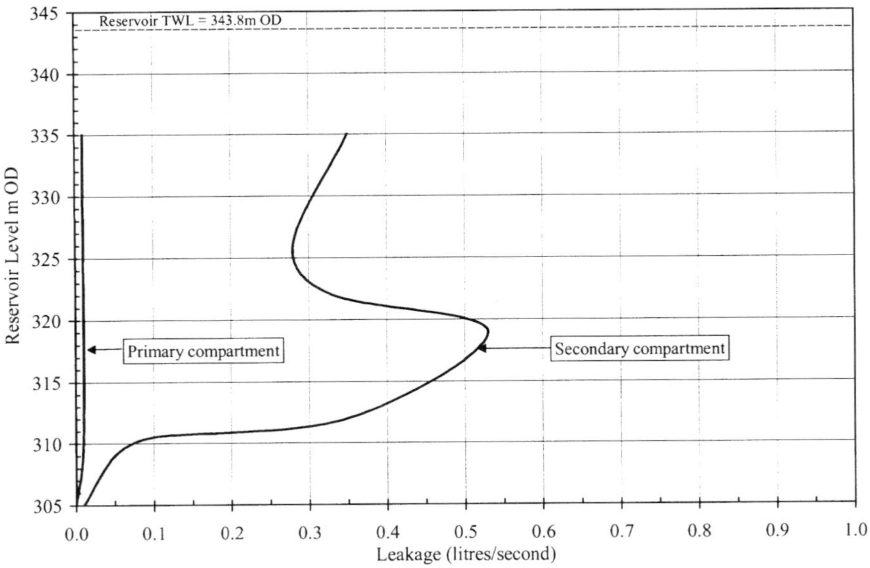


Figure 9 Leakage through geocomposite liner

Seepage through the dam foundation

Pulses of flow into the system, which correspond to rain percolating through the downstream face are evident from the records. The effect varies depending amongst other things upon the micro-catchment area of each compartment. A lag between rainfall and peak outflow of about 24 hours is apparent, which suggests an intrinsic permeability of about 10-3 m/s.

Figure 10 shows the response of the underdrains to refilling once these temporary effects have been removed.

Prior to the remedial works the bulk of the seepage flow originated from drainage zone 4 in the valley bottom. In particular, more than 60% of the total flow drained into manhole chamber SMMH7. At the end of February 2002 this zone contributed less than 20% of the combined flow.

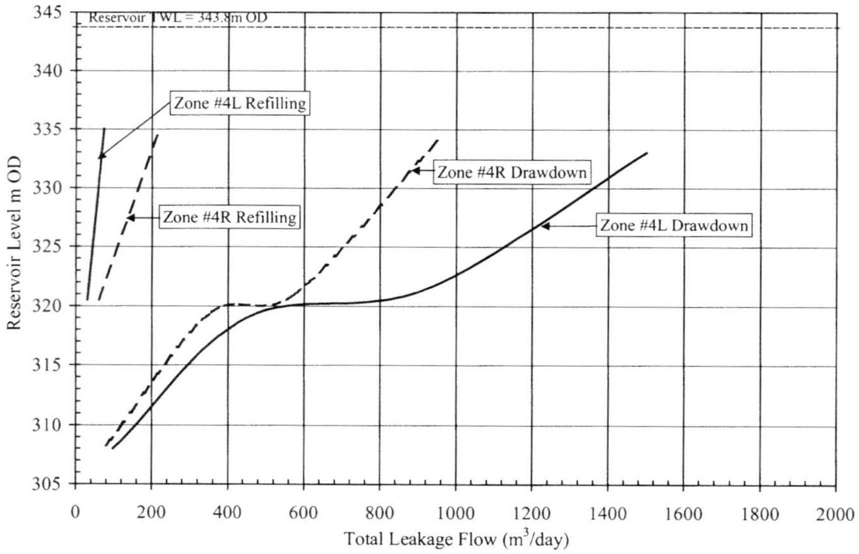


Figure 10 Seepage in to central underdrainage compartment.

The current indications are that total seepage flow has reduced considerably, both against the first successful filling in 1980 and also the situation at the start of 2001. The behaviour can be compared on Figure 11, which suggests an improvement of about 50% against the former and 80% against the latter. In terms of quantities, the estimated daily leakage during January 2001 was in excess of 4,000 $m^3$.

Following reservoir refurbishment and based on extrapolation of current data, the projected leakage at Full Supply Level will be 500 $m^3$ per day (i.e. less than 6 litres per second).

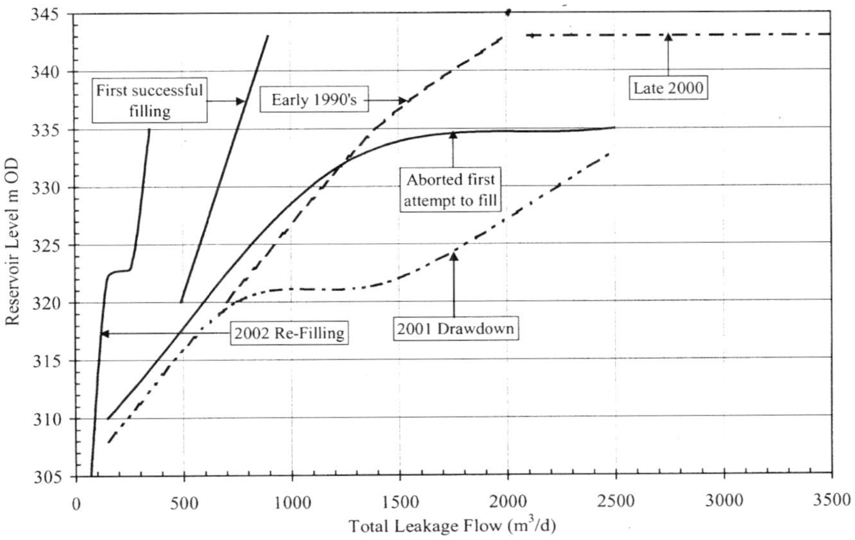


Figure 11 Total leakage/seepage before and after leakage remedial works

## CONCLUSIONS

The investigations confirmed significant ageing and deterioration of the 26-year old upstream asphaltic concrete membrane to Winscar Dam. The evidence suggests that the rate of incidence of defects may increase with time. A type of defect not previously encountered at this site (the 'construction joint' vertical crack) was observed. Major cracks were found that are believed to have developed as a result of settlement of deep fill behind the toe beam. One crack had opened into a hole through erosion.

Drilling, grouting and testing works suggested that the existing curtain remains in a generally good condition, although high grout takes and inter-connectivity between some holes suggests that leaching of the thin cement grout may have occurred in localised areas and/or stratigraphic horizons.

The geocomposite liner that now lies over the original dam face has been found to be very effective in reducing seepage through the dam. Early indications suggest that leakage through the face of the dam at Full Supply Level will not exceed 20 millilitres/second. The geocomposite liner works have enabled the installation of a new leak detection system that will enable defects to be pinpointed accurately and dealt with relative easily at low cost.

The projected daily leakage based on the extrapolation of current data is estimated to be about 20% of that observed prior to the refurbishment works. The gross reduction in water losses over a complete year with the reservoir operating under normal conditions is of the order of 1 $Mm^3$. In terms of the cost of developing an alternative resource of equivalent size, the refurbishment will have repaid the capital outlay of these works within three years.

## REFERENCES

Collins P C M & Humphreys J D (1974) Winscar Dam. *Journal IWES Vol 28,* pp 17-34.

Routh C D (1988) Investigation, identification and repair of the asphaltic concrete facing of Winscar Dam. *Trans. 16th Cong. ICOLD, San Francisco,* pp 655-677.

Carter I C, Claydon J & Robertshaw A C (2000) Winscar Dam: 24 years of manual seepage measurement replaced by automation. *Trans. 20th Cong. ICOLD, Beijing,* pp 1211-1220.

Wilson A C & Robertshaw A C (1998) Winscar Dam: Investigations and repair to asphaltic concrete membrane *Proc. 10th Conf British Dam Society, Bangor,* pp 292-302.

Aguiar E & Blanco M (1990) Experience in connection with performance of plasticized PVC sheeting in Tenerife basin sealing. *Proc. Conf. on Research and Development in the Field of Dams, Crans-Montana, Switzerland,* pp 361-374

## ACKNOWLEDGEMENTS

The authors wish to thank Yorkshire Water Services for their permission to publish this paper. The authors gratefully acknowledge the contributions made by their colleagues, notably, Ian Scholefield, Andrew Robertshaw and Lindsay Deuchar at YWS; Alan Harrison, John Hutchinson and Dean Robinson of MWH; Guy Wilson, Jon Thompson, Dave West and Richard Bakewell of Morrison Construction; Alberto Scuero and Gabriella Vaschetti of Carpi Tech; and Phillip Ball at Skanska Cementation Foundations.

Figure 12 The refurbished upstream face of Winscar Dam

# Turlough Hill – Upper Reservoir: condition of the lining after 30 years

B HAUGH, ESB International, Ireland

SYNOPSIS: Due to the development of cracks and increasing leakage, concerns have arisen in recent years regarding the integrity of the asphaltic concrete lining in the sloping sides of the reservoir at Turlough Hill Pumped Storage Scheme. Prior to April 1995 the leakage was negligible but subsequently started to increase and this was the starting point for the current phase of investigations into the integrity of the lining. This paper gives details of the lining and outlines it's operating history. The various investigations, repairs, possible causes of cracking and remedial options considered are briefly summarised. The policy adopted to deal with the problem and the success of this policy to date is also outlined.

## INTRODUCTION

Turlough Hill Pumped Storage Scheme is located in the Wicklow Mountains approximately 60km south of Dublin. The scheme was designed and is owned by the Electricity Supply Board (ESB) and has 4 No. 73 MW pump/turbine units. The main visible features of the scheme are a natural lower reservoir and an artificial upper reservoir (see Fig.1), which were constructed between 1969 and 1973.

The embankments forming the upper reservoir are founded on and constructed from sound and weathered granite. The "no voids" rockfill embankment dam was constructed by placing and compacting 1.3 x $10^6$ cubic metres of the material to achieve a density of 95% of the modified Proctor value. The maximum height of the embankment is approximately 34 metres and its crest length is approximately 1,445 metres. The surface area of the reservoir at high water level is approximately 14.2 x $10^4$ square metres and the capacity is 2.3 x $10^6$ cubic metres. A one metre thick drainage layer of 10-200mm crushed granite covered by a blinding layer of 35 to 55mm is provided on the inner face of the embankment (see Fig.2). A ring drainage gallery of reinforced concrete is connected to the floor and embankment drainage zones and has 75mm dia. openings in the walls at 2m centres to indicate leakage zones.

*Reservoirs in a Changing World*, Thomas Telford, London, 2002, 431–443

Lining

The embankment is sealed on its upstream face by an asphaltic concrete lining. The total area of lining is 160,000 square metres with 70,000 square metres on the $30^0$ slopes of the embankment.

Fig. 1 Turlough Hill Upper & Lower Reservoirs

The composition of the lining on the slopes of the embankment, which was laid in 3.5m wide strips, is shown in Fig.3.

The composition of the dense asphaltic concrete layer is as follows:

| | | |
|---|---|---|
| Crushed Diorite: | 2 to 12mm | 40% |
| Crushed Dust | 0 to 3mm | 29.8% |
| Natural Sand | 0 to 3mm | 14% |
| Filler | | 8 % |
| Asbestos | | 1 % |
| Bitumen | 60/70pen. | 7.2% |

The asbestos was added at the request of the contractor who maintained that it would allow him to add more bitumen for greater flexibility without loss of stability on the slope.

The specification of the asphaltic concrete lining for the upper reservoir at Turlough Hill was based on defining performance and quality standards, as follows:

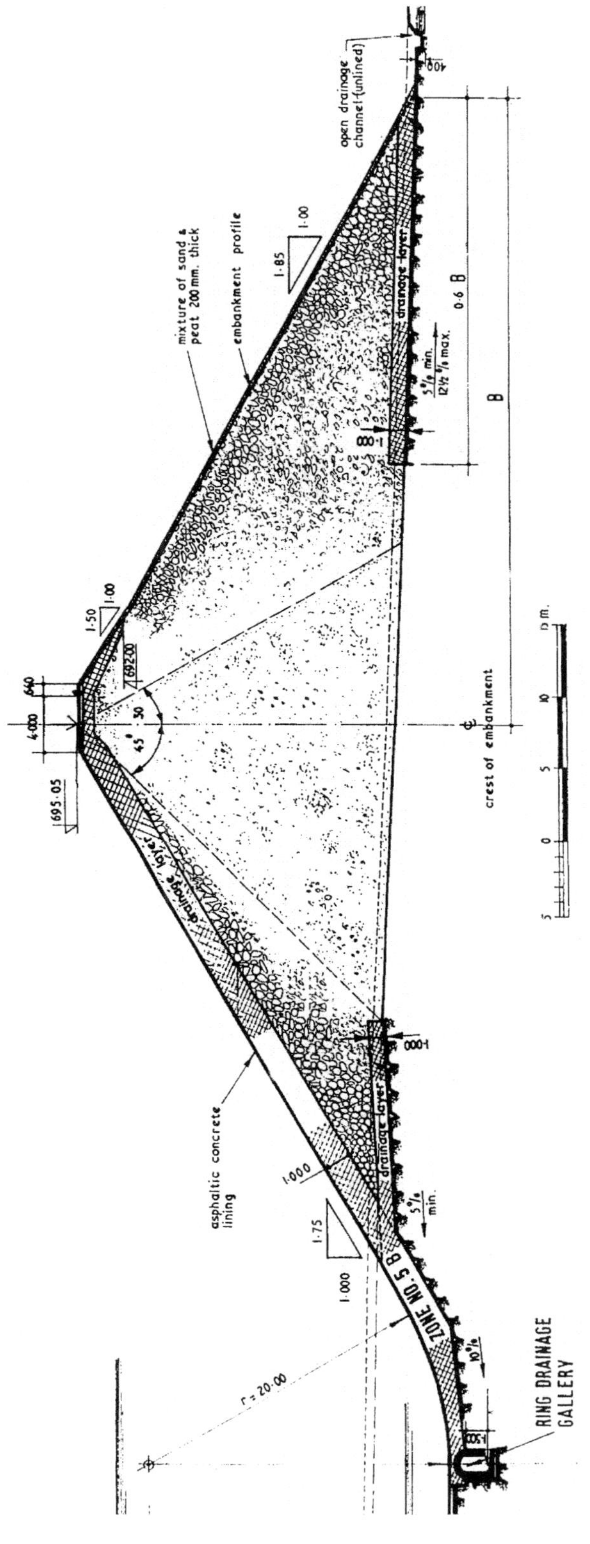


Fig. 2 Typical Cross Section of Embankment

- The lining should, by plastic deformation, bridge any cavities in the underlying drainage layer and accommodate movements of the dam without loss of watertightness, should not deteriorate as a result of weathering, should be stable on slopes at temperatures up to 70°C and should not be liable to cracks, loosening or disintegration at temperatures down to -20°C.

- The lining should be almost completely watertight. The total seepage from the reservoir should not exceed 6 litre/sec (l/s) and the seepage from any area should not exceed 1 litre per 25,000 square metres and the coefficient of permeability of the finished lining should not exceed $1 \times 10^{-9}$ cm/s. The water absorption of a sample of the finished lining should be less than 2 %.

Operating Regime

The reservoir is in use daily, The top water level varies from the maximum operating level to 1m below this and the lower level generally varies between 10m and 15m below the maximum level. Therefore the embankment is subject to a cyclical loading pattern possibly leading to minor reversible movements of the crest and with possible resultant fatigue effects on the lining.

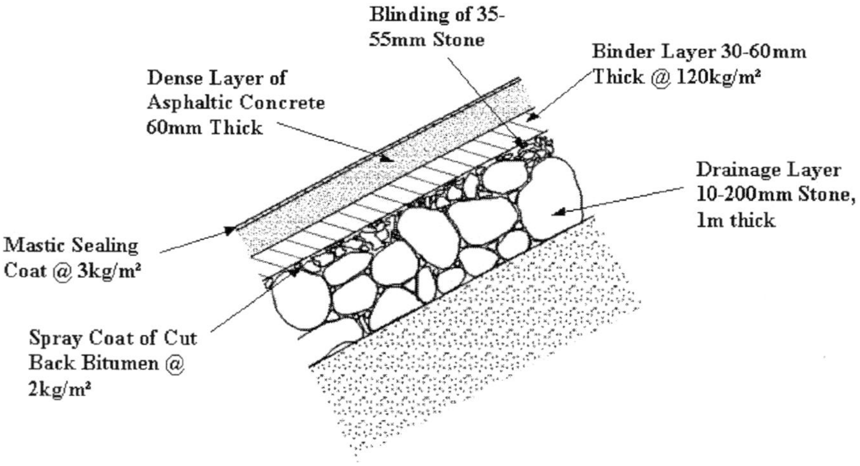


Fig. 3 Lining Details on Slope

OPERATING HISTORY OF LINING

1973-1975

In the first year of operation (1973) there were a few problems with the lining. Blisters appeared on the south west slopes as a result of moisture retained in small balls of matted asbestos fibres, which had not been properly dispersed. There were about a hundred of these 20 mm diameter blisters. Repairs were carried out by gouging out the balls of fibres and filling the resultant cavity with heated asphalt.

Other blisters on the east slope, caused by water trapped between the dense and binder layers of asphaltic concrete, were repaired as they were encountered. There were about ten of these, varying from 100 to 500mm in diameter and they were repaired by drilling a 16mm dia. hole through both layers into the drainage layer, from the lowest part of the blister. A short hardwood plug was driven into the dense layer to the interface with the binding layer. At least 40mm above the plug was filled with heated asphalt rammed into place and smoothed with a heated trowel. These plugholes were covered over with a flat stone and a raised dome of asphalt, 25mm thick. Up to the 1990's no further problems were encountered.

1990-1993

In the early 1990's the top mastic sealing coat, which had been applied hot, was observed to be severely weathered over the top 10 metres of the slope ( 5m vertically). Due to the operating regime of the station the top 4m on the slope is never underwater and the remaining 6m is underwater for a limited period, hence this area is the most exposed to direct sunlight.

In 1993, after seeking advice from the original lining contractor – Teerbau GmbH, a bituminous emulsion in a water base was sprayed on the top 10m of the inner slope by that contractor. This sealing coat differs from the original in that there is no sand content as opposed to 32% in the original coating and was applied cold. There was some initial blistering but most of these subsequently disappeared.

1995-Date

In April 1995 the total seepage from the ring drainage gallery increased to 0.33 l/s from the previous range of 0.15 to 0.25 l/s thus activating the alarm which was set at 0.3 l/s.

Following the very hot summer of 1995, the situation worsened in 1996 with new cracks in the lining and flows in gallery drainage pipes increasing to a new maximum of 1.0 l/s. Most of the leakage and cracking was on the east side of the reservoir and was coming from the embankment side of the gallery.

In March 1997 the flow peaked at 3.5 l/s when the reservoir was at a high level for a continuous 16 hour period. The alarm was increased to 4.0 l/s which is less than the specified limit of 6.0 l/s for total gallery flow for the entire reservoir. Later in 1997 the maximum leakage increased to 8 l/s for a short period. At this stage it was decided to carry out a detailed survey of cracks in the lining and to investigate methods for trial repairs of the cracks.

## INVESTIGATIONS & REPAIRS OF CRACKS

### Crack Surveys

In September 1997 the first detailed baseline crack survey of the lining on the sloping sides was carried out while the reservoir was empty. A total of 88 cracks with a total length of 209m were found, varying in length from 0.3m to 15m. Most of the cracks were less than 3m long and varied in width up to 4mm. ESB/ESBI speculated at this stage that the combination of age, temperature and UV light was causing the lining to become brittle.

Since 1997 surveys have been undertaken approximately twice yearly with the reservoir at an intermediate level (see Table 1). In 1998 the number and total length of cracks increased by over 25%. During 1999 and 2000 the rate of increase in crack development slowed considerably. However in 2001 the rate increased again, partially due to low reservoir levels allowing a greater surface area be examined than was previously possible. The surveys show that nearly all of the cracks run down the slope. The majority are in the top 10m of the slope (5m vertically) and the greatest concentration are on the east side. The majority of the cracks occur along the laying joints and the rate of cracking may be influenced by a range of temperatures occurring over a period.

Table 1 – Results of Crack Surveys

| Date | Number | Percent | Length (m) | Percent |
|---|---|---|---|---|
| Oct.1997 | 88 | 40 | 209 | 56 |
| May 1998 | 40 | 18 | 64 | 17 |
| Feb.1999 | 17 | 8 | 14 | 4 |
| July 1999 | 4 | 2 | 5 | 1.5 |
| Oct.1999 | 8 | 4 | 9 | 2.5 |
| May 2000 | 2 | 1 | 1.5 | 0.5 |
| Apr.2001 | 41 | 18 | 38.5 | 10.5 |
| Sep.2001 | 20 | 9 | 30 | 8 |
| **TOTALS** | **220** | **100** | **371** | **100** |

### Experiments

It was considered that the rate of leakage was related to high reservoir levels as most of the cracks are located in the upper section of the lining. In April 1998 after the rate had increased to 5.5 l/ it was decided to experiment with maximum reservoir levels (see Fig.4) Therefore the maximum operating reservoir level was reduced from 693.1 m.O.D. to 692.2 m.O.D. in 300mm stages over a period of time. Reductions in total gallery flow occurred at each stepped reduction in level. At 692.2 m.O.D. the total gallery flow reduced to 0.33 l/s approximately. This experiment confirmed that at high reservoir levels leakage increased progressively.

Trial Repairs

In September 1997 trial repairs on sections of cracks on the east side were carried out. A number of proprietary repair/sealing materials as specified by international suppliers, were applied to cracks in accordance with the manufacturers instructions by a specialist contractor. None of these were successful. In addition trial repairs were also carried out by preparing a V-notch along lengths of crack and filling it with hot applied roofing grade asphalt. This method of repair was successful but is slow and requires specialist personnel and equipment to carry out the works.

Following the experiments of reducing reservoir levels, repairs were carried out by local ESB personnel in June 1998 on some of the cracks, particularly on the eastern side of the reservoir, which was the area contributing most to the leakage. These repairs involved placing a nylon reinforcing mesh (Flamenco scrim), 500mm wide, over the crack and brushing on to the mesh two coats of a bitumen based thixotropic sealing material (Super Flamenco supplied by MMP International Ltd.). Following these repairs total gallery flow reduced to 0.2 l/s. It was subsequently decided to restore the maximum reservoir level to 693.1 m O. D. (see Fig. 4). The leakage remained at the reduced level and the repairs were monitored over the following years.

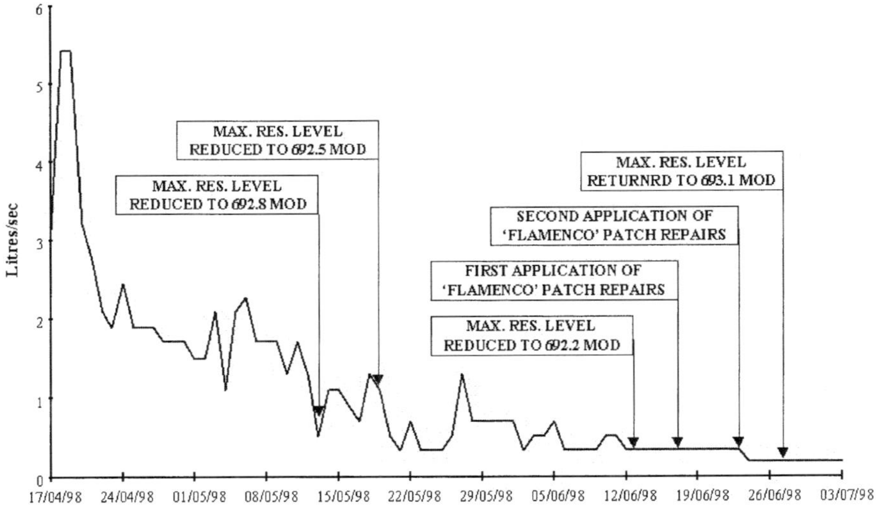


Fig.4 – Maximum Daily Total Gallery Flow

## CONDITION ASSESSMENT OF LINING

In September 1997 nine 150mm diameter cores were taken from the lining on the east and south slopes. The majority of cores were 150mm deep and most were taken at crack locations on the laying joints. Three cores had surface cracks, two had partial depth cracks and three had full depth cracks. In general the condition of the cores was good, however there was no bond between the dense and binding layers for two of them.

In 1998, Walo Bertschinger AG, a Swiss asphalt lining contractor inspected the lining and took two of the 1997 cores to their laboratory for examination and testing. Tests were carried out for binder content, bulk density, air voids and penetration. The results of the laboratory tests are summarised in Table 2 below.
Walo produced a report which agreed that the cracks derived mainly from the ageing and brittleness of the bitumen in the asphaltic concrete lining and indicated that there could be a progressive increase in the number and widths of cracks within the next few years. They recommended that in the short term information on cracks and leakage should be collated at least twice a year. In the long term they recommend that consideration would have to be given to the renewal of the outer layer of the lining on the slope if the condition of the asphalt continued to deteriorate to the point where leakage became a problem. They noted that almost all the visible cracks occurred on or near the laying joints.

In 1999 six 150mm diameter cores were taken from uncracked areas of lining away from laying joints and at roughly equal spacings around the circumference of the reservoir. Four of the cores were 150mm deep and two were 250mm deep. All cores were intact and in good condition. Four of these cores were again sent to Walo to carry out the same range of tests as previously. The results of these later tests are also summarised in Table 2

Table 2: Results of core tests

| **Test** | **Specified** | **1997 Cracked** | **1999 Uncracked** |
|---|---|---|---|
| Binder Content (%) | 7.2 | 7.8 to 7.9 | 7.35 to 8.1 |
| AirVoids (%) | 3 | 6.9 to 8.7 | 1.2 to 1.8 |
| Penetration (1/10 mm) | 60/70 | 16 to 28 | 31 to 37 |
| BulkDensity (kg/m$^3$) | 2350 | 2185 to 2254 | 2368 to 2433 |

The following conclusions can be drawn from the above results:

Binder Content: All values are greater than the specified 7.2%.
Air Voids: The voids for the uncracked cores (1999) are less than the specified 3% while the values for the cracked cores (1997) are much greater.
Bulk Density: The uncracked cores were denser than the cracked cores.
Penetration: The values for the uncracked cores indicate a loss of 50% approximately from the specified 60/70 pen.

Generally the results of the tests indicate that the cracks derive mainly from the ageing and brittleness of the bitumen in the asphaltic concrete lining. It appears that most of the cracks have occurred at the laying joints between adjacent strips of the lining, where some degree of reheating of the lining would have occurred from placing hot asphaltic concrete in contact with the lining already in place, resulting in a probable loss of flexibility. Cores taken from uncracked areas, away from laying joints, display a considerably greater degree of flexibility. Penetration values for cores, where the full depth of the lining was cracked, were in the range 16 to 28, values for cores in uncracked areas were in the range 31 to 37. Therefore, it appears that the uncracked areas of the lining are generally in good condition at this time and would have to become considerably less flexible before serious cracking might be likely to occur there.

## REVIEW OF OPTIONS

In 1999, ESB carried out a review of options to deal with to deal with the deterioration of the lining based on the experience of works carried out up to then by ESB and current international practice.

### Maintain Current Practice Regarding Monitoring and Repairs

Currently twice yearly crack surveys, generally in Spring and Autumn are carried out on the lining. Furthermore, material tests have been carried out on cores taken from the lining. These have revealed that the bitumen in the asphaltic concrete is considerably more brittle and less flexible in cores taken through cracks than in cores taken in uncracked parts of the lining. The tests also indicated that the general condition of the asphaltic concrete is sound and is unlikely to deteriorate suddenly.

Repairs carried out with the nylon mesh and the bitumen based sealing material are still in good condition. The reduction in seepage flows and their maintenance at previous low levels indicates that the repairs have successfully sealed the cracks. There is ongoing monitoring of leakage and observations for signs of washed out material. There is an expectation that the repairs are likely to keep leakage under control as long as the lining material remains sound, even if it is becoming brittle. Variations on this monitoring and repair policy have been applied successfully to other dams of similar ages, e.g. Marchlyn Dam at Dinorwig Pumped Storage Scheme.

### Milling the Asphaltic Concrete at Cracks and Replace

All identified cracks could be repaired by milling off the existing asphaltic concrete at either side of the crack and replacing it with fresh asphaltic concrete. This method was applied by Vorarlberger Illwerke AG(VIW) to one of their dams in Austria, but only after the failure of more simplified repair methods.

While this method of repairing the cracks, effectively removes all the cracks identified at the time of the works, it involves the placing of hot asphaltic concrete against existing asphalt concrete. This reheating effect may further add to the ageing of the existing asphaltic concrete. This possible further loss of flexibility in the existing asphaltic concrete may eventually lead to cracking on either side of the repair. Indications are that this is beginning to happen in the case of the Austrian lining above. This method does not prevent the ageing process from continuing in the remainder of the lining.

Replacement of the Dense Layer on the Side Slopes

This approach would involve milling off the existing dense and mastic layers on the side slopes of the reservoir and replacing it with a new dense layer of fresh asphaltic concrete and a mastic sealing layer. While the asphaltic linings of many dams are now reaching an age where ageing is causing defects such as cracking, there are not many reported cases of this approach having been adopted.

One example of this is at the Saint Cϑcile d'Andorge Dam in France (Herment, Huynh & Jensen,1997). On this dam the milling process was carried out to a depth of 70mm at a rate of 320 $m^2$ per day. This was replaced with 60mm of light coloured asphaltic concrete. This light colour should reduce the effects of heat on the lining.

A second example is at the Hardap Dam in Namibia (Schewe, 1997), where 70 mm of asphaltic concrete was milled off down to low water level and replaced with new dense and mastic layers over an area of 22,000 $m^2$. However, shortly after completion the entire repair works failed, with cracks developing over the entire area. This lead to further rehabilitation works having to be carried out, involving the complete removal and replacement of the recently placed asphaltic concrete. These repairs are reported to be successful. From the problems experienced at the Hardap Dam, it can be seen that there is potential for problems in the initial period, particularly relating to adhesion between the new dense layer and the original binder layer.

Application of Geomembrane

Winscar Dam in the UK is of similar vintage to Turlough Hill, having been constructed between 1972 and 1975, and has a similar lining. The main difference in the lining is that Winscar has two layers of dense asphaltic concrete, 40mm and 80mm thick, while Turlough Hill has only one. In 1996 repairs were carried out on cracks, blisters and other defects at Winscar. It was noted during the repair works that at most of these repair locations there was no bond between the two layers of material (Wilson & Robertshaw, 1998).

In 2001, when leakage had increased considerably it was decided to apply a 2.5mm thick PVC geomembrane over a geotextile membrane, instead of a new asphaltic concrete lining. Carpi laid 25,000 $m^2$ of the geomembrane in 2001. The predicted maximum leakage rate is 1 l/s (NCE, 2001).

## POLICY ADOPTED BY ESB

Most of the above methods involve specialist contractors and considerable costs combined with disruption to the station operations. Having reviewed the above range of options the ESB has decided to continue the current practice of monitoring and repairing cracks using local ESB staff. Repairs carried out by ESB are proving adequate in sealing cracks and keeping the rate of seepage at acceptable levels and cause no disruption to station operations.

While deterioration of the lining is likely to continue, it is envisaged that with the current policy outlined above the lining will continue to give good service for a considerable period. Further, more elaborate, repair and maintenance strategies will only be considered if the current repair regime becomes ineffective.

In 1989, the current ESB dam safety organisational structure was instigated. This included the appointment of an External Dam Safety Committee (EDSC). This committee of international experts is an independent board which oversees the safety of ESB dams.

During the 10-year EDSC Inspection of Turlough Hill upper reservoir in June 2000, the lining was examined. The EDSC considered that the repairs carried out at Turlough Hill are very successful and recommended that the current approach of monitoring and repairing cracks as soon as they appear be continued as long as this proves successful in controlling leakage. This approach allows time for future developments in technology to come up with cost-effective solutions to the problem of ageing asphaltic concrete reservoir linings. They also advocated the installation of a water sprinkling system at the crest of the embankment. The activation of this system at preset temperatures would help to cool the lining and reduce the adverse effects of sunlight and high temperatures. The installation of such a system is being actively pursued.

## CURRENT STATUS

After the leakage was successfully controlled in 1998 there was no major problem with leakage until 2001 even though little repair works were carried out in the meantime. However coinciding with the increase in the number of cracks in the lining up to May 2001, peak values for the total gallery flow increased to approximately 2 litres/sec. Most of these cracks and the associated leakage were on the east side. Following repairs carried out by ESB in May 2001 and August/September 2001 flows again reduced to 0.2 l/s.

In December 2001 there was another increase in flow to 3.5 l/s. The leakage on this occasion was in the south of the gallery, thus emanating from the north facing slope of the lining and the highest flows coincided with the coldest days. It was also noted that the flows peaked approximately one hour before the maximum reservoir level. Therefore the highest flows appear to be coinciding with the coldest temperature of the lining, just prior to being warmed by the reservoir water and daytime temperatures. The low temperatures may be causing contraction and consequent opening of joints/cracks in the lining, which are closed by the temperature of the reservoir water as the reservoir fills. Further investigation is necessary to confirm this. These cracks will be repaired in late spring or early summer 2002 as soon as weather conditions are suitable for repair work.

## CONCLUSIONS

- In general, from the results of investigations and assessments, the main body of the lining at Turlough Hill upper reservoir is in good condition.
- The majority of cracks correspond with laying joints in the lining.
- Cracking of the lining results from the ageing and consequent loss of flexibility in the asphaltic concrete.
- The majority of cracks are concentrated on the east side, which is the side most exposed to sunlight and therefore to ultra violet light.
- Nearly all of the cracks run down the slope with few extending beyond 10 metres from the top.
- The operating regime of the reservoir causing daily cyclical loading and unloading possibly contributes to fatigue in the lining.
- Leakage on the south side in winter could be caused by low temperatures causing contraction and consequent opening of cracks.
- Leakage is related to high water levels in the reservoir, with minor reductions in level causing large reductions in flow. This may change as the ageing progresses.
- The ongoing repairs being carried out by local ESB personnel using a bitumen based sealing material and a nylon reinforcing mesh are proving successful at controlling leakage and will be continued as necessary.
- While the rate of crack development in 1997 and 1998 was quite rapid it had slowed considerably in 1999 and 2000, but increased again in 2001. Extreme weather conditions may affect the rate of crack development in the future.

## ACKNOWLEDGEMENTS

The author is grateful to the Electricity Supply Board for permission to publish this paper and to his colleagues in ESB International for their assistance in it's preparation.

REFERENCES

Herment R, Huynh P & Jensen A (1997). Renovation of Saint Cecile d'Andorge dam facing with white asphalt. Proc. 19th ICOLD Conference, Florence, Vol.1, pp909-938.

Schewe L (1997). Successful remedy for the failed initial refurbishment at Hardap dam. Hydropower & Dams, issue 6,1997 pp57-62.

Wilson A & Robertshaw A (1998). Winscar dam: Investigations & repairs to asphaltic concrete membrane. Proc. 10th Conf. British Dam Society, Bangor, pp292-302.

Search for the seal. New Civil Engineer, 15th November 2001.pp24-25.

# Colliford and Roadford dams: performance of the asphaltic concrete membranes and the embankments

J K Hopkins, Earth Tech Engineering. Ltd
P Tedd, Building Research Establishment. Ltd
C Bray, South West Water

SYNOPSIS. The use of upstream asphaltic concrete membranes as the sealing element on embankment dams has been limited to six dams in the UK. This paper reviews the performance of Colliford and Roadford dams, the two most recently constructed and compares their performance with other asphaltic membrane dams in the UK and Eire. Although constructed more than 10 years ago, there have been relatively few problems with the embankments or upstream asphaltic membranes.

## INTRODUCTION

Two of South West Water's strategic reservoirs, Colliford in Cornwall impounded in 1983 (Johnson & Evans, 1985) and Roadford in Devon impounded in 1989 (Johnston et al, 1995) have asphaltic concrete membranes as the waterproofing element on the upstream face. Colliford Dam is 29m high, 520m long, 254m AOD and impounds 29,100Ml. The embankment is formed by sandwaste from local china clay operations. Roadford dam is 41m high, 430m long, is 126m above sea level and impounds 37,000Ml of water. The embankment is constructed from low grade rockfill quarried from within the reservoir basin, (Wilson & Evans, 1990).

There are only six dams in the UK with upstream asphaltic concrete membranes despite their wide use in Europe. Details of the other dams constructed in the UK have been summarised by Wilson & Robertshaw (1998). This paper reviews the monitoring, performance and maintenance of the upstream membrane of Colliford and Roadford dams and compares their performance with other upstream asphaltic membrane dams.

## COLLIFORD AND ROADFORD DAMS

The membrane, at both dams, consists of two layers of bituminous material laid over a drainage blanket placed on the face of the embankment fill, (Evans & Wilson, 1994). The lower binder layer is 70mm thick and provides a foundation for the 80mm thick, dense virtually impermeable layer. The binder layer has a permeability greater than 1 x $10^{-4}$m/s to ensure that any seepage through the dense layer reaches the underlying drainage

*Reservoirs in a Changing World*, Thomas Telford, London, 2002, 444–455

blanket. The detail provided at the connection to the upstream toe is shown in Fig. 1.

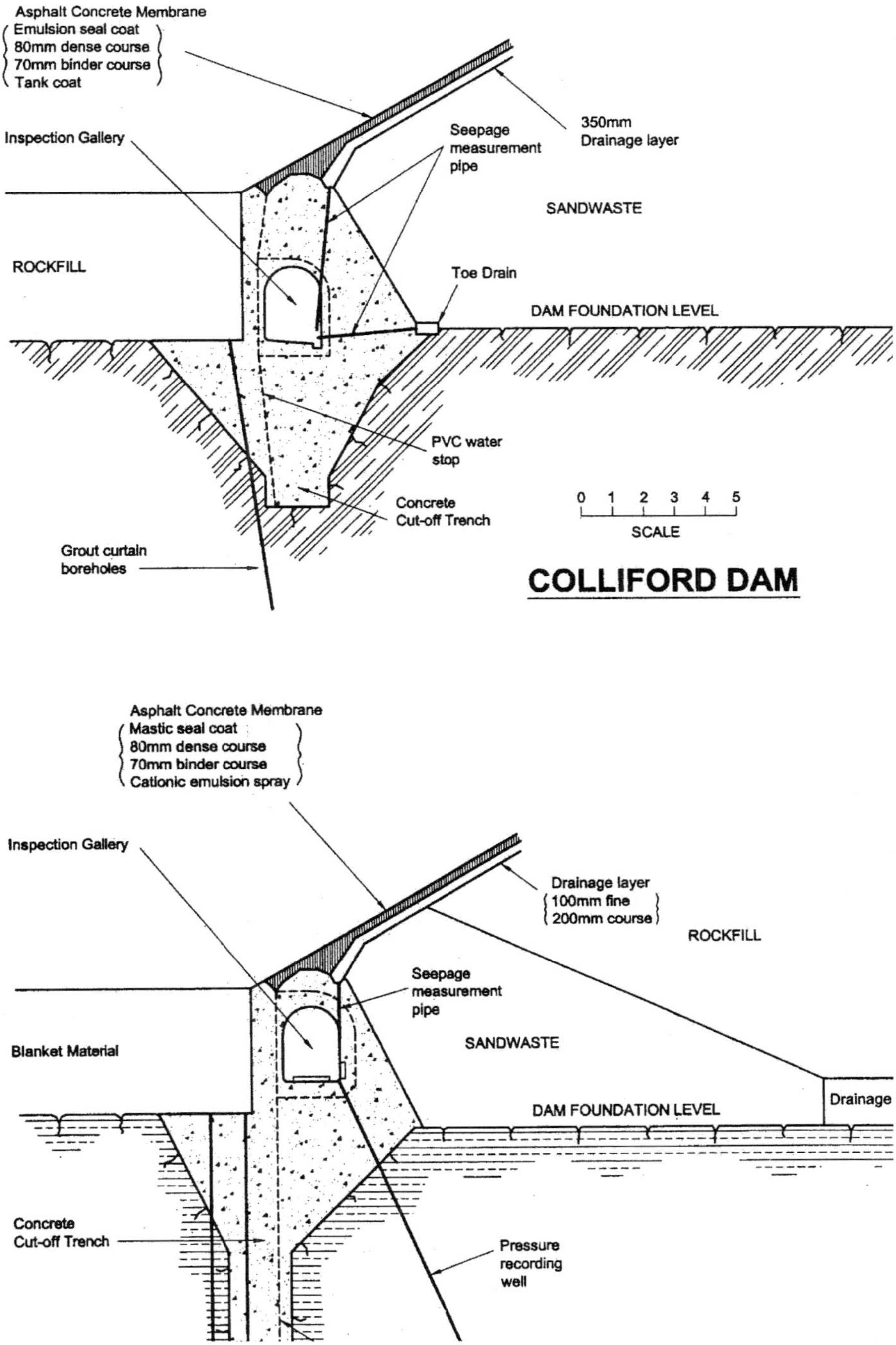


Fig. 1. Detail of connection to the cut-off (after Evans & Wilson, 1994)

The membrane performance criteria, at both dams, for flexibility, slope stability, permeability and durability was specified. The bitumen content of the dense material was 4.7% at Colliford and 4.0% at Roadford. Testing indicated an average air void content of 1.6% at Colliford and 1.1% at Roadford against 3% specified.

The seal coat used at Colliford was a two layer Flintkote cold emulsion applied by spray with the first layer containing mineral fibres. At Roadford hot applied mastic was laid with a squeegee spreader.

The connection with the concrete cut-off was designed to minimise differential settlement and create a satisfactory seal. A large curved and grooved contact area, together with an inclined downstream face were used. Very small settlements occurred at Colliford where sandwaste had been used for construction of the embankment. At Roadford, a zone of the same material was placed between the toe structure and the more compressible rockfill to reduce the relative movement of the asphaltic membrane.

To give added protection to the toe, a Hypofors bitumen sheet, reinforced with nylon and polyester fabrics, was sealed to the face of the asphalt at Colliford. At Roadford an additional 50mm layer of dense asphalt was used, which contained 0.5% more bitumen to increase flexibility.

## MEMBRANE PERFORMANCE AND REPAIRS

During 1987 a broad line was observed just below the top water line at Colliford. The Flintcote emulsion seal coat had pimpled, 1-2mm high and 5-10mm across, and discoloured over a width of 300mm. This had been caused by algae that had died and shrunk following draw-down causing the bituminous material to pimple. The coat had been spray applied which resulted in a thicker rougher coat than necessary. The damaged area was cleaned off and repaired with a brush applied primer followed by one coat of Flintcote no.3 and one of Flintcote no.5.

Small blisters were again found in the seal coat on the exposed upper part of the membrane, 15.5m down slope, during 1996 together with a few roller cracks in the asphaltic concrete and some damage at the Hypofors connection with the concrete gallery. Repairs were carried out when the reservoir was drawn-down during 1997. A primer coat was spray applied over the whole area. A single coating of Bituproof no.3 was squeegee applied over the asphaltic area without previous sealing. A new seal in two layers of Bituproof no.3 squeegee applied across and down slope was applied over the whole area. This was still in good condition in 2001. Minor repairs to joints were carried out at the same time. Minor repairs to top up the mastic seal between the membrane and precast concrete wave wall units had previously been carried out in 1993.

Similar lines of algae have been observed at Roadford on a number of occasions, but no damage to the seal coat has been observed.

Fig. 2. Colliford dam: damage to seal coat in 1996, new seal coat being applied in 1997

## DRAINAGE FLOWS

At both Colliford and Roadford any seepage through the membrane is collected in the underlying drainage layer and discharged into the inspection gallery where the flow can be monitored (Fig. 1). In addition, at Colliford there is a toe drain immediately downstream of the cut-off, which can collect water from the drainage layer due to the high permeability of the sandwaste fill. At Roadford there are a series of pressure recording wells, which can act as shallow drains for the foundation immediately downstream of the cut-off.

### Colliford

Very little flow was noted from the membrane drains during impounding at Colliford. During the period December 1985 to June 1986 flows occurred from two drains on the west side, accompanied by increased flow from the adjacent toe drains. In general there has been only small flows from the toe drains until the reservoir is within 3m of TWL, with little or no flow from the membrane drains. Flows into the gallery are generally related to reservoir level although they also respond to high rainfall, see Fig 3. The main flows enter via the toe drains where seepage flow from the flanks is intercepted. There is little or no flow in the central section where seepage flow drains into the underlying drainage blanket. Flows from the toe drains are accompanied by large quantities of iron and manganese rich ochre. This tends to block the drains and regular cleaning is required.

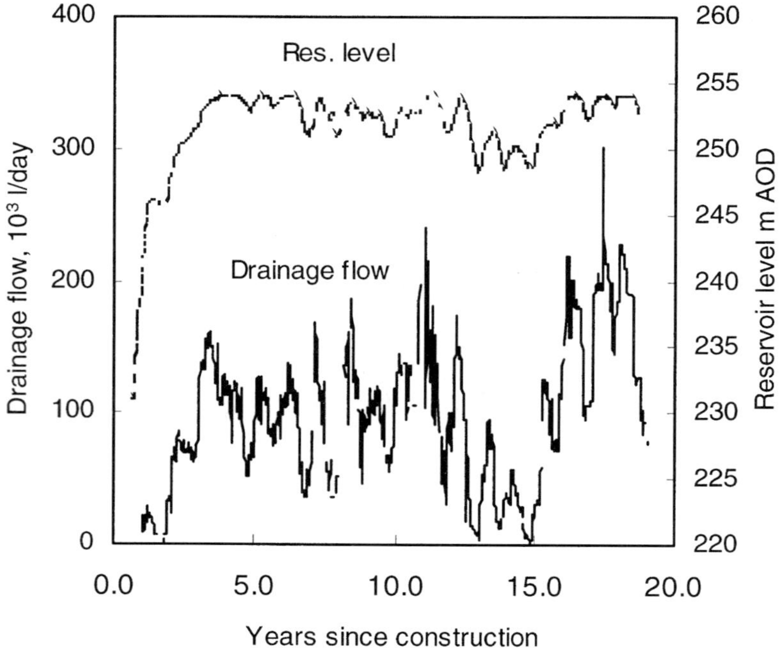


Fig. 3. Drainage flows at Colliford dam

Roadford

At Roadford small flows were observed from the membrane drains shortly after impounding. During the first winter 1989/90 membrane drain flows increased significantly coincident with rising water level. This virtually ceased during the next summer. During winter 1990/91 when the reservoir approached full, flows increased significantly, with flows the following summer virtually ceasing. This trend has continued although winter drainage flow peaks have generally been on a gradually reducing trend with some correlation with water level. Water temperatures show a variation between approximately 18°C in summer, corresponding to virtually no flow in the membrane drains to 7°C in winter when flows are at a maximum, see

Fig 4. The majority of the membrane drain flow originates from the area between the access tunnel and the centre of the dam.

Brauns (1994) suggested that the leakage may be a function of the thermal expansion/contraction of the asphaltic concrete when local imperfections open during cold weather with a corresponding increase in leakage. The gradual decline in maximum membrane leakage rates at Roadford suggests that these local imperfections self seal with time.

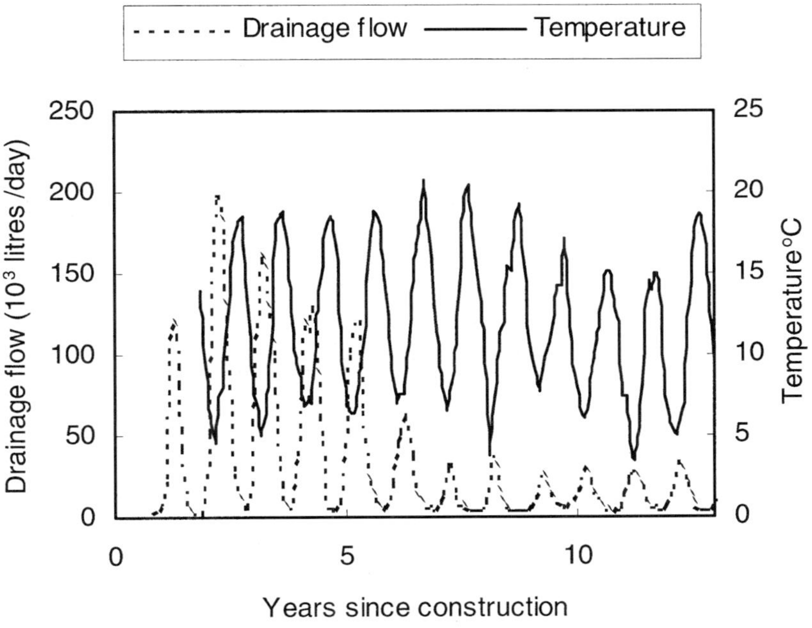


Fig. 4. Membrane drainage flows at Roadford dam.

## SETTLEMENT

### Colliford

At Colliford overflow settlement cells were provided under the asphalt to measure any deflections. Readings have generally been within the range 40mm settlement to 20mm heave, and are probably due to error range of the instrument rather than actual settlement. These readings have now been discontinued. Crest levels at Colliford confirm that very small movements in the range of +/- 15mm have occurred which was expected from the low compressible sandwaste. No long term movements have been measured since impounding. Generally net settlement has occurred near the flanks and slight heave at the centre of the dam.

Roadford Dam

Overflow settlement cells were also used at Roadford but there was little confidence in the observations and they have been discontinued.

Crest coping levels at Roadford have indicated a reducing settlement trend with time. Maximum settlement at the crest copings since impounding in October 1989 has been 425mm and is well within the freeboard design allowance of 981mm, see Fig 5. This compares well with the settlement measured by the vertical magnet settlement gauge for the same period. An additional 110mm was measured by the settlement gauge prior to impounding.

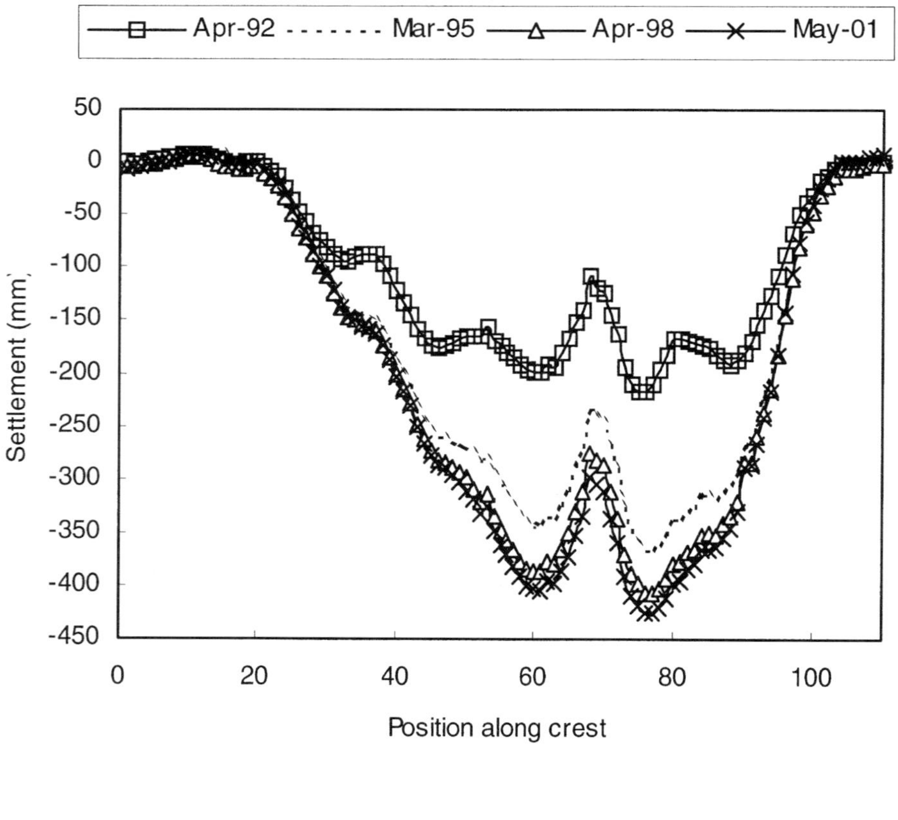


Fig. 5. Crest settlement at Roadford dam

To minimise the deflection of the membrane close to the cut-off structure, sand waste, similar to that used at Colliford which has a compressibility approximately one third that of the sandstone/mudstone/shale rockfill was placed as shown in Fig. 1 to form a transition zone. To monitor the deflections of the membrane at this transition zone an electro-level system was installed immediately under the membrane at three locations along the dam. It consisted of electro-levels built into a series of stainless steel articulated box sections 1m and 0.5m long. The shorter lengths were placed closest to the cut-off structure. At the highest section of the dam, the system extended 14m up the face of the dam from the toe while those on the abutments extended 11m. Details of the system and observed deformations for the first five years are described by Tedd et al (1991, 1995).

The instruments were installed in April 1989 and impounding of the reservoir began in October 1989. Readings are still being taken in 2002. The deflections of the membrane normal to the slope and relative to the toe of the dam at the highest section are shown in Fig. 6 for different times after impounding and various depths of water. The largest deflections occurred between the toe and 1m up the slope (17mm per m), and at the sand waste/rockfill interface (12.5mm per m). It can be seen that there is very little differential deformation over the area where the sand waste was placed. Figure 6 also shows a simple prediction of the membrane deflections based on the compressibility of the sand waste and rockfill from 1m diameter oedometer tests. The prediction is based on full reservoir head and can be compared with observed deflection when the reservoir was first full (plot 2). The agreement is reasonably good close to the toe but the prediction underestimates deflections further away from the toe.

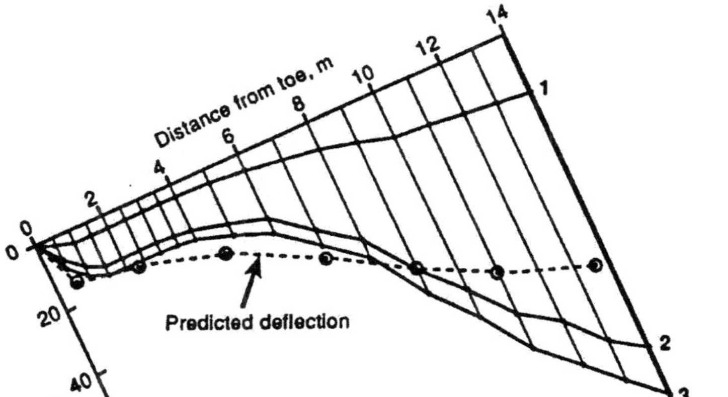


| Date | Res head m |
|---|---|
| **1** Nov 89 | 14.6 |
| **2** Apr 91 | 33 (full) |
| **3** Apr 01 | 33 (full) |

Fig. 6. Deflection of the asphaltic membrane at Roadford dam

Figure 7 shows the deflections measured with time at two locations at the deepest section B, at 11m (15B) and 14 m (18B) from the cut-off, and at the top of the two abutment sections, A and C. At the deepest section, deflections increased with increasing reservoir head to approximately 110mm at 14m up the slope at the highest section when the reservoir was first filled. Deflections are time dependent with increasing deflection occurring when the reservoir level did not change. Reducing reservoir level caused a reduction in deflection, but only after a time lag of approximately 3 months. Cyclic changes in defection, shown in Fig. 7. between 1992 and 1997 correspond to cyclic changes in temperature varying between approximately 18°C in summer to 7°C in winter as shown in Fig. 4. However, these cyclic changes in deflection have not been measured since the beginning of 1999, which may reflect the effect of only small reservoir changes during this period.

The readings from the electro-levels at Roadford dam appear to have given sensible deflection measurements both in the short and long term.

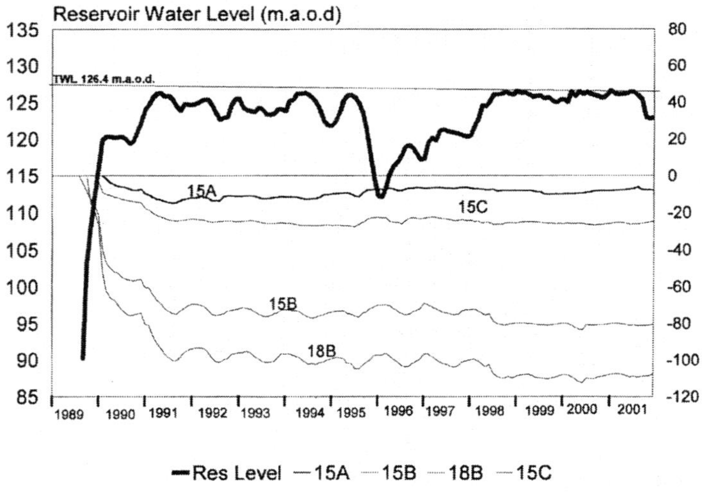


Fig. 7. Reservoir level and deflection of asphaltic membrane at Roadford dam as function of time

## PERFORMANCE OF UPSTREAM ASPHALTIC MEMBRANE DAMS

Some details of the other dams in the UK and EIRE are summarised in Table 1. Relatively minor maintenance works have been carried out on the asphaltic membranes in the UK with the notable exception of Winscar

(Routh 1988, Wilson and Robertshaw 1998) which has had major repairs on a number of occasions and has recently had a geocomposite liner installed on the upstream face (Carter et al, 2002).

Brauns (1994) has discussed the problems of blisters on asphaltic membranes. Many hundreds can occur and they are often filled with water. They only occur where the reservoir is drawn down and investigations have shown that water in the blisters has under gone boiling confirming its entrapment during construction. Repeated frost cycles are thought to weaken the bonds between asphalt layers allowing the blisters to form.

Fielder et al (1996) describes the performance of Porabka-Zar pump storage scheme which was completed in 1979. Thousands of blisters were observed 2 years after construction in the 20m zone where the reservoir is drawn down.

## CONCLUSIONS

The performance of asphaltic membranes at Colliford and Roadford dams continues to be very successful. Minor repairs and replacement of the exposed protective seal coat have been required at Colliford due to deterioration of the spray applied emulsion. The squeegee applied mastic seal coat at Roadford continues to perform well 12 years after construction. The electro-level system installed at Roadford dam continues to work well and indicates little long term movement of the membrane relative to the toe.

The occurrence of blisters and cracks on asphaltic concrete membranes seems inevitable, particularly where water has been allowed to become entrapped during construction. Regular maintenance to the seal coat particularly in the zone of frequent drawdown where the membrane is exposed to UV and frost appears necessary with many dams.

## ACKNOWLEDGEMENT

The Authors would like to acknowledge the permission of South West Water Ltd to publish this paper.

## REFERENCES/BIBLOIGRAPHY

Brauns J (1994). Blisters in an asphalt membrane study – a case study. Discussion 8th British Dam Society. Conference. Reservoir Safety and the Environment, pp 43-51

Carter I C, Claydon J R & Hill M J (2002). Improving the watertightness of Winscar reservoir. Reservoirs in a changing world. Proceedings of the 12th Conference of the British Dam Society. Dublin. pp 400-415

Charles J A & Penman A D M (1988), The Behaviour of Embankment Dams with Bituminous Watertight Elements, Transactions of 16th International Commission on Large dams, San Francisco, vol 2, pp 693-705.

Eccles P G & Kaye K A (1984), Sulby Reservoir, Isle of Man, Journal Institution of Water Engineers, Vol. 38, Jun, 197-216.

Evans J D & Wilson A C (1994), The asphalt membranes at Colliford and Roadford Reservoirs, 'Reservoir Safety and Environment', Thomas Telford, London, 1994, pp1-11.

Fiedler K, Ptaszynski Z & Szymanski A (1996). Repair of asphaltic concrete sealing of the reservoir. Proc of Conf Repair and upgrading of dams. Stockholm pp83-90.

Haugh B (2002). TurloughHill – Upper reservoir: condition of the lining after 30 years. Reservoirs in a changing world. Proceedings of the 12th Conference of the British Dam Society. Dublin. Thomas Telford, pp 416-428

Johnston T A & Evans J D (1985). Colliford Dam sandwaste embankment and asphaltic concrete membrane, Proceedings of the Institution of Civil Engineers, Vol.78, Aug. 689-709.

Johnston T A, Wilson A C & Evans J D (1995). Roadford Reservoir: design, construction and performance. Proc. ICE, Wat. Marit. & Energy, vol. 112, pp 316-325.

Kast K & Brauns J (1996). Ageing and remediation of an Asphaltic conctere faced embankment dam Proc of Conf Repair and upgrading of dams. Stockholm, p185-194

Poskitt F F (1972).The Asphaltic Lining of Dungonnell Dam, Proceedings Institution of Civil Engineers, Vol. 51, Mar, 567-579.

Routh C D (1988). The Investigation, Identification and Repair of the Asphaltic Concrete Facing of Winscar Dam, Transactions of 16th ICOLD, San Francisco, vol 2, pp 655-677.

Smith R (1998). The use of Asphalt in Dam Maintenance Work, Dams & Reservoirs, Vol. 8 No. 3, Dec, 25-31.

Szling Z and Szymanski A (1991). The origin and damage to sealing shields. Transactions of the 17th ICOLD Vienna Q65 R6.

Tedd P, Price G, Evans J D & Wilson A C (1991), Use of the BRE electrolevel system to measure deflections of the upstream asphaltic membrane of Roadford Dam. Proceedings 3rd International Symposium on Field Measurement in Geomechanics, Oslo.

Tedd P, Charles J A, Evans J D & Macdonald A (1995). The use of electro-levels to monitor dam deformations. Proc. of symposium on Research and Development in the field of Dams. Crans-Montana, Switzerland. pp883-890.

Wilson A C & Robertshaw A C (1998). Winscar Dam: Investigations and Repairs to Asphaltic Concrete Membrane, The prospect for reservoirs in the 21st century. Proceedings of the 10th British Dam Society, Bangor, pp292-302.Thomas Telford, London

Wilson A C & Evans J D (1990).The use of low grade rockfill at Roadford Dam, Proceedings of the 6th British Dam Society, The Embankment Dam, Nottingham, pp21-27.

Table 1. Asphaltic membrane dams in the UK and Eire

| Dam | Date built, height (m) | Construction details | Repairs to membrane |
|---|---|---|---|
| Dungonnell | 1968-70<br>17 | Double seal coat<br>2 x 50mm thick dense asphaltic concrete<br>125mm porous bituminous macadam<br>75mm asphaltic concrete underseal<br>Surface dressing | None known |
| Turlough Hill | 1969-1973<br>34 | Mastic seal coat<br>60mm dense asphaltic concrete<br>30-60mm binder layer<br>Spray coat<br>Drainage layer | 1975: Repair of blisters due to trapped moisture<br>1993: Top sealing coat repaired<br>1997 – 2000: Repair of cracks |
| Winscar | 1972-75<br>53 | Double mastic seal coat<br>80mm dense asphaltic concrete<br>40mm dense asphaltic concrete<br>60mm asphaltic concrete layer<br>Bitumen emulsion spray<br>Fine granular regulating layer | 1980: Repair of cracks over culvert<br>1996: Joint seal with parapet and cracks in membrane<br>2001: Repair with geocomposite lining |
| Marchlyn | 1976-79<br>72 | Seal coat<br>80mm dense asphaltic concrete<br>60 mm asphaltic concrete binder<br>Cationic bitumen emulsion spray<br>Crushed rockfill 100mm max size | 1991: Joints repaired and new sealing coat applied |
| Sulby | 1979-82<br>25 (2nd stage)<br>60 (total) | Hot mastic seal coat<br>80mm dense asphaltic concrete<br>60mm asphaltic concrete binder<br>Cationic bitumen emulsion spray<br>Drainage layer | None known |
| Colliford | 1981-1984<br>29 | Two coat cold emulsion<br>80mm dense asphaltic concrete<br>70mm asphaltic concrete binder<br>Bitumen emulsion spray<br>Granular drainage layer | 1987: Pimpled zone near top water level repaired<br>1993: Joint seal with parapet<br>1997: Repair to seal coat |
| Roadford | 1987-89<br>41 | Hot mastic seal coat<br>80mm dense asphaltic concrete<br>70mm asphaltic concrete binder<br>Bitumen emulsion spray<br>Granular drainage layer | None |

# Breaclaich Dam – upstream face joint bandage sealant and wavewall refurbishment works.

K J DEMPSTER, Scottish and Southern Energy plc, UK.
N LANNEN, Scottish and Southern Energy plc, UK

SYNOPSIS. Breaclaich is a 26 m high, 434 m long concrete faced rockfill dam. Areas of the concrete panel joints were progressively deteriorating and leakage monitoring confirmed a clear upward trend with leakage on occasions exceeding the previously determined maximums established under the Reservoirs Act. The wave wall was noted in the original form of record as being of indifferent appearance and alignment, both of which had continued to deteriorate with time to an unacceptable level. Recent design calculations demonstrated that the wavewall would be unstable under design wave surcharge conditions.

The paper describes the investigation and works carried out on the adopted bandage sealant system that was applied to all of the upstream face joints to reduce leakage and the augmentation works carried out to the wavewall. Other completed refurbishment work carried out on the guard and control scour valves and to the tunnel emergency gate, following an uncontrolled closure are also described.

## INTRODUCTION

The works covered by the project at Breaclaich Dam were required to address issues which had been raised during Supervising Engineer Reservoir Inspections carried out under the Reservoirs Act 1975 (1) and which were investigated further by Mott MacDonald prior to the 10 yearly statutory inspection of the reservoir carried out in June 2000 by Mr John Cowie. This statutory inspection confirmed and emphasised the concerns over the dam, with matters raised in the interests of safety to (1) Reduce the leakage through the joints in the concrete upstream face by the application of an approved sealing system and (2) Repair or replace the wave wall to ensure an adequate factor of safety against overturning under design conditions.

## DESCRIPTION OF RESERVOIR AND DAM

Breaclaich Reservoir is situated approximately 2.5 km east of Killin and was completed in 1960. The reservoir is a headpond reservoir storing water for power generation at Lednock, St Fillans and Dalchonzie power stations before eventually discharging to the River Earn a short distance upstream of Comrie in Perthshire.

*Reservoirs in a Changing World*, Thomas Telford, London, 2002, 456–468

The reservoir is formed by the construction of Breaclaich Dam across the Allt Breaclaich raising the water level of the original loch by 19.8 m. The dam is 434 m long at crest level and is of rockfill construction with an upstream face of reinforced concrete panels that provide the impermeable membrane to the dam and was completed in 1960. The dam comprises of a central section and two wing sections at a slight angle to the central section. A concrete toe beam formed in a trench up to 3 m deep supports the upstream concrete facing. A precast concrete wave wall provides protection against waves overtopping the dam. The reservoir is at an altitude of 442.88 m Ordnance Datum (OD) at top water level and has an impounded capacity of 6,750,000 $m^3$. A typical cross section through the dam is shown as Figure 1 and general layout plan of the dam is shown as Figure 2.

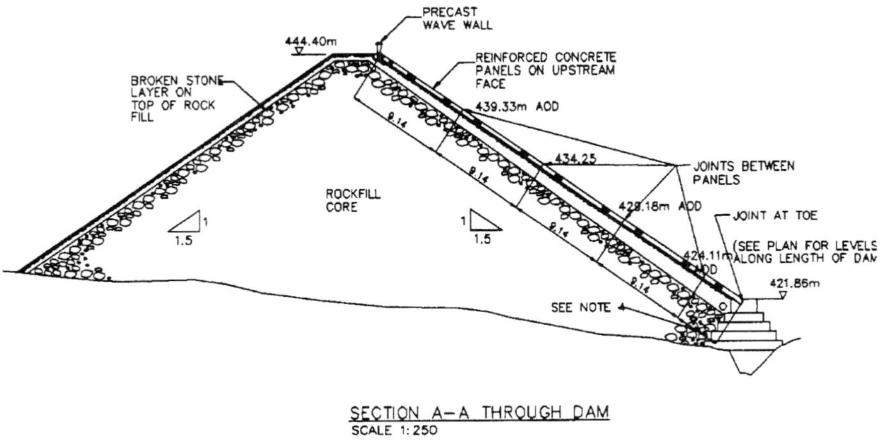


Figure 1. Typical Cross section through Breaclaich Dam

Spill facility from the dam is provided by a 15.24 m long finger spillway at the right hand end which discharges to a spillway channel around the flank of the dam before joining the original watercourse downstream of the dam.

The draw off works are located in the lower section of a twin concrete culvert running from the downstream toe of the dam to a point close to the upstream face. The draw off consists of a 915 mm scour pipe with a screened bell mouth entry that is controlled by means of an upstream guard gate and a needle dispersal valve discharging into a culvert running through the downstream rockfill embankment.

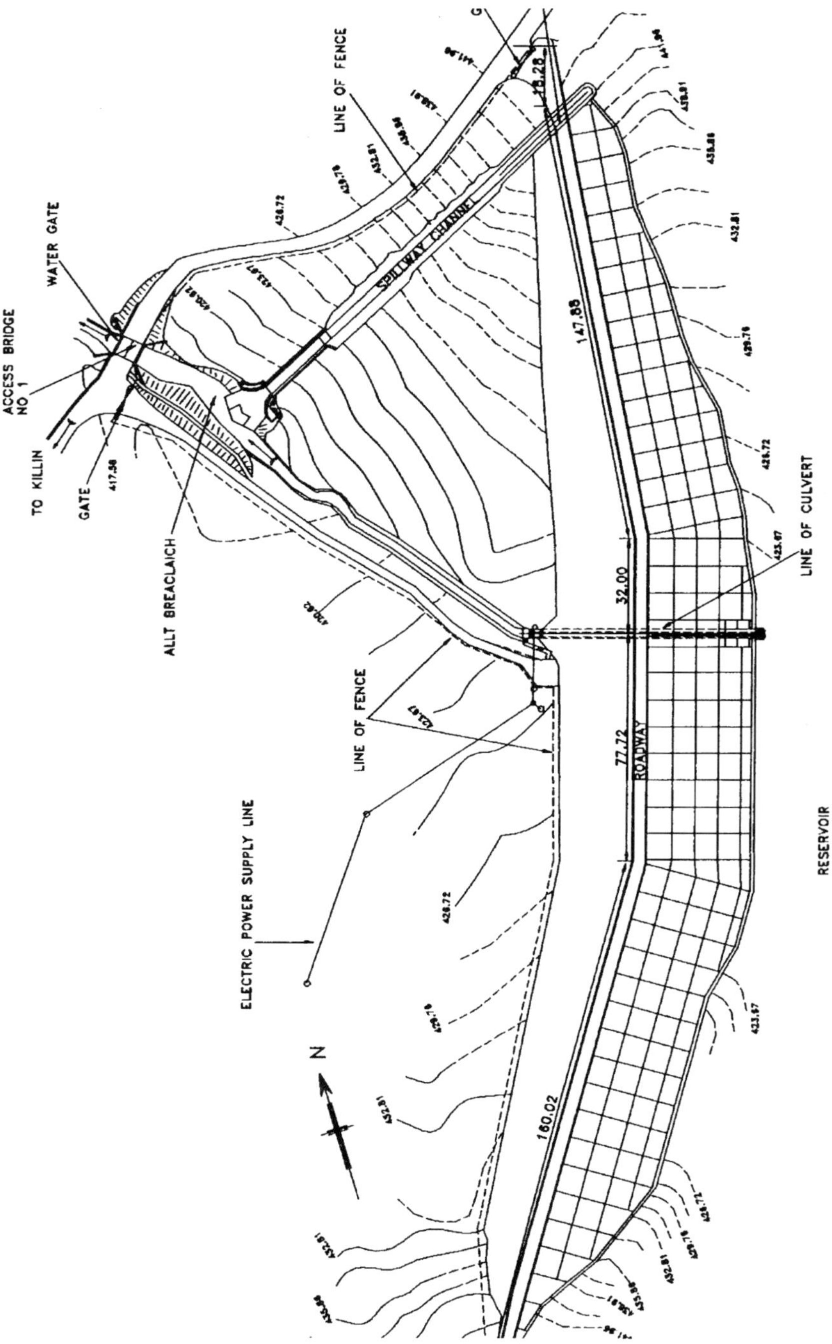


Figure 2. General Layout Plan of Breaclaich

An intake tower located at the western edge of the reservoir transfers water from Breaclaich through a 2.1m diameter concrete lined tunnel to the 3 megawatt (MW) Lednock Power Station.

## INVESTIGATION

### Dam Upstream Impermeable Membrane and Leakage

The upstream face of the dam is covered with reinforced concrete panels to provide an impermeable membrane. The slabs are generally 9.1 m square except in the bottom sections where they are shaped to fit the toe level. They increase in thickness from 380 mm at the top to 450 mm at the base of the dam and are reinforced. The panels are supported on a layer of broken stone with the joints located over concrete ribs cast level with the stone. The joints between panels have a 12.5 mm gap filled with compressible filler in the lower parts and a wedge shaped chase 125 mm deep by 50 mm wide, filled with a hot poured bitumen seal in the top face. A copper water bar with central joggle extends into the adjacent panels at the bottom of the chase formed from 16 gauge annealed sheet.

Leakage through the upstream face of the dam is collected in 225 mm and 300 mm diameter porous pipes, which run behind the toe wall. The leakage is measured at two V notch boxes that are located at the upstream end of the scour culvert at the outlets from the toe drains. These drains collect leakage from the north and south sides respectively as well as rainfall percolating through the dam from the crest and downstream face, the run-off from a small area local to the upper parts of the dam site and also any deep seated leakage through the grout curtain.

In the original "Description of Works" for the dam a leakage limit of 100,000 gallons per day (18,942 litres per hour) was set. This was further emphasised in the 1990 Inspecting Engineers report as a measure to be taken in the interests of safety and the need to refer any increase in leakage above this level to a Qualified Civil Engineer within the meaning of the Act.

A review of the records relating to the dam revealed leakage problems since the dam's initial construction and more latterly repairs to specific sections of the dam. The main problems and the repairs that have been undertaken are as follows: -

> *1960-61*: Sharp increase in leakage caused by cracked concrete panels around the scour culvert due to settlement and movement at joints exceeding design limits. The cracked sections were cut out and replaced with additional joints incorporated.
> *1961-62*: During re-impounding further leakage occurred at the copper water bar of the repaired joints. The water bar was found to have been pulled out and were repaired by concreting small sections at a time with a small steel plate covering the joint
> *1961-62*: A fine crack developed all along the upstream toe caused by settlement of the rockfill beneath the panels and consequent lack of support. The intended hinging action of the concrete panels did

not occur due to the space behind the toe joint being filled with concrete rather than soft bitumen and compounded by casting the bottom end of the panel support ribs directly on the toe concrete. The cracks were not found to be leaking but were repaired by cutting a chase along the line of the crack, filling with bituminous sealant and capping with a fillet of waterproof mortar.
*1987*: Repairs to some expansion joints and the upstream face with replacement bituminous sealant and waterproofing
*1989*: Significant increase in leakage, a length of bitumen sealant at slab S1\S2 was replaced and shutter bolt holes were also sealed.
*1993*: Repairs to joint sealants and cracking in the panels were carried out, as required by the 1990 Inspecting Engineer, using cold applied bituminous filler in the joints and a polymer modified cementitious material in the cracks.

A review of the leakage data for the reservoir showed that on occasions the total recorded leakage did exceed the specified limit. Figure 3 shows the general trend from 1985 to 1998 and shows that the mean annual leakage rate does not exceed the specified limit but that leakage rates increased rapidly from 1994 to 1996. From 1996 to 1998 the leakage rate continued to rise, through at a much lower rate. If leakage continued to increase at the same rate of 3% per year then the mean annual flow rate will exceed the limit before 2004. The rate of increase is not steady and the risk of exceeding the limit before 2004 could not be discounted.

Further analysis of the water level and leakage records confirmed: -

- All trend lines indicated that the rate of leakage from both north and south sections increase with reservoir level although at different rates.
- The leakage in the southern section has increased with time whilst the northern is approximately constant over the same period.
- Leakage from the southern section is generally far greater than the northern, which has had more repairs completed.

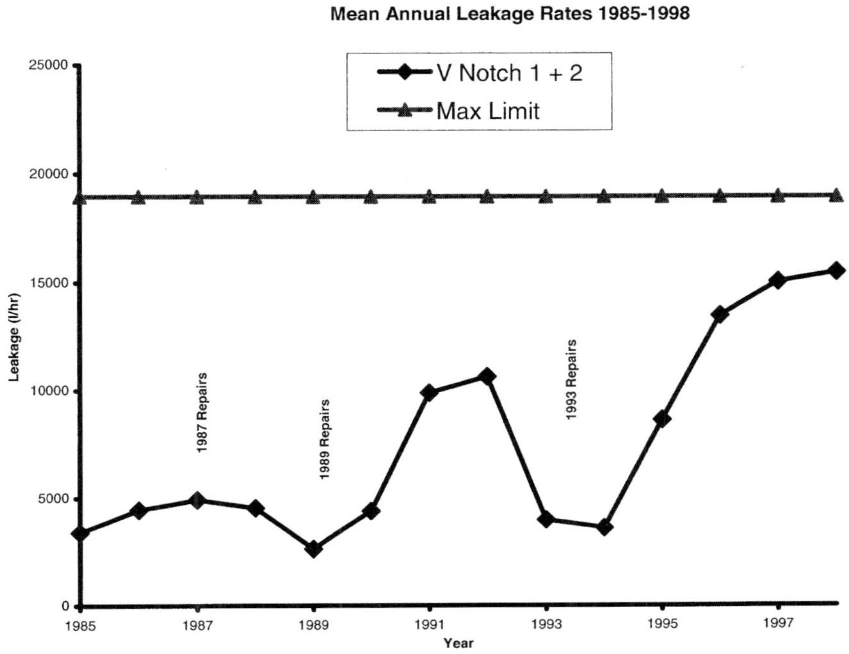


Figure 3. Mean Annual Leakage Rates 1985 to 1998

Wave wall

Precast concrete wave wall units run along the length of the upstream crest protecting the crest from wave overtopping. The units are each 2.43 m long and are independent of one another with no connection or seal between them. The units have a curved upstream face to reflect direct waves and are seated in a recess between the top of the reinforced concrete panels and the broken stone on the crest. The original description of works noted that a cope was added to the top of the wave wall due to the contractors inability to set the units to line and level which itself was noted as of somewhat indifferent appearance. The wave wall was therefore badly aligned and out of level and had continued to deteriorate since its original construction. Deterioration to the majority of units taking the form of delamination of surface mortar, compression spalling at joints, frost damage and inadequate bedding of copes and a number of units tilting towards the upstream face. Whilst still in a condition to perform its function the condition of the wave wall had reached a point where refurbishment was required.

As part of the review carried out prior to the Statutory Inspection flood studies and wave surcharge calculations were carried out in accordance with the Guide to Floods and Reservoir Safety (2) and the Flood Studies Report (3). The worst case wave level was found to be with a 1 in 200 year wind speed combined with still water level at spillway level and a maximum water level of 445.21 mOD calculated, as shown in Figure 4.

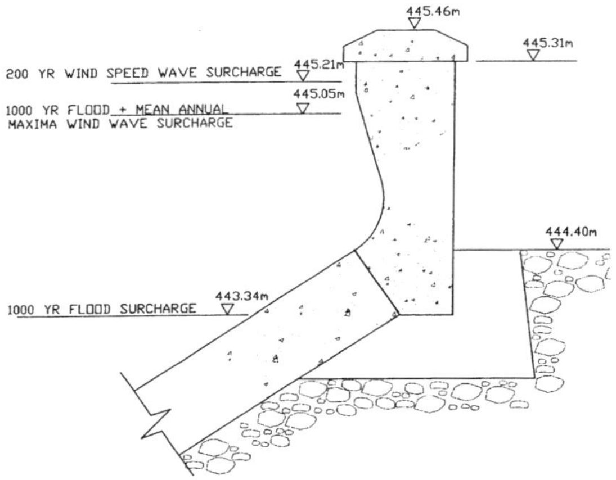


Figure 4. Wave wall flood and wave surcharge

The calculations undertaken demonstrated that the wave wall arrangement was satisfactory with regards physical level for the static and wave levels however the maximum level predicted was significantly higher than the previously adopted combined flood and wave surcharge level of 444.39 mOD. It was therefore considered prudent to have checks on the wave wall stability carried out.

A check was carried out of the existing wall stability using wave impact loading based on BS 6349: Maritime Structures (4). From this, the factor oof safety against overturning was found to be approximately 0.21 compared to a target value of 1.1. It was clearly concluded that the wave wall would be unstable under wave loading and improvement was required

Scour Valves

The guard and control valve on the scour culvert were both in reasonable condition but had never been refurbished nor examined in detail to determine if any problems existed since their installation. Leakage through the gland seal around the operating spindle of the guard valve was noted as increasing and adding to recorded leakage from the toe drainage system.

Tunnel Intake Emergency Gate

The intake emergency gate is 2.13 m by 2.44 m bulkhead gate with the operating winch common to both it and a double fine screen arrangement. Control is via intermeshing drives with dog clutches to select park and drive operations from an upper platform level. During a routine operation the emergency gate dog mesh disengaged because neutral was achieved via the mechanism whilst attempting to re-select the park position after a successful gate operation test. In neutral the drum shaft is free to rotate, resulting in an uncontrolled fall of the gate into the intake shaft. The gate was retrieved and fortunately suffered no obvious structural damage to its main frame but damage was evident to the roller train, guide wheels, seals and secondary elements of the gate frame.

## PURPOSE OF THE PROJECT

A refurbishment strategy was developed with the overall purpose to: -

- Significantly reducing leakage through the concrete slab joints.
- To provide a wave wall to current design standards.
- Refurbish the leaking scour valves.
- Repair and refurbish the intake emergency gate, including provision of fail safe operation.

For operational reasons the works were programmed to coincide with an outage at St Fillans Power Station to replace the machine runner. This work required a 12 week outage during late summer 2000 and in order that minimal water was lost both Lednock and Breaclaich reservoirs would require to be drawn down. This provided an optimum period to carry out work on the upstream face and implement the projects simultaneously.

## OPTION REVIEW

### Dam Upstream Impermeable Membrane and Leakage Options

Providing a membrane over the whole of the upstream face was considered to be prohibitively expensive and due to the generally good condition of the concrete panels would not reduce the leakage much more than joint repairs. Two options to repair the upstream face concrete panel joints were considered. Either replacement of the existing bituminous joint sealant or a bandage sealant over the existing joint were considered in relation to reduction in leakage, avoidance of damage to the existing copper waterbar, long term durability and cost effectiveness. Each of which were assessed and are summarised in Table 1.

Table 1 Summary of upstream face options

| Impact | Importance | Replace sealant | Bandage sealant |
|---|---|---|---|
| Reduce leakage | Essential | Medium | High |
| Avoid waterbar | High | Medium | Best |
| Durability | High | Medium | Best |
| Installation | Medium | Medium | Best |
| Programme | High | Medium | Best |
| Cost factor | High | 1.15 | 1 |

Both options would achieve the primary objectives although a bandage repair was considered to offer a better long term solution, in that it provides a more watertight membrane over the surface of the joint, as opposed to the replacement sealant. This combined with some existing doubts regarding the integrity of the copper waterbar over its full length and the need to rake out the bitumen and expose the waterbar to possible damage in the joint replacement option, lead to the selection of a bandage sealant option. Figure 5 below details the typical joint repair detail.

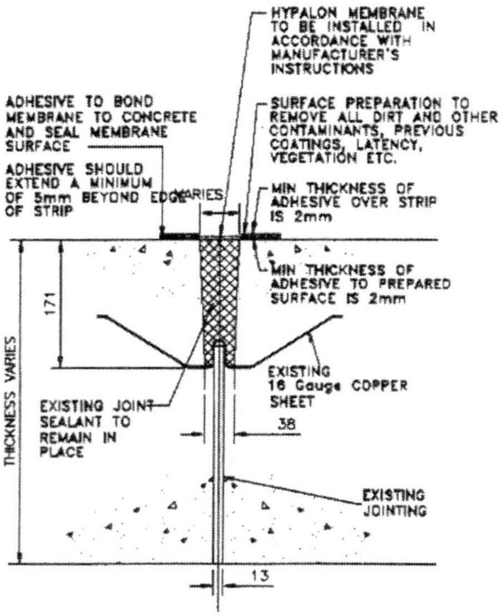


Figure 5. Typical joint repair detail

<u>Wave wall Options</u>

Four options to repair/replace the existing wave wall were considered, namely: -

1. Anchor wall to the upstream slabs and repair the face.
2. Concrete over clad the existing wall.
3. A foundation extension and repair the face.
4. Replacement wave wall.

The options were each considered in relation to stability of the wave wall, aesthetic appearance, long term durability and cost effectiveness. Each of which were assessed against the anticipated degree of their relative importance and are summarised in Table 2 below.

Table 2. Summary of wave wall options

| Impact | Importance | Option 1 | Option 2 | Option 3 | Option 4 |
|---|---|---|---|---|---|
| Stability | Essential | Achieved | Achieved | Achieved | Achieved |
| Durability | High | Worst | Medium | Worst | Best |
| Alignment | High | Achieved | Achieved | Achieved | Achieved |
| Appearance | High | Worst | Medium | Worst | Best |
| Maintain crest width | Medium | Achieved | Not achieved | Not achieved | Not achieved |
| Avoid face panel work | Medium | Not achieved | Achieved | Achieved | Achieved |
| Avoid fill excavation | Medium | Achieved | Not achieved | Not achieved | Not achieved |
| Programme | Medium | Best | Medium | Medium | Worst |
| Cost factor | High | 1 | 1.05 | 1.21 | 1.83 |

Replacing the wave wall would provide the best solution in terms of durability and appearance but would be significantly more expensive than remedial works to the existing wall and could not be justified in economic terms. Stabilisation and repair of the wave wall by anchoring was identified as the cheapest option but provided poor long term durability and would necessitate extensive localised repairs and mortar cladding the wall. Concrete over cladding the existing wave wall provided a robust and durable solution to the problem whilst maintaining cost effectiveness and was determined to be the best option. Figure 6 below details the over clad to the wave wall that was adopted.

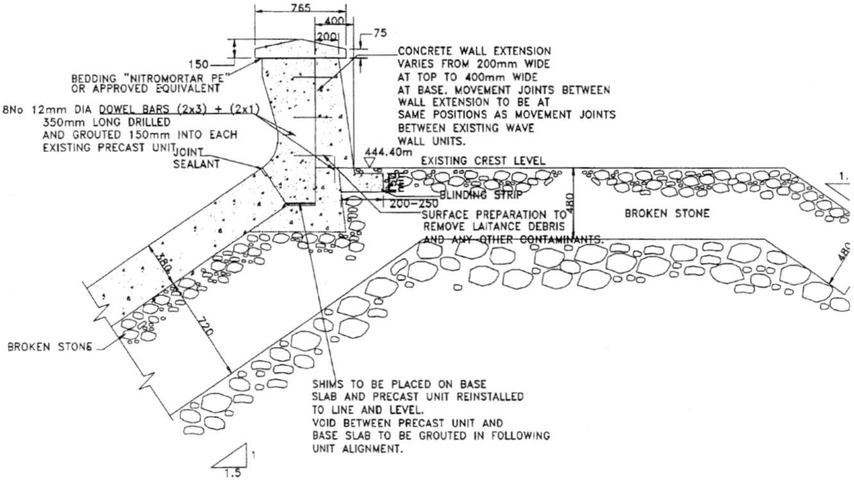


Figure 6. Wavewall over clad details

Scour Valve Refurbishment

The works planned on the scour guard and needle valves was relatively routine and included stripping down into their component parts, dressing all cogs, bearings, gears and teeth. The sealing faces were polished to take out minor irregularities following NDT inspection. The valve bodies were blasted primed and repainted. A new "rotork" actuator was installed on the neddle valve. With the installation of a temporary bulkhead, the opportunity was also taken to blast and repaint the internal sections of the scour pipeline upstream of the valves. Total refurbishment costs were £70,000.

Tunnel Emergency Gate Repair

Following retrieval of the gate from the gate shaft its condition was assessed and found to be remarkably good, considering its freefall. The main concern was to avoid repetition of the incident and the design of the winch and drives was considered in detail. Rather than relying on a single winch a second winch was installed and linked directly to one of the fine screens, the second screen was removed. The existing winch and drive mechanism were refurbished and permanently linked to the gate, thus avoiding the need to change drives and greatly simplifying the operation of the gate. Work to the gate itself included replacement main pulley wheels, main roller coasters including bolts, split pins, new stainless steel spindles and deva bearings to free running roller wheels, new gate seals and new starter and control panels. Refurbishment and enhancement works costs were £110,000.

## SITE WORKS

Dewatering of the reservoir was commenced in March 2000 to allow maximum utilisation of the stored water for power generation prior to opening the scour valves over the final 4 m. The target draw down level to below the toe beam was achieved by August 2000 for work to commence as planned. The upstream bandage works and wave wall were generally progressed in parallel with sequencing of works to avoid wave wall construction above men on the downstream face. Although during wet weather efforts tended to concentrate on the wave wall due to access and bandage application conditions being less favourable, this allowed some flexibility to resource deployment during such periods. Due to the remote location all concrete was batched on site, as were precast coping units, using an air entrained mix. The works were completed in November 2000 and impounding started immediately. The total civil costs were £315,000.

## POST CONSTRUCTION LEAKAGE MONITORING

Following completion of the bandage sealant works, reservoir filling and subsequent operational period leakage monitoring has demonstrated an overall reduction in leakage of approximately 60%. Figure 7 details the leakage monitoring results to July 2001.

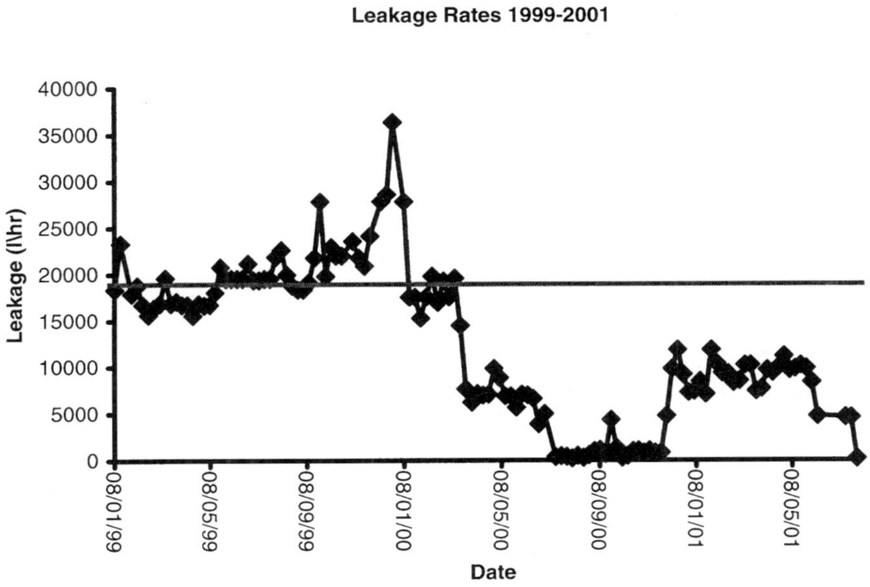


Figure 7. Leakage rates 1999 – 2001

Whilst there may still be an element of leakage through the upstream concrete panels and their joints the majority of leakage is now considered to be from rainfall percolating through the dam from the crest and downstream face, the run-off from a small area local to the upper parts of the dam site and also perhaps some deep seated leakage through the grout curtain.

## CONCLUSIONS

The project followed proactive investigations prior to the planned Statutory Inspection and allowed refurbishment works to be implemented quickly and to the satisfaction of the AR Panel Engineer.

Deficiencies were addressed at the reservoir by adopting a holistic approach and not piecemeal repairs as were done in the past. The leakage has been significantly reduced and the wave wall now meets current design standards as well as a dramatic improvement in its long term durability.

The scour valves were refurbished and should require little or no major works for a further 15-20 years.

The tunnel emergency gate operational safety has been significantly improved, the risk of operational failure reduced and it has been returned to service.

## REFERENCES


(1) Reservoirs Act, 1975, HMSO, London 1975

(2) The Institution of Civil Engineers, (1996), Floods and Reservoir Safety Third Edition, Thomas Telford, London.

(3) Natural Environmental Research Council, (1975) Flood Studies Report, Natural Environmental research Council, London.

(4) BS 6349: Maritime Structures

# Safety and risk

# Tailings dam incidents and new methods

A D M PENMAN, Geotechnical Engineering Consultant, Harpenden,
Chairman ICOLD Committee on Tailings Dams and Waste Lagoons.

SYNOPSIS. There is sufficient geotechnical engineering knowledge today to enable safe tailings dams to be constructed, yet they have been failing at an average rate of 1.7 per year for the last 30 years. Some examples are described and discussed and attention drawn to Bulletin No. 121 "Tailings dams: Risk of dangerous occurrences. *Lessons learnt from practical experience,*" published by the International Commission on Large Dams, which gives 221 examples of incidents and failures.

## TAILINGS DAMS.

Rich deposits of metals in seams and veins have long since been worked out and we have to get our metals now from their hiding places, distributed within the mass of, in many cases, good sound strong rock at concentrations lower than half a percent by weight. In the Andes, hard rock is quarried in the traditional ways of drilling vertical holes behind a quarry face to take explosives that blow off the next section of rock. The resulting fragmented rock is carted to a processing plant where it is crushed, ground and milled down to sand and silt sizes, then passed through a wet process to extract copper and the remaining slurry of sand and silt is discharged from the tail end of the plant as waste tailings. In Ireland, underground mining is used to obtain rock containing zinc, which is put through similar process of crushing and grinding. Because of the small content of metals, there are large amounts of waste tailings that have to be put somewhere. The water content is usually high enough for the slurry to be made to flow though pipelines so that it can be carried far away from the processing plant to a site that may be out of sight and out of mind. Any moneys spent on this waste material reduces the profit from the mine, so every effort is made to keep the cost of disposing of this material to a minimum. An indication of the volumes involved may be gleaned from a report in the Mining Journal (September 2000) of the development on the Indonesian island Sumbawa of the Batu Hijau mine. The planned production is for an average of 600 000 tonnes/day for the 15 year life of the mine, with the aim of winning 830 tonnes of copper and 0.06 tonnes of gold per day, leaving at least 599 000 tonnes a day of waste fine rock for disposal.

It is usual to store this waste material in large impoundments retained by dams. This fine derivative from a sound rock can form an excellent construction material, but if it were to be placed in layers and compacted to form an

*Reservoirs in a Changing World*, Thomas Telford, London, 2002, 471–483

embankment, its water content would have to be seriously reduced. In the early days when crushing and milling was less efficient, tailings often contained a range of particle sizes from coarse sand of 0.5 to 0.1 mm to fine silt of 0.06 to 0.006 mm. It was clearly of advantage to use the coarsest sizes for dam construction. Separation of sizes for one of the first methods of tailings dam construction, referred to as the upstream method, was effected by use of a beach. Tailings discharged from the dam crest flowed slowly down a fairly flat beach, allowing the coarsest sizes particles to settle out first, with ever finer material being deposited further down the beach leaving only the finest particles to be carried into a pond, where they settled out, and clear water was decanted, often down a vertical pipe tower for discharge though a culvert under the dam. The overflow into the pipe tower was raised as the level of the tailings in the impoundment rose, by adding sections of pipe or stoplogs depending on the type of tower and the crest of the dam was slowly raised to keep ahead of the impoundment level and moved upstream to maintain a suitable downstream slope for the dam. Drainage and drying produced some suction in the pore water between the particles near the surface, tending to pull them together, but this was the only form of compaction commonly available. Tailings that had to be pumped through the delivery pipeline could come out with some velocity. This was turned to advantage by the development of simple hydro-cyclones to separate out the larger size particles. The cyclone consists of a steel cylinder with funnel shaped base and an entrance pipe coming in tangentially near the upper end of the cylinder. The rate of flow of the tailings causes it to rotate within the cylinder, throwing the larger sizes to the outside, leaving the finest material in the middle. A vertical pipe dips down from the upper end plate and the fine tailings flows through it to be discharged well into the impoundment. The coarse fraction falling from the funnel is used for dam construction, and in some cases can be machine compacted. Adjustable nozzles on the end of the funnel can be used to control the sizes of the particles included in this coarse fraction, but the volumes that can be used depend on the proportion of coarse particles in the tailings as delivered from the processing plant, and the volume required for dam construction has to be sufficient to keep it high enough to retain the impounded tailings. Even without a sufficiency of coarser material, the cyclone has the effect of reducing the water content of the tailings.

Terzaghi described the principle of effective stresses in English for the first time in 1936, although he gave it in German in 1925. Geotechnical engineers are now very well aware of this principle that controls the strength of particulate materials and so are able to design tailings dams that can permanently retain these impoundments of waste tailings. A major problem is the control of water and the water content of the tailings forming the dams. The climate of the area for the impoundment plays a vital role. It is much easier to construct stable dams in regions of low rainfall and high evaporation as found in parts of South Africa, Australia and Central America, than where there is the reverse of low evaporation, high rainfalls and the possibility of frost.

Because of its greater density, tailings can cause much greater damage than water. The uncontrolled release of water from a reservoir due to dam failure can cause damaging floods, but it can flow through and around buildings without destroying them and persons may be rescued before they drown. But the release of liquefied tailings from an impoundment can destroy buildings and crops and cause the deaths of persons. The cost to a mining company of the failure of one of their tailings dams, in lost production, repair or construction of another means for tailings storage, repair to third party damage, compensation for loss of crops, livelihood and even life itself, is so great that it would not be expected for any mining company to take the risk. Engineering knowledge today is quite adequate to enable of the safe design, construction and maintenance of tailings dams. Yet throughout the world, they have failed at an average rate of 1.7 per year for the past 30 years. In many cases failure has been due to silly mistakes: a lack of full attention to detail. An exception may be due to the violent forces caused by earthquake, but even in highly seismic regions, types of construction and special provisions can be made to minimise the risk of major damage.

EXAMPLES OF FAILURES.

There were many impoundments retained by dams built by the upstream method associated with the copper mines in Chile. That country is subject to earthquakes and failures were not uncommon. A well known example is that of the El Cobre old dam, built to a height of 33 m between 1930 and 1963, with a downstream slope of 1 on 1.2 to 1.4. Two years after its construction stopped, the area was struck by the 1965 La Ligua earthquake, which occurred during daylight. Eye witnesses said dust clouds came up from the dam, obstructing it from view as it failed, releasing the liquefied tailings to flow down the valley, engulfing the miners' village and continuing for a further 5 km. Many lives were lost. This failure and others in Chile, have been described by Dobry and Alvarez (1967).

The release of dust is typical of the failure of dry loess slopes and is caused by the volume reduction on shearing that occurs in loose particulate materials. Air from the voids is expelled, carrying dust with it. The downstream face of the dam had clearly been relatively dry before it disintegrated allowing the release of the unconsolidated tailings slurry. Because of the potential volume reduction, what little effective stress existed within the mass of the saturated tailings would be entirely transferred to the pore water, causing the material to loses all shear strength and becomes a liquid.

In Japan, the Mochikoshi impoundment was being built in a hollow near the top of a hill to store gold mine tailings and was retained by three tailings dams. These were being built by the upstream method from very sound starter dams made from the local volcanic soil. The dams were raised by building dykes from the volcanic soil placed on the beach and compacted. The impoundment was subjected to a ground acceleration of 0.25g from the Iso-Oshima earthquake of magnitude 7.0 that occurred on 14th January 1978. The highest of the three dams failed during the main shock, releasing 80,000 $m^3$ of tailings contaminated by sodium cyanide, through a breach 73 m wide and 14 m deep. The tailings flowed 30 km into the Pacific Ocean. The second highest dam failed next day, 5 hrs. 20 mins. after an aftershock, releasing a further two to three thousand cubic metres of tailings through a breach 55 m wide and 12 m deep. These and other earthquake related failures led to recommendations that the downstream method of construction should be used in earthquake areas, rather than the upstream method. In this method, coarse material, possibly from cyclones, is placed on the downstream part of the dam where effective drainage measures can be employed and the fill can be compacted. Alternatively the dam can be built from borrowed fill, as with water retaining dams.

Stava. At 12.23 on 19th July 1985 two tailings dams, one above the other and both built by the upstream method, collapsed. A total quantity of 190,000 $m^3$ of tailings slurry was released and flowed, initially at a speed of 30 km/hr down into the narrow, steep sided valley of the Rio Stava, demolishing much of the nearby small village of Stava and continued, at increasing speed, estimated to have been 60 km/hr to another small town, Tesero about 4 km downstream, at its junction with the Avisio River in northern Italy. The only surviving eye-witness, a holiday maker, had the horrifying experience of watching the disaster from the hillside and saw the hotel where his family were taking lunch being swept away by the torrent of tailings. This failure caused 269 deaths.

The tailings dams, as indicated by Fig. 1, were for a fluoride mine that was begun in 1962 and were sited on a mountain side slope of 1 on 8. The decant was in the form of a concrete culvert laid up the sloping floor, with coverable openings about every 0.5 m vertical rise. Water from the pond decanted into the openings, which were covered, one by one, as the level of the tailings rose. The lower dam was built by the upstream method to a slope of 1 on 1.23. When it reached a height of 19 m, the second dam was begun at the upstream end of the impoundment and built to a slope of 1 on 1.43. When it reached a height of 19 m, further planning consent was required. This was given on the condition that a 5 m wide berm was constructed at that level and permission given for the dam to be built to a height of 35 m. Construction continued at the same slope of 1 on 1.43 and the failure occurred when it was 29 m high. The cause is thought to be due to a combination of blockage and leakage from the culvert under the toe of the upper dam, thereby raising the phreatic surface sufficiently to cause a rotational slip, as indicated by Fig. 1.

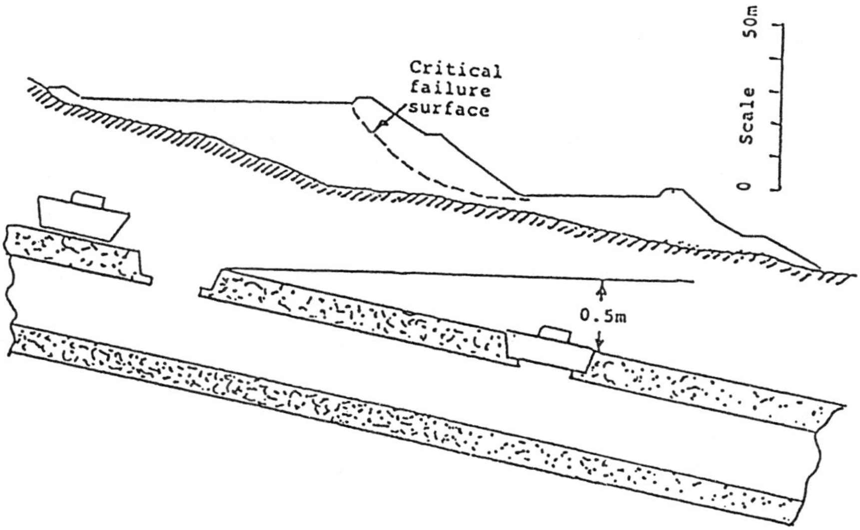


Fig. 1. Stava tailings dams. Detail shows concrete culvert with coverable openings for decanting.

Six months before the failure, a local slide occurred in the lower portion of the upper dam on its right side, in the area where the decant pipes pass underneath the dam, due to freezing of the service pipe during a period of intense frost, according to Berti et al (1988). For the next three months water was observed seeping from the area of the slide. A month before the failure, the decant pipe underneath the lower impoundment fractured allowing the free water and liquid mud from the pond to escape towards the Stava river, creating a crater above the point of fracture. A bypass pipe had to be installed through the top portion of the lower dam, and the broken decant pipe blocked to restore use of the system. During this operation the water level in the upper impoundment was lowered as far as possible, then just four days before the failure, both ponds were filled and put back into normal operation. 53 minutes before the failure a power line crossing below the impoundments failed, then only 8 minutes before, a second power line failed. The tailings from the failure reached Tesero about 4 km distance, within a period of 5 to 6 minutes. As a result of this failure, the strict Italian law governing the design and construction of water retaining dams, according to Capuzzo (1990), is being extended to include tailings dams.

Merriespruit. The Virginia No 15 tailings dam had been built by the 'paddock' method that is used extensively in South Africa's gold mining industry. It was a long dam encircling and retaining an impoundment of 154 ha holding 260x106 $m^3$ of gold mine tailings containing cyanide and iron pyrite. The foundation soil was clay and drainage was required under the dams. General experience was that drains were often blocked by iron oxides and other residue. The impoundment formed one of several similar impoundments of the Harmony Gold Mine near Virginia in the Orange Free State. The suburb of Merriespruit containing about 250 houses had been built near the mine in 1956. Virginia No 15 lagoon was begun in 1974 and a straight northern section of the dam nearest the suburb was placed only 300 m from the nearest houses. Dam construction and filling of the lagoon continued until March 1993, when the section of the dam closest to the houses was 31 m high.

The summer of 1993/4 in the Orange Free State had been particularly wet and on the night of Tuesday 22nd February 1994 there were violent thunderstorms over Virginia and a cloudburst when 40 mm of rain fell in a very short time. The water level in the lagoon rose due to direct catchment: there was no stream or other natural source of water that came into the lagoon, which while operational, had a launder system that removed the transportation water decanted from the tailings slurry that had been delivered into the impoundment. During the early evening at about 19.00, water was found running down the streets and through gardens and an eye witness saw water going over the crest of the dam above the houses. The mining company and contractor were informed, but when their representatives reached the site it was already dark. One of the contractor's men rushed to the decants and found water lapping the top rings but not flowing into the decants. He removed several rings to try to get the water flowing, but the main pool was next to the north dam crest with no direct connection to the decants. At the same time, another contractor's man was near the downstream toe of the dam, and saw blocks of tailings toppling from a recently constructed buttress that had been built against a weak part of the dam. An attempt was made to raise the alarm, but before anyone had been contacted, there was a loud bang, followed by a wave of liquefied tailings that rushed from the impoundment into the town. Cross sections of the dam during early stages and during failure, given by Blight (1997) are shown by Fig.2.

A breach 50 m wide formed through the dam, releasing 2.5x106 $m^3$ of tailings that flowed for a distance of 1,960 m, covering an area of 520x103 $m^2$. The flow passed through the suburb where the power of the very heavy liquefied tailings demolished everything in its path, houses, walls, street furniture and cars, carrying people and furniture with it. According to newspaper reports, people already in bed at about 21.00 hours when the mudflow struck, found themselves floating in their beds against the ceiling. 400 survivors spent the night in the Virginia Community Hall, a kilometre

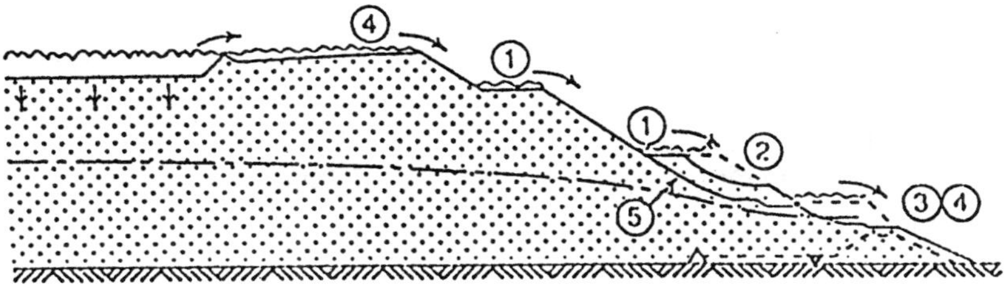


1. Berms overtop after thunderstorm.
2 Loose tailings infill to earlier failures on lower slope erodes.
3. Tailings buttress starts to fail.
4. Pool commences overtopping and erodes slopes and tailings buttress.
5. Unstable lower slope fails and failed material is washed away.

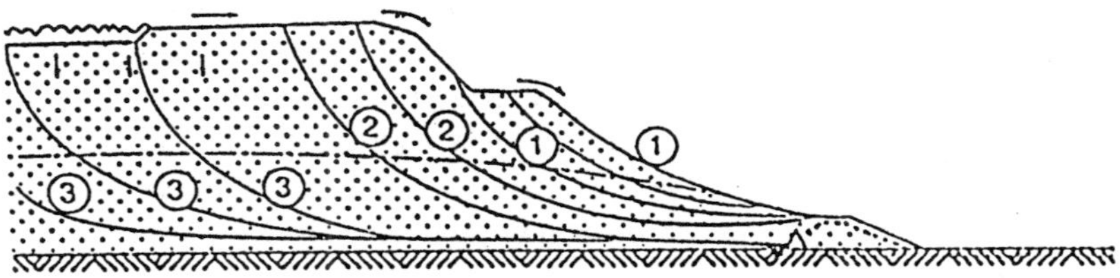


1. Lower slopes fail and are washed away.
2 Domino effect of local slope failures which are washed or flow away.
3. Major slope failures with massive flow of liquid tailings engulfing town.

Fig. 2. Section of Virginia No.15 dam. (a) Critical section of dam during early stages of failure. (b) Critical section of dam during failure. (from Blight 1997).

away. Hetta Williamson said that her husband had gone back in daylight to their former home and found nothing but the foundations. It is remarkable that only 17 people were killed.

Apparently this north section of the enclosing dam had been showing signs of distress for several years, with water seeping and causing sloughing near the toe. A drained buttress constructed from compacted tailings had been built against a 90 m long section, but continued sloughing had caused the mine to stop putting the normal flow of tailings into the impoundment more than a year before the failure, i.e. the impoundment had been closed. At that time the freeboard was, according to the contractor, a respectable 1 m. But sloughing at the toe continued, and construction of the buttress was continued. Not long before the failure, slips had occurred in the lower downstream slope just above the buttress. In fact, although the placement of tailings had been stopped, wastewater containing some tailings continued to be placed and the water overflowed into the two decants. Unfortunately there formed a sufficient deposit of further tailings to cut off the decants and cause the main pool of water to move towards the crest of the north part of the dam, leaving a freeboard of only 0.3 m, and water was still being pumped into the impoundment from the mill on the night of the failure. Evidence of what had been going on since supposed closure was provided by satellite photography. A Landsat satellite passed over the area every 16 days and the infrared images revealed the positions of the tailings and the water pool.

Government Mining Regulations that had come into force in 1976, required a minimum freeboard of 0.5 m to be maintained at all times for this type of impoundment, to enable a 1 in 100 year rainstorm to be safely accommodated without causing overtopping. Evidence of the level of tailings in the Virginia No 15 lagoon showed that the tailings had been brought up to within 15 cms of crest level prior to abandoning this storage in March 1993. Had the government regulations required inspection of the dam, particularly at closure, the very small freeboard would have been noticed and a further raising of the dam crest enforced to prevent overtopping in the event of a maximum probable precipitation. The more modern approach to freeboard is to consider free volume, which should be sufficient to accept a 1 in 100 year rainstorm and, according to the shape developed in the upper part of the impoundment, may require a much larger vertical freeboard below the lowest point of the dam crest.

A failure at Baia Mare. The expanding city of Baia Mare in Romania was beginning to encroach on old mining areas where there were disused impoundments of tailings. Removal of these impoundments and their retaining tailings dams would both release valuable land for city development and allow extraction of remaining metals from the old tailings. The scheme at Baia Mare involved construction of a new impoundment and a new efficient processing plant that would accept tailings removed from the old impoundments. Initially three were to be reworked and pipelines were laid out to transmit water from the new impoundment to be used for powerful jets that would cut into the old tailings, producing a slurry that would go to the new processing plant for extraction of remaining metals, with the tailings from it flowing to the new impoundment. The system used the same water going round and round with no interference with the environment.

The site for the new impoundment, well away from the city, was on almost level ground, with its main axis 1.5 km long, sloping down only 7 m from NE to SW with a width of about 0.6 km, as indicated by Fig.3. An outer perimeter bank 2 m high with 1 on 2 side slopes, as shown by Fig.4, was built from old tailings, and the whole area of about 90 ha, lined by HDP sheet, anchored into the crest of the perimeter bank. Drainage was installed to collect any seepage, that would be pumped back so that there should be no escape of contaminated water into the environment. About 10 m inside the perimeter, starter dams were built, also with 1 on 2 side slopes, to heights of about 5.5 m along the

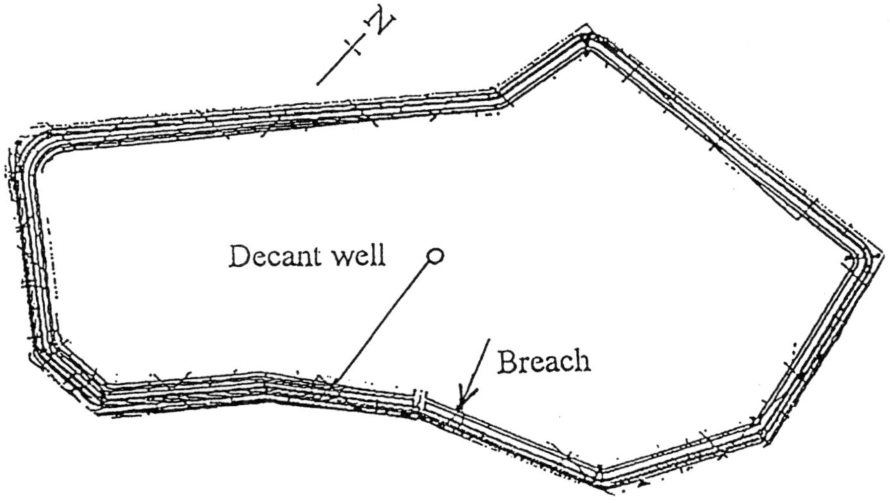


Fig.3. Baia Mare. Plan of new impoundment showing position of eventual overtopping and breach.

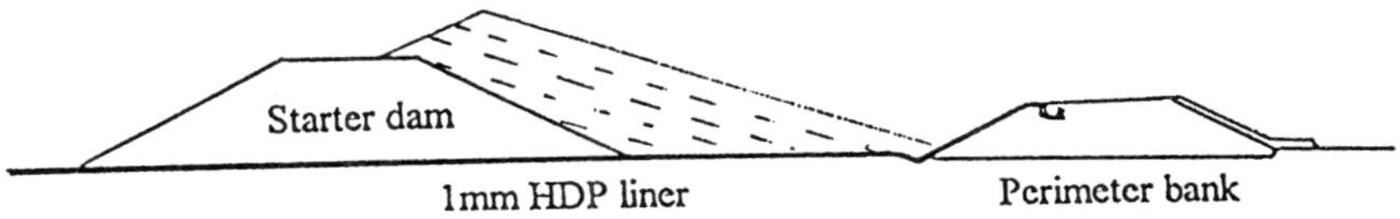


Fig.4. Section of perimeter bank and starter dam showing position of the HDP sheet liner.

SW lower edge of the impoundment, tapering down to 2 m height about half way along the sides, with the remainder around the NE end of the impoundment, about 2 m high. Cyclones mounted along the crest of the SW starter dam and part way along the side starter dams accepted the tailings piped from the new processing plant, discharging the coarser fraction on to the downstream side to fill the space to the parameter dam, and raise the whole dam, with the main volume of fine tailings slurry being discharged into the impoundment. Collected water was discharged into the central decant, drained through a 450 mm diameter outfall pipe embedded under the HDP liner and pumped back to operate the monitoring jets in the first of the old impoundments, 6½ km away, and close to the city. Cyanide was used in the new processing plant for the extraction of gold, so that the tailings and water in the new impoundment contained considerable amounts of cyanide. No water should leak from the pipe work circuit, although the water used in the cutting jets flowed over the unlined floor of the old impoundment where it could soak into the ground. First discharge into the impoundment was in March 1999, and during the summer everything worked well, particularly during June, July and August when the average evaporation was 142 mm per month, although the delivered tailings did not contain quite as much coarse material as had been envisaged and the rate of height increase of the dams was lower than intended. During the winter, however, conditions became greatly changed. The temperature fell below zero on 20 December and remained low during most of January, freezing the cyclones and producing a layer of ice over the impoundment, which became covered by snow. Tailings from the processing plant was warm enough to keep the operation working, but there was no further height increase for the dams because the cyclones were out of action. Precipitation during September to January averaged 71 mm per month and fell as rain and snow on both the whole area of the impoundment but also on the old tailings impoundments that were being worked. This extra water was stored in the impoundment causing the level to rise under the now thick layer of ice and snow.

On 27th January there was a marked change in the weather. The temperature rose above zero and there was a fall of 37 mm of rain. The ice and snow covering melted and the dams, half way along the sides of the impoundment, where they were only starter height, were lower than the developing water level. At 22.00 hours on 30th January 2000, a section overtopped, washing out a breach 25 m long that allowed the escape of about 100 000 $m^3$ of heavily contaminated water that flowed following the natural slope of the area, towards the river Lapus. This in turn fed into the rivers Somes, Tisa and Danube, eventually discharging into the Black Sea. A very large number of fish were killed with serious consequences for the fishing industry for a time. The Hungarian authorities estimated the total fish kill to have been in excess of one thousand tonnes. Water intakes from the rivers had to be closed until the plume of toxic contaminates had passed and for some time afterwards until the purity of the water could be confirmed. The cyanide plume was measurable at the Danube delta, four weeks later and 2000 km from the spill source.

The concept of a closed system in which none of the process water should escape into the environment should have been excellent, with the new tailings impoundment completely lined with plastic sheeting and provision for the collection of any seepage. Unfortunately no provision had been made for the additional water that would accrue from precipitation, nor had the problems of working at low temperatures been addressed. The scheme was one that could have worked well in the hot and dry conditions found in some parts of Australia and South Africa.

NEW METHODS.

Thickened tailings. As a complete alternative to the use of tailings dams, the tailings can be de-watered until it becomes a non segregating material that when deposited from one point, will form a slope of 2 to 6%. In such thickened tailings all the particles stay together forming a homogeneous material with a high capillary suction. The removal of much of the process water from the tailings is achieved by passing the tailings through high compression thickeners. The system was first used in 1972 at the Kidd Creek copper/zinc mine in Ontario, Canada, for dealing with 12,000 tons a day. The disposal area of 1420 ha, with an average diameter of 1.5 km, is on a topographical high, surrounded on three sides by a river. Although discharge was initially from a series of spigots from a central ramp, later single point discharge was placed at the north end of the area with the intention of moving it progressively to the south to create a ridge that can be reclaimed while deposition continues. The slopes formed are between 2.5 and 3%, and the only retaining dam required is 19 m high across a small valley.

Another example of the use of thickened tailings is at the Mount Keith nickel sulphate mine in Western Australia, that produces 11.5 x $10^6$ tonnes of tailings a year. The storage area is 1700 ha, with an average diameter of 4.6 km on land with a very slight fall of only 12 to 14 m from west to east. A perimeter bund 14 km long surrounds the site to prevent the spread of any materials that might be carried by rainfall run-off. There is a central riser pipe surrounded by 8 other risers 35 to 45 m high and a fully automated three-train, two stage pumping station able to deliver the thickened tailings to any riser at a rate of 3,000 $m^3$ per hour. Under drainage was installed in the ground surrounding each riser, plus open drainage to collect decanted water. The facility was commissioned in December 1996. This information was supplied by the mine owners, WMC Resources Limited, Perth.

Yet a third example is at the Peak Gold Mine, Cobar, New South Wales. This underground mine, established in 1992, produced gold, copper, lead and zinc, causing 0.3 x $10^6$ tonnes of waste tailings a year. The absence of any significant amount of waste rock from this underground mine, made the idea of an impoundment requiring no large dams particularly attractive, and the central discharge avoids the considerable work of having to move discharge points as required during the construction of other types of tailings dams.

Paste tailings, wet cake tailings, dry cake tailings and dry stack tailings.
These four methods have much in common. Paste tailings is essentially thickened tailings with a hydrating additive such as Portland cement. It can be pumped by displacement pumps. Wet cake tailings have been de-watered to the extent that pumping is impossible, but the cake is virtually saturated, while dry cake tailings has an even lower water content and is unsaturated, requiring transport by conveyor belt or some type of vehicle on road or rail.

Dry stack material has had water extracted by filtering using pressure or suction force. The most common filtration plants use drums, horizontally or vertical stacked plates or a system using horizontal belts. An example of the use of filtration to reduce the water content of pulverised fuel ash (PFA) was give by Martin et al (1994) and quoted in ICOLD Bulletin No. 106. Fly ash from power stations was wetted so as to be able to pump it to the site for an impoundment, where some of it was de-watered to an optimum condition and used to construct the downstream shoulder of the tailings dam by placing, spreading and compacting the material, using waste colliery shale to form a core, with the remaining wet tailings being pumped straight into the impoundment.

These new methods are more expensive than the tradition methods of tailings storage, but have the advantage of requiring less volume to accommodate a given weight and presenting less stability problems and in general, a lack of risk of failure and no problems of liquefaction. In many cases this must make the methods attractive to mine owners.

## CONCLUSIONS.

Geotechnical knowledge with engineering experience can enable safe tailings dams to be designed and constructed, but the current rate of tailings dam failures that have averaged 1.7 a year during the past 30 years shows that this knowledge and experience has not been applied in every case. A failure can stop production, and clean-up operations, compensation for damage and in some cases death, plus finding new storage for tailings, can be so extremely expensive for a mine that it would be expected that every care would be taken to avoid failure. It would appear that management fail to engage staff able to understand tailings dams sufficiently to be able to detect deficient design or construction procedures, so that dangerous conditions can be allowed to persist until failure occurs. In order to improve the situation the International Commission on Large Dams with the United Nations Environmental Programme have recently published ICOLD Bulletin No. 121 that gives 221 examples of incidents with tailings dams and discusses causes, with the aim of helping those in charge of tailings dams to understand some of the simple mistakes that continue to occur.

## REFERENCES.

Berti G, Villa F, Dovera D, Genevois R and Brauns J (1988). The disaster of Stava/Northern Italy. *Proc. Specialty Conf. "Hydraulic fill structures" GT Div/ASCE, Fort Collins, CO, pp 492-510.*

Blight G.E. (1997). Insight into tailings dam failures – technical and management factors. *Proc. Intn. Workshop on Managing the Risk of Tailings Disposal, Stockholm, Sweden, May, pp 17-34.*

Capuzzo D. (1990). A brief description of the main causes underlying the collapse of Stava tailings dams, Trento - Italy. *Proc. ICOLD Intn. Symp. Safety and Rehabilitation of Tailings Dams, Sydney, Australia, vol 2, pp.64-66.*

Dobry R. and Alvarez L. (1967). Seismic failures in Chilean tailings dams. *Jour. Soil Mech. and Fdn. Eng. A.S.C.E. vol 93, no. SM 6, pp. 237-260.*

ICOLD (2001). Bulletin 121. Tailings dams: Risk of dangerous occurrences. Lessons learnt from practical experience. *International Commission on Large Dams, 151, Boulevard Haussmann, 75008 PARIS. 144 pps.*

Martin P.L., Vaciago G, Somers N.R. and Burbridge H.T. (1994). Additional lagooning on Stage 1 of the Gale Common Ash Disposal Scheme: Investigation, design and preparatory works. *Proc. 1st Int. Congress on Environmental Geotechnics, Edmonton, Canada, pp 325-330.*

Report, Cyanide Spill at Baia Mare, UNEP/OCHA Assessment Mission Report, April 2000. *Available at www.natural-sources.org/environment/BaiaMare*

# The IMPACT Project - Continuing European Research on Dambreak Processes and the Failure of Flood Embankments

M W MORRIS, HR Wallingford, UK

SYNOPSIS. The CADAM project (*Concerted Action on Dambreak Modelling*) highlighted areas where our understanding of dambreak failure and extreme flood processes was poor and which contributed greatly to uncertainty in the prediction of potential flood conditions resulting from the failure of a dam or flood defences. The IMPACT project (*Investigation of Extreme Flood Processes and Uncertainty*) focuses research in three of these areas; namely *Breach Formation*, *Flood Routing* and *Sediment Movement* with the objectives of improving understanding of these basic processes and developing tools and methodologies for their more accurate and reliable prediction. Work also focuses on identifying the uncertainty associated with the prediction of these processes and subsequently the implications for end user applications, such as emergency planning.

## INTRODUCTION

In recent years there have been a series of extreme flood events within the UK that have raised public awareness of the impact and hence risk of flooding. During these extreme events a range of flood defences have been overtopped and some (flood embankments) have breached. Whilst no dams have failed during these events, there have been cases where embankments have been overtopped, raising the issue of what might happen in the event of a failure. With growing public awareness has come an increased pressure to identify and mitigate against the risk of failure and flooding.

Predicting the potential impact of failure of a dam or flood defences is not yet a legal requirement in the UK. However, legislation is moving towards such an approach. Guidance for the *Risk Management of UK Reservoirs* (CIRIA, 2000) has recently been produced and there is currently ongoing DEFRA funded research investigating probabilistic approaches to dam and reservoir risk assessment.

Fundamental to the assessment of risk from dam or defence failure (whether for emergency planning or asset management) is the prediction of the failure conditions themselves. A framework for risk assessment and management will provide a mechanism for identifying and ordering risks, but the procedure is still dependent upon the accuracy with which the various failure or event processes can be determined. The reliability and flexibility

*Reservoirs in a Changing World*, Thomas Telford, London, 2002, 484–493

of the tools and methodologies typically used in the UK for predicting these conditions can, at best, be described as limited when compared to simulation tools used now for river flow modelling or construction design.

Over the past 20 years some dam and defence owning companies have undertaken dambreak analyses for their structures – typically using the DAMBRK software package, which also contains the NWS BREACH model for predicting failure of an embankment. These models were developed during the 1980's and in recent years have been shown to have limitations and deficiencies (Mohamed, 1998). Fear of public reaction to potential flood risk from dambreak has also resulted in many of these studies remaining strictly confidential. This confidentiality has probably had the effect of limiting the review, assessment and development of new and better predictive tools. If a topic is not openly discussed then there is no perceived need for such tools!

The need to improve knowledge and develop better tools was identified during the CADAM project. With growing trends towards placing flood risk data openly in the public domain, the pressure to provide accurate and appropriate dambreak predictions becomes greater, although of course, the significance for end uses, such as for emergency planning, remains as important as ever.

This paper presents a brief overview of key issues identified during the CADAM project, which are now being addressed within the IMPACT project. The implications of this work for dambreak / extreme flood prediction are also presented.

## THE CADAM PROJECT

CADAM was funded by the European Commission as a Concerted Action Programme that ran for a period of two years ending in January 2000. The project promoted the comparison of dambreak and breach modelling performance and practice across Europe. Key outputs from the CADAM project included:

- Proceedings from 4 workshops covering a variety of dambreak topics
- Test case data for various dambreak laboratory tests as well as field data
- Guidance document (complementing ICOLD Bulletin 111 (ICOLD, 1998)) entitled *Dambreak Modelling Guidelines and Best Practice* (Morris & Galland, 2000)
- Project report document providing an overview of the project work and summarising key findings, including recommendations for future R&D

All publications from the CADAM project may be accessed via the project website at www.hrwallingford.co.uk/projects/cadam. Paper and CD-ROM copies are also available. A summary of findings may also be found in the BDS Bath Conference proceedings (Morris, 2000).

Conclusions from the CADAM Project

A total of 31 specific conclusions are identified within the CADAM project report. Selected conclusions from this report are presented below:

*Conclusion 9: The accuracy of numerical models in predicting general hydrodynamic conditions is relatively good in comparison to other aspects of a dambreak study. Our ability to model complex flow conditions (such as flow in urban areas) is relatively poor.*

*Conclusion 10: The issue of accuracy is caught in a loop. Modellers are reluctant to define accuracy since there are so many unknowns and assumptions in the modelling process. Equally many end users are unclear as to what conditions they should work towards – particularly where legislation does not exist. A true assessment of accuracy will not occur until clear guidance on the required level of accuracy is given.*

*Conclusion 11: It proved impossible within the scope of CADAM to identify a single best model or type of model appropriate to any or all dambreak flow conditions. A more detailed and in depth analysis of data and model performance is required if this is to be achieved...*

*Conclusion 12: Uncertainties within the breach modelling process may be the greatest contribution to uncertainty within the whole dambreak analysis process.*

*Conclusions 13 & 15: Our current ability to predict the rate and location of breach growth is quite limited. Breach model accuracy is very limited. An estimate of ±50% for predicting peak discharge is suggested, with the accuracy of predicting the time of formation considerably being worse.*

*Conclusion 17: Currently, there is no single recommended breach model. Whilst the NWS BREACH model is widely used it has significant limitations. A number of researchers are currently working on the provision of improved breach models. There is a clear need to integrate knowledge from both the hydraulics and soil mechanics disciplines in order to advance expertise in this field.*

*Conclusions 20 & 21: Large scale movement of debris and sediment is likely to occur during a dambreak event leading to large variations in valley topography, particularly near to the dam... likely to significantly affect predicted water levels. Consideration of these effects should therefore become a part of dambreak analysis.*

*Conclusion 23: Research suggests that high-density debris and sediment flow significantly affects the flood wave propagation speed.*

*Conclusion 28: There are many aspects to impact assessment of dambreak floods that require more detailed investigation and research to provide reliable data for use in risk assessments.*

THE IMPACT PROJECT

The IMPACT project team comprises 9 partners from 8 different countries and a variety of academic and industry based organisations. The objective of the project is ultimately to aid the assessment and reduction of risk from extreme flooding caused by natural events and / or the failure of dams and flood defences.

Research Topics

Research is focussed around 3 key areas, and combined by an overall approach to identifying uncertainty associated with the prediction of processes in these areas (Fig. 1). The following sections offer an overview of this work.

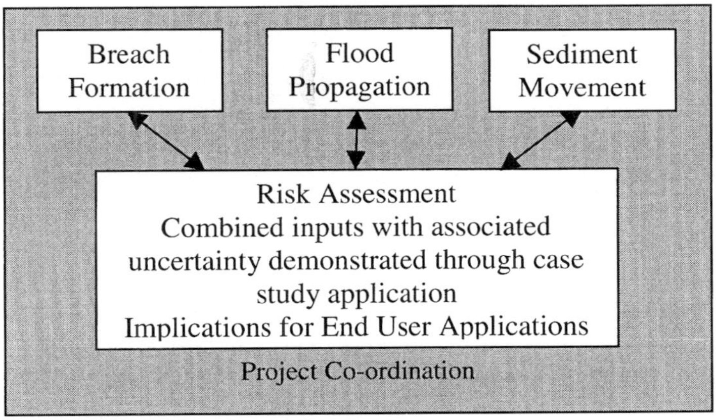


Fig. 1 Research topics within the IMPACT project

Breach Formation

The object of research under this area of the project is how to reliably predict the breach formation process through a flood embankment, or embankment dam. Existing models such as the NWS (DAMBRK) BREACH model do not offer a 'reliable' prediction tool (Mohamed, 1998) hence a programme of work to aid both understanding of the fundamental processes and the development and validation of improved modelling tools has been established. This work includes:

- Breach formation field work – controlled failure of 6m high embankments to identify and monitor key formation processes
- Breach formation laboratory work – physical modelling of embankment failure to identify and monitor key formation processes
- Numerical model development and comparison using field and lab data
- Breach location – development of a methodology / prototype tool for a risk based approach to identifying potential breach location

Field Modelling A test site has been established in Norway where a test embankment some 6m high and 30m wide may be constructed across a river valley, allowing the creation of a temporary reservoir retaining up to 100,000m$^3$ water. A total of 5 embankment failure tests will be undertaken, allowing close monitoring of the breach formation process.

Photo courtesy Dr. Ing Kjetil Arne Vaskinn (Statkraft Grøner)

Fig. 2 Location of embankment breaching test site (Norway)

A reservoir upstream of the site provides the opportunity of controlling additional inflow into the test reservoir whilst a further lake area downstream allows attenuation of the breach flood flow. Remoteness of the site and the nature of the topography permits minimal environmental impact of the test work.

Photo courtesy Dr. Ing Kjetil Arne Vaskinn (Statkraft Grøner)

Fig. 3 An embankment failure test at the Norwegian test site

With the number of field tests limited by cost, the choice of test conditions (i.e. type of material, geometry etc.) is difficult. By combining the field test programme with a laboratory test programme of 25 failures, it is anticipated that the value of the field data may be extended to cover a wider range of failure conditions. Field tests for 2002 have been established as:

Test 1 Homogeneous earthfill embankment (non cohesive) - overtopping
Test 2 Homogeneous earthfill embankment (cohesive) – overtopping
Test 3 Composite (2 layer) embankment - overtopping

Laboratory Modelling To complement the field test programme a series of 25 laboratory tests are planned. These tests will be divided into three series of 8 tests, scheduled before, during and after the 2 series of field tests. The objective of the laboratory tests is to compare and validate the scaling of test data and to extend the range of data for different embankment geometries, material and failure mechanisms.

The first 8 lab tests prior to / interactive with the first series of field tests will comprise:

| | |
|---|---|
| Tests 1-3 | Direct replication of field tests |
| Tests 4-6 | Varying embankment geometry from Tests 1-3 |
| Tests 7-8 | Varying embankment material parameters for 1 of Tests 1-3 |

Laboratory testing will be undertaken within the Flood Channel Facility housed at HR Wallingford, which offers a flume some 10m wide, 50m long, 0.8m deep and with an inflow capacity in excess of $1m^3/s$ (Fig. 4).

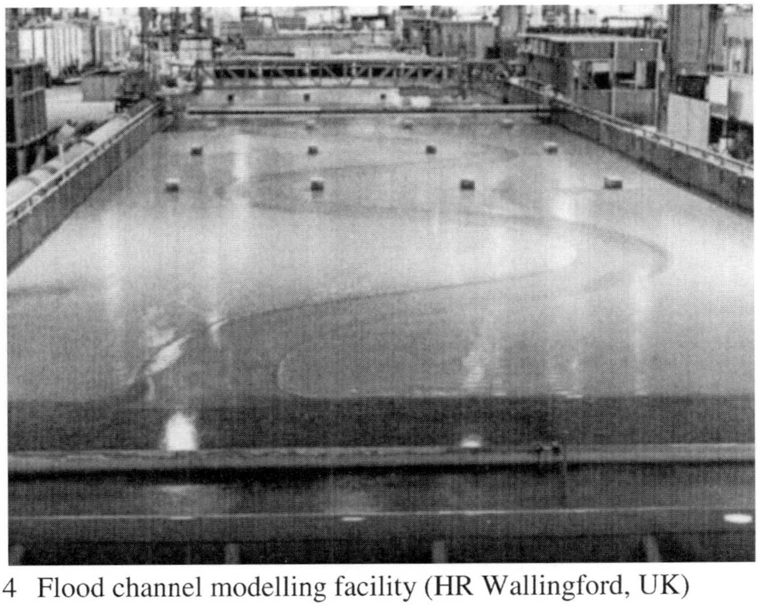

Fig. 4 Flood channel modelling facility (HR Wallingford, UK)

Numerical Modelling Field and laboratory data will be used to test, validate and develop numerical breach models. Five partners within the IMPACT team plus additional organisations from Europe and the US will also participate in model testing and development. A single breach model will not be developed but a comparison of various modelling approaches and performance will be made to identify the most effective modelling tool(s).

Breach Location A separate module of work will be undertaken by HR Wallingford and Statkraft Groner looking specifically at factors contributing to breach location, with the aim of developing a tool or methodology for identifying the (relative) risk of a breach occurring at a specific location. This is of particular importance when considering the risk of breach formation through long lengths of flood defence embankment or through a bunded reservoir.

Flood Propagation

The focus of research here is to produce and extend reliable modelling methods for the propagation of catastrophic flood flows. The overall objectives are:

- To identify dam-break flow behaviour in complex valleys, around infrastructure and in urban areas
- To compare different modelling techniques and identify the best approach, including the assessment of modelling accuracy
- To adapt existing, and develop new modelling techniques for the specific features of extreme 'dambreak' floods

The approach adopted is to:

- compare different mathematical modelling techniques
- identify the best approaches, including the assessment of implementation methods in industrial software packages
- check the accuracy and appropriateness of various methods by validation of the models against results from physical experimentation
- validate the different modelling techniques adopted, both existing and newly developed, against field data obtained from actual catastrophic flood events

The research has been divided into two distinct work packages looking firstly at urban flood propagation and then flood propagation in natural topographies. The development and validation of appropriate techniques for modelling flood flow through urban areas has become particularly relevant in recent years. Currently such modelling is undertaken in a variety of ways, however no formal comparison and validation of these different methods has been undertaken.

Sediment Movement

This area of research considers the need to improve (or even include) the prediction of sediment movement in association with catastrophic flood modelling. Case study data from dambreak events in the US has shown that significant quantities of sediment and debris can move during extreme floods. This in turn can affect the rate of flood wave propagation as well as floodwater level due to potentially large variations in bed level (Fig. 5).

Fig. 5 Sediment movement - before and after a flood event

The nature of the problem is different from that found during normal flood flows in that the quantity and size of sediment will be much greater during catastrophic flood flows. This is an important issue for the accurate prediction of downstream conditions since:

- the river bed elevation can vary by tens of meters
- the river can be diverted from its natural course (as for the Saguenay river tributary – the Lake Ha!Ha! dam failure, Canada, 1997)

Research in this area is divided into two work-packages that address the (near field) sediment flow during dam-break conditions and the (far field) geomorphological changes that can occur in a valley during dambreak flows. The approach adopted is to combine physical model experiments, designed to improve our physical understanding of these cases, with the development and testing of mathematical modelling methods for simulation of these flow conditions.

Risk and Uncertainty

The research outlined in the sections above focuses on processes that are currently poorly understood or poorly simulated by predictive models. An important aspect of any process that contributes towards an overall risk assessment (i.e. prediction of flood risk) is an understanding of any uncertainty that may be associated with prediction of that particular process. For example, a flood level may be predicted to reach, say, 20m and an emergency plan developed to cope with this. However, if the uncertainty associated with this prediction is, say, ±2m rather than ±0.2m, then different measures may be taken to manage this event.

The problem to be solved is therefore to quantify the uncertainty associated with each process contributing to the risk assessment and to demonstrate the significance of this for the specific end application. This may be in the form of uncertainty associated with flood level prediction, flood location, flood timing or flood volume – depending upon the particular application of data. Uncertainty will be quantified by working closely with the team researchers and demonstrated through a number of case study applications. The procedure for combining the uncertainty associated with different data will depend upon the process itself. This may require multiple model simulations or combination through spreadsheet and / or GIS systems as appropriate.

Having identified process uncertainty and the effect that particular processes may have on the end application, it will also be possible to identify the importance (with respect to the accuracy of a risk assessment or end application) that each process has and hence the effort that should be applied within the risk management process to achieve best value for money.

Interaction With other National and International Research

Whilst the IMPACT project has been funded according to a specific project team and work plan, considerable effort is being taken to ensure that the work interacts with other National and International initiatives to promote maximum dissemination and benefit. Within the UK, the project is supported by the joint DEFRA / Environment Agency R&D programme, and specifically links with an initiative entitled "Reducing the Risk of Embankment Failure under Extreme Conditions". Other IMPACT partners have linked the research work with specific national initiatives, so widening the network of expertise and knowledge. Internationally, it is likely that additional (but unfunded) partners, such as Electricité de France (EDF) and the US Bureau Reclamation and US Department of Agriculture (Research Service), will also participate in model testing and comparison, so increasing the range and scope of the research work.

## CONCLUSIONS

The tools used for modelling breach formation and dambreak in the UK have not changed significantly during the past decade. However, research during the last few years has highlighted potential problems with some of these tools (such as the NWS DAMBRK and BREACH models) and has identified the need for targeted research and development to both improve our understanding of basic flood processes and to provide more reliable predictive tools.

The IMPACT project runs from December 2001 until November 2004 and is focussing research in three key areas, namely breach formation, flood propagation and sediment movement. The focus of this work will be to investigate fundamental processes but also, most importantly, to develop better predictive tools for industry and, through maintaining a risk based philosophy, to quantify uncertainty within the prediction process.

As work continues on this project information will be posted to the project website. For more information visit:

**www.impact-project.net.**

ACKNOWLEDGEMENTS

The financial support offered by the European Community and by DEFRA / EA in the UK in support of this work is acknowledged.

The IMPACT project team comprises Universität Der Bundeswehr München (Germany), Université Catholique de Louvain (Belgium), CEMAGREF (France), Università di Trento (Italy), Universidad de Zaragoza (Spain), Enel.Hydro (Italy), Statkraft Grøner AS (Norway), Instituto Superior Technico (Portugal), and HR Wallingford Ltd (UK).

REFERENCES

CIRIA (2000). Guide to Risk Management for UK Reservoirs. CIRIA Publication C542.

Department of the Environment (DoE) (1991). Dambreak flood simulation program: DAMBRK UK.

ICOLD (1998). Dambreak flood analysis: Review and recommendations. ICOLD Bulletin 111.

Mohamed (1998) Informatic tools for the hazard assessment of dam failure. MSc thesis for Mohamed Ahmed Ali Mohamed, IHE Delft, May 1998.

Morris M W (2000). CADAM – A European Concerted Action Project on Dambreak. Proceedings of the biennial conference of the British Dam Society, Bath.

Morris M W, Galland JC (EDS) (2000). Guidelines and Best Practice for Dambreak Modelling: Conclusions from CADAM. Drawn from a range of conclusion papers produced during the CADAM project and published by the European Commission.

# A historical perspective on reservoir safety legislation in the United Kingdom

J A CHARLES, Building Research Establishment Ltd, UK

SYNOPSIS. Popular demand for reservoir safety legislation arose in the middle of the nineteenth century as a result of two dam failures which caused major loss of life. Factors that led to the long delay in introducing legislation and significant influences on the form taken by the reservoir safety legislation enacted in 1930 and 1975 are identified.

## INTRODUCTION

At a meeting of the Institution of Civil Engineers (ICE) in 1856, I K Brunel announced the death of J M Rendel and, after praising his former colleague, commented that *'He was always considered a safe man'* (Lane, 1989). This description was then amplified in a way calculated to remove any idea that it was intended as a compliment: '*- one who would seldom do anything that he had not, or others had not done before – and this is, strange to say, a recommendation with men of business, who do not understand engineering. But it is quite clear this quality is not the characteristic of a Smeaton, a Stephenson, or a Watt, or the world would make no progress.*'

It would seem that Brunel viewed safety and innovation as mutually exclusive alternatives. This view may have had some validity in those areas of civil engineering in which Brunel and Rendel were engaged, but the construction of embankment dams in the nineteenth century was to provide ample evidence that a lack of innovation did not prevent serious failures.

During the first half of the nineteenth century the demand for unpolluted water supplies for the rapidly expanding industrial towns led to a major increase in reservoir construction. The comments of two eminent dam engineers in the middle of the nineteenth century give some insight into dam construction in that period:

- Henry Conybeare (1858) affirmed that *'It was much to be regretted, that the Minutes of Proceedings of the Institution contained so little information regarding works for impounding water, especially as many admirable works of that class had been recently executed by different Members, for the water supply of Liverpool, Manchester, Edinburgh, and other large towns'*. Nearly all the dams built in Great Britain in the nineteenth century were embankment dams of a traditional type of construction with a narrow central core of puddle clay and doubtless the paucity of papers was in part due to the perceived lack of innovation in this type of construction.

- In a discussion at the ICE in 1859, Robert (later Sir Robert) Rawlinson made the less than encouraging claim that *'In regard to the form and mode of constructing embankments for reservoirs, he could not admit that modern engineers were worse, in these respects, than their predecessors'*. He went on to criticise loose tipping of fill from waggons and the practice of carrying outlet pipes or culverts through the fill. Rawlinson's comments suggest that standards of construction, if not actually deteriorating, were not improving.

A significant number of embankment dams failed during construction. These dams "failed" in the geotechnical sense that there was a slip in one of the embankment slopes caused by inadequate shear strength possibly due to excessive construction pore pressures. The slip might be contained within the fill or might involve both fill and foundations. Although this type of construction incident could require major remedial works, it occurred when there was little or no water in the reservoir and public safety was not at risk. In the rest of the paper the term "failed" is restricted to cases where the dam is breached with a consequent uncontrolled release of reservoir water.

When on 23 August 1848 a small dam failed by overtopping during a storm at Darwen in Lancashire with the loss of 12 lives, the verdict of the jury at the inquest was that *'... all the deaths inquired into occurred by an accidental cause, that cause being the excessive rains of Tuesday night and Wednesday morning, by which the reservoir at Bold Venture Lodge overflowed and washed away the embankment ...'* (Anon, 1848). The jury recommended steps to be taken in the event that the dam was rebuilt.

A few years after the failure at Darwen, popular demand for reservoir safety legislation arose as a result of two dam failures in Yorkshire which caused major loss of life. The failures of Bilberry and Dale Dyke were not associated with extreme floods and human culpability seemed more obvious. Resistance to introducing safety legislation arose from concerns over proposals for a government inspectorate and fears that safety legislation would lessen the dam owner's responsibility. Some 427 lives have been lost in 14 failure events which are listed in Table 1.

BILBERRY 1852

The collapse of Bilberry dam at 1am on Thursday 5 February 1852 caused 81 deaths and much property damage in the Holmfirth area. Construction of the dam commenced in 1839 and was beset with problems. The reservoir owners, the Holme Reservoirs Commissioners, did not wish to incur the expense of frequent site visits by George Leather, the engineer who designed the dam, and supervision of construction was inadequate. The dam was overtopped and breached following a long period of leakage and settlement caused by internal erosion. There had been heavy rainfall in the area and there were a number of eye-witnesses of the failure. The reservoir emptied in 30 minutes (Anon, 1910). The unsafe state of the dam was recognised by those living closest to it and some of these people escaped the flood which engulfed the inhabitants of Holmfirth lower down the valley.

Table 1. British dam failures causing loss of life (after Charles, 1993)

| Dam | Height (m) | Reservoir volume ($10^3 m^3$) | Date built | Date failed | Deaths |
|---|---|---|---|---|---|
| Tunnel End | 9 | | 1799 | 1799 | 1 |
| Diggle Moss | | | 1810 | 1810 | 5 |
| Whinhill | 12 | 262 | 1821 | 1835 | 31 |
| Welsh Harp | 7 | | 1837 | 1841 | 2 |
| Glanderston | | | | 1842 | 8 |
| Darwen | 5 | 20 | 1844 | 1848 | 12 |
| Bilberry | 20 | 310 | 1845 | 1852 | 81 |
| Dale Dyke | 29 | 3240 | 1863 | 1864 | 244 |
| Rishton | | | | 1870 | 3 |
| Cwm Carne | 12 | 90 | 1792 | 1875 | 12 |
| Castle Malgwyn | | | | 1875 | 2 |
| Clydach Vale | | | | 1910 | 5 |
| Skelmorlie | 5 | 24 | 1861 | 1925 | 5 |
| Coedty & | 11 | 320 | 1924 | 1925 | 16 |
| Eigiau | 10 | 4500 | 1911 | 1925 | |

This was the first dam disaster in Great Britain to draw major attention to reservoir safety. The Home Secretary, Sir George Grey, arranged for Captain R C Moody of the Royal Engineers to inspect the remains of the dam and give expert evidence at the inquest. Moody drew attention to the failure to properly control fill placement and ensure that the more cohesive fill was placed next to the puddle clay core with the more granular fill in the outer slopes. He remarked that a sinkhole in the crest was located above the culvert *'and is no doubt due to the washing away of the bad puddling over and above the culvert where it passes through the puddle wall below'*. A private act for the reconstruction of the reservoir contained some provisions relating to safety, including the appointment of J F Bateman as the engineer.

At the inquest, the verdict of the jury on Friday 27 February contained strong criticism of the Holme Reservoirs Commissioners, and pointed to the need for legislation: '*... and we regret that the reservoir, being under the management of a corporation, prevents us bringing in a verdict of manslaughter, as we are convinced that the gross and culpable negligence of the Commissioners would have subjected them to such a verdict had they been in the position of a private individual or firm. We also hope that the* *legislature will take into its most serious consideration the propriety of making provision for the protection of the lives and properties of Her Majesty's subjects exposed to danger from reservoirs placed by corporations in situations similar to those under the charge of the Holme Reservoirs Commissioners.'*

The possibility of a quick legislative response to the catastrophe must have been reduced by changes in government that occurred in 1852. Sir George Grey had been Home Secretary since 1846 in the Whig administration of Lord John Russell, but on 28 February 1852, the day after the jury's verdict, a Tory government under the Earl of Derby took office. This only lasted a few months and was replaced on 28 December 1852 by a Peelite coalition led by the Earl of Aberdeen with Viscount Palmerston as Home Secretary.

When Palmerston became Prime Minister of a Liberal government in 1855, Grey again became Home Secretary. Although not a brilliant speaker, he showed much practical ability during the Chartist riots and Fenian activity in Ireland. While Grey was Home Secretary for a third time, from 1861 to 1866, legislation was introduced in response to the Bilberry failure. There were nine clauses in the Waterworks Clauses Act of 1863 on the security of reservoirs. Any interested person could make a complaint to two justices of the peace that a reservoir was in a dangerous state. If satisfied that the complaint was well founded, the justices had powers to order the lowering of the reservoir and the execution of works to remove the cause of the complaint. It seems that these provisions were not repealed until the Water Act 1945 and the Water (Scotland) Act 1946 (Agnew, 1984).

DALE DYKE 1864

Dale Dyke failed during first filling of the reservoir on the night of Friday 11 March 1864 at 11.30pm. Earlier that day a crack had been observed along the downstream slope near the crest of the dam, and an attempt was made to blow-up the overflow weir, but the dam collapsed before this was achieved (Harrison, 1864). The reservoir was far larger than that at Bilberry, and the catastrophe claimed 244 lives in the vicinity of Sheffield. This is the most serious dam failure that has occurred in Great Britain and the country's worst civil engineering disaster. John Towlerton Leather, the engineer for Dale Dyke, had been responsible for five other dams for the water supply of Sheffield and was a nephew of George Leather, the engineer for Bilberry. The Home Secretary, Sir George Grey, appointed two civil engineers, Robert Rawlinson and Nathaniel Beardmore, to assist in the enquiry into the cause of the catastrophe (Amey, 1974).

The verdict of the jury on Thursday 24 March included the statement: *'that, in our opinion, there has not been that engineering skill and that attention to the construction of the works, which their magnitude and importance demanded; that, in our opinion, the Legislature ought to take such action as will result in a governmental inspection of all works of this character; and, that such inspections should be frequent, sufficient, and regular;'*

A leader in *The Times* on 17 March 1864, argued that those threatened by reservoirs could not be expected to defend themselves and needed protection. There was pressure for a government inspectorate, but the Home Office wanted to pass responsibility to the Board of Trade (Simpson, 1984).

In their report to the Home Secretary, Rawlinson and Beardmore were critical of both the design and construction of the dam. They believed that failure was most likely to have been caused by leakage from a fractured outlet pipe, but the design and construction of the embankment was also criticised: '*... the puddle-wall is much too thin, and the material placed on either side of it is of too porous a character, .... No puddle-wall should ever be placed betwixt masses of porous earth, as puddle, under such conditions, will crack, and is also liable to be fractured by pressure of water.'*

Rawlinson and Beardmore did not endorse the recommendation from the jury for government inspection: *'We cannot, however, recommend it for adoption. Any approval of plans or casual inspection of waterworks embankments cannot ensure ultimate safety in such works. The responsibility must remain, as at present, with the engineer and persons immediately connected with the works. Magistrates have jurisdiction under clauses inserted in recent Waterworks Acts.'*

Sheffield Waterworks Company retained leading engineers, including Hawksley, Bateman, and Simpson. Contrary to the findings of Rawlinson and Beardmore, these men considered that the dam collapsed as a result of a landslide and was an unavoidable accident. Sheffield Corporation engaged nine engineers, including Sir John Rennie and James Leslie, to report individually on the failure and they agreed with Rawlinson and Beardmore that there had been faulty construction. The failure mechanism has continued to be disputed (Binnie 1978; 1981).

## SELECT COMMITTEE ON THE WATERWORKS BILL 1865

On 23 February 1865, Sir George Grey announced that a draft bill was being prepared, but that the matter was the responsibility of the Board of Trade (Simpson, 1984). The bill was referred to a select committee of the Commons. The Select Committee on the Waterworks Bill 1865, with J A Roebuck, Member of Parliament for Sheffield, as chairman, reported on 23 June 1865 and made the following recommendations (Sessional Paper 401):

> 1. That in all cases in which it is proposed to construct a large reservoir, the undertakers should submit to the Home Office or Board of Trade plans and sections of the site selected for such reservoir, and of the works to be erected, and also descriptions of the mode of construction.

2. That in all cases it should be the duty of the Home Office or Board of Trade to send a competent person to such site in order to verify and report upon such plans and sections and descriptions.
3. That it should be the duty of the Home Office or Board of Trade to submit such plans and sections and descriptions, together with the observations of the person so sent, to any select committee appointed by either House of Parliament to consider any Bill by which powers are to be given to construct the contemplated works.
4. That it shall be the duty of the Home Office or Board of Trade to send some competent person to inspect and report on the works that are to be constructed, so that if there be any glaring deviation from the rules laid down by any private Act under which the reservoir inspected may be constructed, or by any general Act respecting waterworks that may be passed, the deviation may be made known to the Home Office or Board of Trade. And it seems important, in order that competent inspectors should be employed by the Government, that they should receive ample remuneration for their labour and services.
5. That when any reservoir inspected as above is completed, the undertakers should be bound to give due notice of such completion to the Home Office or Board of Trade, and that such reservoir shall not be filled with water until after the expiration of a specified time. And that during such time it shall be competent to the Home Office or Board of Trade to prohibit the undertakers from allowing water to be let into such reservoir. But if within the specified time no such prohibition is issued, then it may be competent to the undertakers to fill the reservoirs with water.
6. And further, the Committee having found that large reservoirs have at times been allowed to decay and become dangerous, they suggest that an adequate supervision over all large reservoirs should be maintained by the Home Office or Board of Trade; and to that end, from time to time, competent persons should be sent to inspect and report upon such reservoirs.
7. The Committee do not intend to diminish in any degree the responsibility of undertakers of waterworks contemplated by the Bill referred to them, to pay all such damage as may result from the water stored by them. In the opinion of the Committee, the provisions by which such responsibility is imposed upon such undertakers should be stringent and completely unambiguous.

A new bill was introduced in 1866, which largely followed the recommendations of the Select Committee, including a clause imposing strict liability, and which was restricted to reservoirs impounding over a million cubic feet of water ($28 \times 10^3 m^3$) (Simpson, 1984). However, the Waterworks Bill of 1866 ran out of time and did not become law. There was a change in government in July 1866 when a Conservative administration took office. In February 1867, the new President of the Board of Trade announced in the Commons that there was no intention of reintroducing the bill.

SKELMORLIE 1925

Following heavy rainfall, the Skelmorlie lower reservoir in south-west Scotland failed at 2pm on Saturday 18 April 1925. The flood water, which was probably sufficient to overtop the embankment, was augmented by water from a quarry which, having partially filled due to a blocked culvert, suddenly emptied when the blockage cleared. The embankment was breached over a 9m length and the reservoir emptied in 15 minutes (Davidson, 1996). Five people were killed in the village of Skelmorlie.

The failure was attributed to a grossly deficient overflow and inadequate freeboard. Coutts (1934) quotes from the *Glasgow Herald* of 16 June 1925 the verdict of the jury at an enquiry held at Kilmarnock Sherriff Court as follows: *'The disaster was caused by absence of any regular skilled supervision and inspection.'*

EIGIAU AND COEDTY 1925

The Eigiau and Coedty dams are situated in north Wales. Eigiau was a small mass concrete gravity dam built between 1907 and 1911 which impounded a reservoir of $4.5x10^6m^3$. It breached at 08.45pm on Monday 2 November 1925 (Walsh and Evans, 1973). The failure was attributed to a blow-out of a section of the dam wall which was 5m high at that location. Some $1.5x10^6m^3$ of water were discharged through the breach in the hour following failure, lowering the reservoir by 1.5m.

The Coedty embankment dam, which was situated 2.5 miles downstream of the Eigiau dam, impounded only $0.3x10^6m^3$ of water. The water flooding in from the Eigiau reservoir filled the almost full Coedty reservoir within minutes. The dam was overtopped and the concrete core-wall collapsed. A wave of water and mud hit the village of Dolgarrog at 09.15pm. Providentially most of the villagers were watching the weekly film show in the Assembly Hall, situated on high ground. However, ten adults and six children died in the disaster.

Technical evidence at the inquest was given by Ralph Freeman to the effect that the foundation of the Eigiau dam had not been sufficiently deep. The jury returned a verdict of accidental death: *'caused by the bursting of the dam under the wall in consequence of the wall lacking a proper foundation.'* The coroner's jury recommended regular government inspection.

EDWARD SANDEMAN AND THE TIMES

Edward Sandeman was a leading dam engineer in the early part of the twentieth century. He was born in 1862, two years before the Dale Dyke failure, and died in 1959 at the age of 96. Sandeman was a founder member of the Institution of Water Engineers (IWE) and its President in 1911-1912. A long involvement with reservoirs led to his 1925 letter to *The Times* in which he expressed views based on discussions with fellow engineers in private practice and in government. There was a convention of public anonymity of professional persons and his letter was signed 'Civil Engineer'. An initial version including the full recommendations of the 1865 Select Committee was rejected by the newspaper as being too long. The shortened version was published on Friday 4 December 1925:

> Sir, - In view of the loss of life following upon the failure of the dams of reservoirs in Scotland and North Wales recently, I beg to call your attention to the report made by the Select Committee on the Waterworks Bill 1865 (Sessional Paper 401), after they had considered the evidence as to the failure of the earth embankment of the Bradfield Reservoir on March 11, 1864, when 245 lives were lost. Shortly, this Committee recommended (*inter alia*) that -
>
> (1) Plans of proposed large reservoirs should be submitted to the Home Office or Board of Trade, together with a description of the mode of construction.
>
> (2) The plans should be reported upon by a competent person, and the report submitted to any Select Committee appointed by Parliament to consider the Bill by which powers were to be obtained to construct the works.
>
> (3) The works should be inspected to ascertain if they complied with the rule laid down by any private Act authorizing them.
>
> (4) As large reservoirs have been allowed to decay and to become dangerous, periodical inspection should be made of them.
>
> It is unfortunate that some such control as was then suggested has not been brought into effect. Those of us who are engaged in work connected with the water supply of our towns are not infrequently impressed by the fact that there is a vital need for an examination and enquiry into the present condition of the reservoirs of the country, from the point of view not only of their sufficiency to withstand the pressures to which they are subjected, but also to ascertain whether, in view of the exceptionally heavy rainstorms which have occurred locally in the last ten or 15 years, they are provided with overflows of a capacity requisite to deal with the floods following upon similar rainstorms.
>
> Three of these cases of heavy rainfall occurred in areas of low annual rainfall in the counties of Norfolk, Somerset, and Lincoln, when 6in. to 9in. fell in short spaces of time; had one of these occurred upon a watershed draining to a reservoir deficient in respect of overflow capacity in one of the thickly populated valleys in the northern part of England, the loss of life might conceivably exceed anything previously experienced in this country.
>
> Within my own experience (and probably other engineers will be able to give similar instances) I have within the last five years warned two water authorities that the earth embankments of their impounding reservoirs – made 40 or 50

> years ago – had sunk in the centre so as to be dangerous. Both these cases have now been rectified. In other instances I have found the overflows – which are the safety valves of reservoirs – interfered with and altered to a dangerous extent. These instances alone are sufficient to show that inspection of existing reservoirs is necessary. In the interests of the public safety it is of importance that the matter should receive attention without delay.
>
> I am, Sir, yours faithfully,
> CIVIL ENGINEER

At the Thirtieth Winter General Meeting of the IWE on 4 December 1925, the Secretary read out the letter published that day in *The Times*. The following resolution, passed by the Council on the previous day, was put to the members and was carried: *'That, in view of the number of reservoirs in the country used for the impounding of water, the desirability of holding a full enquiry into the circumstances of the recent disasters at Dolgarrog, in north Wales, and Skelmorlie, Scotland, be represented to the appropriate Departments, having regard to the uneasiness caused generally by the occurrence of such events.'* It was proposed to send copies of the Resolution, embodied in a letter, to the Prime Minister, the Home Office, the Ministry of Health, the Board of Trade, the Electricity Commissioners, and the Scottish Office.

At a meeting at the Home Office on 11 December 1925, crucial issues related to whether there should be Government supervision and what was required for existing reservoirs, as opposed to new works. A draft report was sent to the Ministry of Health on 31 December 1925. A reply from the Ministry dated 4 January 1926 questioned the reasonableness of requiring an annual inspection of all reservoirs containing more than $1 \times 10^6$ gallons ($4.5 \times 10^3 m^3$). An interdepartmental conference with representatives from the Home Office, the Ministry of Health, the Ministry of Transport, and the Board of Trade reported in December 1926. It was asserted that the main legal safeguard against defective reservoirs was the Common Law liability of reservoir owners. Another four years were to pass, with a change of government in 1929, before legislation was introduced.

## RESERVOIRS (SAFETY PROVISIONS) ACT, 1930

The Act which became operative on 1 January 1931 is described as: *'An Act to impose, in the interests of safety, precautions to be observed in the construction, alteration, and use of reservoirs, and to amend the law with respect to liability for damage and injury caused by the escape of water from reservoirs.'* Some of the main features of the legislation were as follows:

- applicable to reservoirs holding more than $5 \times 10^6$ gallons ($22.7 \times 10^3 m^3$)
- design and supervision of construction by a qualified civil engineer

- reservoir not to be filled until a qualified civil engineer has issued a certificate
- inspection by a qualified civil engineer at intervals of not more than ten years
- qualified civil engineer defined as a civil engineer who is a member of a panel constituted for the purposes of the Act
- undertakers to keep a record of water levels, leakages and settlements.

In Section 7 of the Act it is stated that where damage or injury is caused by the escape of water from a reservoir constructed after the passing of the Act under powers granted after that date, the fact that a reservoir was so constructed does not exonerate the undertakers from any proceedings to which they would otherwise have been liable.

With the advice of the President of the ICE, the Secretary of State constituted two panels of qualified civil engineers:

- Panel A being engineers qualified to design and supervise the construction or alteration of, and to inspect and report on, all reservoirs to which the Act applied
- Panel B being engineers qualified to inspect and report upon reservoirs under Section 2(5) of the Act only (this section deals with inspection).

In 1946, a four panel system was introduced. Engineers in all the panels could now act as the construction engineer as well as inspecting reservoirs. The differences related to the type of reservoir:
Panel I – all reservoirs
Panel II – non-impounding reservoirs
Panel III – non-impounding reservoirs holding less than $50 \times 10^6$ gallons ($227 \times 10^3 m^3$)
Panel IV – non-impounding reservoirs constructed of brickwork, masonry or concrete.

The 1930 Act did not meet with universal approval as shown by a critical letter to the editor of *Engineering* (Griffiths, 1930): *'The immediate question is whether water engineers are to accept this new form of departmental despotism without protest, but the far wider and more important question inevitably follows: namely, is it not obvious that, with such a precedent actually established, the principle of "panels" will be applied to other branches of engineering work, and thus our profession will gradually slide into the control of Government Departments?'* Seventy years later, it can be affirmed that this fear has not been realised. Another criticism in this letter was that admission to the panels was not determined by a body of experts, nor in accordance with any standard of technical qualifications, but rather at the discretion of the Home Secretary after consultation with the President of ICE.

A discussion on the Act took place at the Annual Winter General Meeting of IWE on 5 December 1930. The report (Anon, 1930) that *'This was open to members of the Institution only, and no report is published'* suggests that concern among engineers over some aspects of the Act was not confined to the letter from Griffiths (1930) to *Engineering*. Three years later at the Annual Summer General Meeting of IWE in May1933, the retiring President, J K Swales, affirmed that the Act was a government measure and that the Council of IWE did not approve of its terms. Much of the disapproval centred on the role of the President of ICE who was rarely a water engineer. IWE had proposed to ICE that, with regard to panel appointments, the President should be assisted by an Advisory Committee with representatives from a number of interested bodies including IWE.

The ICE President did set up a small informal advisory committee, but it was 35 years after the Act came into force before the then President, Sir Robert Wynne-Edwards, proposed that advice to the Secretary of State on the composition of panels should come from a special committee set up by the Institution, as provided under Section 8(1) of the 1930 Act. This was formed as "The Institution Committee under the Reservoirs (Safety Provisions) Act 1930". The rules and composition of the Committee were approved by the ICE Council in January 1965. At least six members of the committee were to be on panel I.

One feature of the legislation that can cause surprise is that neither the 1930 Act, nor the 1975 Act which superseded it, contains any technical standards or guidance. Such guidance has been provided in other forms. Shortly after the 1930 Act became law, the ICE set up a committee, with W J E Binnie as chairman, to determine the maximum intensity of flood for which provision should be made. Their conclusions were published in 1933 as the Interim report of the Committee on Floods in Relation to Reservoir Practice. The maximum intensity of rainfall was related to the catchment area and a formula was given for calculating the run-off. It was recommended that the catastrophic flood be estimated as at least twice the normal maximum flood.

## INSTITUTION OF CIVIL ENGINEERS REPORT 1966

Following the implementation of the 1930 Act, there have been no dam failures causing loss of life, and, in view of the link between catastrophic failures and the demand for legislation, further legislation might have seemed unlikely. However, serious incidents and a number of failures have occurred. Wright (1994) listed 10 failures in the period 1960-1971 (including some in Northern Ireland). There was also the effect of major failures abroad and in a report on reservoir safety (Institution of Civil Engineers, 1966) it was stated that:

*'Recent reservoir disasters abroad have focused attention on the legislation which governs the design, construction and operation of reservoirs in the*

*countries which have regulations on these matters. The first step towards a review of existing legislation in Britain was taken following a discussion on 19 February 1963, of a paper presented to the Institution of Civil Engineers by Edward Gruner entitled "Dam disasters". Comments about the operation of the Reservoirs (Safety Provisions) Act 1930 were invited from engineers associated with reservoir design, and a review of these, as collected and analysed by Mr H F Cronin, CBE, MC, Past President of the Institution, was made available to this Committee. International discussion on the safety of reservoirs during the Eighth Congress on Large Dams held in Edinburgh on 4-8 May 1964, showed that several countries were reviewing their regulations. Following that Congress, Mr J Guthrie Brown, CBE, MICE, as President of the International Commission on Large Dams, suggested to the Council of the Institution of Civil Engineers that they should consider setting up a special committee to submit proposals to the Home Office for a revision of the Reservoirs (Safety Provisions) Act 1930.'*

An ad hoc committee, chaired by R W Mountain with Arthur Penman as Technical Secretary, was set up in 1964 and gave detailed recommendations for amending the law relating to reservoir safety (Institution of Civil Engineers, 1966). Three years later the administration of the 1930 Act was transferred from the Home Secretary to the Minister of Housing and Local Government (subsequently the Secretary of State for the Environment) and the Secretaries of State for Scotland and Wales (Ellis, 1975).

RESERVOIRS ACT 1975

In 1970 the Government announced that they intended to introduce new legislation and five years later the Reservoirs Act 1975 received Royal Assent. At the BNCOLD Symposium held at Newcastle in September 1975, Ellis (1975) stated that *'The Reservoirs Act 1975 repeals the Reservoirs (Safety Provisions) Act 1930 and re-enacts the legislation on reservoir safety in a strengthened and more effective form'*. He went on to assert that *'The Act received Royal assent on 8 May 1975 and it is expected that most of its provisions will be brought into force some time before the end of 1975'* This expectation proved to be optimistic. It was on 25 April 1983, during a debate in the House of Lords on the Report of the Select Committee on Science and Technology on the water industry, that Lord Skelmersdale stated that the Government had decided to follow the Select Committee's recommendation and to implement the Reservoirs Act 1975.

The Reservoirs Act was implemented between 1983 and 1987. The Act applies to "large raised reservoirs", that is reservoirs designed to hold or capable of holding more than 25 000$m^3$ of water as such above the natural level of any part of the land adjoining the reservoir. This volume is about 10% greater than the 5 million gallons specified in the 1930 Act. The Act together with the associated Statutory Instruments provides the legal framework within which qualified civil engineers make technical decisions relating to the safety of reservoirs. In effect, the legislation imposes a system of safety checks on reservoir construction and operation. The Act recognises four types of person or organisation with distinct functions and responsibilities:

- undertakers (duties as owner or operator),
- enforcement authority (to ensure compliance with legislation),
- qualified civil engineer (to advise on safety),
- Department of the Environment [now DEFRA] (legislator).

The 1975 Act kept the same approach to reservoir safety as the 1930 Act, maintaining the panel system of engineers, but creating the new role of Supervising Engineer. The enforcement role was given to local authorities. However, in his report to the Secretary of State for the Environment on the failure during construction of Carsington embankment dam, Coxon (1986) recommended that consideration should be given to centralisation of key records relating to certification and inspection of dams.

The new Act embodied most of the recommendations of the 1966 ICE report with the following significant exceptions:

- the size of reservoir was not raised to $10x10^6$ gallons ($45x10^3m^3$)
- a requirement for undertakers to have third party insurance was not introduced
- the Act was not extended to structures storing liquids other than water
- central control of records and enforcement was not introduced, but enforcement was delegated to local authorities (these authorities were given the duty of compiling a register of reservoirs in their area).

FURTHER DEVELOPMENTS

In response to a recommendation by the House of Lords Select Committee, the Department of the Environment instituted a programme of reservoir safety research in 1983. This was linked to the decision to implement the Reservoirs Act 1975 and the realisation that many dams would require remedial works to satisfy modern safety standards. The research programme (Wright *et al*, 1992) has been directed towards the safety of existing dams rather than new construction. The objective is to promote adequate and consistent safety standards for large reservoirs at minimum cost. A Guide to the Reservoirs Act 1975 (Institution of Civil Engineers, 2000) has been published as part of the research programme.

Further developments in reservoir safety legislation may take place. In November 2000 a consultation paper was issued on a Draft Water Bill which included two provisions concerning reservoir safety. The first would transfer the role of enforcement authority in England and Wales to the Environment Agency; the second would give the appropriate authorities in England and Wales the power to direct undertakers to prepare a plan to control or mitigate the effects of flooding consequent on an escape of water from the reservoir. These provisions have not yet become law.

CONCLUSIONS

1. The two dam failures which caused the greatest loss of life occurred in 1852 and 1864 respectively, but no comprehensive reservoir safety legislation was introduced until 1930. A number of reasons for this delay can be identified.

a. There were concerns that legislation would not increase safety and that the formation of a government inspectorate would lessen the responsibilities and liabilities of engineers and owners for their dams.
b. Changes in government during the 1850s and 1860s occurred at critical moments when legislation appeared to be imminent.
c. Although the failures of Bilberry and Dale Dyke were seen as national disasters, they did not enter the national consciousness as the Tay Bridge disaster did in 1879 despite the death toll of 75 in the bridge collapse being smaller than at either of the dam failures.

2. When the record of dam failures and consequent loss of life in the 100 years preceding the 1930 legislation is examined (Charles, 1997; Charles *et al*, 1998), it is clear that the situation was unacceptable by modern standards and fully warranted the legislation that was enacted.

3. Under the 1930 Act, and the 1975 Act which superseded it, the responsibility for reservoir safety lies with owners. The legislation provides a framework within which independent qualified civil engineers make technical decisions relating to reservoir safety and great reliance is placed on the competence of these engineers.

4. The safety record since 1930 has confirmed the soundness of the British approach to reservoir safety legislation. Each time safety legislation has been improved, there has been an improvement in the dam safety record (Wright, 1994).

5. Future safety depends on maintaining the professional and technical standards of qualified civil engineers and on continuing vigilance at every reservoir. This latter requirement includes providing appropriate levels of surveillance and, where necessary, upgrading old dams to acceptable standards. While there has been major expenditure on overflow works to meet improved flood standards, there has not been any comparable effort to upgrade old embankment dams with regard to the hazards of internal erosion and slope instability (Charles, 1998).

ACKNOWLEDGEMENTS

This paper has been prepared as part of the DEFRA Reservoir Safety Research Programme. Mr Michael Kennard kindly lent some papers of

Edward Sandeman which cast light on the events which led to the 1930 Act. Mr Keith Swettenham provided information on the Darwen failure. Dr Arthur Penman and Dr Paul Tedd made helpful comments on the paper.

REFERENCES

Agnew S C (1984). Reservoirs safety legislation. Water Management: a review of current issues. Proceedings of Symposium, London, pp 4.1.1-4.1.11. Institution of Water Engineers and Scientists, London.

Amey G (1974). The collapse of the Dale Dyke dam 1864. Cassell, London. 212pp.

Anon (1848). Dreadful accident at Over Darwen. *Blackburn Standard*, 30 August.

Anon (1910). The Holmfirth flood caused by the bursting of the Bilberry reservoir. *Holmfirth Express*. (Reprinted by Kall-Kwik Printing, Huddersfield, 1991). 87pp.

Anon (1930). Safety of reservoirs. Transactions of Institution of Water Engineers, vol 35, p274.

Binnie G M (1978). The collapse of the Dale Dyke dam in retrospect. Quarterly Journal of Engineering Geology, vol 11, no 4, pp 305-325.

Binnie G M (1981). Early Victorian Water Engineers. Thomas Telford, London. 310pp.

Charles J A (1993). Embankment dams and their foundations: safety evaluation for static loading. Proceedings of International Workshop on Dam Safety Evaluation, Grindelwald, vol 4, pp 47-75.

Charles J A (1997). Acceptable risk and reservoir safety. Dams & Reservoirs, vol 7, no 2, July, pp 7-12.

Charles J A (1998). Lives of embankment dams: construction to old age. 1998 Geoffrey Binnie Lecture. Dams & Reservoirs, vol 8, no 3, December, pp 11-23.

Charles J A, Tedd P and Skinner H D (1998). The role of risk analysis in the safety management of embankment dams. The prospect for reservoirs in the 21st century. Proceedings of 10th BDS Conference, Bangor, pp 1-12. Thomas Telford, London.

Conybeare H (1858). Description of the works, recently executed, for the water supply of Bombay in the East Indies. Minutes of Proceedings of Institution of Civil Engineers, vol 17, session 1857-58, pp 555-575.

Council of Institution of Water Engineers (1925). Safety of reservoirs. Transactions of Institution of Water Engineers, vol 30, pp 78-80.

Coutts J C (1934). Reservoirs (Safety Provisions) Act, 1930: impressions and experiences of an inspecting engineer. Water and Water Engineering, June, pp 380-386.

Coxon R E (1986). Failure of Carsington embankment. Report to the Secretary of State for the Environment. HMSO, London. 176pp.

Davidson W (1996). The Skelmorlie reservoir disaster. The Scots Magazine, vol 144, no 4, April, pp 393-398.

Ellis L E (1975). Reservoirs Act 1975. Inspection, operation and improvement of existing dams. Proceedings of BNCOLD Symposium, Newcastle-upon-Tyne, paper 1.1.

Griffiths P (1930). Letter to the Editor on Reservoirs (Safety Provisions) Act, 1930. Engineering, vol 130, no 3370, 15 August, p211.

Gruner E (1963). Dam disasters. Proceedings of Institution of Civil Engineers, vol 24, January, pp 47-60.

Harrison S (1864). A complete history of the great flood at Sheffield. *Sheffield Times.* (Reprinted by Evans and Longley Associates, Dewsbury, 1974.) 160pp.

Institution of Civil Engineers (1933). Interim report of the Committee on Floods in Relation to Reservoir Practice.

Institution of Civil Engineers (1966). Reservoir safety. Report of the ad hoc committee set up by the Institution of Civil Engineers to submit proposals for a revision of the Reservoirs (Safety Provisions) Act 1930.

Institution of Civil Engineers (2000). A guide to the Reservoirs Act 1975. 209pp.

Lane M R (1989). The Rendel Connnection – a dynasty of engineers. Quiller Press, London. 224pp.

Rawlinson R (1859). Discussion on Water supply to Melbourne. Minutes of Proceedings of Institution of Civil Engineers, vol 18, session 1858/59, pp 391-397.

Simpson A W B (1984). Legal liability for bursting reservoirs: the historical context of Rylands v Fletcher. Journal of Legal Studies, vol 13, June, pp 209-264.

Swales J K (1933). Transactions of Institution of Water Engineers, vol 38, pp 22-27.

Walsh P and Evans J (1973). The Dolgarrog dam disaster of 1925 in retrospect. Quest, Journal of City University, London, no 25, pp 14-19.

Wright C E (1994). UK reservoir failures and safety legislation. Dams & Reservoirs, vol 4, no 3, October, pp 20-21.

Wright C E, Coats D J and Charles J A (1992). The reservoir safety research programme of the Department of the Environment. Water resources and reservoir engineering. Proceedings of 7th BDS Conference, Stirling, pp 411-424. Thomas Telford, London.

# Risk assessment and the safety case in dam safety decisions

D N D HARTFORD, BC Hydro, Canada, and Trinity College, Dublin
R A STEWART, BC Hydro, Canada

SYNOPSIS. This paper outlines in a conceptual way some thoughts on how complex dam safety decisions might be made in terms of a "Safety Case". The ideas are based on the guidance provided by the UK Health and Safety Executive and are consistent with the guidance on the application of the analytic-deliberative approach put forward by the US National Research Council. The conceptual proposal incorporates relevant aspects of experience in other industries and situations where decisions concerning the management of risk engender significant public concern. The role of effective communication between the dam owner's engineers, the owner, the regulator and the risk bearers is discussed.

## INTRODUCTION

Dam owners, their engineers and dam safety regulatory authorities have, for many years, been grappling with the problem of defining how "safe" dams should be and demonstrating that dams are in fact "safe". While the generally excellent safety record achieved by the dams community serves as testament to the success of these efforts and despite significant research efforts on the part of some major dam owners, the underlying problem of defining a "safe" dam has remained essentially intractable.

Although the debate about safety criteria for dams within the dams community continues (e.g. Hydrovision, 2002), the underlying and possibly more important issue of how defensible dam safety decisions can be made in a transparent way receives little or no attention. Dam owners who are liable for the consequences of dam failure and who also require the consent of the public for continued operation of their dams are now faced with the problem of meeting these evolving societal expectations in respect of openness and transparency in public safety management. The remainder of the paper explores how dam owners might achieve these objectives.

## CONTEXT

While the generally good safety record of dams can be held as testament to the effectiveness of established dam safety practice, a great deal of dam safety decision-making is neither transparent nor objective. Recent investigations into applications of risk assessment have revealed it is heavily

*Reservoirs in a Changing World*, Thomas Telford, London, 2002, 510–519

reliant on arbitrarily selected deterministic standards, and the values and opinions of regulators, owners and their engineers, often but not always working within a loosely defined framework of accepted practice. Expert opinion or engineering judgement has long been and remains a cornerstone of dam safety practice, with little in the way of requirements for the scientific validity of dam safety assessments to be demonstrated. There are exceptions, but in most cases the values and opinions of engineers, be they retained by regulators or dam owners, expressed as engineering judgements dominate the decision-making process.

This situation is no longer in keeping with societal norms. There is a need to explore and reveal the considerations involved in the exercise of engineering judgement. The recent safety assessment of the Forth Railway Bridge (HSE, 1996) serves as an excellent example of how this can be achieved. Further, there is a need to broaden the decision-making process to ensure that the values and expectations embodied in the decision-making process properly reflect those of the various parties affected and are not restricted to the values and judgements of those deeply involved. Court rulings, regulatory instruments, existing health and safety laws and proposed legislation pertaining to corporate killing all point to a need for objectivity and scientific validity for defensible decision-making concerning matters of public safety. Public trust in experts continues to diminish and there are increasing requirements for transparency. The report of the World Commission on Dams (WCD, 2000) provides a broad statement of public concern about all aspects of the role of dams in society.

According to Miss J. Bacon, former Director General of the Health and Safety Executive; "It is the nature of risk that, frequently, those who create the risk do not bear its consequences nor its wider costs. So the market does not function properly as a distributive mechanism. The State must intervene to regulate risk." (Bacon, 1999). Concerning dams, which often have the potential to cause a large number of fatalities in a single event, Harold Laski's (Laski, 1930) observation that "we must ceaselessly remember that no body of experts is wise enough, or good enough, to be charged with the destiny of mankind" seems particularly appropriate.

## DAM SAFETY STANDARDS SETTING – INDICATORS OF CHANGE

Traditionally, the dam engineering community, comprising government and private dam owners and their engineers, safety regulators where they exist, learned societies and professional bodies have combined and have implicitly been charged with setting and implementing standards for dam safety assurance. The focus of these efforts has been the pursuit of zero risk by preventing dam failures through defensive "design-in-depth" for normal operating conditions and, in the case of natural hazards through the concepts of the Probable Maximum Flood (PMF) and the Maximum Credible

Earthquake (MCE). The result was a technocratic approach to dam safety decision-making whereby a dam was declared to be safe in an absolute sense if it met these standards.

However, this essentially simple and straightforward approach to decision-making has not proved to be completely successful for several reasons including:

- Concerns about the manner in which these standards were derived and the costs of meeting these standards emerged from within this same community in the 1970's and these concerns have increased ever since.
- The dam safety community has not yet developed a generally acceptable scientifically based risk assessment methodology.
- The responsibilities and interests of the dam safety regulator and the dam owner are quite different, as are the responsibilities and interests of the engineers involved. "Regulator capture" whereby the regulator begins to view things from the perspective of the regulated should be avoided.
- There is a growing need to address the interests of the affected public in a deliberative way (NRC, 1996) through effective communication.
- Dam owners are increasingly aware of their legal liabilities with an attendant need for their decision process to be based on reliable knowledge. This brings requirements for scientific validity of the engineering analysis and transparency of the engineering judgements that pervade the process.
- Health and Safety Law has made significant advances in recent years and the "goal setting" approach to risk regulation has emerged, particularly in the United Kingdom where a 'permissioning' approach applies to higher hazard industries (HSE, 2000).
- The state's regulator has to confront some basic issues: most notably, the need for economic, social and technological progress compared with "zero risk" or "guaranteed safety". (Bacon, 1999, ibid.).
- The dam safety decision-making has analogies in the other higher hazard industries, and the decision-making processes in these other high hazard industries set precedents for dam safety decision-making.

## THE SAFETY CASE CONCEPT AS IT MIGHT APPLY TO DAMS

We have found that the goal setting approach to risk regulation developed in the United Kingdom by the Health and Safety Executive to provide a useful framework for dam safety decision-making. There are many similarities between the decisions that have to be made for dams and the safety decisions made by the Health and Safety Executive. While we recognise that the Safety Case is a legal requirement in the United Kingdom, we see benefits in embodying the essential features of a Safety Case in dam safety decisions.

The primary responsibility in dam safety decision-making rests with the dam owner, the creator of the risk, and the dam owner is responsible for effective management of the risk. This is essentially the same as the guiding principle upon which the UK regulatory regime is based. This regulatory regime comprises three essential elements; basic standards for risk reduction set in goal setting legislation; preventive action to secure compliance, concentrating on high risk areas; and "enforcement on complaint" in low-risk areas - which relies on a degree of acceptance and risk toleration by workers and the public.

The requirements imposed by this risk-regulatory regime are that duty holders are required to identify the hazards, assess the risks and prepare a safety case demonstrating how the risks will be prevented or otherwise controlled. They are also required to set out a safety management system showing how the safety case will be implemented and maintained. Thus in a general way, the requirements of the risk-regulatory regime are comparable with accepted dam safety management procedures which include hazard identification, various forms of risk assessment, and documented rationale for the risk control measures. Although we have yet to document a formal safety case, the essential elements of the safety case are embodied in our dam safety management system (Figure 1).

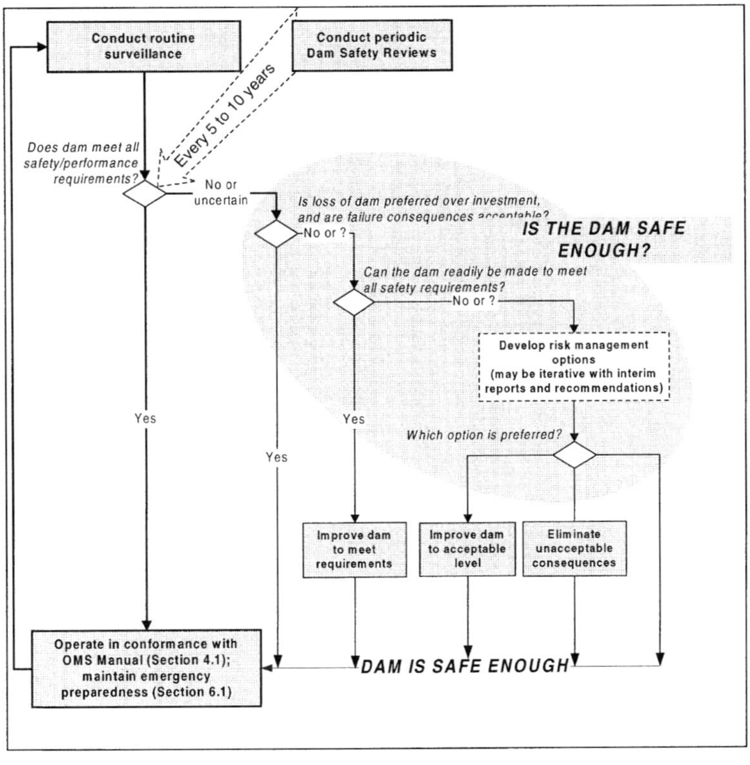


Figure 1. BC Hydro's Dam Safety Management System (BC Hydro, 2002)

The area of safety management that we are presently developing and applying pertains to the As Low As Reasonably Practicable (ALARP) principle. We understand that this involves consideration of three further important elements: "justification" for incurring risks at a particular level; "trades-offs" between costs and benefits; and "proportionality" of response to risk.

The control measures and their costs must be proportionate to the degree of risk, and the uncertainties surrounding the calculation of the risk. The tolerability of the resultant risk will reflect the benefits from the risky activity and the alternative risks associated with not incurring some risk, and that sometimes this will be a political rather a regulatory decision. Ironically, making this ALARP process formal, logical and transparent is proving to be immensely.

## GUIDANCE FROM OTHER INDUSTRIES

Clearly, dams are not the only high hazard facilities that are beneficial to society and yet impose involuntary risks and which have the potential to cause multiple fatalities in single events. In the United Kingdom, the offshore oil and gas industries, onshore major chemical hazards, the nuclear industry and the railways are all subjected to a higher degree of regulation than other industries, because they engender greater societal concern. The decision-making processes in these industries and the way that they are regulated provide guidance for dam safety decision-making and in meeting the expectations of regulators of higher hazard industries. Amongst those guidance that we have found most useful is the decision context framework of the UK Offshore Operators Association (UKOOA, 1999).

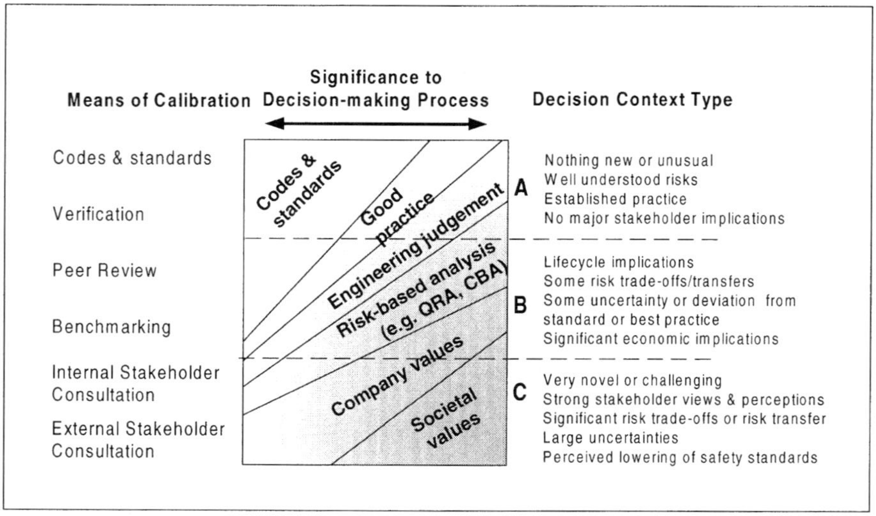


Figure 2. The UKOOA Risk Decision Framework (UKOOA, 1999)

## GENERAL FRAMEWORK FOR APPLICATION

As a matter of principle, dam safety policy making shoúld be considered independently of dam safety analysis issues. Stakeholder interests should also determined independently of policy-making and decision-making although there are obviously important interfaces between these activities. This clear separation of these related activities is consistent with the findings of the Health and Safety Executive's report Policy Risk and Science (OXERA, 2000).

The purpose of a risk assessment is to inform the decision process not to make the decision. Thus the decision context determines the nature and scope of the risk assessment (Stewart, 2000). Risk assessment methods are not considered to be an alternative to established approaches to dam safety decision-making, rather that established approaches that lead to effective risk control are subsets of a more general risk assessment framework. If it is necessary to implement a risk assessment approach, be it qualitative or quantitative, it should be done with a clear understanding of how the results relate to the decision framework.

When using quantitative approaches, it is wise to impose strict rules on the risk assessment process as it is well known and that simple approaches to characterising the uncertainties that pervade complex behaviours (often described as judgmental probabilities) are likely to give an erroneous impression of the nature and extent of uncertainties (NRC, 1996, ibid., Cooke, 1991, Morgan et al., 1991). Hence, decisions reliant on such estimates may be prejudiced, and in retrospect following an incident or failure may be judged to be poor, or in the extreme negligent. Generally, the wise approach is to use risk assessment to inform the decision process, but not used to make the decision.

## PERFORMANCE GOALS

Development of a safety case requires coherent and comprehensive consideration of a number of related factors including: good design; physical (hardware) risk control measures that reflect current authoritative good practice; risk mitigation focused operational and maintenance procedures; effective and properly resourced management systems that are transparent and auditable to ensure that they are working properly; appropriate consideration of human factors that reflect how people behave in reality; and emergency arrangements that mitigate the effect of an accident if all else fails (HSE 2001).

## SAFETY AND RISK ASSESSMENT

One process for characterising the acceptability of the performance of the dam for extreme floods and the need for dam safety improvements is illustrated in Figure 3.

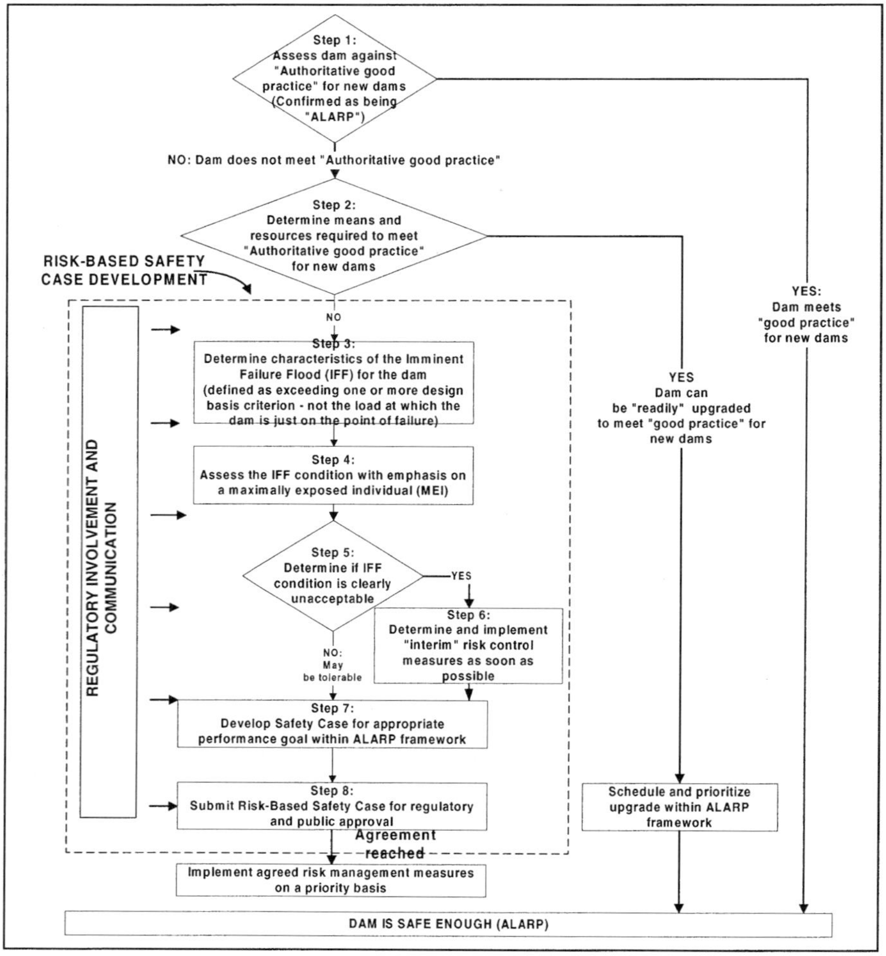


Figure 3. Outline of a Procedure for Selecting Performance Goals for Existing Dams for Extreme Floods.

BC Hydro has been investigating the potential application of these concepts in dam risk management. Examples include Aberfeldie Dam (1997), Coursier Dam (1998), Elsie Dam, (1999) and Coquitlam Dam (2000). The analytical process is illustrated in Figure 4. The border between unacceptable and tolerable set here at $10^{-4}$ is our suggestion. Any decision as to its acceptability is a matter for discussion with the regulatory authorities.

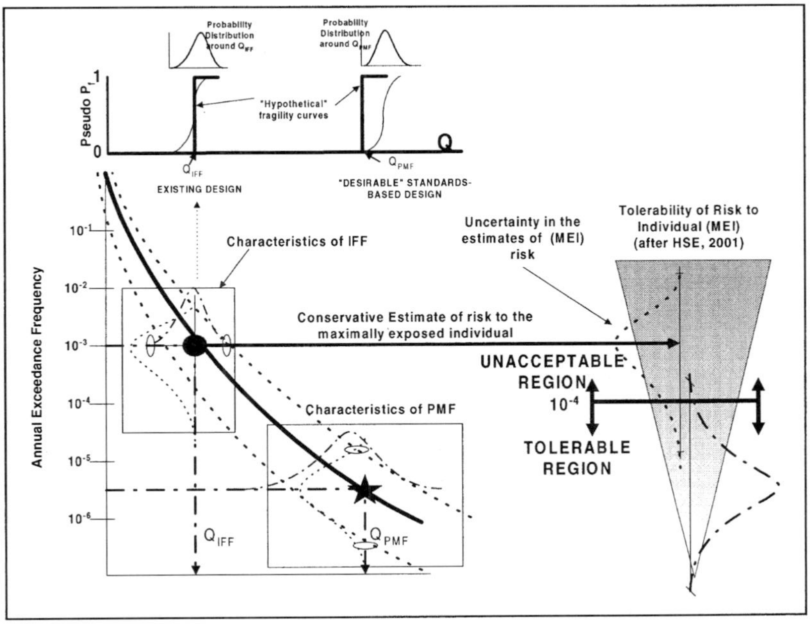


Figure 4. Analytical illustration of the concept for floods set in the framework of the HSE's tolerability of risk framework.

## COMMUNICATION

Open, transparent and effective communication of dam safety and risk management issues with those affected may be of central importance to the process. Deregulation and "corporatisation" of dam ownership has started to changed the relationship between the public and dam safety decision-making. This open communication approach which we have adopted for almost ten years in dam safety has proven to be very beneficial to all parties involved, independent of the size of the community. For example, the small community of Bull River in South Eastern British Colombia wrote the following letter of thanks in the regional newspaper: "The residents of Bull River wish to thank BC Hydro & staff for their concern & professionalism in handling the potential Aberfeldie Dam problem. You could not have done more to help the Community & relieve their concerns. We all feel that we have gained new friends." (Kootenay Advertiser, 1999).

Similarly, it was the success of the communication that provided for the enduring support of the local communities during the investigation and remediation of the W.A.C. Bennett Dam sinkhole in 1996, (Stewart and Garner, 2000). Communication within the professional community is an important part of this process as it assists in demonstrating that the safety management is subject to external and independent technical scrutiny.

## CONCLUSIONS

The purpose of this paper has been to present some ideas about how the Safety Case concept might be applied in dam safety decision-making with the view to having these concepts debated in more detail by the dam engineering profession and the wider scientific and public safety policy communities.

The following conclusions are drawn:

1. The proposed process enables dam safety decisions to be made in a manner consistent with other high hazard activities, thereby permitting the safety of dams to be managed in terms of the societal norms that apply to other higher hazard activities.
2. By focusing on the performance capability of existing dams in the context of what would be achieved through modern design practices, the process provides a basis for judging the acceptability of the safety of dams whose design and construction are not in keeping with modern practices.
3. The concept facilitates making the process of dam safety decision-making transparent.
4. A great deal of knowledge concerning how to handle decision problems of this nature reside outside the dams community and there are benefits to be gained by incorporating this knowledge as appropriate.

We envisage that the greatest challenge facing the dam safety community will be in demonstrating that the risks have been reduced As Low As Reasonably Practicable;

- that incurring risk at a particular level is justified;
- that the "trades-offs" between costs and benefits are appropriate;
- and that the response to risk are "proportionate" to the degree of risk.

Miss J. Bacon (Bacon, 1999, ibid.) summarised the challenge succinctly as follows:

"The task of the risk regulator - and of the scientific and engineering communities - is to *reassert* the concepts of justified risk and of "safe enough"; to demonstrate the effectiveness of good science and technology in providing robust systems of risk management and control; and to make *transparent the* processes undertaken for arriving at scientific judgements and engineering decisions".

This challenge should not be taken lightly and we expect that we will have to devote considerable effort and resources to transform dam safety decision-making from a technocratic exercise dominated by subjective engineering judgement to a transparent process that justifies the taking of risk at a particular level.

ACKNOWLEDGEMENTS
In developing this approach, we have relied heavily on the published work of the Health and Safety Executive and on the advice and guidance of Dr. J. McQuaid.

REFERENCES

Bacon J H (1999). Categories and Structures of Man-made Risks and Related Basic Problems A risk-regulator's perspective. Forum Engelberg, 23/24 March.

BC Hydro (2002). Dam Safety Management Manual (revised annually).

Cooke R M (1991). Experts in Uncertainty. Oxford.

Health and Safety Executive (1996). An assessment by HSE of the structural integrity of the Forth Rail Bridge. HSE Books.

Health and Safety Executive (2001) (revised). Reducing Risks, Protecting People. HSE Books.

Health and Safety Executive (2000). Regulating higher hazards: Exploring the issues.

Kootenay Advertiser (5 July 1999).

Laski H (1930). The Limitations of the Expert. Harper's Magazine.

Morgan M G & Henrion M. Uncertainty: A Guide to Dealing with Risk and Uncertainty in Quantitative Risk and Policy Analysis. Cambridge University Press.

NRC (National Research Council, United States) (1996). Understanding Risk: Informing Decisions in a Democratic Society. Stern P S & Fineberg H V (eds.). National Academy Press.

OXERA (2000). Policy, risk and science: Securing and using scientific advice. Contract Research Report 295/2000. Health and Safety Executive.

Stewart R A & Garner S J (2000). Performance and Safety of WAC Bennett Dam, A Seven Year Update. 53rd Canadian Geotechnical Conference, Montreal.

Stewart R A (2000). *Dam Risk Management – the ICOLD Lecture* Proc. GeoEng 2000, Balkema.

# Risk assessment – its development and relevant considerations for dam safety

J McQUAID, University of Sheffield, UK

SYNOPSIS. Although risk is a way of describing an uncertain outcome, it is often presented as an objectively-determined parameter. Objectivity and uncertainty are unnatural bedfellows. The non-objective judgements that are inevitable in handling uncertainties need to be explained and, in some areas of risk regulation, it is a condition of operation that this explanation should formally be set out in a safety case. A key question is how to ensure the robustness of expert input on risk questions. The paper outlines developments in risk assessment and discusses issues that are considered to be relevant to dam safety.

## INTRODUCTION

The evolution of risk assessment as an underpinning feature of the UK approach to industrial safety regulation effectively began with the report of the Robens Committee(1972). The report laid the foundation by advocating reform to an essentially risk-based regulatory regime. As a consequence, the processes of risk assessment and of decision-making based on proportionality to risk have progressively replaced the prescriptive and inflexible legislative requirements of former times. Prior to the Robens Committee's work, the approach to regulation had been based largely, though not entirely, on the need to respond to experience. The Robens Committee recognised the need for anticipatory action and observed: 'The safety system must look to future possibilities as well as to past experience'. Thus the view of the future based on experience of the past was recognised for its limitations in a time of rapid technological change introducing new hazards and also giving opportunities to reduce existing ones. The goal-setting regulatory regime established by the Health and Safety at Work Act of 1974 addressed the problem by compelling the regulator and the regulated to think about what could go wrong and to base control measures on the results of that thinking. The 1987 EC Framework Directive on Health and Safety and its implementation by the 1992 Management of Health and Safety at Work Regulations sharpened the focus. Risk assessment came to occupy and continues to occupy centre-stage in the UK regulatory system for industrial safety, including risks to the public from work activities.

*Reservoirs in a Changing World*, Thomas Telford, London, 2002, 520–533

## EARLY DOMINANCE OF HAZARD ASSESSMENT

As already remarked, regulation of industrial safety in the pre-Robens era was not entirely based on past experience. That was certainly the case in coal mine safety. At a time of rapid mechanisation, there was a strong recognition of the need to identify what could go wrong and hence to develop measures to prevent the untoward from happening. There was much research on the hazards of new methods of mining. But the emphasis was entirely on the assessment of the hazards in terms of the consequences if things went wrong. The chance of it happening did not explicitly influence the thinking and the priority was to avoid disasters from happening. This was an example of the application of the Precautionary Principle i.e. taking action in advance of experience or the availability of scientific certainty. There was little if any of the distinction in terms of precaution between hazard (the consequences) and risk (the combination of consequences and chance) that is now part of the language of safety. This was also the case in the early 1970s in the assessment of the safety of large-scale transportation of liquefied natural gas (LNG) by sea. Again, the interest of the regulator was only in the consequences of postulated accidents. Worst-case scenarios were of particular concern irrespective of the chances of them happening.

## THE CHANGE TO RISK ASSESSMENT

The Canvey Island Investigation (HSE,1978) was profoundly important. It was a comprehensive investigation of the safety of several large scale chemical installations in a populated area. The investigation brought into the reckoning, for the first time, the assessment of the chance of severe accidents. For many of the scenarios, the assessment was based on informed judgement rather than systematic analysis. The judgements on chance became an explicit consideration in decisions on what was regarded as 'safe', changing the whole basis of assessment from hazard to risk. The Canvey Island Investigation drew heavily on the expertise of the nuclear industry which had rather earlier adapted from sole attention to a 'maximum credible accident' approach to safety judgements. The Advisory Committee on Major Hazards (1976 et seq) in the wake of the Flixborough disaster of 1974 provided an additional stimulus to the practice of risk assessment, especially in its quantitative form.

However, during this period, the world around was also changing. Society and the political system increasingly demanded that serious problems should be distinguished from apparent ones, however interesting and challenging they might be as a matter of scientific and technical investigation. The need was, then, for a better ordered basis for the technical input to the exercise of choosing between conflicting options. Many of the problems of concern brought with them uncertainties of various kinds, e.g. about the protection afforded by equipment or systems and by the actions of human operators interacting with those systems, as much as about the uncertainties in assessing the consequences in the event of failure of the protection. Risk is the natural form for the expression of the uncertain outcomes of events or activities. The information on risk can then be coupled with other influential factors - benefits, convenience, contribution to quality of life, and other aspects of public preferences - in making the choice. Where the analytical basis for the assessment of risk was weak due to uncertainties about the consequences or the chances or both, the regulator must proceed with due caution that takes account of the quality of the judgements applied to the assessment of risk, on which more will be said below.

## INCREASING DIVERSITY OF INPUTS TO RISK DECISIONS

A feature of the development of the risk-based approach has been the growing diversity of the risk community. For a long time, engineers and scientists provided the dominant input, based on the perception that the objectivity of their approach provided all that was needed for decision-making on risk. There is now a multiplicity of other players engaged - government, industry, workers, the public and their representatives. This broadening of the range of participants is a response to the recognition that judgements on risk issues are about both science and the values held by individuals and society as a whole.

Science and technology continue to play a great part in decisions. The risk debate at the technical level embraces the nature of hazards, the consequences and degree of risk, the control measures needed and the relative balance of risks and costs. In relation to the latter, economists are increasingly influential, in view of the need to balance the costs and benefits of risk control measures and the need to unravel these since their totality often includes costs and benefits that are borne by or accrue to parties other than those being regulated and those directly exposed to the risks.

The public is now more frequently involved, with the public attitude often represented as aiming for zero risk. More sympathetically, though, it can be seen as a questioning of the interpretation of the concept of risk as no more than a quantifiable physical quantity, ignoring the importance of public fears, values and beliefs (Watson, 1981). Very often there is a questioning too of the actual framing of an issue in terms only of the determination of the technical measure of the risk, thus pre-empting the debate (Stern&Feinberg,1996).

Finally, there are the psychological and sociological dimensions of risk communication, which has broadened from attention only to the conveyance of information to the active participation of stakeholders at all stages of the decisionmaking. This trend has progressed to the extent that it is now argued (Stirling,1998) that public participation is as much a matter of analytical rigour as it is of political legitimacy. However, there is still a considerable range of views on the extent to which risk perceptions should influence public policy decisions and the manner in which it should be done

(Pidgeon,1998). It is apposite to recall here the advice of Edmund Burke to the public in 1774:
'Your representative owes you, not his industry only, but his judgement; and he betrays, instead of serving you, if he sacrifices it to your opinion.'

## SOME ASPECTS OF RISK ASSESSMENT

Risk assessment provides the essential anticipatory element which underpins compliance with a regulatory regime based on goalsetting. It comprises a structured and systematic examination of the likelihood of harmful events and the consequences should the events occur. It may be based on an extrapolation of past experience, with an in-built assumption that the past is representative of the future. In many circumstances, this accumulated experience will have influenced the development of standards and codes of accepted good practice and these provide the starting point for the consideration of whether anything further needs to be done. In the circumstances of particular interest to this conference, there may be little or no past experience, either direct or plausibly relevant. The risk assessor must then construct a description of how events may develop from a given initial state. Such a description is conventionally referred to as a model of the behaviour of the different elements of the system under stated influences.

Risk assessment provides a powerful means of improving discipline in the deliberation by experts on uncertain phenomena. It serves to improve communication on the qualitative and quantitative aspects of risks and to inform the dialogue on the measures needed to control the risks in accordance with a declared philosophy of safety. The presentation of expert views in the form of a risk assessment framework allows greater transparency to be achieved in respect of:

- the realism with which the experts have represented practical circumstances;
- the way in which the experts have exercised judgement at different points in the risk assessment;
- the extent to which the uncertainties in the physical processes have been taken into account, and
- the sensitivity of risk control decisions to the modelling assumptions.

These desirable attributes reflect the quality or fitness-for-purpose of the risk assessment and its presentation. Fitness-for-purpose therefore relates to the facilitation of the dialogue as well as the innate suitability of the assessment to the problem at hand. The quality of the dialogue surrounding the acceptance of the assessment by the peer review process and by the regulator will be influenced by the perception of quality. There are two particular aspects worthy of further discussion – the role of models and the way expert judgement enters into risk assessment.

## THE ROLE OF MODELS IN RISK ASSESSMENT

There are many aspects to the appraisal of models used in risk assessment (see McQuaid,1998) and only an outline can be given here. It is useful to distinguish between two types of model since different quality considerations apply. The two types may be called 'science' models and 'predictive' models. The predictive models may, and usually do, incorporate science models but go beyond them in having to deal with issues that cannot be subjected to the procedures of science.

A science model is a means of representing the state of knowledge or 'science' concerning a phenomenon. It provides an interpretation in mathematical terms of what is currently known or accepted as physical descriptions of the phenomenon. The quality of a science model, or the 'soundness' of the science, is assured by the normal processes of science. These operate on the way knowledge is acquired by experiment or practical observation, the robustness of the assumptions used in framing the mathematical description and the form of validation of the final result.

The domain of science models is the relatively easier one in which to quality assure, with much reliance on the rigour of peer review in the open literature. This is a powerful means to root out lack of care, ignorance of relevant information or neglect of evidence running counter to the hypothesised model, and promotion of commercial interest by selective disclosure and biased examination of evidence as distinct from advancement of knowledge.

Predictive models represent a conjecture of what might happen under stated assumptions, for example if an earthquake were to occur at a particular location with associated population distribution, location of vulnerable structures, etc. Such a predictive model would incorporate science sub-models describing the progression to the defined failure state. The hypothesised progression will be identified by some systematic technique such as fault tree analysis.

A predictive model is a tool of risk assessment and incorporates assumptions and judgements about the effects of particular practical circumstances. Such assumptions and judgements will not be testable by the methods of science (otherwise the model would be a science model by the above definition). Where judgement has to be exercised, there is a need for conformity to some principles as will be discussed later.

A predictive model is not necessarily a mathematical representation of the conjectures underlying the model. An example is the safety assessment of the Forth Rail Bridge (HSE,1996). This assessment needed a predictive model to be developed in terms of a weighing of the evidence for the continuing structural integrity of the bridge. The result was, in effect, a prediction of the future safety state of the bridge, conditioned by measures to be put in place to correct identified deficiencies in the management of the bridge's safety.

An important factor in quality assessment of predictive models is representativeness. Predictive models are necessarily idealisations incorporating approximations to reality. The presentation of a predictive model needs to be clear about:

- what features of the practical situation are chosen to be represented and why?
- what features are judged not to need representation and why?
- what features cannot be represented and why?

Transparency is key to quality assessment to enable independent judgements to be made.

Validation of models is the process of establishing the extent to which the application objectives of a model are attained, defined in terms of both the average or expected behaviour and the variability about the expected results. Validation of science models will obviously be provided by comparisons with relevant data. Validation of predictive models can rarely be completed in the same way, since experimental data will not be available or even not practically possible to acquire. Sometimes it is possible to obtain validatory evidence from observations following accidental events. Where validation of one kind or another is limited, it is essential to test the sensitivity of the model predictions to plausible variations in the input assumptions.

The main predisposing factors for ensuring model quality are complete openness and diversity of expert inputs. These provide security against the operation, inadvertent or otherwise, of bias which can arise from effects such as 'group think' and slavish conformity to scientific and risk assessment fashions.

## THE EXERCISE OF JUDGEMENT

One of the outstanding features of public risk debates is a lack of understanding of the nature of the science involved in risk decisions. The characteristic of risk decisions is that they are by definition based, to a greater or less extent, on incomplete data and understanding. This contrasts with the prevailing public and media belief that science is about facts and incontrovertible proofs. In actuality, the science in risk assessments is pervaded by judgements and assumptions influenced by scientific values and fashions. They may also be influenced, unfortunately, by commercialism driven by the market value of scientific information and methodologies and this can affect the strength of adherence to a particular scientific view.

The distortion in public and media understanding of the nature of expert advice can easily extend to those who have to make the ultimate decisions on what to do, how much to spend and how to cope with public and media reactions. There is ample evidence of a lack of clarity about the nature of the expert advice on risk issues and a suspicion that extraneous social value judgements are allowed to influence what should be and are expected to be position statements on the engineering and scientific dimensions of risk issues. Traditionally, the engineer dealing with risks has used judgement based on experience in deciding on the degree of conservatism to adopt and has reflected this in a safety factor. This has the benefit of stability in decision making, important in risk regulation, and clarity of a kind about the extent of the engineer's ignorance of the underlying phenomena. But it may also incorporate an undeclared allowance for the engineer's own aversion to risk, with the danger that this trespasses on the decision maker's remit. The nature of goalsetting regulation provides scope for responding to advances in understanding and the process of risk assessment provides a powerful tool for probing these engineering judgements on conservatism.

Much has been written about the different types of uncertainty but the point is that the handling of uncertainty in all its forms requires expert assessment. The process of expert assessment covers a spectrum. At one end, the assessment can be made on the basis of formal and transparent analysis using models that are based on established physical laws and mathematical relationships. The models are supported by relevant and validated data. Moving across the spectrum, the role of formal analysis diminishes to the point where modelling may provide only a partial answer. There may be various ways of representing the features of the problem, each consistent with the available, usually sparse, data. Expert judgement figures increasingly in the conduct of the assessment. The quality of the judgement depends on depth of relevant experience and availability of evidence from similar circumstances accumulated over time or across different disciplines. Eventually the point is reached where the uncertainties and lack of understanding are such that the assessment must be based largely or wholly on expert opinion. The quality of the assessment then depends on such characteristics as credibility, standing and independence of the assessors.

The spectrum as described is in decreasing order of tractability, by which is meant the ease with which the reasoning behind the conclusions can be drawn out and explained. The process of drawing out or eliciting judgement has received much attention, particularly in nuclear safety studies. Formal methods of elicitation have been developed, covering the manner of numerically representing the judgements of individuals and of aggregating them into a combined view using standard statistical techniques. But much expert or professional judgement is elicited informally and the weighing-up of the consensus view can be largely subjective. The conclusion may simply emerge or appear to emerge.

Risk management decisions need to take due account of the nature of the assessment The degree of precaution to be observed or the safety measures that have to be put in place will depend on where the assessment lies on the spectrum. Decision makers need to understand the basis of the assessment on which the advice they are given is based. The compression of what is often a complex mixture of analysis, judgement and opinion into a single numerical expression of risk can lead to an impression of certainty based on science as it is commonly understood.

There are several questions arising from this:
- how should advisers be selected?
- what is the nature of the advice?
- what form of consensus is offered?
- what is meant nowadays by an independent adviser?

The questions point to a need for:
- procedural guidelines for decision makers so that they can better organise the provision of the advice taking account of the nature of the decision and the extent to which they may have to defend it – a consideration particularly relevant to Government;
- recognised protocols for eliciting judgements from experts and for combining those judgements into a consensus view properly defined;
- the judgements to be made tractable in the sense defined above and, most importantly
- the advice to be seen as a living entity, with the advice to be revisited in a preordained way as new information emerges which may challenge the basis of the original advice.

REGULATING HIGHER HAZARDS

In the UK goal-setting regulatory regime, all duty holders – those with duties under the law – are required to identify hazards and assess risks. Some industries are in a high hazard category with a potential for catastrophic accidents, albeit that they may be low risk as demonstrated from either actual experience or from a risk assessment. In such cases, the requirements on the duty holder are enhanced under regulations specific to each industry. Currently, there are regulations in place for four industries – nuclear sites, onshore chemical plants (above a certain size in terms of hazard potential), offshore exploration and production rigs, and railways. The general requirement is that the duty holder has to demonstrate in a safety case or a safety report how the risks will be prevented or else controlled to a level that is as low as reasonably practicable. In addition, the safety case must set out how safety will be managed so as to achieve the objectives of the control regime. The safety case must be notified to the regulator (HSE) who will assess it against published principles and conduct a dialogue with the duty holder as to its validity in terms of the requirements of the law and, if necessary, seek further evidence in cases of doubt.

The regulations impose more rigid frameworks on the regulator and require more positive regulator engagement than other approaches to regulation. Nuclear installations can only operate under licence from the regulator who can impose conditions, revoke or refuse a licence. For onshore chemical plants, the regulator has power to prohibit operation if insufficient information is provided and must prohibit operations if serious deficiencies are identified in the measures taken. For offshore installations and railways, the regulator must formally accept the safety case and it is an offence to operate without an accepted safety case. The import of these commitments on the regulator has been to bring the tag 'permissioning' into use in collectively describing the different regulatory regimes.

A comprehensive evaluation of the permissioning regimes has been published by HSE (HSE,2000). This summarises their rationale as being to give society an added level of confidence that duty holders are capable of discharging their legal responsibilities to control the risks. They provide the regulator with additional levers which can be developed in the light of the industry's performance. The duty holder's documentation provides part of the basis for targeting regulator intervention. However, it is emphasised that permissioning does not provide a guarantee of safety in the operation of the duty holder's arrangements.

The requirements of permissioning regimes compel the duty holder and the regulator to engage in a dialogue at a high technical level about every aspect of the assessment and control of risks and of the costs of options for risk reduction. The requirements are used as a basis for discussion and diagnosis rather more than for enforcement and sanctions by the regulator. A permissioning regime can only operate effectively if the regulator's staff are very well informed scientifically and technically and have the experience to apply standards consistently so that they are able to conduct the dialogue with the duty holder on an equally well-informed basis. They must be able to defend their judgements on whether risks have been reduced as low as reasonably practicable beyond which expenditure to achieve further reductions would not be justifiable. This brings with it the danger of a perception that responsibility is transferred from the duty holder to the regulator. The avoidance of this danger requires rigorous traceability of the basis of all decisions. The permissioning regimes all have arrangements, differing in detail, for communicating with the workforce and the public about the content of safety cases which by their nature make clear that responsibility for risk control lies with the risk creators.

## THE NEED FOR TRANSPARENCY AND EXPLANATION

The intense public interest in risks is a pervasive aspect of modern living to which anyone involved in the assessment and management of risks, especially large scale risks, must pay due regard. There are incessant demands for reassurance. Controversies are readily fuelled by highly publicised disagreements between scientists. There is a natural concern that technological progress can be compromised by unfounded fears. There is an equally natural concern that decisions should take account of the best available science and, where the science is in doubt or strongly disputed, that there should be a presumption in favour of precaution. Overall, there is a recognition of the need for a better public understanding of risk issues.

All of these concerns give rise to a quest for greater openness and transparency in the science underlying policies on risk. This quest is an attempt to substitute for the lack of objectivity in matters pertaining to risks on the argument that any inclinations to take a biased or less than fully informed view will thereby be exposed for scrutiny by peers and stakeholders. The issue has been cogently stated by Weinberg (1972) thus: 'In a democratic society, the public's right of access to the debate in the sense of being informed about it and participating in it is as great as the public demands it to be. Especially where experts disagree, the public has little choice but to engage in the debate at an earlier stage than the experts themselves find convenient or comfortable.'

This need has been addressed by guidelines issued to UK government departments by the Chief Scientific Adviser (OST,2000). The guidelines are structured around key principles:

- early identification of research needs should always be the aim;
- government departments need to have a developed network of sources of intelligence;
- evidence pertaining to an issue should be subjected to sufficiently questioning review from a wide ranging set of view points.
- collaboration and involvement of all those with an interest should be fostered, and
- data should be made available as early as possible to enable a wide range of research groups to tackle the issue.

Adherence by government departments to the guidelines is monitored by a Ministerial committee. Much of the scientific advice to government that generates public concern relates to risk issues. In turn, much of the public concern relates to the manner and content of communication by government departments on the issues. Guidelines on risk communication for government departments have been developed (HSE,1998). The guidelines recognise that effective communication is a two-way process and requires arrangements for the engagement of stakeholders in the deliberation on courses of action.

The overall objective of the various actions - openness of the research process, rigour in the formulation of advice and effectiveness of communication - is to improve the level of trust and confidence in risk regulation. Beyond that, there is a need also to achieve a proper balance between regulation by government and allowing the public to make their own decisions - a perennial problem in a democratic society.

## IMPLICATIONS FOR DECISION MAKING

The developments that have been reviewed above have implications in various ways for the approach to decision making by the regulator and the regulated. The regulatory regime for industrial safety confers considerable discretion on the regulator. The processes of risk assessment (as a means to ensure proportionality of action) and of cost benefit analysis (as a means to ensure economic efficiency of those actions) provide important sources of defence against arbitrariness in the exercise of that discretion in making regulatory decisions. They serve to expose the thinking of the regulator to scrutiny in the course of consultation on legislative developments. In addition, in a goal setting framework of regulation, the regulator has to make judgements on compliance by the regulated with the requirements of the law. An important step was the recent publication (HSE,2001a) of the principles by which HSE judges whether risks have been reduced as low as reasonably practicable. But these steps are not in themselves sufficient and increasingly this is so in the light of greatly increased public concerns and need for reassurance across the whole spectrum of risks. As a consequence, there has been an evolution of measures to secure that reassurance. The principal measures to note are as follows:

- the adoption and promulgation of principles of good regulation by the Cabinet Office (Better Regulation Task Force,1998). The principles are summarised under the titles of transparency, consistency, targeting, proportionality and accountability;
- the publication by all government departments and agencies of explanations of the influences they take into account in their decision making on risks within their jurisdiction. For example, HSE has published (HSE,2001b) the generalisation of the Tolerability of Risk framework (HSE,1992) to the complete range of HSE's risk decision making.
- the intention of the Government to publish a Statement on Risk as part of its implementation of the recommendations of the Phillips Inquiry into BSE (Phillips, 2000).

All of these measures, welcome as they are, do not represent a significant departure from the top-down model of risk communication which assumes that public concerns arise from a deficit of knowledge about risk issues and how they are handled. The need for other measures besides the supply of more information has been increasingly pressed in recent years. This has resulted from sometimes intense media attention and the growth of single-interest groups able to command and organise sources of counter information. The public can be understandably confused and has tended in opinion polls to reflect a preference for information from non-official sources. The end result is a pressure for inclusionary forms of dialogue between the regulator and stakeholders, involving participation and engagement at all stages of the decision making process. The involvement needs to begin at the earliest stage when issues are being framed since the framing can pre-empt the form of the solution and exclude options that, in the view of protagonists, need to be retained throughout the process of deliberation. Not to do so results in alienation of those with a particular interest and opposition to the eventual solution however 'rational' it might appear to its proponents. However, the need to get a balance between the different considerations has not yet resulted in any general agreement on the way that inclusion should be achieved.

A number of participative structures have emerged in recent years, including focus groups, consensus conferences and citizens juries. These complement and do not displace moves towards more openness in the normal institutional mechanisms such as advisory committees. However, the effectiveness of these evolving structures has still to be tested, in terms of the quality or perceived quality of the decisions as measured by success in damping down problems and controversies. It seems logical that the choice of participative process should be matched to attributes of the problem at hand and of the decision required.

The need for innovation in this area has been due in part to the way in which discussion in the public arena can be influenced, sometimes grotesquely so as for example with GMOs, by the attentions of the media. There is as yet a poor understanding of the reasons why some risk issues can attract great media attention and public outrage while other risks, of greater importance by any objective measure, are virtually ignored. Any moves to seek improvement through participative structures need to be properly informed on this phenomenon of the social amplification (or attenuation) of risk (Petts et al,2001).

## FURTHER DEVELOPMENTS

The evolution of risk-based decision making and of the social and political climate in which it is conducted will continue to present challenges to the regulator (Rimington,1993). Technological innovation and demands from society for cleaner, healthier and safer conditions will have the potential to generate conflicts. If these mutually desirable aspirations are to be achieved, some improvements in the fundamental underpinning of society's capacity to cope will be necessary. I will conclude by mentioning three areas relevant to this conference.

First, there is the need at university level for a greater awareness and understanding of risk and risk control concepts to become an accepted part of the education of engineers. Allied with this would be an exposure of undergraduates to explanatory frameworks for the exercise of professional judgements in making decisions involving uncertainties and incomplete information. The welcome trend to output-based or competence standards should be beneficial towards these ends. A necessary requirement will be that the universities should be able to draw on the experience of regulators and industry to provide illustrative material to enable the integration of concepts into the teaching of individual subjects.

Second, there is a need to promote to the public that the main benefit of risk assessment arises from the discipline it imposes on the process of anticipation in order to counter the perception that its purpose is to provide an exclusive and technically-oriented fix to issues where public attitudes and values may be paramount.

Third, there is a strong argument for extending the use of safety case regimes to other areas of government regulation where risks affect the public or give rise to public concerns, with full exposure of the principles of assessment and public availability of the safety case information. The ordered recording of the basis for decisions would do much to engender public confidence.

## REFERENCES

Advisory Committee on Major Hazards (1976) 1st Report. (1979) 2nd Report. (1984) 3rd Report. London. HMSO.

Better Regulation Task Force (1998). Principles of Good Regulation. London. Cabinet Office.

HSE (1975). Canvey: An investigation of potential hazards from operations in the Canvey Island/Thurrock area. London. HMSO.

HSE (1992). The Tolerability of Risk from Nuclear Power Stations. London. HMSO.

HSE (1998). Risk Communication: A Guide to Regulatory Practice. London. HSE.

HSE (2000). Regulating higher hazards: exploring the issues. London. HSE. Available at www.hse.gsi.gov.uk.

HSE (2001a). Principles and guidelines to assist HSE in its judgements that duty holders have reduced risk as low as reasonably practicable. Available at www.hse.gov.uk/dst/alarp1.htm>

HSE ( 2001b). Reducing risks protecting people.

McQuaid J (1998). Quality considerations in predictive modelling. Proc. Fire and Explosion III Conf. I. Mech. E., London,pp1-6.

OST (2000). The Use of Scientific Advice in Policymaking. London. DTI.

Petts J, Horlick-Jones T & Murdock G (2001). Social Amplification of Risk: The Media and the Public. Contract Research Report. Sudbury. HSE Books.

Phillips Lord (2000). The BSE Inquiry. HMSO.

Pidgeon N (1998). Risk perception versus risk analysis. Reliability Engineering and System Safety, Vol.59,pp1-4.

Rimington J D (1993). Coping with Technological Risk: A 21st Century Problem. The 1993 CSE Lecture. London. Royal Academy of Engineering.

Robens Committee (1972). Safety and Health at Work. London. HMSO.

Stern P C & Feinberg H V (Eds) (1996). Understanding Risk: Informing Decisions in a Democratic Society. Washington DC. National Academic Press.

Stirling A (1998). Risk at a Turning Point? J. Risk Research, Vol.1,No.2, pp97-109.

Watson S R (1981). On risks and acceptability. J. Soc. Radiological Protection Vol.1, No.4,pp21-25.

Weinberg A M (1972). Science and trans-science. Minerva, Vol.10,pp209-222.

# Multi-attribute performance monitoring for reservoir systems

J W HALL, University of Bristol, UK
J W LE MASURIER, University of Canterbury, New Zealand
E A BAKER, University of Bristol, UK
J P DAVIS, University of Bristol, UK
C A TAYLOR, University of Bristol, UK

SYNOPSIS: A methodology for managing the performance of complex dam systems is described. The system is represented hierarchically, so that high level business decisions and more detailed operational decisions can be supported by the same methodology. Performance of each sub-system is captured by a set of performance indicators. Evidence of performance is assembled from all available sources. Uncertainty in the available evidence is modelled using Interval Probability Theory, providing a commentary on sources and implications of uncertainty in the decision.

## INTRODUCTION

The aim of the research described in this paper has been to develop new decision support techniques to enable performance-based management of complex civil engineering infrastructure systems. The work is focussing on the group of economically important and safety-critical infrastructure systems, including dams, flood and coast defences and engineered and natural slopes, with the following characteristics:

- The physical failure mechanisms are complex and site-specific. Available models of the failure mechanism have significant deficiencies.
- The structural behaviour is spatially and temporally varied. This is often associated with natural variability in loading regime (wind, wave, rainfall, seismic) and geotechnical conditions.
- Monitoring information tends to be scarce and can be expensive to obtain.
- Because of both the scarcity of quantitative information and the complexity of the physical processes there has traditionally been a major element of expert judgement in condition monitoring.
- Condition assessments are characterized by uncertainty. Consequently monitoring and remediation resources can be misdirected.

*Reservoirs in a Changing World*, Thomas Telford, London, 2002, 534–550

## ANALYSIS OF THE PROBLEM DOMAIN

In the UK there are roughly 2500 dams of a capacity exceeding 25000m$^3$, making them subject to the provisions of the 1975 Reservoirs Act. These dams are subject to statutory safety checks, which are based largely on the experience and engineering judgement of the Inspecting Engineer. The individual responsibility that the Act places on engineers involved in safety assessment is regarded as an enduring strength of the now aging legislation. However, despite the tradition of tight regulation, there is some concern that the prescriptive approach to safety in the dam sector may not have kept pace with developing approaches to risk and hazard management. So, for example, provisions for surveillance of dams are based on fixed intervals rather than on concepts of risk-based surveillance. For hazards that have in the UK only become a concern in relatively recent years, notably seismic hazard, there is a great deal of inconsistency of approach. Recent developments in hydrological analysis of spillway capacity have also caused confusion (Hartford, 2000). There is an ongoing tendency to invest in dam improvements, but the evidence of beneficial impact of that investment on asset performance is inconclusive. Fundamental to the contemporary problems of dam safety assessment in the UK (as well as in many other countries) is the aging nature of the dam stock. Concomitant is the loss of expertise in dam design and loss of corporate memory of dam behaviour.

There is some reluctance in the dams sector to adopt widely reliability methods. Amongst the rather complex reasons for this reluctance (Blockley, 1999a) are the recognized limitations of existing models of failure mechanisms and scarcity of data.

Civil infrastructure asset management decisions are a multi-disciplinary endeavour involving a complex set of technical, economic and environmental issues (Hsieh and Liu, 1997, Chowdhury et al, 2000, Hastak and Abu-Mallouh, 2001). Individuals are engaged in cycles of decision-making in their own domain, which contribute to key points of resource commitment in the collective process (Mintzberg et al, 1976, Boland et al, 1990). Yet, particularly in the public sector, it can be unclear when and by whom these decisions are actually taken, being the result of negotiation processes that take place in parallel at several levels within the organization.

The processes of options analysis and evaluation can involve assembling and manipulating vast quantities of evidence. The evidence will appear in a range of formats, including dense numerical model results, textual evidence in technical reports, analogous cases, expert judgements, and perceptions and value judgements from the wider stakeholder group. In other words the evidence appears at very different levels of granularity and does not lend itself to being compressed into a single format. Whilst there may be a large volume of information relating to a decision, it is on the whole only of partial relevance, incomplete and sometimes conflicting.

A further characteristic of the domain is the growing awareness of the impacts of uncertainty and the need for improved decision-making. The quest for improved decision-making is driven by intensifying organizational and cultural change. Decision-makers in both the public and private sectors are under great pressure to use resources efficiently. There are also increasing demands to identify and mitigate adverse impacts of asset management decisions. At the same time in-house expertise has been reduced due to down-sizing and out-sourcing of technical services. As a consequence, and also due to greatly improved communication and modelling technologies, decision-makers are facing intense information processing demands (Hall and Davis, 2001).

In the light of the analysis of current practice reported above, the following needs for decision support were identified:

1. to assemble evidence about asset condition and performance from diverse sources and represent it in a common and coherent model;
2. to externalize expert judgements;
3. to provide a commentary on sources and implications of uncertainty in the evidence;
4. to provide a platform for testing the implications of alternative asset management options (including data collection options);
5. to facilitating dialogue between experts and other decision stakeholders.

## KEY PRINCIPLES

The challenges outlined above have been addressed by the development of generic principles and a software tool that helps the non-expert user to implement those principles in a straightforward way. The following key principles are proposed:

1. The infrastructure system of interest is described *hierarchically*.
2. The hierarchy is constructed by considering the *processes* that the system enacts.
3. Performance of all systems and sub-systems is described by a *figure of merit*, which is a non-dimensional measure, on a 0 to 1 scale, of how the system is performing against objectives.
4. The figure of merit is calculated by assessing evidence of performance from either or both of two sources:
    - the figures of merit of sub-systems that are below the system of interest in the hierarchical system model,
    - and performance indicators that are associated with the system of interest.

5. Evidence of performance is assembled from all available sources, ranging from monitoring measurements and inspection records, design calculations and model studies to expert judgements, analogous cases and accounts of past failures. All of these types of evidence may be used as performance indicators.
6. Performance targets are expressed as *value functions*, which map from the (usually dimensional) scale of the particular performance indicator to a non-dimensional scale of performance relative to objectives.
7. Uncertainty in performance indicators, value functions and figures of merit is handled explicitly.
8. Asset managers may be interested in specific aspects of performance, for example cost, safety or environment, as well as the overall figure of merit, so it is possible to isolate system performance and performance indicators that relate to these aspects.

The main elements of the proposed modelling approach are illustrated in Fig.1. The photograph on the bottom left hand side of the diagram represents the 'real' system of interest, in this example a reservoir system. Abstracted from this are measurements of performance (where the term 'measurement' is used in its most general sense, as discussed above) and a hierarchical system model. The performance indicators are associated with one or more relevant sub-systems in the hierarchical model. Value functions are based on organisational values and objectives, codes of practice and company and regulator standards. Performance indicators are projected through value functions and weighted to generate a figure of merit for each sub-system. A revised set of weightings is used to generate figures of merit for specific aspects of system behaviour. The conceptual and theoretical structure of the proposed approach is described in detail by Hall et al, 2002.

Process

A process is a purposeful activity, in the sense that it enacts a transformation in a controlled manner. Process is a fundamental construct in the analysis of purposeful systems like civil engineering infrastructure systems. Fig. 2 illustrates the primary transformation that the process enacts and the second order control activities that ensure that the process continues to deliver the required performance in a dynamic environment. The transformation between demand and response may apply to physical loads, but equally well to more general demands placed on the system. The process may thus be thought of as taking some resources, be they physical or information resources, and transforming them into an output.

The transformation is enacted by some sub-system within the overall system under consideration. There is therefore a link between the processes that delivers a desired response and the sub-system that enacts that process. The sub-system may be 'hardware', 'software' or 'bioware' (Wymore, 1993). The concept of process therefore combines in generic terms 'hard' and 'soft' aspects of general systems, helping to overcome the traditional and rather unproductive distinction (Blockley, 1999b). A process perspective emphasizes the dynamic nature of infrastructure systems, focussing on how they deliver performance.

In the current research the aim has not been to develop workflow description of the process, but rather an overview of the system at a range of levels of resolution. A hierarchical process-oriented view of infrastructure systems has therefore been adopted. The concept of process provides a rational criterion for decomposing the system.

Hierarchical representation of processes

The system of interested is represented by a hierarchical model of nodes and links. Higher levels represent more abstract descriptions of the system, whilst lower levels will be more closely related to elements of the physical system. The extent and structure of the network is at the discretion of the user, but can be extended up to the highest-level processes within the organisation and down to whatever level of detail is appropriate. The aim is to structure the problem in a way that forms the basis for logical thinking, rational debate and new insights into the decision problem. The notion of hierarchical description of systems is fundamental to systems thinking (Haimes, 1977, Haimes and Jaing, 2001, Checkland, 1981) and well established in practice for infrastructure management (see for example Ezell et al, 2000a, Ezell et al, 2000b, Hastak and Abu-Mallouh, 2001).

The hierarchy is in the first instance built 'top-down', though model building will usually be an iterative process. Having identified the high level processes within a system, the user asks themself what are the sub-processes that are required to enact the high level processes. Note that the model is not merely an inventory of the physical elements of the system. It can include human sub-systems, and should be structured to represent the processes that the sub-systems enact, rather than necessarily representing their physical proximity or connectedness.

As the example in Fig. 1 illustrates, the hierarchy is a directed acyclic graph rather than being a strict tree structure i.e. each sub-system can be connected to more than one super-system. The only constraint on model structure is that it has to be hierarchically layered.

There will inevitably be an element of judgement in the development of the model structure, since the criteria for model decomposition are open to different interpretations. However, experience with this approach to hierarchical process modelling (Davis and Hall, 1998) has demonstrated that by using a combination of group sessions and guidance from the process modelling expert it is possible to develop reasonably stable model structures.

Performance

Performance can be thought of as those aspects of system behaviour that are relevant to meeting objectives. A system may enact several processes, and there will be several perspectives on any given process. However, only the behaviours and perspectives that are relevant to the objectives embody the performance of the system (Fig. 3).

Traditionally, performance has been treated in rather crude terms in engineering systems. At the most basic level, performance has been constrained to describing the system as either 'failed' or 'not failed'. This is extended to consider perhaps two performance criteria, serviceability and ultimate limit states. More recently, the move towards performance-based engineering is leading to multi-attribute descriptions of performance under a whole range of loading conditions (SEAOC, 1995, Hirano et al, 2000, Crandall and Freeborne, 2001, Sato and Ogi, 2001). This more general notion of performance has been adopted.

Performance indicators

Evidence about performance of a system or sub-system is provided by a set of *performance indicators*. As illustrated in Fig. 3, the performance indicators represent a sub-set of the system state variables that are relevant to the system objectives. Objectives are themselves derived from organisational and stakeholder values.

High level performance indicators may, for example, be the probability of catastrophic failure, the probability of interrupted service to customers or the Net Present Value of an asset. Lower level indicators may, for example, include rates of structural decay and maintenance costs. Evidence of performance is assembled from all available sources, ranging from monitoring measurements and inspection records, design calculations and model studies to expert judgements, analogous cases and accounts of past failures. In the systems being studied, monitoring information tends to be scarce and can be expensive to obtain, so no evidence is excluded by pre-specifying a particular (for example probabilistic) format. Performance indicators may be derived from monitoring activities or from model predictions of future behaviour, expressed for example as probabilities of failure or fragility curves. The measured value of a performance indicator may be a numerical value, but could also be a linguistic statement (e.g. "poor" or "very good").

A performance indicator may provide evidence about the performance of more than one sub-system. A performance indicator may be directly measured, or it may be the result of a computation of other performance indicators. For example the average age of plant, illustrated in Fig. 4, is the average of each of the age performance indicator for each individual item of plant.

Valuing performance: value functions and multi-attribute weighting

A given sub-system will usually have a range of performance indicators associated with it, which may be measured against different dimensions. Therefore, in order to generate an overall estimate of the performance of a given sub-system it is necessary to:

1. Map all of the performance indicators onto a common (non-dimensional) scale;
2. Compare each of the performance indicators with targets that represent acceptable performance, or, in general, functions that value performance;
3. Weigh the relative importance of the various performance indicators.

The first two tasks are achieved by mapping the performance indicator through a value function that represents how the user rates different levels of performance and maps the performance indicator onto a non-dimensional scale (French, 1988, Wymore, 1993). The third task is achieved by applying a weighting to each of the performance indicators.

Fig. 5 illustrates the general shapes of value functions. Codes of practice or regulatory thresholds are equivalent to a stepped value function (Fig. 5b) – performance one side of the threshold is 'acceptable' and given a score of 1, whilst performance on the other side is 'unacceptable' and given a score of 0. Under some circumstances it will be desirable to 'smooth' the stepped value function, which is the S-shape function (Fig. 5f). Economic benefits will conventionally be represented by a linear (Fig. 5a) or decreasing marginal value (concave) function (Fig. 5c).

The value function has to be chosen by the user in the light of organisational objectives, codes of practice, regulatory standards etc. The user defines the value function by specifying:

1. The type of function.
2. The upper and lower bounds on the score (i.e. bounds beyond which performance is scored as either 0 or 1).
3. A further parameter for curved functions, which sets the amount of curvature.

Propagating evidence through the hierarchy

Besides the evidence from the performance indicators associated with a given system, the performance of its sub-systems also provides evidence of performance. In other words, as well as being *measured locally*, evidence about performance is *propagated* through the hierarchy. The amount of evidence provided by sub-systems depends on the criticality of those sub-systems to the performance of the system. The lowest-level sub-systems in the hierarchy will just have measured evidence of performance, but all other systems will have propagated evidence and will usually also have locally measured evidence. Propagation of evidence is achieved using the uncertain inference mechanism of Interval Probability Theory (Hall et al, 1998) and is discussed in Hall et al, 2002.

Specific aspects of system attributes

The figure of merit provides a summary of system performance against a range of objectives. Asset managers may be interested in seeing how the system is performing against specific objectives, for example, cost, safety or environment. This is achieved by applying a different weighting to each performance indicator according to the aspect of system performance under consideration. So, for example, the number of days lost due to accidents on a given system can apply to the safety aspect, but perhaps also to cost and operations. By weighting the performance indicators in this way it is possible to generate multiple views of the system for a range of different aspects, as illustrated by the multiple sheets on the upper right hand side of the diagram in Fig. 1.

## HANDLING UNCERTAINTY

Performance indicators and figures of merit will inevitably have uncertainty associated with them. The uncertainty will be both due to variability in measurements and predictions and due to epistemic uncertainties in their dependability.

The figure of merit can be interpreted as a measure of belief in the hypothesis that the given system is performing satisfactorily. In other words it can be interpreted as a subjective probability of the truth of the hypothesis. A probability interval is used to represent the uncertainty in this belief measure. In the current research the aim was to use a relatively straightforward approach to evidential reasoning, called Interval Probability Theory (IPT) (Cui and Blockley, 1990, Hall et al, 1998), which retains the desirable properties that have made evidence theory more attractive than conventional Bayesian approaches.

## IMPLEMENTATION

The method has been implemented in a Windows-based software tool. The tool comprises of a hierarchical systems model linked to a database of performance indicators, with the following key elements:

1. A graphical tool for drawing hierarchical models.
2. A model manager, to navigate large models and switch between alternative special views.

3. A database of performance indicators, which is intended to be compatible with an organisation's database and intranet systems.
4. A graphing tool for illustrating how performance indicators have varied with time.
5. A library of parameterised value functions, which can be chosen and adapted by the user.
6. An inference engine for implementing IPT.

Graphical model construction

Each system in the model has a coloured bar (an 'Italian flag') associated with it (shown beneath system name), which is a graphical representation of the interval-valued figure of merit (see Fig. 1). The green proportion (shown light grey on the left of the bar) represents the evidence for satisfactory performance, the red proportion (shown dark grey on the right of the bar) represents the evidence against satisfactory performance and the white portion between the two represents the uncertainty. The Italian flags provide an immediate overview of system performance, enabling the user to identify areas of poor performance and their implications. Further interrogation takes the user to the relevant entries in the database of performance indicators (see below) and the value functions through which they have been projected.

Database of performance indicators

Every performance indicator is held in a database (which in many organisations already exists). In the database the following is held:

1. The name of the performance indicator.
2. Its current value.
3. Its dimensions.
4. The sub-system(s) it provides information about.
5. A default value function.

Where available, a time series of performance indicator measurements is archived and can be viewed with a graphing tool.

Higher-level performance indicators will often be aggregations of performance indicators at lower levels in the process model. It is therefore possible to sum or average over other performance indicators in the database. To do so, the arithmetic operation and database references are entered in the relevant field.

A special case of uncertainty is when performance indictors are recorded as linguistic values, for example on a five-word scale from 'very poor' to 'very good'. If information on confidence in the linguistic judgement is also obtained, again on a five-word scale, then it is possible to use these two pieces of information to generate in interval measure, again on a [0, 1] scale (Fig. 6). So, for example, a judgement of 'good' performance made with 'medium' confidence yields an interval probability of [0.62, 0.87]. Linguistic performance indicators are elicited directly from the expert as a judgement of performance relative to objectives so there is no need to project the performance indicator through a value function.

IPT inference engine

Having established the model structure and captured the performance indicators, the final step is to enable propagation of evidence through the hierarchy. At each level in the hierarchy other than the lowest level, the user enters:

- 'necessity' and 'sufficiency' numbers, which represent the criticality of the performance of the sub-systems to the performance of their super-system.
- 'dependency' numbers, which represent the strength of dependency between each of the sub-systems

These are then automatically combined with the figures of merit of the sub-systems to generate the propagated estimate of the figure of merit. Weights are entered to enable the merger of measured and propagated figures of merit to calculate the value of the merged figure of merit for display in the graphical model and propagation up the hierarchy.

## APPLICATION

The method outlined above has been implemented in a case study for Scottish and Southern Energy (SSE), described in detail in Hall et al, 2002. The hierarchical model of the SSE system is shown in Fig. 7. The high level systems within the hierarchy reflect the company structure and its sub-division into generation, transmission and distribution processes. The focus of the case study was on modelling the processes within the hydro generation department, and this aspect of the model was developed in more detail. In order to fit the model on a single page, only part of the hierarchy is illustrated, however, Fig. 7 does show the full height of the hierarchy, from low-level sub-systems that form part of a specific reservoir to high level business processes. Note how each system and sub-system in the model has an 'Italian flag' associated with it, which is the graphical representation of the interval-valued figure of merit.

In group decision-making situations the model can be actively used to test alternative changes and watch the impact propagate through the hierarchy.

## CONCLUSIONS

Descriptive analysis of current asset management practice within the dam sector in the UK has demonstrated a need for improved ways of assembling and representing diverse evidence about system performance in order to improve decision-making. A decision support methodology has been introduced that aims to address this need by ordering the large number of processes and quantity of information that the decision-maker is expected to assimilate, providing an overview of the system and insights into those areas of the decision complex that are most influential.

The approach has the potential to improve communication, forming the basis for discussion and negotiation and enhancing transparency in decision-making. The model provides a guide to where attention and intervention in a system should be concentrated. It thereby provides a mechanism for prioritising asset management actions. Decisions can be readily justified because evidence upon which they are based is recorded in an auditable way. Finally, the approach provides a common platform for individuals engaged in different aspect of the system to work upon. Local technical decisions and higher level business decisions can be explored and justified using the same model, and the influence between these different types of decision-making can be illustrated.

ACKNOWLEDGEMENTS
The research described in this paper was funded by the UK Engineering and Physical Science Research Council, Scottish and Southern Energy and the Environment Agency. Data for the case study were provided by Scottish and Southern Energy. Dr Hall's research is funded by a Royal Academy of Engineering post-doctoral research fellowship.

REFERENCES
Boland, R. Greenberg, R., Park S., and Han I. (1990). "Mapping the process of problem reformulation, Implications for understanding strategic thought." Mapping Strategic Thought, A.S. Huff, ed., Wiley, Chichester.

Blockley, D.I. (1999a). "Risk based structural safety methods in context." Structural Safety, 21, 335-348.

Blockley, D.I. (1999b). "Process modelling from reflective practice for engineering quality". Civil Eng. and Env. Syst., 16, 287-313.

Checkland, P. (1981). Systems Thinking, Systems Practice. Wiley, Chichester.

Chowdhury, M.A., Garber, N.J., and Li, D. (2000). "Multiobjective methodology for highway safety resource allocation". J. Infrastructure Syst., ASCE, 6(4), 138-144.

Crandall, J.H., and Freeborne W. (2001). "Performance-based damage assessment and design verification of U.S. housing in extreme earthquake and hurricane events." NIST Special Publication SP963, Gaithersburg, MD, 343-362.

Cui W.C., and Blockley, D.I. (1990). "Interval probability theory for evidential support." Int. J. Intelligent Systems, 5, 183-192.

Davis, J.P., and Hall, J.W. (1998). "Assembling uncertain evidence for decision-making." Proc. 3rd Int. Conf. on Hydroinformatics, Balkema, Rotterdam, 1089-1094.

Ezell, B.C., Farr, J.V., and Wiese, I. (2000a). "Infrastructure risk analysis model". J. Infrastructure Syst., ASCE, 6(3), 114-117.

Ezell, B.C., Farr, J.V., and Wiese, I. (2000b). "Infrastructure risk analysis of municipal water distribution system". J. Infrastructure Syst., ASCE, 6(3), 118-122.

French, S. (1988). Decision theory: An introduction to the mathematics of rationality. Ellis Horwood, Chichester.

Haimes, Y.Y. (1977). Hierarchical analysis of water resources systems: Modelling and optimisation of large scale systems. McGraw-Hill, New York.

Haimes, Y.Y. and Jiang, P. (2001). "Leontief-based model of risk in complex interconnected infrastructures." J. Infrastructure Syst., ASCE, 7(1), 1-12.

Hall, J.W., and Davis, J.P. (2001). "Sources and implications of uncertainty for coastal managers in the UK. Water and Env. Mgt. J. 15(2), 103-108.

Hall, J.W., Le Maurier, J.W., Baker E.A., Davis, J.P. and Taylor, C.A. (2002) "A decision-support methodology for performance-based infrastructure management", submitted to J. Infrastructure Syst., ASCE.

Hartford, D. (2000). "Dam risk and the owner's dilemma." Int. Water Power and Dam Construction, 52(9), 18-23.

Hastak, M., and Abu-Mallouh, M.M. (2001) "MSRP: Model for station rehabilitation and planning". J. Infrastructure Syst., ASCE, 7(2), 58-66.

Hirano, Y., Gojo, W., and Omi, T. (2000). "Framework for reforming building regulatory systems with 'performance-based' approach -A comparative study of English, New Zealand, Australian and Canadian systems." J. Architecture Planning and Environmental Engineering, 531, 221-228.

Hsieh, T.-Y., and Liu, H.-L. (1997). "Multistage heuristic approach for solving infrastructure investment decision problems." J. Infrastructure Syst., ASCE, 3(4), 134-142..

Mintzberg, H., Raisinghani D., and Théorêt, A. (1976). "The structure of 'unstructured'decision processes." Administrative Sci. Quarterly, 21, 246-275.

Sato, H., Ogi, K.-i. (2001). "Study on performance-based design methods for wind effects on highway bridges." NIST Special Publication SP963, Gaithersburg, MD, 107-114.

SEAOC. 1995. Vision 2000 - A framework for performance based design, Structural Engineers Association of California, Vision 2000 Committee, Sacramento, California.

Wymore, A.W. (1993). Model-Based System Engineering, CRC Press, Boca Raton.

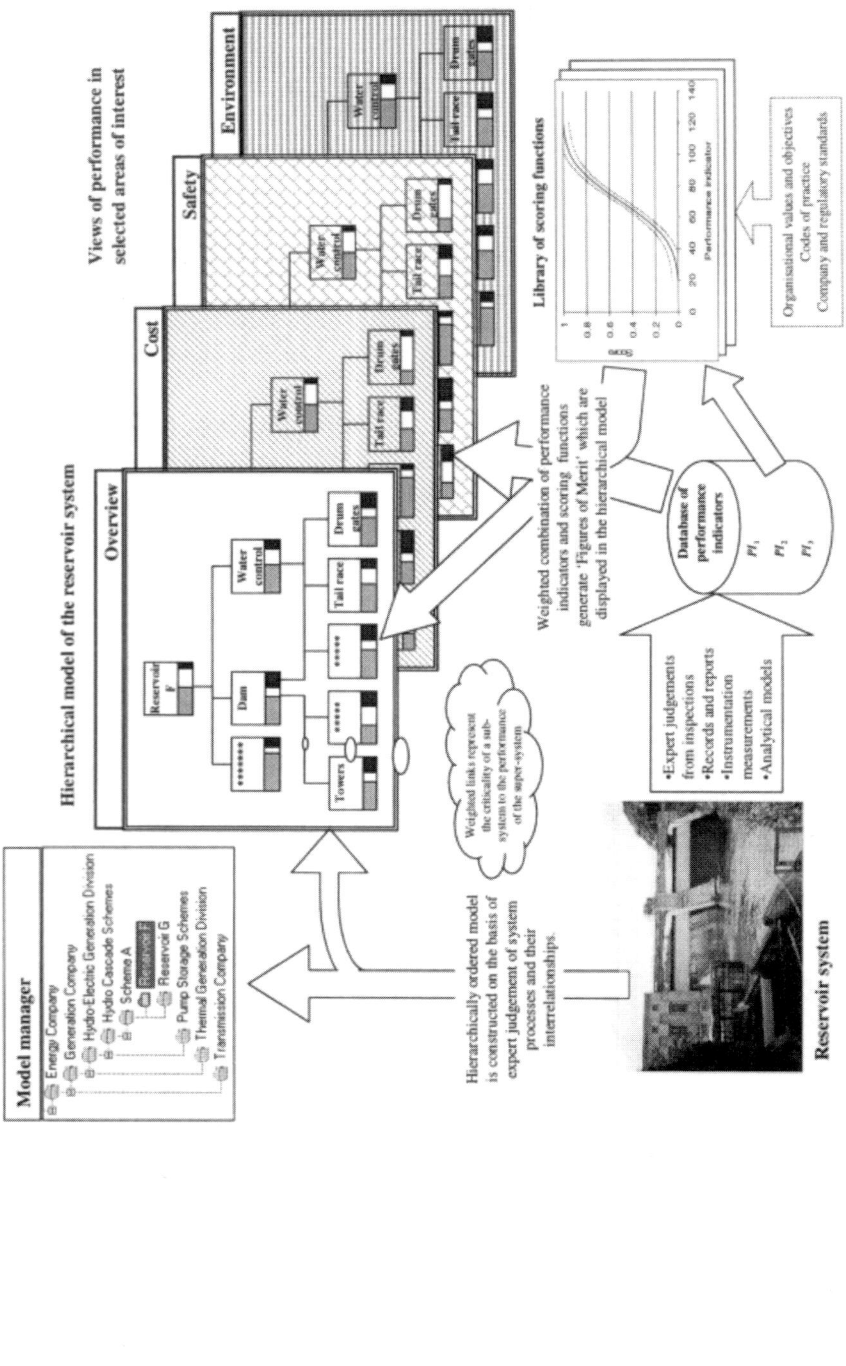


Fig. 1. Schematic of the model

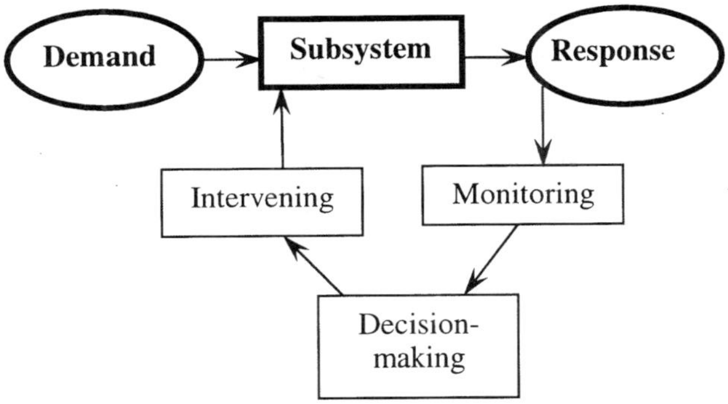


Fig. 2. Generic process

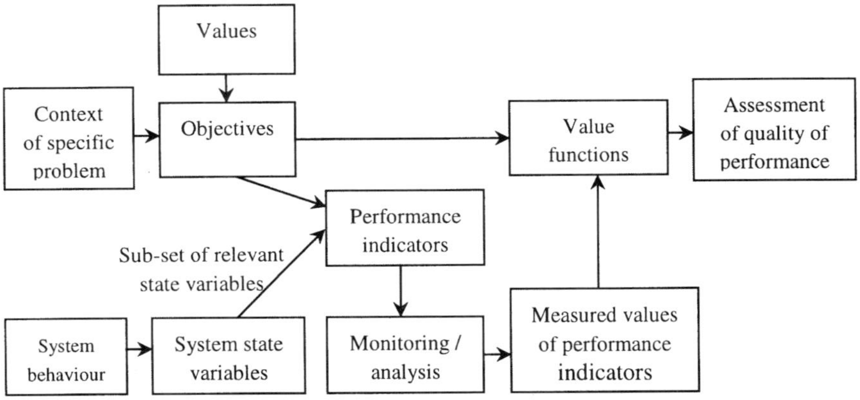


Fig. 3. Derivation of performance indicators

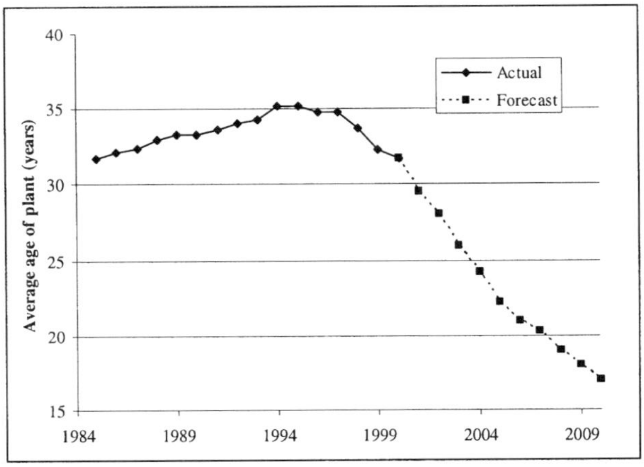


Fig. 4. Time series average age of plant (after Smith 2000)

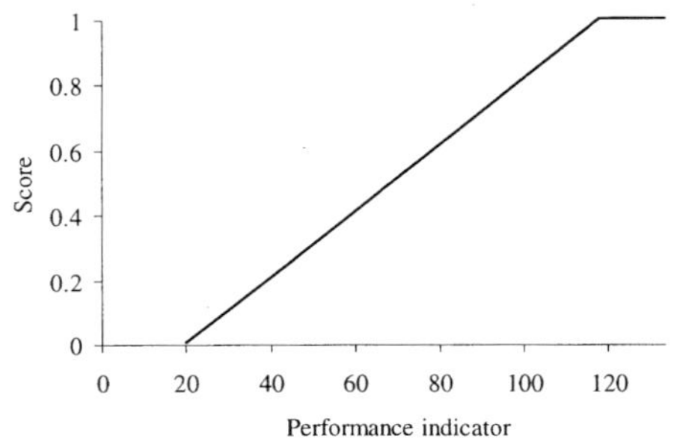


a) Linear

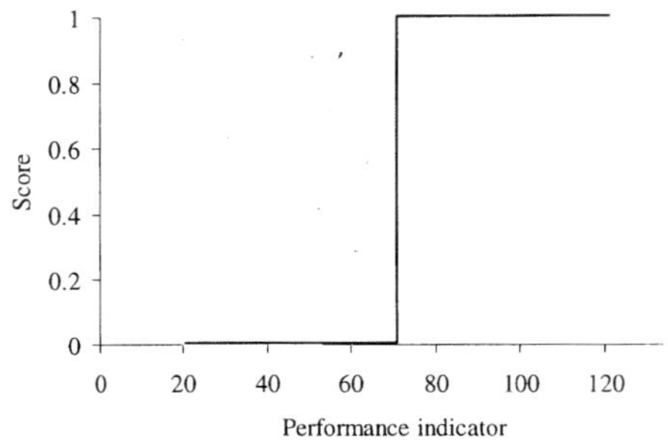


b) Stepped

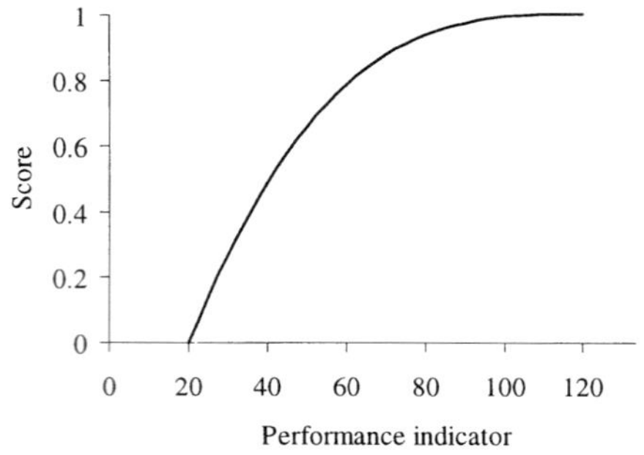


c) Convex

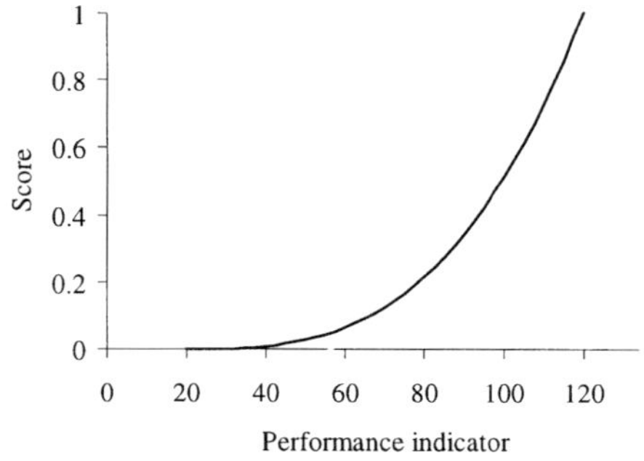


d) Concave

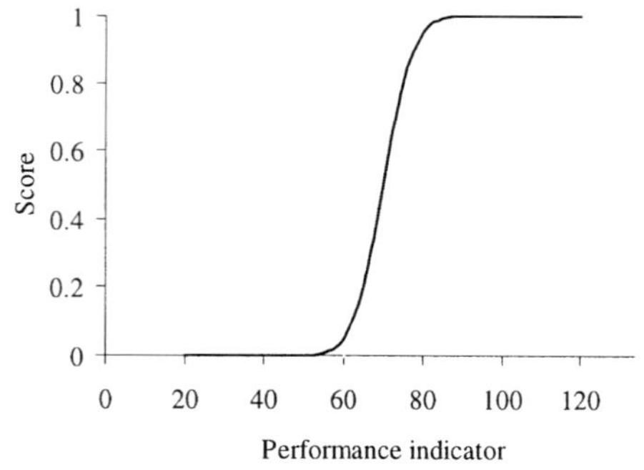


e) s shaped

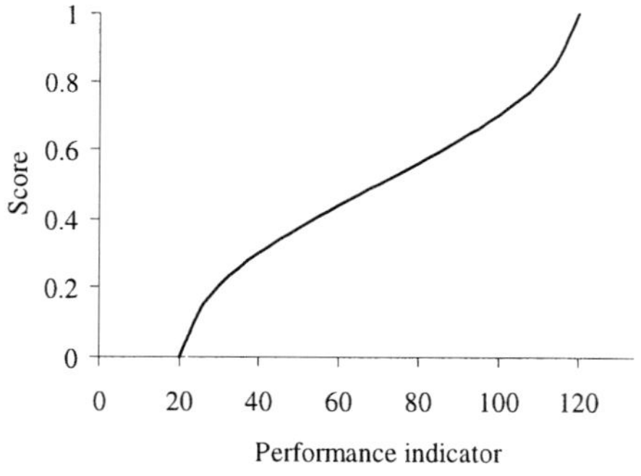


f) z shaped

Fig. 5. Shapes of value function curves

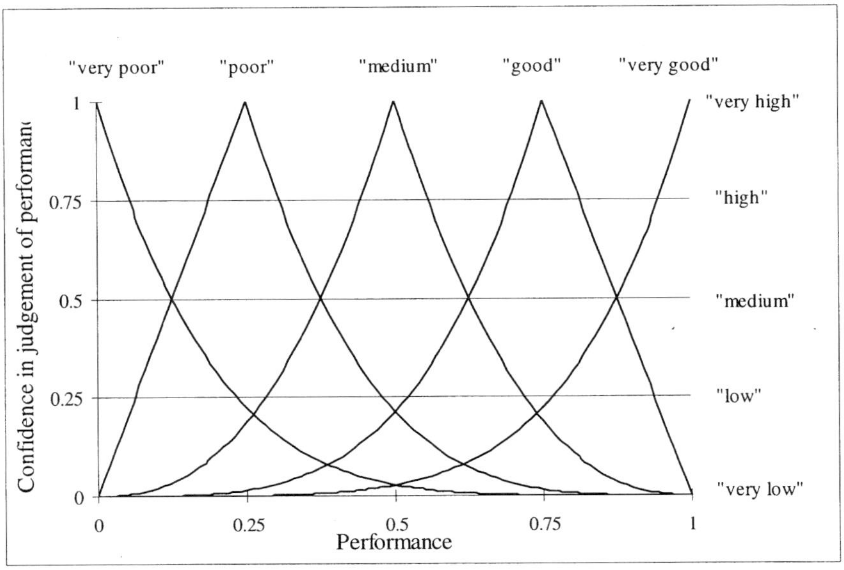


Fig. 6. Mapping from linguistic descriptions of 'performance' and 'confidence in judgement of performance' to interval values

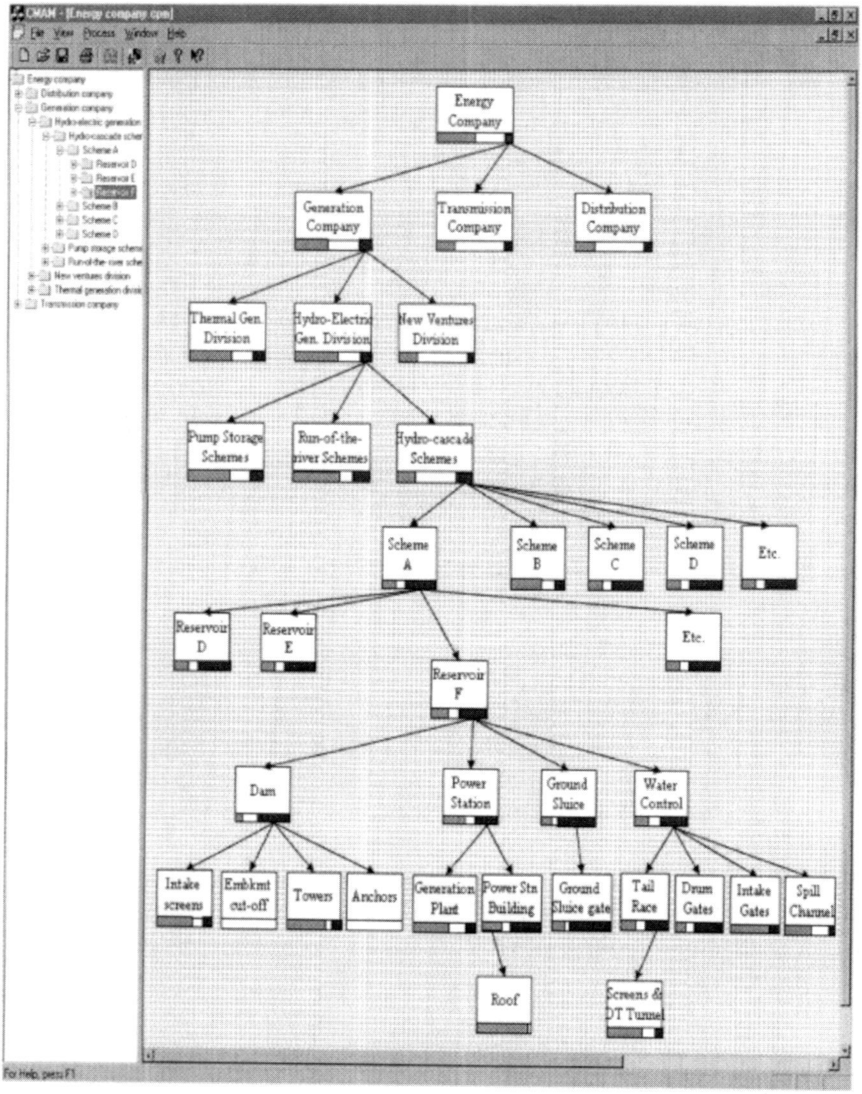


Fig. 7. Hierarchical model of the SSE system

# Reservoir risk assessments in the north of Scotland

FIONA TARRANT, Binnie Black & Veatch Ltd, UK
JOHN ACKERS, Binnie Black & Veatch Ltd, UK
NIGEL GRAHAM-SMITH, North of Scotland Water Authority, UK

SYNOPSIS This paper describes the experiences of Binnie Black & Veatch (BBV) in carrying out reservoir risk assessments for the North of Scotland Water Authority (NOSWA), using the guidance given in the recent CIRIA Report C542 '*Risk management for UK reservoirs*'. It provides a general account of the methodology and the particular approach adopted, and highlights some of the issues raised and lessons learned by the BBV team of assessors. It is hoped that this paper will make a contribution to the development of reservoir risk assessment methodology.

## INTRODUCTION

NOSWA is undertaking a programme of risk assessments of its reservoirs (including converted natural lochs), most of which come within the ambit of the Reservoirs Act 1975. The purpose of the present project is to carry out appraisals of the risks associated with NOSWA's dams and other reservoir structures, in order to prioritise the development of integrated contingency plans for the communities situated downstream of the reservoirs.

The project covers 75 dams and reservoirs, which range in capacity between about 10Ml and 5500Ml and in height from <1m to >15m. They are located in all six operating regions of NOSWA. Table 1 summarises the reservoirs by region and type.

Table 1 Numbers of reservoirs in project by region and type

| Region | Embankment dams | Service reservoirs and raw water tanks | Concrete and other dams | All reservoirs |
|---|---|---|---|---|
| Grampian | 5 | 5 | 0 | 10 |
| Highland | 10 | 2 | 22 | 34 |
| Orkney | 3 | 0 | 5 | 8 |
| Shetland | 0 | 0 | 4 | 4 |
| Tayside | 3 | 1 | 2 | 6 |
| Western Isles | 1 | 0 | 12 | 13 |
| Totals | 22 | 8 | 45 | 75 |

*Reservoirs in a Changing World*, Thomas Telford, London, 2002, 551–562

The risk appraisals, which were carried out in accordance with the guidance given in the recent CIRIA Report C542, '*Risk management for UK reservoirs*' (Hughes *et al*, 2000), may be divided into three stages:

1. ***Potential failure impact assessment***, based primarily on an assessment of the potential magnitude of the failure flood passing down the downstream valley and a field appraisal of the resulting impacts in terms of risks to life and property;
2. ***FMECA selection***, in which the results of the Stage 1 impact assessment are used to decide whether and what type of Stage 3 assessment is required ('*FMECA*' means 'failure modes, effects and criticality analysis'); and
3. ***FMECA assessment***, in which possible failure modes of the dam and other pertinent reservoir structures are considered and analysed.

The Stage 1 field appraisals were carried out between January and April 2002, in a wide variety of weather conditions, ranging between snowstorm and bright sunshine. High winds in the Western Isles in mid-February resulted in the cancellation of all flights and ferries to Barra for several days, with the result that the appraisals for the two NOSWA reservoirs on Barra were undertaken in the comfort of a B&B dining room in South Uist and relied more heavily than others on information from the maps and previous inspection reports.

*Fedderate dam and spillway (Grampian Region)*

Stage 3 assessment were required for about 20% of the reservoirs. It was originally anticipated that the results of Stage 1 would be presented and discussed before a decision was made on which reservoirs were to be subject to Stage 3 assessments. However, the decision made at Stage 2 is based entirely on an impact score derived in Stage 1, so it was agreed that the Stage 3 assessments should follow on directly from the field appraisals.

Although the methods of appraisal and assessment generally followed the guidance in CIRIA Report C542, there were a number of points where the methodology needed to be clarified, to ensure that a consistent approach was adopted by the members of the assessment team. This paper is intended to highlight some of the issues and experiences encountered by the BBV team of assessors.

## STAGE 1: IMPACT ASSESSMENT

The impact assessment is intended to provide a relatively rapid method of assessing the consequences of a possible dam failure, resulting in an 'impact score' which can be used both to categorise dams (Stages 2 and 3) and also prioritise the need for further more detailed studies.

The CIRIA report divides the process into five steps:

1 Collation of information and site visit
2 Prediction of dam failure discharge
3 Prediction of downstream flood levels following dam failure
4 Assessment and scoring impacts
5 Combining the scores and identification of likely consequences of dam failure

The BBV approach was to carry out the site visit after steps 2 and 3, so that the assessor was aware of what discharges could apply at various points in the downstream valley and could base the field assessment of the impacts (step 4) on the likely flood depths and velocities. In some cases, as a result of observations during the site visit regarding the form of the dam and the valley topography (in particular the shape of the river channel and floodplain), some adjustments were made to the calculations in steps 2 and 3 during or after the field visits.

The following information was provided by NOSWA for most of the reservoirs in advance of the field visit:

- a limited amount of basic data on the dam height, storage capacity and appointments of inspecting engineer and supervising engineer, contained in a database kept by NOSWA;
- a copy of the most recent inspection report under Section 10 of the Reservoirs Act 1975 and the supervising engineer's annual statements under Section 12, normally going back for a period of five years and
- 1:10 000 scale maps extending 5km downstream and 1:25 000 scale maps extending 30km downstream of the reservoir, derived from NOSWA's GIS system.

Because of shortcomings in some of the GIS-derived maps, including normally a relatively wide contour interval of 10m, the standard Ordnance Survey 1:25 000 'Pathfinder' published map sheets were also procured. These mostly have the advantage of a 5m contour interval in lowland areas and were found to be easier to read and interpret, although in many cases the

GIS-derived maps had the advantage of including recent development not shown on the published map sheets.

As far as practicable, the BBV assessor arranged to meet and discuss the reservoirs with the appropriate supervising engineers, prior to making site visits. Where necessary for safety or security reasons, NOSWA personnel accompanied the assessor.

Dam failure flood prediction

The CIRIA guide provides two distinct methods of discharge prediction, depending on the type of dam:

- embankment dams; and
- concrete dams.

For this study these methods were coded into spreadsheets, with the approach for concrete dams being further subdivided, giving three types:

Type A Embankment dams;
Type B Concrete gravity dams; and
Type C Concrete arch, multiple arch and buttress, or gravity arch dams.

The CIRIA guidance for flood prediction from the failure of a service reservoir is not explicit, suggesting that it is likely to fall somewhere between the values determined using the formulae for embankment and concrete dams. It goes on to recommend that it should be derived 'after due consideration of the structural design and the potential discharge, assuming it to comprise either an embankment or concrete structure'.

The usual form of service reservoir encountered in these studies comprises a concrete tank surrounded by embankments. The view taken was that the possible failure of such concrete walls is likely to be inhibited by support from the surrounding embankments, so that the use of the concrete dam formula is likely to give an unrealistically high discharge and consequently too short a timebase for the release of the stored volume. We therefore normally used the formula for an embankment dam in these cases.

Flood prediction downstream of the dam

The CIRIA method divides the downstream floodpath into two parts:

- the 'near valley', which extends to 5km downstream of the dam; and
- the 'far valley', which extends from 5km to 30km downstream.

A simplified form of flood routing and hydraulic calculations are used to estimate the attenuation of the flood peak as it passes down the valley and to estimate the corresponding flood depths. The flood attenuation calculations are carried out by dividing the near and the far valley into four zones each. The zone lengths can be unequal and were generally chosen so that zone boundaries occur at points where the valley floor level could be ascertained by the presence of a contour running across it.

*Loch Ordie dam, near Dunkeld (Tayside Region)*

For valleys in which the river channel is small, the river channel is disregarded in the analyses, effectively representing the passage of the entire flood down a valley in which any river channel has been is filled with inert material. An alternative way of looking at this case is to assume that the channel is already running bankfull as a result of antecedent runoff. In cases where the river channel is large in relation to the flood discharge (for example the failure flood from a small reservoir in a minor tributary valley), so that the flood would be expected to be contained within the river banks, the river width is used in the analyses. In cases where the river channel is deeply incised and the flood is likely to be fully contained within it, then again it is the incised channel geometry which is used in the flood routing calculations.

It must be recognised that the representation of the near and far valleys each as four zones, within each of which the geometry and roughness are taken to be uniform, requires a great deal of simplification. In many cases this is likely to result in an underestimation of the potential flood storage and flood peak attenuation which would really result from the continuously varying channel and valley geometry. This may be expected to result in an increasing degree of conservatism with regard to peak discharges as the flood progresses down the valley. On the other hand, flood depths in the upstream part of the valley may be slightly underestimated.

CIRIA Report C542 provides Tables A and B for presenting the hydraulic and flood routing calculations. We developed the formats of these tables, presenting them as sheets of a spreadsheet, so that the calculations would be carried out automatically. An additional row giving the mean flow velocity was included, as this is relevant for assessing the risks to life and property resulting from the flooding.

Although Tables A and B refer to the near and far valleys respectively, in many cases the boundary between them was taken at a distance other than 5km, for various reasons:

- to suit a convenient point for a zone boundary, generally taken where a contour line crosses the valley; and
- if the overall downstream valley length is much less than 30km, for example for reservoirs near the coast, Table A may cover only part of the near valley, with the rest included with the far valley in Table B.

CIRIA Report C542 draws attention to the limitations of the approach, pointing out that the method 'is only intended to give an order of magnitude estimate to allow the scale of potential flood impact to be assessed'. It goes on to point out that it should not be taken as a substitute for full dambreak modelling. BBV concur with that warning and would recommend that full dambreak analysis be undertaken for those reservoirs posing the greatest risk to downstream communities. In some cases, a particular recommendation has been made for a dambreak study, for example if the assessor felt that the CIRIA methodology may not adequately represent the severity of the downstream floodwave.

Flood prediction for cascades of reservoirs

Where the failure of a dam results in flows into a downstream loch or reservoir, or indeed into a series of water bodies, the approach outlined above is not usually appropriate. CIRIA report C542 includes guidance on how to approach cases where there is the possibility of a 'cascade' failure of a series of reservoirs in a single valley, together with a test case involving three such reservoirs. The approach recommended is to route the failure flood hydrograph from the first reservoir through the storage in the second reservoir, triggering failure of the second reservoir when its spillway capacity is exceeded and flow passes over the crest of the dam, and so on for any further reservoirs.

For many of the reservoirs in the Western Isles, the Orkneys and coastal areas of the Highlands, many of which comprise raised natural lochs, the dam breach flood passes through one or more downstream lochs, often with small differences in elevation between the lochs. The peak breach floods involved are relatively modest and result in peak outflows from the downstream lochs which are unlikely to breach the natural rim or any outlet weir structure and result in the escape of additional water from the downstream lochs. For these cases BBV developed a simple 'cascade' spreadsheet technique for carrying out the routing of the failure flood through the downstream lochs. In most of these cases the routing through the downstream lochs is responsible for virtually all of the flood peak attenuation, the attenuation in the channels between the lochs being insignificant because of the long timebase of the flood hydrograph.

Downstream impact assessment

The downstream impact assessments were undertaken by field visits, guided by the 1:10 000 scale GIS-derived mapping and the 1:25 000 Ordnance Survey maps. In order to assess the practical limits of the flood envelope on the ground and the likelihood of structural damage and risks to life, the assessors took account of the flood depths and velocities estimated in Tables A and B. In addition, they made rapid assessments of the effects of potential constrictions such as bridge openings and raised embankments on the passage of the peak discharge and the associated risks.

CIRIA Report C542 considers the impacts under seven headings essentially by land use type, and for each of these an impact score of between 0 and 4 is defined depending on the numbers of people or properties affected or on a qualitative assessment of the risk. For the first four impacts another parameter is defined, the 'PAR' or 'persons at risk'. The table below is an example of one of these tables, covering Impact 4. There is an obvious difficulty in the consistent application of this table, which is (for example) whether to score 50 people affected as 2 or 3. We therefore modified the scoring system slightly, raising the bottom of the ranges for scores of 1 to 4 by 1 person.

Table 2 Impact scores for recreational sites

| Disruption | Number of people affected | Score | PAR |
|---|---|---|---|
| None | 0 | 0 | 0 |
| Minor | 0 to 10 | 1 | 10 |
| Appreciable | 10 to 50 | 2 | 50 |
| Significant | 50 to 100 | 3 | 100 |
| Major | More than 100 | 4 | Estimate |

*Loch Smalag, South Uist*

Tables giving guidance on the selection of appropriate scores are included in CIRIA Report C542, which also includes two standard Tables – C, and D – to use for recording the impacts in the near and far valleys respectively. The general format suggested in CIRIA Report C542 was followed for these tables, except that they were expanded to include the information on the scoring system, both to aid rapid completion and to give the reader an immediate reference to the guidance against which the assessment has been made. In order to obtain a consistent approach to impact assessment from all our assessors, potential problems of interpretation found in the early part of the fieldwork were discussed and agreement reached between the assessors on how to treat a number of issues.

Combined impact score

CIRIA Report C542 assigns weightings to each of the impact scores for the near and far valley. The PAR values for the near and far valley are totalled separately, then the following equations applied:

$$\text{Near valley potential loss of life} = 0.5 \times \text{PAR}$$

$$\text{Far valley potential loss of life} = \text{PAR}^{0.6}$$

The weighted impact scores and total PAR values for the near and far valley respectively are then combined to give a total reservoir impact score, after applying various further factors. CIRIA Report C542 includes a standard form – Table E – to carry out these weightings and score combinations. We included this in the same spreadsheet file, in order to capture the impact scores and PAR values automatically from Tables C and D.

## STAGE 2: FMECA SELECTION

Stage 2 categorises the reservoir according to the downstream risks, reaches a decision on whether a Stage 3 risk assessment is required and, if so, whether it should be a Level I or Level II assessment. This decision depends on the Stage 1 impact score, as set out in the table below.

Table 3 Classification and risk assessment selection criteria

| Total impact score | Grade | Risk assessment required | Indicative flood standard |
|---|---|---|---|
| >750 | *** | Level II | A |
| 175–750 | ** | Level I | A/B |
| <175 | * | No further | C/D |

## STAGE 3: FMECA ASSESSMENT

As noted earlier, 'FMECA' stands for 'failure modes, effects and criticality analysis. It is intended to represent a systematic approach to:

- analysing and listing the ways in which a system can fail (the failure mode);
- for each failure mode, assessing the effects of that failure; and
- to identify how critical that mode of failure is to the operation of the system.

CIRIA Report C542 (Hughes *et al*, 2000) commends the approach in the following manner:

> 'FMECA offers a balance between the two extremes of relying solely on engineering judgement and the rigour (and expense) of full probabilistic analysis. It also provides the flexibility to deal with varying levels of knowledge of the performance and reliability of different dam components. The approach does not require specific probabilities to be attributed to the failure of a specific component. Instead, it requires a qualitative assessment as to the probability, likelihood of detection and consequence of failure.... In this way the system draws on the assessor's knowledge without causing difficulty by demanding specific probabilities for each component within the failure tree.'

The Stage 3 risk assessment has two levels, depending on the total impact score, as in the table above:

- Level I Only complete failure of the dam is considered
- Level II Partial failure modes, such as a gate failure, are considered in addition to complete failure

LCI diagrams

A key part of the Stage 3 risk assessment is the LCI ('location, cause, indicator') diagram. The LCI diagram has a tree-like structure, starting with the locations at which failure may occur, for each of these identifying a series of failure causes, then for each cause listing the indicators which might suggest a risk of such a failure developing. For each cause/indicator element, a 'criticality score' (described below) is calculated, which is intended to give 'a measure of the hazard that the *cause/indicator* element creates for the overall dam safety'.

Consequence, likelihood and confidence scores

These three factors are considered for each cause/indicator element in the LCI diagram and are each assigned scores of between 1 and 5. For some of the LCI diagrams, the CIRIA project resulted in guidance being offered on appropriate values for some of the consequence scores and included in the LCI diagrams, but these may not always be appropriate. In all cases it is for the assessor to decide on the scores to assign to the factors.

*Consequence:* How directly is failure of a given element related to failure of the dam?

1 = low: failure of the element is unlikely to lead to dam failure

5 = high: failure of the element is highly likely to lead to dam failure

*Likelihood:* What is the likelihood of the failure of this particular element?

1 = low: there is a low probability of the element failing

5 = high: there is a high probability of the element failing

*Confidence:* What is the assessor's confidence in the reliability of the values assigned to these consequence and likelihood scores?
5 = low: the assessor has a low confidence in the predictions
1 = high: the assessor is very confident in the predictions

With regard to the likelihood scores, these are relative and a high likelihood still has a low absolute probability. The confidence scores are not as might be intuitively expected from a comparison with the scores for the other two factors, but were designed so that a high score (low confidence) represents another indication of potentially high risk. Thus, high scores for all three factors represent a higher degree of potential risk, allowing the values of the factors to be combined (as described below) by simple multiplication.

*A very small dam (but within the ambit of the Reservoirs Act), Loch Mhic Gille-bhride, North Uist*

The confidence score is intended to allow the assessor to take account of whichever of the following issues are applicable:

- the detectability of failure (whether the available instruments can give appropriate prior warning);
- whether knowledge of material quality, workmanship and construction history is available and taken into account in the likelihood score;
- whether routine maintenance can be relied upon to happen;
- the quality of the records used to determine the likelihood score; and
- incompleteness of knowledge (for example where structural components and/or interactions are difficult to assess).

Criticality and risk scores

These scores are intended to provide the means of prioritising the high-risk elements of the LCI diagram. The criticality score ranging between 1 and 125, is defined as:

$$\text{Criticality score} = \text{Consequence} \times \text{Likelihood} \times \text{Confidence}$$

Because of the inclusion of the confidence score in this product (representing a poor knowledge of or confidence in the consequence and/or likelihood scores), CIRIA Report C542 considers the criticality score not to represent a 'robust' identifier for high-risk elements. The report therefore also considers a combined score which excludes the confidence score, thus:

*Consequence × Likelihood score,* with a range from 1 to 25.

The risk score is defined as:

*Risk score = Reservoir impact score (from Table E) × Criticality score*

As this combines the impact score (the possible effect on downstream communities) with the risks associated with features of the reservoir, it provides a simple measure by which prioritisation might be decided between different reservoirs.

Ranking and prioritisation

CIRIA Report C542 sets out procedures by which the scores obtained for each cause/indicator element are used to rank the elements for review, prioritisation and action. The rankings cover:

- *Criticality score*;
- *Consequence × Likelihood score*; and
- *Confidence score.*

The aim of the prioritisation procedure is to provide the reviewer with suitable information to make an informed decision on appropriate action to take. In general terms it would be expected that:

- for elements marked high for *Consequence × Likelihood*, remedial works should be considered; and
- for elements with high *Confidence* scores (that is, low confidence) consideration should be given to the advisability of investigative work which would reduce the uncertainties and clarify the need (if any) for remedial works.

Presentation of Stage 3 assessments

CIRIA Report C542 provides a series of standard tables for documenting the process and presenting the results of the Stage 3 study. These are:

Table F LCI diagram score justification
Table G Risk summary table
Table H Summary of highest risk elements

BBV prepared spreadsheet versions of these tables, designed so that the relevant information is carried forward as necessary and also to allow the necessary sorting of the information to be performed.

Calibration of methodology

The categories, factors and scores suggested for the methodology of the Stage 3 risk assessment, as formulated during the CIRIA study (Hughes *et al*, 2000), were based on 'a combination of engineering judgement, database analysis, test case application and feedback from workshop and steering group sessions'. The report stresses that they are 'not intended to offer a definitive methodology or calibration, but [to] act as a starting point from which the risk assessment methodology may be developed'. It is anticipated that this series of reservoir risk assessments, in particular those involving a Stage 3 assessment, will make a contribution to the development of the methodology.

## RESULTS OF STUDIES

The project was not complete at the time of writing this paper, so the results of the studies will be presented at a later date.

## ACKNOWLEDGEMENTS

The authors wish to thank NOSWA and BBV for permission to publish this paper. The authors also wish to thank Bob Bryce (NOSWA) and his team, including the supervising engineers, for their assistance in the course of the project. They also acknowledge the work of the other members of the BBV team of assessors, namely Chris Scott, Andy Rowland, Dominic Molyneux, and Aled Hughes, with the support of Katy Davies.

## REFERENCE

Hughes, A, Hewlett, H, Samuels, P G, Morris, M, Sayers, P, Moffat, I, Harding, A and Tedd, P (2000) Risk management for UK reservoirs, Report C542, CIRIA (London)

# Lake Sarez Risk Mitigation Project

L J S ATTEWILL, JacobsGIBB LTD, UK
L SPASIC - GRIL, JacobsGIBB LTD, UK

SYNOPSIS. Lake Sarez was created after an earthquake induced landslide formed a natural dam across the Murgab river in the Pamir region of Tajikistan in 1911. The volume of the lake has now grown and poses a threat to the inhabitants downstream of the lake and to the entire Amu Darya plain. The paper describes the studies and investigations into the dam's vulnerability to internal erosion, erosion by overtopping waves and the threat posed by lake shore instability, earthquakes and floods. The risk mitigation measures, which comprise both a monitoring/early warning system and long term structural measures are described as well as their economic justification.

INTRODUCTION

Lake Sarez in Tajikistan was formed in 1911 when a massive earthquake-triggered landslide buried the village of Usoy under a 650 m high obstruction of Murghab River. The resulting 60 km long lake, with a surface area of 85 km$^2$ containing over 17 km$^3$ of water, is located behind the so-called Usoy natural dam in the Pamir range (Figure 1).

Fig. 1. Location of Lake Sarez

*Reservoirs in a Changing World*, Thomas Telford, London, 2002, 563–574

It is feared that the Usoy dam could fail for various reasons. The failure of one of the several potential landslides could create a large wave causing overtopping and destruction of the dam. Failure of Usoy dam would be catastrophic and it would be a major disaster for the Aral Sea Basin, Central Asia and the World community. It would create a mud wave along the Bartang, valley and inundate areas in Tajikistan, Afghanistan, Uzbekistan and Turkmenistan. Valuable agricultural land, areas, structures, vegetation and animals would be destroyed. Although the safety of the Lake has been studied over many years, significant gaps and inconsistencies exist in the data available. Despite this lack of data, however, it is clear that the risk to the downstream population living in the Bartang and Panj river valleys remains considerable.

Under Aral Sea Dam Safety & Reservoir Management Project a preliminary assessment of the safety of Lake Sarez carried out by GIBB[1]. The work was followed, in October 2000, by the Lake Sarez Risk Mitigation Project (LSRMP). The LSRMP with its three components aims to address and mitigate the risk. LSRMP components are as follows:

- Component A - The design and implementation of a monitoring system and early warning system
- Component B – Social Component that identifies safe havens and trains the vulnerable downstream communities under the disaster preparedness programme
- Component C - Preliminary studies of long term solutions

Components A and C are funded by the Swiss Secretariat for Economic Affairs (Seco) grant and GIBB are assisting the Swiss consultant Stucky in both components. Component B is funded by USAID and other grants and is being carried out by international humanitarian agencies.

This paper describes the work undertaken within the Components A and C of the LSRMP.

## STUDIES AND INVESTIGATIONS

Numerous studies were undertaken and reports and papers have been written on the Lake Sarez. The work carried out on Sarez could be broadly classified into three groups:

- Work carried out before 1999
- Work carried out in 1999-2000
- Work carried out under LSRMP

### Work carried out before 1999

The vulnerability of the area to landslides was noted and described by Russian scientists as early as the 1880's. After the 1911 earthquake a series of expeditions were made to prepare topographical and geological maps of the dam.

The potential landslide on the right flank of the lake was identified in 1968[2] and since 1984 the work was focused on the Right Bank Landslide. Between 1985-1987 Geological Department of Moscow State University undertook a geological investigation of the Right Bank Landslide[3] as well as the mathematical and physical modelling of overtopping of the Usoy dam due to possible Right Bank Landslide with a total estimated volume of 0.9 $km^3$

Despite many studies and reports, no reliable records of horizontal or vertical movements of the dam and the Right Bank Landslide are available. Similarly, despite several geological investigations almost no reliable subsurface data are available.

Work carried out between 1999 and 2000

The potential danger of Lake Sarez and Usoy dam was brought to the attention of the Secretariat for the International Decade for Natural Disaster Reduction (IDNDR) who in 1999 organised a mission to Sarez and subsequently reported on the findings.[4]

Under Aral Sea Dam Safety & Reservoir Management Project a preliminary assessment of the safety of Lake Sarez was carried out by GIBB in 1999-2000. This work was primarily a desk study that provided the first opportunity for western engineers to translate and study Russian language reports.

Work carried out under LSRMP

*Field Inspections*

The work under LSRMP started in August 2000 and field missions to Sarez were organised in September and October 2000 to examine the Usoy dam and the Right Bank Landslide. Possible locations of tunnel exit portals were also studied.

The geological and tectonic structure of the landslide was studied, which confirmed that the Usoy collapse was mainly controlled by tectonic features.

The canyon that developed in the downstream slope of the dam was also examined. The canyon is some 25m deep and its head has been irregularly migrating upstream: the last movement was in 1994 when a backward erosion of 45m in length was caused by an increase in the annual maximum discharge from $70m^3/s$ to $170m^3/s$. Since that time no further erosion has been recorded. There are about 45 springs in the canyon, most of them are emerging at approximately at the same elevation indicating that the lower part of the dam has much lower permeability than the upper part where the seepage takes place. The present situation appears stable for all normal flows but it can be destabilised by internal and surface erosion.

The Right Bank Landslide (See Fig 2) was also studied during the field mission.

Fig. 2. Right Bank Landslide

Observations made during the visits and the subsequent studies have lead to a revision of the estimates of the potential volume of the unstable mass of the Right Bank Landslide, from 0.9km³ (Akulov) to a provisional best estimate of 0.5km³. The reduction is based on the fact that there is no firm evidence to show that the potential slip is deep seated as previously reported.

It has therefore been proposed and agreed with the Client to carry out the following supplementary investigation:

Usoy dam:

- Complementary geological mapping for determining the optimal positioning of the monitoring instruments
- Preliminary GPS survey to establish a precise geodetic reference network for determination and referencing displacements and for studying the feasibility of permanent observation schemes
- Dye tracing analysis to check the filtration regime through the dam

Right Bank Landslide:

- Complementary geological mapping in order to remove hypothetical and interpretative elements from the factual data in the existing geological maps
- Determine position of 4 deep boreholes for the inclinometers to be installed in the framework of the Monitoring System (see below)
- Preliminary GPS survey to establish a precise geodetic reference network for determination and referencing displacements and for studying the feasibility of permanent observation schemes

The proposed supplementary investigation works were scheduled to take place in September 2001 but the works were postponed until May 2002 due to security situation in Tajikistan.

The second field mission comprised a visit to Bartang Valley and was related to the Component A studies and a study of possible routes for the access road. During the visit the information on the existing EPP and EWS as well as existing power supply and communication capability was examined.

*Overtopping desk study*
With a freeboard of over 40m, neither an incoming flood nor the displacement resulting from a landslide falling into the lake is likely to raise the lake level sufficiently to cause overtopping. The only conceivable mechanism that can cause overtopping is a landslide generated wave. Previous studies carried out by Russian scientists assumed that the landslide would be very large (volume > $2km^3$) and fast moving (velocity > 20m/s). Model tests indicated that such a landslide would generate a wave of over 150m in height that would overtop and, thus, probably destroy the natural dam.

In the desk study wave heights resulting from a $0.5km^3$ landslide were predicted using theoretical relationships (Noda[5], Slingerland & Voight[6]) and by reference to the physical model of Morrow Point dam in USA[7] and the tsunami that devastated the coast of Flores Island, Indonesia in 1992[8], both of which had a similar Froude number to Sarez, with the following results:

| | Maximum wave height (m) |
|---|---|
| Noda | 72 |
| Slingerland & Voight | 17 |
| Analogy with Morrow Point model | 40 |
| Analogy with Flores Island | 38 |
| Average | 42 |

Following a similar procedure the average and maximum wave heights generated from a $1km^3$ landslide would be 54m and 91m respectively.

The potential for a landslide generated wave causing external erosion of a scale sufficient to threaten the integrity of the dam depends on how the wave is propagated within the lake. Based on the studies carried out on the Tajford waves[9], the height of the wave arriving at the dam has been estimated by applying a reduction factor of 0.77 to the initial wave height, giving a wave height in the range 32-55m for the $0.5km^3$ landslide and 41-70m for the $1km^3$ landslide.

The volume of water that would overtop the dam has been estimated from an empirical formula of Muller[10]

$$V/V_0 = (1-f/R)^{2.2}$$

Where: $V_0$ = incident wave volume
$V$ = overtopping volume
$f$ = freeboard
$R$ = run up

The volume of the incident wave is derived from the Morrow point dam tests and is estimated as

| Landslide volume ($km^3$) | case | Wave height (m) | Specific Wave volume ($m^3/m$) | Overtopping volume ($Mm^3$) |
|---|---|---|---|---|
| 0.5 | average | 32 | 9,295 | 0.5 |
| | maximum | 55 | 28,000 | 5 |
| 1.0 | Average | 41 | 15,000 | 2.5 |
| | maximum | 70 | 44,500 | 25 |

These estimates, which are very approximate, demonstrate the sensitivity of the overtopping volume to landslide volume. It is reasonable to suppose that the integrity of the dam would not be threatened by the average wave height resulting from a 0.5$km^3$ landslide. A 1$km^3$ landslide however could result in an overtopping volume 25 $Mm^3$ and it is doubtful whether the dam could survive this.

*Internal erosion*

There is no evidence of internal erosion despite the high flow velocities. The dam material is very blocky and the cavities or water passages that have been eroded appear to be stable. However peak lake levels continue to rise which, with a constant inflow implies that the outflow does not vary with head.

Raised lake levels, whether by flood, or landslide would expose new material to through flow which might be less stable.

Landslides caused by earthquakes could shorten the seepage path and increase the hydraulic head.

## EARLY WARNING SYSTEM (EWS)

Review of the existing early warning system

The first EWS was installed for the Lake Sarez in the late 1980's. It was supposed to be triggered by sensors detecting high floods of the Bartang Valley in a section located 36 km away from Usoi dam. The message would then be sent by satellite to Moscow and via computer network to Dushanbe.

Since the independence of Tajikistan, this warning system has not been in operation due to lack of maintenance and a lack of funds for renting the satellite channels. In 1999, under LSRMP Component B, ECHO, Agha Khan Foundation and FOCUS in close co-operation with the Ministry of Emergencies in Dushanbe developed and implemented a new temporary EWS which will be in operation until the Component A EWS is installed. The existing EWS will be fully integrated into the Component A EWS.

Emergency preparedness plans

The objectives of developing the EPP for the Bartang Valley is to protect people from the threat related to Usoy dam overtopping. The protection relies on:

- Early Warning System that will warn people and the rescue teams about an emergency (developed under Component A)
- Preparedness Plan on which some communities in the valley have already been trained under Component B (definition of escape lanes and safe places, storage of basic food and blankets)
- Training and preparedness of Rescue teams in Khorog (Component B)

Monitoring system

The following Monitoring System (MS) that will be a part of the Lake Sarez Early Warning System (EWS) will be installed at:

- Usoy dam – lake level monitoring devices (2No), settlement cells (12No sets), GPS(3No), gauging stations at the toe of the dam (1No), manual measurement of turbidity increase , Strong Motion Accelerometers (SMA)(3No)
- Right Bank Landslide – GPS(4No), inclinometers(4No), piezometers (4No), geophones (6No), SMA (1No)
- Irkht monitoring stations (south extremity of the lake) – gauging station (1No), meteorological station (1No), lake level monitoring devices (2No)

The above instruments will be grouped in eleven Monitoring Units (MU1 – MU11). The signals from the Monitoring Units will be transmitted to and collected in the Central Unit (CU) based in a house that will be constructed on the Usoy dam. The staff in charge of the maintenance of the MS and EWS will also be stationed in the house.

The monitoring signals collected from the MUs will be compared with trigger values that will be related to the Four warning levels (1"abnormal situation", 2"get ready", 3"escape", 4"back to normal"). In addition, if the operators detect an abnormal situation or event, not detected by the survey system, or in case of this one is not operating normally, they can activate manually the sirens in the villages, level 2 or 3, and transmit this information to Dushanbe via one satellite way. The level 4 ("back to normal signal") is only activated from Dushanbe.

Flood inundation mapping and emergency preparedness

As mentioned in the introduction, flood inundation mapping and emergency preparedness plans are being produced under LSRMP Component B by USGS and FOCUS and are not the subject of this paper.

STRUCTURAL MEASURES

There are two main categories of long term solutions:

- Measures designed to reduce the hazard of a failure of the Usoy dam
- Measures designed to reduce the effects of such a failure

Either set of measures would require the reconstruction of the existing road up the Bartang valley or the construction of a new access from the east: either would be very expensive.

Measures designed to reduce the hazard of a failure of the Usoy dam

The risk of external erosion can only be reduced by increasing the freeboard which can be achieved either by lowering the lake level or by raising the dam crest. The risk of internal erosion can be reduced either by lowering the lake level (to reduce the hydraulic gradient) or by reinforcing the dam to resist erosive forces. Thus the only measures, or combination of measures, that address both internal and external erosion are:

- Lowering the lake level (solution A)
- Raising the dam crest in combination with strengthening the dam (solution B)

The extent of freeboard increase

The existing freeboard is approximately 40m. It is considered that a 60m increase to give a total freeboard would provide sufficient protection against any conceivable landslide wave.

*Solution A*

Four methods of lowering the lake level were considered:

- Diversion of the main inflow into an adjacent catchment upstream of the lake
- Outlet tunnel through Usoy dam right abutment
- Outlet tunnel through Usoy dam left abutment
- Pumping from the lake

Of these options, the diversion of the inflow and the pumping were eliminated on the grounds of cost. Of the outlet tunnels the left abutment option is to be preferred on account of less difficult geology. The works would comprise two 8m diameter tunnels, one connecting Lake Sarez with Lake Shadau and the second from Lake Shadau to the river downstream of the Usoy dam.

*Solution B*

This solution, which aims at converting Usoy from a natural dam into an engineered structure capable of withstanding the hazards by which it is threatened, comprises the following elements:

1. Raising the Usoy dam crest to increase the minimum freeboard to 100m to eliminate the risk of overtopping.

2. Construction of an overflow surface spillway to arrest the long term increase in lake level.

3. Strengthening the vulnerable area of the downstream face of the dam (the canyon) to provide greater security against internal erosion.

In combination all these measures would provide a level of protection against external and internal erosion that is equivalent to lowering the lake level.

Solution C: Measures designed to reduce the effects of such a failure

These measures represent an alternative approach to risk mitigation but it must be appreciated that they are not strictly comparable to the measures described above. Measures to reduce the impact of failure, unlike those measures described in the foregoing sections, do nothing to reduce the probability of failure and will only be effective for a relatively limited or partial failure that would result in the uncontrolled release of a relatively small volume of water.

The measures comprise the construction of a downstream storage dam and flood defence work in the Bartang valley.

The only uninhabited part of the downstream valley where emergency storage could be provided without interfering with people's lives or livelihood is the section between Barchadiv and Usoy dam. It is evident that storage of $5Mm^3$, sufficient to retain the maximum estimated overtopping volume from a $0.5km^3$ landslide, could be created by the construction of a dam of approximately 20m high.

Costs and benefits

The estimated costs of the three solutions are summarised as follows:

| **Element** | **Estimated cost (US$m)** | | |
|---|---|---|---|
| | **construction** | **access** | **total** |
| *Solution A: lowering the lake by 60m* | | | |
| Left bank diversion | 32 | 100 | 132 |
| Right bank diversion | 22 | 90 | 112 |
| *Solution B: engineering the Usoy dam* | | | |
| Usoy dam crest raising | 75 | | |
| Spillway | 55 | 90 | 370 |
| Dam strengthening | 150 | | |
| *Solution C: Flood damage mitigation* | | | |
| Downstream retention dam | 10 | | |
| Flood defences | 20 | 75 | 105 |

The following conclusions can be drawn from these estimates:

1. The cost of all solutions are dominated by the cost of access
2. It is cheaper to increase the freeboard by lowering the lake than by raising the dam and providing a spillway
3. the lowest cost solution is likely to be flood mitigation, although this does not provide the same level of security as solution A or B

The risk mitigation benefits of these structural measures are difficult to quantify. In the Bartang valley cost of the potential damage is probably small since the value of the infrastructure that would be destroyed is low. The primary benefit would therefore be a reduction in the probability of the population at risk losing their lives: the previous paper showed that after the installation of the Early Warning System the population at risk is estimated to be 1000, and that the structural measures would reduce the probability of failure from $10^{-3}$ to $10^{-5}$. The cost per life saved, defined as

$$\frac{\text{annualised cost of the structural measures}}{\text{no of population at risk * reduction in probability of failure}}$$

is estimated at $12 million. The risk mitigation benefits alone are therefore insufficient to justify the structural measures, and other benefits must be identified if the risk of failure is to be reduced.

Hydropower generation

Of the various long term solutions described above only the schemes designed to lower the lake level possess any commercial potential. Of these the left bank diversion seems to offer the most benefits at lowest incremental cost.

For any of the solutions it will be important to reduce or severely eliminate the filtration through the dam so as to maximise the flow through the turbines. This can be achieved either by maintaining the lake level below 3160masl or by carrying out engineering works, such as grouting or blanketing the Usoy dam, in order to reduce the filtration if the scheme is to be operated with a lake level above 3160masl. Because the likelihood of constructing an effective grouted cut-off or blanket at an affordable cost is rather low, only the former alternative has been considered.

Power and energy

As the lake level drops, the rate of filtration through the dam decreases and the maximum allowable flow through the turbines increases. The combined effect of the variations in flow and head is that the power that can be generated rapidly increases to a peak value of 220 - 250MW, depending on the maximum total outflow from the dam. Once the lake level is drawn down to its long term level – assumed to be 3160m – the power reduces to 125MW, as shown in Figure 3. These estimates are based on an overall efficiency of 75% and a plant factor of 100%. Once the lake is draw down the station could be operated to generate peaking power with a plant factor of about 50%.

The annual production peaks at about 2000GWh/year in the period when the lake level is being drawn down, and settles at a long term value of 1100GWh/year,

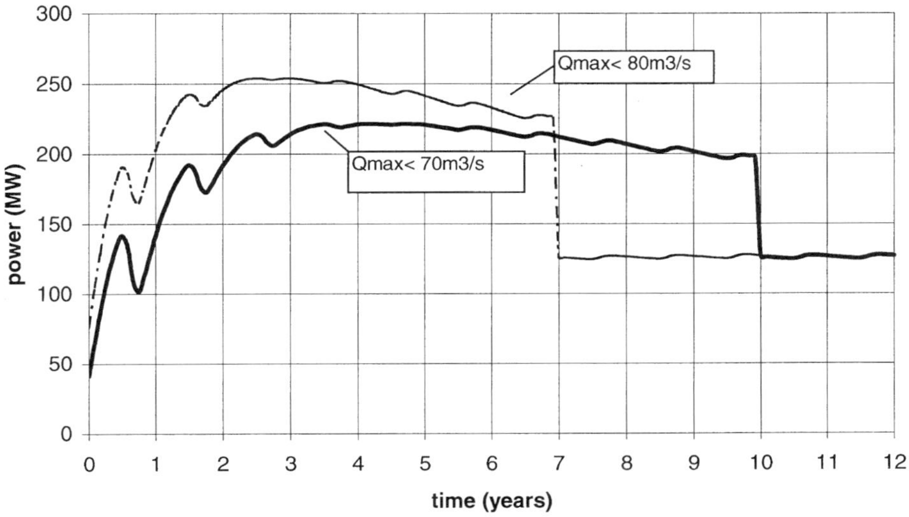


Figure 3. Power availability

Electricity demand

Tajikistan currently imports 4000GWh of electrical energy from Kyrgystan and Uzbekistan. Thus the production from Sarez could easily be absorbed by the domestic market.

Transmission

A 300km transmission line would be required to connect Sarez with the existing Tajikistan grid. However this connection could form a major part of the planned interconnection with China. For this study a double circuit, single 200mm$^2$ conductor, 500 kV transmission line is recommended.

Cost & Benefits

The total cost of the left bank diversion/hydropower scheme is estimated as follows:

| | US$ million |
|---|---|
| - civil works | 53 |
| - equipment | 67 |
| - access | 100 |
| - transmission & switching | 105 |
| total | 325 |

O&M costs are assumed to be $4million/year

The net annual energy benefit, valued at the current import price of 3 cents/kWh, peaks at $54 million in year 5 and thereafter diminishes to its long term value of $29 million.

At a discount rate of 12% this scheme is economically feasible: in other words, by the development of hydropower the risk mitigation benefit would be obtained at zero cost.

ACKNOWLEDGEMENTS
The authors are grateful to the Sarez Agency in Tajikistan for permission to publish this paper and to their colleagues in Tajikistan and Switzerland for shared ideas.

REFERENCES

[1] GIBB Ltd (2000). Lake Sarez Safety Assessment Report

[2] Sheko A.I. et al (1967 – 1969). Ministry of Geology of the USSR. - Reconnaissance inspections of lake's banks

[3] Akdodov Y, Kozak Y (1985-1987). Geological Department, Integrated engineering-geological studies in the Sarez Lake region for evaluation of stability of slopes and the forecast of process development for 1985-1987

[4] United Nations (2000). Usoi Landslide dam and Lake Sarez

[5] Noda E (1970). Water waves generated by landslides. Journal of the Waterways, Harbors and Coastal Engineering Division, ASCE

[6] Slingerland & Voight (1979) Occurrences, properties and predictive models of landslide generated water waves. Rockslides & Avalanches – vol 2 Elsevier Science Publishing

[7] Pugh C A & Harris D W (1982): Prediction of Landslide Generated Waves 4th ICOLD, Rio de Janeiro

[8] Imamura F (1996): Numerical model for tsunami generation due to subaqueous landslide along a coast, Science of Tsunami hazards, No 14.1 pp 13-28

[9] Harbitz C (1991):Numerical simulation of slide generated waves, Science of Tsunami Hazards, No 17.3

[10] Huber A. (1997): Quantifying impulse waveeffects in reservoirs, 19th ICOLD, FlorenceGIBB Ltd (2000). Lake Sarez Safety Assessment Report

# Where to keep your dam documents?

JEREMY STEWART, COGNICA Ltd, UK

SYNOPSIS. The storage of records is an issue requiring some thought if the most efficient and cost effective solution is to be implemented. The importance of keeping records, data and documents of technical merit has increased significance within water companies. Severn Trent Water has embraced the requirement to hold documentation electronically and in partnership with COGNICA have developed a system that allows them to store and access information quickly and easily. As the requirement to make reservoir information more widely available in the future increases, Severn Trent Water can be safe in the knowledge that their records are held in a manner that can improve accessibility without jeopardising the document security.

## INTRODUCTION

Working with paper can pose some difficult dilemmas. Do you really need to keep the documents? Do you have the physical capacity to store the documents? How often will you need to refer to them? How long will it take you to retrieve them?

When you have an enforced requirement to keep documentation this tends to take these dilemmas away but they are replaced by potentially more difficult decisions. How is it best to store the documents? Can you quickly and easily access the information when it is needed? Can you be safe in the knowledge that these documents are suitably secure?

The Reservoirs Act 1975 has encouraged water companies to store and reliably retrieve certain records and documentation relating to their reservoirs. Lillie and Hitchmough (1999) raised questions about the security of the reservoir archive and suggested that an electronic system would improve the current position. Reservoir owners adhere stringently to the demands of the Act with respect to the dam structures but looking after the records relating to each asset is often seen as a lower priority. The value of the records held in reservoir and dam archives should not be underestimated and the protection of these records is vital.

## INFORMATION STORAGE

Current reservoir archives, in a variety of formats, occupy a large amount of physical space. The larger the archive the more likely it becomes that the records will be held in more than one location. A dispersed archive is more

*Reservoirs in a Changing World*, Thomas Telford, London, 2002, 575–580

difficult to control and unauthorised access is very difficult to prevent. Problems of access are compounded by the fact that legitimate users of only part of the record need to be given access to the whole archive.

As well as taking up space, archive records tend also to be susceptible to damage and deterioration. This is particularly of concern for records that exist solely in paper form such as old typed reports and large-scale drawings on a variety of media including paper, linen and film. Catastrophic loss caused by flood or fire is fortunately rare but deterioration caused by handling, exposure to sunlight and photocopying will lead to a gradual degradation and eventual loss of archive material. The logical assumption is that conversion to a properly managed electronic storage system would eliminate this risk to the records.

In addition to eliminating risks, electronic systems address security through access control. Password protection not only limits who can see records, if necessary down to the individual document level, but it can also prevent the data being changed without correct procedures being followed. The data owner could allow different users to have different access privileges. A typical user could be limited to reading records or alternatively to reading and commenting on records, whereas an administrator would be able to only distribute records. A Reservoir Engineer would have responsibility for adding new records and linking related documents but would not be able to edit existing records. Particularly stringent controls could be placed on document deletion. The ability to print data can also be controlled through considered deployment of an electronic system.

Password control has other benefits and can for example allow tracking of document usage against specific users or user groups. Statistics could detail how many times a document is used, when and by whom. Modification tracking as part of the same password protected system, linking changes to specific users, can also provide the basic information for a change control strategy.

## ACCESSING INFORMATION

A secure and well-protected archive is of course only useful if the information held is accessible to those who need it. This means that some up-front effort in structuring and indexing the documentation will be required. The benefits of this activity increase over time, reducing the effort required to find documents on future visits to the archive.

In a standard paper archive it is only possible to categorise documents in one way. An electronic system allows the same document to be viewed from many different starting points. Examples include views by Year written, by Author or by Reservoir name. Grouping documents in this way means that related documents can be viewed together and important data are not overlooked.

Another way to relate common documents is to apply intelligent linking to the records. This allows instant access to referenced works acting as a live bibliography. Another benefit of this approach is that new reports need not reproduce information from referenced work; a link to the existing data is sufficient to support the new report.

Rapid access to the correct information can reduce response times and prevent possible incident escalation. Electronic systems can provide this access both through intelligent indexing and by the use of high-level searches. A high level search based on keywords and titles in an electronic archive means that a document can be found, quickly and efficiently, without the need for intimate knowledge of where it has been filed. An intelligent keyword-searching algorithm can bring together common documents, ensuring that nothing is missed on the topic of interest. Using keywords provides more economical access in terms of processing time than a traditional document database word-search, and so would be particularly effective in an emergency situation.

In an emergency situation it is vital that all parties are using the same information, that all relevant records are available and that the records are up to date. Storing records electronically on a central server that can be accessed across a network makes it easier to ensure that everyone is viewing the same data. Centralised storage means that the administration function can also be centralised and it is therefore easier to keep information up to date. It is no longer necessary to spend time travelling to the archive as it can be accessed from a remote site using a portable computer.

Updating of the reservoir archive with new material can also benefit from an electronic approach; for example where the 1975 Reservoirs Act requires that records be submitted in a 'prescribed form'. The use of an electronic template of the prescribed form of record ensures that all the relevant information is correctly recorded, that all fields are completed before saving and that reports are in a consistent format.

## SEVERN TRENT WATER CASE STUDY

The document archive at Severn Trent Water amounts to approximately 100,000 A4 pages and over 2,000 drawings. These are stored in filing cabinets and drawing files at two different sites, the documents stored for security purposes at the second site are intended as an exact copy of those stored at the first site. The archive grows by around 1,000 pages a month. The drawing archive growth is less easy to predict.

The Severn Trent archives are located, by necessity, in fully serviced office space reflecting both the need for constant access and the susceptibility of the archive material to excessive heat, cold or damp. The space occupied by the filing and drawing cabinets at each of the two locations is the equivalent of that which could possibly be utilised by up to 10 desk-based staff.

Severn Trent Water reservoirs are centrally managed as assets within the engineering division by a dedicated team of 12. The Networks or Water Supply divisions control the day-to-day operation of the reservoirs. However, these divisions must have reference to the central management team for the majority of the information necessary to perform their function.

Reservoir monitoring data and documents are reported and transmitted by various means such as fax, paper report, e-mail and telemetry. The electronic information received is currently printed and added to the paper archive. As more and more information is transferred and stored electronically the question was raised as to whether it was sensible to keep printing this information out to add to the already growing pile of paper. There is obviously also a duplication of effort in updating two separate archives and a time lag in updating that results in the archive only rarely being identical.

Severn Trent Water identified these problems and decided that an electronic solution would increase business efficiencies. The company had previously documented the fact that they have developed their own structure for reservoir data storage. Their requirement was therefore to take this structure and use it as the basis of a document management system.

Severn Trent Water investigated a host of document management systems that were available off-the-shelf. Document management solutions are designed to safely store a variety of documentation in a structured way. They are less focused on providing a user-friendly way of accessing the stored information and they certainly don't pay too much attention to the Reservoirs Act.

COGNICA were approached to discuss whether a bespoke solution could be developed to meet the requirements. With experience of similar work elsewhere in the water industry COGNICA were ideally placed to undertake the project and as a result the AQUIR system was developed.

The AQUIR system, now being used by Severn Trent Water, is web browser based and installed on their Intranet. This enables the users, whether they are based at their offices in Warwick or on a remote site, to view information relating to any one of the company's 60 reservoirs. The Intranet deployment also means that once the system has been updated there is no danger of a different version of the data being accessed elsewhere. The system has several levels of access control allowing different users access to different documents dependent on the role they have been assigned. Similarly their updating permissions are dependent on their role.

The electronic archive is structured to allow document access by report section, by author, by date or via the selection of key words. The size of the electronic archive is several gigabytes but the data is only stored once and usage statistics will allow elements of the data to be deployed in a way that matches the number of times they are viewed.

The growth of the archive will be controlled by the use of document reference links rather than document copies and by the reduced storage space requirement of an electronic document over a scanned document.

The method for attaching new information to the archive utilises drop down menus for describing different document types. This enforced naming convention will ensure a generic approach to future storage. The constraint prevents individual users using non-standard descriptors for the information, which would ultimately lead to difficulties in finding information.

Certain users at Severn Trent Water have the ability to add new documents and make links between documents whereas others can only view the documents. Appointed administrators can also control the addition of new users and assign their roles. This enables the Severn Trent Water reservoirs team to have complete control over their records. As these records relate to an estimated £1 billion worth of assets the data management issue is perhaps an area that more companies in a similar position should look to prioritise.

The point and click interface using icons to represent menu options required little training and the system was introduced to the Severn Trent staff in a single two-hour session. As with all electronic systems there will be a period of familiarisation required before accessing the archive becomes second nature and the issue of user enthusiasm should not be underestimated. Early indications are however, very promising and the increased functionality offered to users by the AQUIR system over the paper archive is encouraging users to log on.

The feedback from users within the Severn Trent reservoir team is that the AQUIR system will give them more time to concentrate on the technical aspects of their roles by eliminating the majority of the administration of the archive. As yet there has been little opportunity to collate the views of users outside of the reservoir team but these will be reported later in the year and a review posted to BDS along with a report on the system stability and the impact of any system failures in the interevening period.

The overall benefit to the business will not be able to be assessed for some time but Severn Trent have already seen cost benefits in the removal of the requirement for expensive storage space and the time and material costs associated with the reprinting and copying of information for delivery to users outside the reservoir team.

## THE FUTURE

Alan Johnston's Geoffrey Binnie Lecture 'Taken for Granted' at the British Dam Society Biennial conference on 15th June 2000 concerned risk and reservoir safety. Much of the discussion involved the risk and safety of the dam structures themselves, but he also made mention of the reports and records that describe those structures. He discussed transparency and the fact that information on the risks attached to reservoirs should be available to the public. He advised that it would be no surprise if this became a requirement in the future. This would have a great impact on how reservoir owners are compelled to hold and distribute their records, and how they will make this information available to the public. Currently most reservoir owners hold their records in a number of formats, including paper, electronic, microfiche and a variety of databases. For a reservoir owner to make any documents available to the public they will have to impose very tight controls over who gains access to which documents and what they are able to do with them.

Improving access but controlling access. Disparate aims they may be but they are both tackled more effectively with an electronic data system. Access control is an issue that will become more important in the coming years. It may well be that the question of whether you can afford an electronic system for holding your reservoir data, becomes more a question of whether you can afford not to have an electronic system for holding your reservoir data.

## ACKNOWLEDGEMENTS

The authors would like to thank Severn Trent Water Reservoir Engineers for allowing their case study to be used for the purpose of this paper.

## REFERENCES

Lillie C & Hitchmough M (1999). Are your data less secure than your dams? BDS – Dams and Reservoirs, December 1999.

Extract from Mr T A Johnston's Geoffrey Binnie Lecture at the 2000 Biennial BDS Conference.

# The Characteristics of UK Puddle Clays – a Review

A I B MOFFAT, University of Newcastle, UK

---

SYNOPSIS. The majority of significant British embankment dams completed in the period 1840-1960 were of so-called 'Pennines' profile, with a comparatively slender central core of puddle clay. In dams completed prior to c. 1880-1890 the use of puddle clay extended also to the cutoff trench, many excavated to considerable depth. Satisfactory long-term performance of puddle core and/or trench is critical to embankment integrity. Progressive deterioration has been identified in a number of 'Pennines' embankments, some 75% of which were constructed pre-1930. Assessment of such dams requires an appreciation of the nature and characteristics of puddle clay. The intention of this paper is to provide a concise first review of puddle clays in the UK.

## INTRODUCTION

Some 2,500 UK reservoirs, retained by 2,650 dams, can be regarded as significant in that their capacity brings them within the ambit of the Reservoirs Act 1975 (BRE, 1994). Median reservoir capacity is modest at approximately 125 x $10^3$ $m^3$, and median height of dam similarly so at 7.5 m, with 28% of dams below 5 m height. Median age is relatively high at c. 98 years, with approximately 1,200 dams dating from the $19^{th}$ C or earlier.

A large proportion of older UK reservoirs are concentrated in the former industrial heartlands such as the North West, West Yorkshire or Central Scotland, and are thus located in close proximity to major centres of population. It has been suggested that up to 75% of reservoirs subject to the 1975 Act have, on inspection, been classified as Category A or B in terms of the ICE Floods Guide (ICE, 1996), i.e. they would endanger life and/or cause extensive damage in the event of breaching (Tedd *et al*, 2000).

The several variants of embankment dam account for 83% of the UK dam stock, i.e. some 2,200 dams. Records indicate that 1,600 dams of all types, with the earthfill embankment overwhelmingly predominant, were completed within the period 1840-1960. The latter timespan can be considered to define the era of the 'Pennines' embankment, i.e. an earthfill dam incorporating a comparatively slender central core of puddle clay in combination with a trench-type cutoff, the latter sometimes of considerable depth. Incomplete records preclude giving a definitive figure for the number of 'Pennines' embankment dams, and on a number of occasions a presumed 'Pennines'-type dam has, on investigation, been found not to have

*Reservoirs in a Changing World*, Thomas Telford, London, 2002, 581–601

a clearly defined core. An indicative estimate would suggest that possibly as many as 60% of UK embankments, i.e. c. 1,300 dams, could be of 'Pennines' type. It will be noted that the construction of a large majority of these pre-dated any formalised understanding of geotechnical engineering.

The increased emphasis on monitoring and surveillance consequent upon implementation of the 1975 Reservoirs Act has indicated less than satisfactory long-term performances in a significant number of 'Pennines' type embankments. Excessive seepage, settlement and other signs of progressive deterioration have been identified, defects associated in most instances with degradation of puddle core integrity. This has frequently necessitated technically difficult and costly remedial work.

Assessment and evaluation of 'Pennines' embankments requires an appreciation of the nature and geotechnical characteristics of puddle clays. Within the UK dam community such appreciation has been inhibited by a notable dearth of published information. The purpose of this paper is to present a concise review of puddle clay in context with an understanding of the 'Pennines' embankment dam.

Puddle clay is not, as sometimes believed, a distinctive type of clay. It is, rather, a soil of significant clay content whose natural fabric and structure have been broken down by intensive working and remoulding at a water content such that the resulting clay 'puddle' is a dense, homogeneous, low permeability and plastic material. Its characteristics are thus suited to use in water-retaining barriers, and there is evidence that what was recognisably a puddle clay was used by the Roman Army to line conduits and aqueducts. Employed similarly by the early canal engineers, it was natural that where reservoirs were required to supply lock systems the watertight element of the impounding dam was also constructed in puddle clay, initially in the form of an upstream blanket but latterly and more generally as a central core. The preparation of puddle clay will be addressed in later sections of this paper.

## THE 'PENNINES' EMBANKMENT

### Evolution

The history of embankment dam development in the UK encompasses three principal eras, commencing with the earliest engineered dams constructed for amenity lakes etc. from the mid-18th C., most notably by Lancelot 'Capability' Brown and John Grundy. Many of these were built as homogeneous embankments, but a number made use of rammed or punned clay – not to be confused with puddled clay – to form a central core wall or barrier. This 'pre-Pennines' era was followed, from c. 1840, by the 'Pennines' era, which spanned some 120 years. The 'Pennines' era concluded c. 1960 with the universal adoption of the rolled clay core, signalling the start of the 'post-Pennines' era. Development up to this point is comprehensively described in Skempton, 1989.

Table 1. Evolution of the embankment dam in the United Kingdom

| Period | Representative Dimensions max. height (m) | Representative Dimensions core H/b ratio[1] | Impervious Element | Cutoff Provision | Supporting Shoulders |
|---|---|---|---|---|---|
| **Pre-Pennines Era** | | | | | |
| 18th C | c. 6 | (c.4) | homogeneous profile (or rammed clay core) | fill or rammed clay in shallow key trench (if provided) | random fill |
| 1800-1840 | 25 | 3 to 4 | central puddle core (or thick upstream 1.0–1.5m blanket) | puddle in key trench | random fill |
| **PENNINES ERA[2]** | | | | | |
| Phase 1 1840-1865 | 30 | 3.5 to 6 | slender central clay core; puddled *in situ* | puddle clay : deep trench as necessary | random fill; lifts up to 1.2m; no compaction |
| Phase 2 1865-1880 | 35 | 3 to 5 | slender central puddle clay core | puddle clay : deep trench as necessary | select fill against core (Moody zoning); lifts to 1.2m; toe drainage; compaction incidental |
| Phase 3 1880-1945 | 35 | 3 to 4 | slender central core; 'pugged' puddle clay | puddle clay: deep trench as necessary | select fill against core; lifts to 1.2m; compaction limited |
| Phase 4 1945-1960 | 45 | 3 to 5.5 | slender central core; 'pugged' puddle clay | concrete in deeper trenches and grout curtain | select fill against core; drainage provision etc; controlled compaction |
| **Post-Pennines Era** | | | | | |
| 1960⇒ | 90 | 2.5 to 3 | broad core, rolled clay | rolled clay key trench and/or grouted cutoff | 'engineered', with zoned compacted earthfill and/or rockfill, drainage/ filters etc. |

1. H = max. core height above base
   b = max. core width at base
2. phase dates indicative only; changes generally progressive and not universal

Table 1 traces the development of the embankment dam, identifying the more significant evolutionary changes in earthfill embankment design and construction associated with each era.

A brief selective chronology can usefully supplement Table 1:

- 1730 : first use of 'punned' clay barrier
- 1766 : first foundation cutoff trench
- c. 1795 : puddle clay barrier first employed
- c. 1865 : zoning of shoulder first adopted: danger of excessively slender cores appreciated
- 1876 : first rock grouting of foundation
- 1937 : soil mechanics analyses employed in design modern plant used in shoulder construction

Stages in development

The 'Pennines' era can be divided into Phases 1 to 4, as identified in Table 1. Successive phases reflect major evolutionary changes made in response to specific influences, notably:

- a small number of disastrous dam failures (notably Bilberry (1852), Dale Dyke (1864))
- numerous serious instances of unsatisfactory performance (e.g. Torside (1854), Lower Lliw (1873 *et seq.*))
- the trend to larger and bolder dams, sometimes founded on difficult ground

Figures 1 and 2 show illustrative profiles of embankments of Phases 1 and 3 respectively.

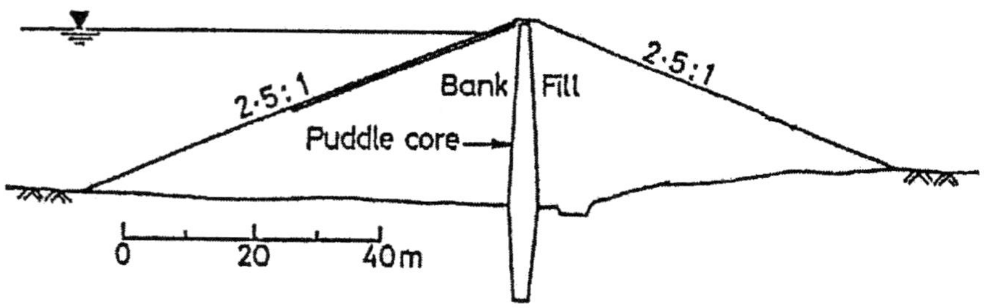


Fig. 1. 'Pennines' embankment: c.1860 (Phase 1) (after Skempton, 1989)

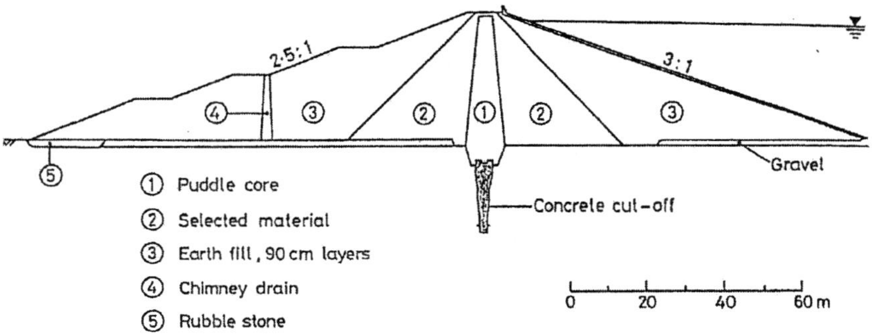


Fig. 2. 'Pennines' embankment: c.1900 (Phase 3) (after Skempton, 1989)

Progressive changes of particular moment included, *inter alia*:

- the move to a form of zoning, with select clayey fill placed adjacent to the core (a simplified version of the zoning recommended by Captain Moody RE in the aftermath of the Bilberry disaster)
- introduction of the steam pugmill, with puddle clay prepared away from the embankment (it may be noted that the placing and working of puddle clay, as opposed to its preparation, was never successfully mechanised)
- increasing use of erosion-resistant concrete rather than puddle clay in deeper cutoff trenches
- outlet works installed in flanking tunnels or in culverts on rock *vice* installation through fill or compressible foundation soils
- application of a modest degree of compaction to shoulder fills
- increasing attention to the control of internal seepage, e.g. by provision of stone pillar drains, drainage blankets etc.

The demise of the 'Pennines' embankment from c. 1960 (Jumbles, the final dam of the type, was completed in 1971) marked a belated recognition of the merits of the modern zoned embankment with rolled clay core, a type already long-dominant in the USA and elsewhere. The reluctance to progress from the 'Pennines' type dam despite its relatively high unit cost, i.e. cost/m$^3$, was attributable to generally satisfactory UK experience with the mature 'Pennines' dams of Phases 3 and 4, where the following design principles had become established (Kennard, 1994):

- core width not less than one third of core height, with minimum of 2 m at crest and batters not steeper than 1 in 12
- select fine fill on both sides of core, preferably with a zone of clay 2-3 m wide immediately either side of core
- coarse fill placed in outer shoulders
- upstream face not steeper than 2:1, with berms in higher dams
- a drainage mattress under the downstream shoulder
- 'slippery or compressible' material in natural ground removed prior to construction

A critique

While overall experience of the 'Pennines' dam design has been generally satisfactory, certain technical weaknesses are inherent in many of the embankments of this type, including:

- slenderness of the core (H/b > 5) with resulting susceptibility to hydraulic fracture and erosion on first filling (This may be self-arresting given a clay downstream zone or shoulder)
- vulnerability of the puddle trench to fracturing and erosion
- lack of effective internal drainage
- shoulders frequently of poorly drained, clayey material
- ongoing consolidation and settlement of the core
- steep face slopes (e.g. 2:1 and steeper)
- susceptibility to drying out and cracking on drawdown

The quality of engineering and workmanship was also extremely variable.

Functions of core and cutoff

The primary functions of the 'Pennines' core and associated trench cutoff can be stated as:

- localisation of head loss, ensuring that seepage does not destabilise the downstream shoulder
- control of water loss
- provision of a more plastic zone which can adjust to deformation without loss of integrity

The inferences with respect to seepage and internal erosion and the provision of drainage and protective filters will be noted.

PUDDLE CLAY

Nature and source soils

The primary purpose of puddling was to destroy the natural structure and fabric of a clay soil while mixing in any natural fine partings of silt or sand, yielding a relatively soft plastic material of uniform consistency which had many desirable attributes as a core fill.

Puddle clay may therefore be defined as a cohesive soil intensively remoulded and worked, or 'puddled', at a water content such as to produce a uniformly dense, homogeneous and plastic soil of very low permeability (c. $10^{-9}$ m/s). Undrained shear strength, $c_u$, of the freshly puddled clay is also low (c. 8.5-10 kPa at placing), while compressibility and ductility are relatively high. In the case of clayey soils of high inherent plasticity with the potential for high shrinkage moderate amounts of sand and/or some gravel were frequently added during preparation to temper or 'tone' the puddled clay before placing.

Table 2 identifies the source clay used in the cores of a range of 'Pennines' dams, demonstrating the suitability of a wide range of cohesive soils to use as puddle clay. It will be noted that while many of the examples refer to soils of low to intermediate plasticity, classified as CL and CI, there are numerous instances of high plasticity source clays being successfully employed, e.g. at Coulter, Monkswood, William Girling *et al.* Also of note is the wide variation in clay fraction, ranging from 18% to 70%.

Table 2. Representative 'Pennines' dams: puddle core clays

| Dam | Location NGR | Height (m) | Completion | Core Clay: BS Classn. | Core Clay: Description |
|---|---|---|---|---|---|
| Aldenham | Herts. TQ 160960 | 7 | 1795 | CV | — |
| Barrow Compn. | Avon ST 541 678 | 12 | 1864 | CH | soft alluvial clay |
| Barrow No.3 | Avon ST 541 678 | 17 | 1899 | CH | soft to firm silty L. Lias clay |
| Blackmoorfoot | W. Yorks SE 099 130 | 14 | 1876 | CH | soft homogeneous clay (45% clay) |
| Burnhope | Co. Durham NY 847 390 | 41 | 1935 | CI | boulder clay |
| Challacombe | Devon SS 697 421 | 15 | 1945 | MI | silty clay (28% clay) |
| Coulter | Strathclyde Reg. NT 037 275 | 24 | 1907 | CH | — |
| Cullaloe[(1)] | Fife Reg. NT 187 873 | 11 | 1876 | CL | soft sandy clay (25% clay) |
| Cwmwernderi | W. Glam. SS 816 903 | 22 | 1901 | CI | silty clay (39% clay) |
| Glencorse | Lothian Reg. NT 222 636 | 23 | 1823 | CL/CI | silty clay (?) |
| Grassholme | Co. Durham NY 955 242 | 34 | 1914 | CL/CI | stiff silty boulder clay |
| Hury | Co. Durham NY 955 242 | 33 | 1894 | CL | stiff silty boulder clay |
| Hallington West | Northumberland NY970 762 | 13 | 1889 | CI | boulder clay (25% clay) |
| King George V | London (Lea Valley) TQ 374 965 | 8 | 1912 | CV/CE | silty alluvial clay |
| Knockendon | Strathclyde Reg. NS 243 519 | 27 | 1948 | CI | boulder clay |

Table 2 (continued)

| Dam | Location NGR | Height (m) | Completion | Core Clay BS Classn. | Core Clay Description |
|---|---|---|---|---|---|
| Lambieletham[2] | Fife Reg. NO 502 133 | 14 | 1899 | CL/CI | silty clay |
| Lockwood | London (Lea Valley) TQ 353 902 | 9 | 1903 | CE | silty alluvial clay (70% clay) |
| March Ghyll | N. Yorks SE 122 511 | 20 | 1906 | CH | — |
| Monkswood | Avon ST 758 710 | 13 | 1893 | CV | Lias clay |
| Muirhead | Strathclyde Reg. NS 260 563 | 23 | 1942 | CI | boulder clay |
| Oakdale Lower[3] | N. Yorks SE 465 964 | 11 | pre 1914? | CHO/CVO | very soft clay, high organic content |
| Ramsden | W. Yorks SE 114 056 | 25 | 1883 | MI/MH | clayey silt (17-42% clay) |
| Rivelin Upper | S. Yorks SK 271 868 | 12 | 1848 | CI/MI | glacial clay |
| Rotton Park | Birmingham SP 043 868 | 14 | 1828 | — | boulder clay |
| Selset | Co. Durham NY 918 211 | 41 | 1959 | CL | boulder clay (18% clay) |
| Silent Valley | Co. Down J 307 219 | 25 | 1933 | CI | boulder clay |
| Staines South | London (Thames Valley) TQ 045 728 | 9 | 1903 | CH/CV | London clay |
| King George VI | London (Thames Valley) TQ 045 728 | 17 | 1947 | CH/CV | London clay |
| Sutton Bingham | Somerset ST 556 115 | 18 | 1955 | CH | Oxford clay |
| Walshaw Dean Lower | W. Yorks SD 966 335 | 22 | 1907 | CI/CH | silty clay (20-40% clay) |
| Walshaw Dean Upper | W. Yorks SD 966 335 | 24 | 1907 | CI/MH | |
| William Girling (Chingford) | London (Lea Valley) TQ 367 941 | 13 | 1951 | CV | London clay |

1. decommissioned and breached c. 1980
2. demolished 1985 following d/s instability
3. discontinued 1991 following serious internal erosion

Illustrative examples of puddle clays deriving from generic source clays, with their consistency limits, are shown in Table 3. Also tabulated are the corresponding liquidity indices, $I_L$, and shrinkage limits, $w_s$. Liquidity index, defined as $I_L = \frac{w - w_p}{I_p}$, expresses the water content, w, of the puddle clay at placing relative to its consistency limits, i.e. liquid limit, $w_L$ and plastic limit, $w_p$, with plasticity index $I_p = w_L - w_p$. It will be noted that many of the values lie in the 40% to 50% range, indicating that the puddle clays generally have a consistency roughly midway between the liquid and plastic limit values. In UK climatic conditions clays with $w_L < 80$ are unlikely to be prone to excessive shrinkage.

Table 3 : Representative generic puddle core clays: consistency limits and shrinkage

| **Generic Core Clay/Dams (BS Classn.)** | **Liquid Limit $w_L$%** | **Plasticity Index $I_p$%** | **Water Content w%** | **Liquidity Index $I_L$%** | **Shrinkage Limit $w_S$** |
|---|---|---|---|---|---|
| **Boulder clays (CL/CI/CH)** | | | | | |
| Muirhead (CI) | 43 | 26 | 28 | 42 | 13 |
| Coulter (CH) | 63 | 39 | 35 | 28 | - |
| Hallington West (CI) | 48 | 24 | 31 | 29 | - |
| Selset (CL) | 30 | 15 | - | - | - |
| Silent Valley (CI) | 38 | 21 | 27 | 47 | 13 |
| **London clay (CV)** | | | | | |
| King George V | 80 | 50 | 50 | 40 | 14 |
| Queen Mary | 75 | 50 | 44 | 38 | 14 |
| William Girling | 73 | 52 | 43 | 42 | 15 |
| **Lias clay (CH/CV)** | | | | | |
| Barrow No.3 (CH) | 66 | 41 | 36 | 27 | - |
| Monkswood (CV) | 85 | 60 | 50 | 58 | 14 |
| **Alluvial clays (CE)** | | | | | |
| Banbury | 132 | 92 | 75 | 40 | 12 |
| King George V | 125 | 80 | 70 | 32 | 11 |

All clayey soils at their plastic limit display comparable undrained shear strengths, with $c_u$ in the range 100-200 kPa. Remoulded, their strength will decrease across the plastic range, from c. 100-200 kPa ($w = w_p$) to c. 1-2 kPa ($w = w_L$). Plasticity index $I_p$ is thus a measure of the change in water content, $\Delta w$, required to change $c_u$ by some two orders of magnitude within the plastic range of the soil. Soils of low plasticity, e.g. CL, CI soils having a low $I_p$, require only a small incremental change in water content, $\Delta w$, to effect a very substantial change in shear strength, $\Delta c_u$. Soils having a high $I_p$ (i.e. highly plastic soils, classification CH, CV etc.) will not stabilise under load without a considerable incremental change in water content, suggesting a possible correlation between plasticity and compressibility.

Liquid and plastic limits, $w_L$ and $w_p$, are a function of clay content and mineralogy. The latter parameters also govern permeability, k, and hence rate of consolidation, implying a correlation between $w_L$ and coefficient of consolidation, $c_v$.

Preparation and placing

In the case of earlier 'Pennines' type dams, e.g. those of Phases 1 and 2, it was customary to puddle the clay core *in situ*. Clay as excavated from the borrow area was placed in the core and water added as necessary to permit effective manual *in situ* puddling by the boots and puddling tools or clay spades of the puddling gang. This frequently led to very wet and soft areas which proved difficult to work effectively into the desired uniform consistency. The gradual introduction from c.1860 of steam-driven pugmills which could process the toughest clays into a homogeneous and uniform puddle was accompanied by two major changes in construction practice:

- clay excavated from the clay fields was stacked or spread and left to weather ('sour'), sometimes for several months, before further processing
- initial clay preparation and 'toning' in terms of water content, with the addition of sand to moderate shrinkage if necessary, was effected in pugmills sited clear of the dam, e.g. at the clay field; the prepared clay was then transferred to the core for placing in shallow layers and further manual puddling *in situ*

These developments led to a more consistent puddle clay and thus to a more uniform core.

Table 4 details examples of clay preparation and placing practice as specified for a number of dams. Indicative contemporary unit costs for a puddle clay core, typically three times the cost of shoulder fill, are quoted in Table 5.

The labour-intensive and costly nature of puddle core construction led to the adoption of relatively slender core profiles, most notably in Phases 1 and 2 of the 'Pennines' era. Incidents associated with such narrow cores induced a later trend to wider cores, as indicated in Table 1, and a core H/b ratio of three came to be widely adopted, particularly in dams below about 12-15 m in height.

Table 4. Puddled clay cores: preparation and placing

| Dam | Completion | Ht (m) | Puddle Preparation: Souring period | Puddle Preparation: Manual (in-situ) | Puddle Preparation: Pugmill | Puddle Placing: Layer (mm) | Puddle Placing: Working-in |
|---|---|---|---|---|---|---|---|
| Toddbrook | 1840 | 24 | n.k. | √ | | 150 | trodden, heeled in |
| Kentmere-head | 1848 | 17 | n.k. | √ | | 200 | soak, cut, crosscut and tread |
| Harperrig | 1859 | 12 | note (1) | √ | | 150 | trodden |
| Barrow No.2 | 1864 | 11 | n.k. | | √ | 150 | tempered, cut, crosscut and tread |
| Barden Upper | 1881 | 38 | yes | | √ | 150 | trodden in |
| Kennick | 1883 | 13 | 3 mths. | | optional | 150 | punned and worked |
| Cant Clough | 1892 | 26 | 6 mths. | √ | | 225 | cut, crosscut, tread and press |
| Hury | 1894 | 33 | 4 mths. | | √ | 150 | trodden, rammed and crosscut(2) |
| Clydach | 1900 | 10 | 1 mth. | | √ | 115 | cut, crosscut, trodden, punned and rolled |
| Grassholme | 1914 | 34 | 1 mth. | | √ | 115 | trodden and punned |
| Burnhope | 1935 | 41 | n.k. | | √ | 100 | cut, trodden and heeled |
| Ladybower | 1945 | 43 | n.k. | | √ | 150 | heeled in |
| Knockendon | 1948 | 24 | yes | √ | note (3) | | see note (4) |
| Cod Beck | 1953 | 23 | n.k. | | √ | 150 | cut, crosscut and heeled |
| Selset | 1959 | 41 | 'if reqd.' | | √ | 100-125 | watered, sliced, trodden and heeled(5) |
| Greenbooth | 1963 | 35 | n.k. | | √ | 100 | heeled in |

1. possible first example of wetting/working up puddle at claypit
2. specification provided for use of steam roller on puddle core, but not carried out
3. not general practice in Scotland to employ a pugmill for 'hill clays' (boulder clays)
4. initially, puddle worked in-situ with spades; with core at half-height, procedure changed to employ in-situ compaction of 'core' by bulldozer
5. attempts to mechanise placing unsuccessful

Table 5.'Pennines-era' clay fills: indicative unit costs

| Dam (height x length) m | Year | Fill Cost (£/m³) | | | Mass Concrete |
|---|---|---|---|---|---|
| | | puddle | select | shoulder | |
| Barden Upper (20 x 479) | 1882 | £0.23 | - | £0.09 | - |
| Kennick[1] (13 x 112) | 1883 | £0.33 | £0.12 | £0.10 | £0.98 |
| Burnhope (41 x 540) | 1935 | £0.49 | - | £0.15 | - |

1. contemporary mass excavation rates: soft £0.10/m³; rock £0.40/m³

Specifications

Early puddle clay specifications were simple in the extreme, with the placing and *in situ* puddling procedures described in very subjective terms. In some cases the source of the raw clay was prescribed, but it was also common practice for the Engineer to require the Contractor to source a suitable material for approval. As the 'Pennines' era developed specifications increasingly tended to reflect an accumulated experience of best practice. They nevertheless ranged from a loose, qualitative specification at one extreme through to the most highly prescriptive specification at the other. In the latter the technique to be employed by the puddling gang for cutting, cross-cutting and treading of the puddle was prescribed in detail, almost in the manner of a military drill.

Extracts from an illustrative specification of 1883 are quoted below:

> *Clay puddle*
> *Cl. (49) Clay suitable for puddle may it is anticipated be found in several places on the land purchased by the Board for the Reservoir, but parties tendering must satisfy themselves on this point, as the Contractor will have at his own cost to procure proper clay however far it may be necessary to go for the purpose.*
>
> *The clay must be dug and turned over on the ground where it is obtained and must there be freed from all stones, turf, peat, sand, shale or other objectionable matter.*
>
> *It must be good strong tough clay unmixed with any spongy, miry or puffy earth and after being dug it must be exposed to the weather in ridges for at least four winter months and immediately before it is required for use it must be worked up into puddle by turning it over, cross-cutting, watering, treading, or machine pugging, into one plastic homogeneous mass of the toughest consistency. All the above operations must be carried out on the ground where the clay is dug, or if this is at a distance, on a site to be assigned near the*

*embankment. As soon as possible after it is worked up into puddle it must be removed to its place in the embankment, and should any of it become from an excess of water too plastic or soft, or from being too long exposed during dry weather after it is worked, cracked and hard, it must be re-tempered and again worked up.*

*In its place in the embankment the puddle must be spread in even layers not more than six inches in thickness, each layer carried out to its full extend trodden, rammed and cross-cut.*

*On no account must any of the working up of the puddle in the clay field, or its deposit in or under the embankment be carried on during frosty weather. [See Section 70 as to rolling.]*

*Puddle wall*
*Cl. (67) The clay puddle must be filled in layers not exceeding 6 inches in thickness, each layer to be horizontal and carried out to its full extent at both ends before another is placed upon it. The embankment must be brought up to keep pace with the puddle as it proceeds and the edges of the puddle wall must be kept clear, neat, and of the true width by planking at the sides which must be drawn as the work proceeds, or otherwise as may be ordered.*

[Hury Dam: J Mansergh]

Further illustrative puddle clay specifications are quoted in Moffat (1999) and Johnston *et al* (1999).

Following the growing appreciation of soil mechanics which developed post-1945 undrained shear strength and permeability were written into some later specifications, thus:

$c_u \nless 7.1$ kPa ) Selset Dam (1955)
$k \ngtr 0.1 \times 10^{-9}$ m/s )

Present-day specifications for puddle clay, as developed by British Waterways Board and by North West Water, are set out in Haider, (1989) and in Johnston *et al*, 1999 respectively.

Testing

A number of simple and essentially qualitative tests for puddle clay evolved in the course of the 'Pennines' era. As an early example, a sample of puddle clay kneaded for three minutes and rolled into a 75 mm ball was not expected to show evidence of disintegration when immersed in water for 48 hours. Later additional tests included those for 'tenacity' as a measure of 'strength', and for 'impermeability' as assessed by 24 hour loss of water from a 'basin' formed of puddle clay.

By the latter part of the 'Pennines' era the testing of a fully prepared and worked-up puddle clay of acceptably low permeability centred upon four index tests:

- pinch index : (susceptibility to spalling and cracking)
- tenacity index : ('strength')
- elongation index : (ductility)
- soaking index : (dispersion/disintegration)

It will be noted that the tests continued to be qualitative in nature; they are described in detail in Head, (1980).

From c. 1950-55 puddle clay index tests were increasingly supplemented by regular laboratory and field determinations of undrained shear strength and permeability, as indicated for Selset.

## PUDDLE CLAY CHARACTERISTICS

### Consistency

Freshly puddled clays have the consistency of a very soft clay, i.e. they can be exuded between the fingers when squeezed. Undrained shear strength, $c_u$, for saturated fresh puddle clay at placing is sensibly constant in the range 8-10 kPa irrespective of the source clay. This equates to the minimum strength required to adequately support the puddling gang; it may be compared with $c_u$ = 60-120 kPa for the typical and much stiffer modern rolled clay core. Puddle clays also generally display some thixotropic increase in strength over the first 12-18 months, as shown in Table 6; clays of higher liquid limit, $w_L$, show the greater thixotropic gain in shear strength. Long-term increase in $c_u$, attributable to core consolidation, is referred to below.

### Shear strength

Undrained shear strength *In situ* and laboratory tests on 'mature' puddle clays confirm that undrained shear strength, $c_u$, is generally much enhanced following long-term consolidation of the core. This is apparent in Table 6, and is further highlighted by the examples of Table 7.

Table 7 also demonstrates that in a number of instances, and within limits set by sampling and testing procedures, $c_u$ increases with depth below core crest, z, in a near-linear relationship. This is essentially a reflection of the long-term consolidation of the core clay as indicated previously; the implications with respect to long-term core settlement will be appreciated. The influence of test method upon recorded shear strength is considerable.

Attention is also drawn to the range of $c_u$ values recorded for some individual dams in Table 7, e.g. at Coulter and Blaennant-Ddu. The extreme variation in $c_u$ may be attributable to localised changes in puddle clay 'quality' at time of construction and/or to later core degradation.

Table 6. Puddled clay cores: long-term gain in undrained shear strength

| Dam (Core Clay (BS Classn.)) | Undrained Shear Strength $c_u$ (kPa) at age: as placed | 1 wk. | 15 mths. | 14 yrs. | 30 yrs. | 40 yrs. |
|---|---|---|---|---|---|---|
| Banbury (alluvial clay (CE)) | [8.5] | - | - | - | - | 16.5 |
| King George V (London clay (CV)) | [8.5] | 10.5 | - | - | - | - |
| (alluvial clay (CE)) | [8.5] | - | - | - | 17.5 | - |
| Monkswood (Lias clay (CV)) | [8.5] | - | - | - | 9.5 | - |
| Queen Mary (London clay (CV)) | [8.5] | - | - | - | 16.5 | - |
| William Girling (London clay (CV)) | 8.0 | 10.0 | 14.0 | 17.5 | - | - |

Table 7. Puddled clay cores: undrained shear strength

| Dam | Compl. | Core Clay BS Classn. | Mature Undrained Shear Strength $c_u$ (kPa) | | |
|---|---|---|---|---|---|
| | | | $c_u$ | triaxial (T) vane (V) | variation $c_u$ with depth z |
| Abberton | 1938 | – | 23 – 33 | V | – |
| Banbury | 1903 | CE | 16.7 | – | – |
| Barrow Comp | 1864 | CH | 8.6 – 10.5 | – | – |
| Barrow No.3 | 1899 | CH | 45 – 60 | T | – |
| Blackmoorfoot | 1876 | CH | 26 | T | – |
| Blaennant Ddu | 1878 | – | 12.5 – 107 | T | 3.5 z |
| Challacombe | 1945 | MI | 20 – 60 | T | (30 + 2z) |
| Coulter | 1907 | CH | 4.8 – 74 (mean 28) | V | – |
| | | | 3.5 – 86 (mean 33) | T | – |
| Cwmwernderi | 1901 | CI | 10 – 65 (SBP (T+10) | T | (20 + 2 z) |
| Hallington West | 1889 | CI | 19 – 50 | T | – |
| Holmestyes | 1840 | – | 25 – 61 | – | – |
| Knockendon | 1948 | CI | 13.6 | – | – |
| Lambielethan | 1899 | CL/CI | 21 - 36 | T | (17+1.5 z) T |
| | | | 30 – 53 | V | (27+1.7 z) V |

Table 7 (continued)

| Dam | Compl. | Core Clay BS Classn. | Mature Undrained Shear Strength $c_u$ (kPa) | | |
|---|---|---|---|---|---|
| | | | $c_u$ | triaxial (T) vane (V) | variation $c_u$ with depth z |
| Lockwood | 1903 | CE | 28 | – | – |
| March Ghyll | 1906 | CH | 10 – 76 | – | (30+0.7 z) |
| Monkswood | 1893 | CV | 9.6 | – | – |
| Oakdale Lower | pre 1914 | CHO/CVO | < 7 | – | – |
| Ramsden | 1883 | MI/MH | 5 – 38 (mean 30) | T | 'no pattern' – |
| Selset | 1959 | CL | 15 (avg. upper core) | – | – |
| Silent Valley | 1933 | CI | 8.6 | – | – |
| Staines South | 1903 | CH/CV | 16 | – | – |
| William Girling | 1951 | CV | 17.7 | – | – |

<u>Shear strength (effective stress parameters)</u> Little data with regard to effective stress parameters c' and $\phi'$ for puddle clays has been published. Table 8 presents information from nine dams, embracing a range of source clays. Silty source clays such as those at Hallington West and Lockwood apart, it is generally realistic to assume $c' = 0$, $\phi' = 20°$ for a first analysis. In the case of boulder clays or glacial tills of lower clay content it may be appropriate to assume $\phi' = 30°$.

Table 8. Puddled clay cores: shear strength parameters (effective stress)

| Dam | Core Clay | BS Classn. | Shear Strength c' (kPa) | $\phi'$ (deg.) |
|---|---|---|---|---|
| Barrow No.3 | soft to firm silty L. Lias clay | CH | 0 | 30 |
| Blackmoorfoot | soft homogeneous clay (45% clay) | CH | 14 | 20 |
| Brent | London clay | CH | 4 | 21.5 |
| Hallington West | boulder clay (25% clay) | CI | 11-17 | 23-29 |
| Lockwood | silty alluvial clay (70% clay) | CE | 23 | 8 |
| Luxhay | – | – | 10 | 30 |
| Rivelin Upper | glacial clay | CI/MI | 3 | 30 |
| Selset | boulder clay | CL | 0 | 24 |
| Staines South | London clay | CH/CV | 4 | 13 |

Permeability

In the case of soils possessing a significant clay fraction permeability is the most variable and difficult of primary characteristics to quantify. Some generic values of coefficient of permeability, k, are shown in Table 9. Single-site datasets of field and laboratory determinations can range over several orders of magnitude due to the influence of local microscale variations in soil fabric and pore structure. It is therefore appropriate that measured values of permeability are regarded as being indicative only; their utility, other than in terms of relativities, is generally limited. Table 10 presents values of permeability coefficient, k, as determined at a number of dams. For puddle clay a coefficient, k, in the range (1-100) x $10^{-9}$ m/s may be regarded as not atypical and acceptable.

Table 9. Generic permeabilities

| **Core Fill** | **Indicative Clay Content (%)** | **Indicative Permeabilities k(m/s)** |
|---|---|---|
| puddled London clay | 50-60 | 1 x $10^{-10}$ |
| rolled boulder clay | 20-30 | 2 x $10^{-9}$ |
| rolled moraine | variable | 2 x $10^{-7}$ |

Table 10. Puddled clay cores: representative permeabilities

| **Dam** | **Completion** | **Core Clay** | **BS Classn.** | **Indicative Permeability* k(x $10^{-9}$ m/s)** |
|---|---|---|---|---|
| Blackmoorfoot | 1876 | soft clay | CH | 0.5 |
| Burnhope | 1935 | boulder clay | CI | 2.0 |
| Cheddar | 1935 | Keuper marl | – | 0.2 |
| Gorpley | 1905 | silty clay | CI | 0.6[(1)] |
| Grimwith[(2)] | 1864 | boulder clay | – | 40 – 3500 |
| Hallington West | 1889 | boulder clay | CI | 600 – 11000 |
| Holmestyes[(3)] | 1840 | silty clay | – | 0.1 – 4000[(1)] (generally 500) |
| Luxhay | 1905 | – | – | 4 – 20[(1)] |
| Selset | 1959 | boulder clay | CL | 0.1 – 0.2[(1)] |
| Staines South | 1903 | London clay | CH/CV | 0.1 |
| Sutton Bingham | 1955 | Oxford clay | CH | 12 – 41 |
| Tamar Lake | 1825 | – | – | 0.6 – 1.5 |
| Usk | 1955 | boulder clay | CL | 0.2[(1)] |

Table 10 (continued)

| Dam | Completion | Core Clay | BS Classn. | Indicative Permeability* $k(\times 10^{-9}$ m/s) |
|---|---|---|---|---|
| Walshaw Dean Upper | 1907 | – | CI/MH | 0.03 – 1.0$^{(1)}$ |
| Alwen No.2 | 1916 | 51% clay | – | 1.3$^{(4)}$ |
| Abercribban Nos.1/3 | – | 58% clay | – | 42$^{(4)}$ |
| Blaen-y-cwm | 1940 | 79% clay | – | 1.3$^{(4)}$ |
| Carno Lower | 1911 | 72% clay | – | 1.1$^{(4)}$ |
| Cyfartho Lake | – | 39% clay | – | 2.0$^{(4)}$ |

* values 'normalised' to units of x $10^{-9}$ m/s
1. field/piezometer value
2. original dam prior to reconstruction: no conclusive evidence of core, but central silty clay zone 'finer'
3. puddle blanket u/s of defective core
4. crude 2.14m 'constant head' test: lab?

Compressibility and consolidation

Little data has been published for puddle clays. Estimated values for compression index, $C_c$, frequently considered more useful than coefficient, $m_v$ on account of the latter's stress-dependency, are shown on Table 11 for a number of dams of volume compressibility. Reported values of $m_v$ are also shown in one instance.

Table 11. Puddle clay cores: representative consolidation characteristics

| Dam | Compl. | Core Clay | BS Classn. | Consolidation Characteristics | |
|---|---|---|---|---|---|
| | | | | $c_v$(m²/yr) | $c_c$ |
| Barrow No. 3 | 1899 | silty L.Lias clay | CH | 0.51 – 0.95 | 0.41 – 0.59 |
| Blackmoorfoot | 1876 | boulder clay | CH | 0.59 – 1.47 | – |
| Hallington West | 1889 | boulder clay | CI | 2.10 – 8.45 | (0.29 – 0.41*) |
| Lockwood | 1903 | silty alluvial clay (70% clay) | CE | 5.10 | – |
| Selset | 1959 | boulder clay | CL | 0.90 | 0.16 – 0.23 |

* $m_v$(m²/kN) = 1.6 – 7.7 × $10^{-4}$

Erodibility

The erodibility of puddle clays is, in general terms, relatively low. Erodibility of natural clays is a most complex characteristic, controlled by the electro-chemistry of the clay mineralogy in combination with the seepage water from the reservoir, i.e. the electro-chemistry and hence erodibility are site-specific.

Some tests for erodibility, such as the 'pin-hole' test (Head, 1980) and others, are not necessarily suited to the assessment of UK puddle clays. Erodibility in the UK context continues to be the subject of study. In the interim a simple cylinder dispersion test, in principle not dissimilar to the early 'soaking index' test referred to previously, has been proposed by Atkinson *et al* (1990).

Table 12 presents some results from ongoing work at the University of Newcastle on clay dispersion as a measure of erodibility.

Table 12. Puddle clay cores: dispersivity

| **Dam** | **Core Clay** | **BS Classn.** | **Test Method[(1)]: Results[(2)]** | | | | |
|---|---|---|---|---|---|---|---|
| | | | **A** | **B** | **C** | **D** | **E** |
| Blackmoorfoot | soft homogeneous clay (45% clay) | CH | D | D | D | C | D |
| Cullaloe | soft sandy clay (25% clay) | CL | ND/D | ND/D | ND | – | ND |
| Lockwood | silty alluvial clay (62% clay) | CE | ND | ND | ND | – | ND |
| Staines South | London clay (50% clay) | CH/CV | ND | ND | ND | C | ND |

(1) Test method:
- A. pinhole test
- B. double hydrometer test
- C. crumb test
- D. cylinder dispersion test
- E. chemical test

(2) D = dispersive
ND = non-dispersive
C = non-dispersive cohesive

## PUDDLE CLAY CORES: PERFORMANCE

Detailed consideration of the performance of puddle clay cores extends beyond the remit of this paper. Reviews of the long-term performance record of UK embankment dams have been presented in Moffat, 1982, and also in Charles & Boden, 1985, and Charles, 1989. Attention is drawn to the three Performance Indices for Settlement ($S_I$), Seepage ($Q_I$), and risk of Hydraulic Fracturing ($HF_I$) (Charles, 1986). These Indices, and in particular Settlement Index $S_I$, have a useful auxiliary role in the monitoring of 'Pennines' era embankments.

## CONCLUSIONS

A high proportion of the UK stock of dams are 'Pennines' type earthfill embankments, characterised by a relatively slim core of puddle clay, the latter often in combination with a narrow and deep puddle clay trench cutoff. The 'Pennines' embankment evolved through a number of identifiable phases or stages, and in its mature form from c. 1880 the type proved satisfactory in service.

Core and cutoff are critical to integrity of the Pennines dam, but the nature and characteristics of puddle clay are less widely appreciated than might be surmised. Puddle clay, an intensively remoulded and tempered cohesive soil whose fabric has been destroyed, is a soft and plastic core fill of low strength and very low permeability. The range of source clays is very wide, embracing natural soils ranging from low to very high plasticity.

The geotechnical characteristics of puddle clays, while relatively consistent across the board at time of placing, subsequently cover a considerable range of values for all key parameters, notably permeability. Data on geotechnical parameters for 'mature' puddle clays is neither as comprehensive nor as readily accessible as is desirable bearing in mind the UK's extensive and ageing stock of 'Pennines' embankments. Valuable research into puddle clay core behaviour has been undertaken in recent years, most notably at Building Research Establishment. Significant gaps in understanding remain, however, and a strong case can be argued for the further work identified below:

1. further research into specific characteristics of certain mature puddle clays which derive from generic soil types, e.g. boulder clays etc.;
2. an enhanced programme to make data on puddle clays and long-term puddle core/cutoff performances more readily accessible; and
3. research into repair materials etc. which are puddle clay compatible.

The data summarised in this paper has been assembled in the course of ongoing research into puddle clays at the University of Newcastle.

## ACKNOWLEDGEMENTS

The data presented in this Paper has been gleaned from an extensive range of sources, not all of which lie in the public domain. Published (and unpublished) work by a number of distinguished engineers, notably Mr M F Kennard, Drs J A Charles, A D M Penman and P Tedd, and by Prof P R Vaughan, together with the work of the late Mr G M Binnie and Profs A W Bishop and A W Skempton, have been of particular assistance. The co-operation and assistance of many other dam engineers is also gratefully acknowledged. Specific references identified below are confined to selected published sources of most immediate relevance to an appreciation of the nature and characteristics of the puddle clay core.

SELECTED REFERENCES

Atkinson J H, Charles J A & Mhach H K (1990). Examination of erosion resistance of clays in embankment dams. Q. Jnl. Engg. Geol., vol 23, No 2, pp103-108.

BRE (1994). Register of British dams. Report BR261, 55pp. Building Research Establishment, Garston.

Charles J A (1986). The significance of problems and remedial works at British earth dams. Proc. 4th Conf. British Dam Society, Edinburgh, pp123-141. BNCOLD, London.

Charles J A (1989). Deterioration of clay barriers: case histories. Proc. Conf. on Clay Barriers for Embankment Dams, London, pp109-129. Thomas Telford, London.

Charles J A & Boden J B (1985). The failure of embankment dams in the United Kingdom. Proc. Symp. on Failures in Earthworks, pp181-202. ICE, London.

Haider G (1989). Reconstruction of Llangollen canal. Proc. Instn. Civ. Engrs., Part 1, Vol 86, Dec., pp1047-1066.

Head K H (1980). Manual of soil laboratory testing. (3 vols.), 1238pp. Pentech Press, London.

ICE (1996). Floods and reservoir safety. 3rd edn., 63pp. Thomas Telford, London.

Johnston T A, Millmore J P, Charles J A & Tedd P (1999). An engineering guide to the safety of embankment dams in the United Kingdom. 2nd edn. Construction Research Communications, London, 102pp.

Kennard M F (1994). Four decades of development of British embankment dams. Geoffrey Binnie Lecture, 8th Conf. British Dam Society, Exeter. In Disen. Vol. 'Dams and Reservoirs', pp1-30, British Dam Society, London

Moffat A I B (1982). Dam deterioration – a British perspective. Proc. 2nd Conf. British National Committee on Large Dams, Keele, pp103-115.

Moffat A I B (1999). Presentation on Puddle clay and the Pennines embankment dam. Report in Dams and Reservoirs, Vol 9, No 1, pp9-15. British Dam Society, London.

Skempton A W (1989). Historical development of British embankment dams to 1960. Keynote address, Proc. Conf. on Clay Barriers for Embankment Dams, London, pp15-52. Thomas Telford, London.

Tedd P, Skinner H D & Charles J A (2000). Developments in the British dams database. Proc. 11th Conf. British Dam Society, Bath, pp181-189. Thomas Telford, London.

# A review of systems used to assess dam safety

A J BROWN, Halliburton KBR, UK
J D GOSDEN, Halliburton KBR, UK

SYNOPSIS. The Reservoirs Act 1975 provides a framework for managing the safety of UK dams, and since the implementation of its predecessor the Reservoirs (Safety Provisions) Act in 1930 there have been no dam failures involving loss of life in the UK. However, in order to ensure best practice continues to be used for managing the safety of UK dams, and in common with other industries, we need to increase our understanding of, and where possible quantify, the risk posed by dams in the UK.

One of the issues that has received prominence recently with the publication of a new method of estimating floods is how dam engineers evaluate the relative importance of the various threats to dam safety, and ensure that appropriate weighting is given to both potential weakness in the event of extreme loading and symptoms of potential problems caused by internal erosion and other internal processes.

The paper reports on the early stages of a research contract for DEFRA for which the terms of reference include consideration of whether "an Integrated System of assessing possible threats to dam integrity could be devised in which standards for reservoir spillway capacity are determined in association with other features of dam design, condition and operation".

The paper identifies and reviews the various systems to assess dam safety that are currently in use world-wide, highlighting the differences and commenting on possible reasons for these differences. It concludes by proposing the requirements for a system that would be appropriate for use as an aid to judgement by a Panel Engineer when carrying out a 10 year inspection under the UK Reservoirs Act.

## INTRODUCTION

This paper presents the results of the early stages of a research contract for DEFRA by Halliburton KBR (formerly Brown & Root) in association with the Building Research Establishment. It reports on literature reviews and informal discussions with practitioners in dam safety in the USA, Canada and Australia. Its purposes are to:

a) present current world-wide best practice in systems of dam risk assessment and comment on the strengths and weaknesses of these systems,

*Reservoirs in a Changing World*, Thomas Telford, London, 2002, 602–618

b) raise awareness and promote debate and consideration of the value of a formalised risk assessment in the preparation of a Section 10 Inspection Report under the Reservoirs Act 1975,
c) suggest a possible specification for the preferred system to assist Panel Engineers and others in the periodic risk assessment of UK dams.

This research contract includes proposing a prototype system and trialling this system on ten dams, which are due for completion in mid 2002. The prototype system and results of the trialling will be reported separately once completed. In parallel with the consideration of systems to assess dam safety reported in this paper, considerable effort was put into interrogating the UK database on dams held at BRE to establish the quality of and conclusions from the available data. The results of this interrogation are reported in a companion paper (Tedd et al, 2002).

## DEFINITIONS

### General

Currently there are significant inconsistencies in the terminology used both between different countries, and between different publications within a country. This is important as until there is a common understanding and agreement of the terminology it will be difficult to compare systems and thus to progress such systems. The Research project has carried out a review of terminology currently in use world-wide and within UK, recommending that the terminology defined in the Research Contract is used as the standard for future work on dam safety in the UK. Some of the key elements of this terminology are reproduced below.

### External threats

External loads, such as floods and earthquake, are random natural events which can be measured and extrapolations made to estimate the magnitude of extreme events that could cause failure of the dam. They are different from the specific mechanisms that can cause degradation of the dam, which are termed mechanism(s) of deterioration. Thus extreme rainfall is a threat, whilst one of the resulting mechanisms of deterioration is that the inflow exceeds the spillway capacity causing the dam to overtop.

### Internal threats

Internal threats relate to mechanisms of deterioration that occur within the body of the dam and are:

- not necessarily random natural events (and thus amenable to statistical analysis),
- often difficult to measure (and thus not amenable to analysis of trend or other time or dose related analysis of measured parameters),
- much less well understood in terms of the mechanism of behaviour.

Event train

Event trains are a useful technique to map the linkage between the potential root cause of a dam failure (threat) and the ultimate mode of failure. Possible event trains originating from one external and one internal threat and associated definitions as shown in Appendix A. The potential complex branching and interlinking should be noted; for example failure by internal erosion can be initiated by extreme rainfall causing a rise in reservoir level, which in turn causes hydraulic fracture. Conversely internal instability may lead to crest settlement, which in turn leads to failure by overtopping in a large flood.

Risk analysis, evaluation, assessment, control and management

The definitions given in Kreuzer (2000) will be used. A risk analysis comprises estimating the overall annual probability of failure of the dam, the consequences if the dam did fail and the consequent risk (probability of failure x consequences). Risk assessment is then taking the results of this assessment and evaluating whether the risk is tolerable. This is illustrated in Fig. 1. By definition if the tolerability of the risk from a dam is to be assessed with any meaningful reliability it is necessary to quantify both the annual probability and consequences of failure of the dam.

Tolerable

A willingness to live with a risk so as to secure certain benefits and in the confidence that the risk is one that is worth taking and that it is being properly controlled (HSE, 2000, page 3)

ALARP (As low as reasonably practicable)

A risk is tolerable only if risk reduction is impracticable or if its cost is grossly disproportionate to the reduction in risk gained.

## RISK ASSESSMENT AND DAMS

Purpose of Risk assessment

The prime purpose of a risk assessment of a dam is to address the issue of "is the dam safe enough". The reasons for deciding to carry out such a assessment are varied, and may include

- periodic risk assessments required under national legislation,
- assessment where an incident has occurred leading to concern over the safety of the dam
- where it has been decided that some form of upgrading is required establishing the magnitude of the upgrading
- prioritising use of funds for safety upgrades within a portfolio of dams with a single owner
- managing the risk of a portfolio of dams, in terms of both third party risk and risk to a business

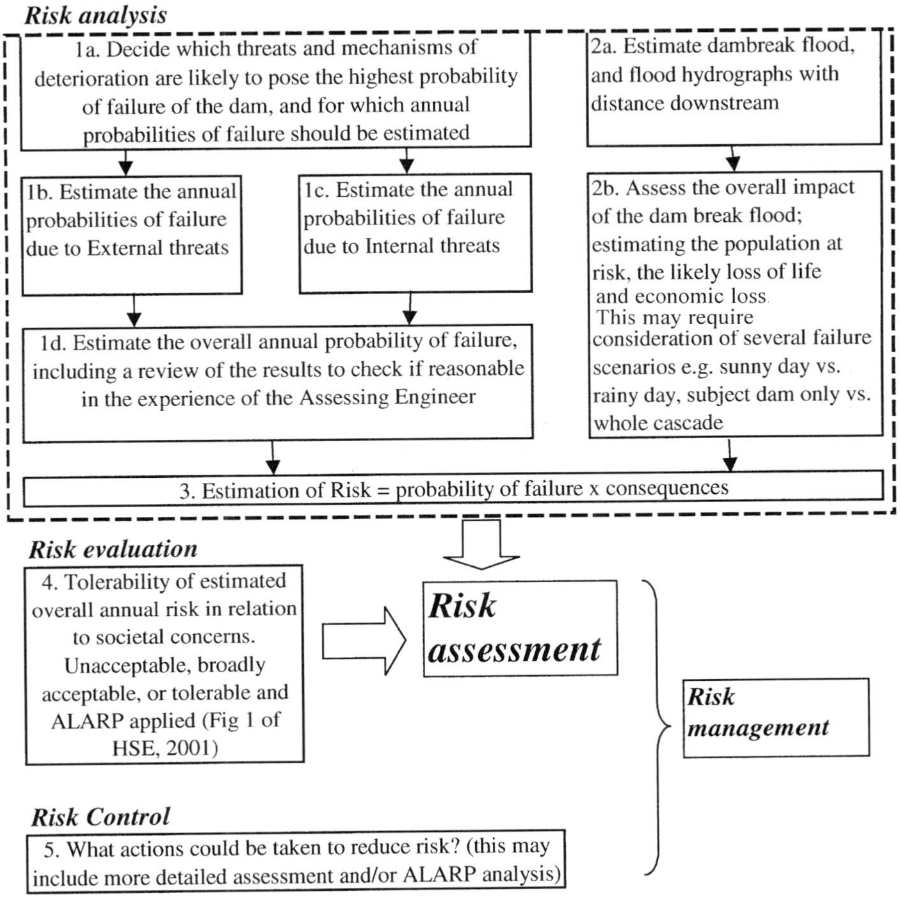


Fig. 1. Main elements in an Integrated system for Risk assessment of dams (based on Kreuzer, 2000)

Levels of Risk assessment

There are several levels of assessment of the safety of a dam, the amount of detail varying with the purpose of the assessment, size of dam, consequence of failure and resources available. McCann (1998) defines five levels of risk analysis, comprising in increasing complexity; (1) scoping , (2) ranking, (3) detailed, (4) comprehensive and (5) full scope. He suggests that the level of analysis is determined by the user (in-house or external), defensibility required, extent to which uncertainty is to be assessed and the consequences of the failure of the subject dam.

In an ideal world the probability of failure and consequences would both be quantified, to allow evaluation of the risk posed against societal acceptance criteria. However, in practice with dams which are high hazard low risk it is not always possible to do this and it may therefore be necessary to accept qualitative, or semi-quantitative measures of some components of the risk assessment.

This will have an impact on the choice of level of assessment carried out, as the value of a very detailed assessment of one component of the risk assessment is questionable where other elements cannot be similarly quantified. It is also important to note that although the overall annual probability of failure is the sum of all individual probabilities that could cause failure (which could be as many as 100 items) it is likely that say 80% of this probability will be governed by only a few mechanisms of deterioration. This is important when considering the level of assessment, as the most value is likely to be obtained by concentrating on estimating the probability of failure of these main mechanisms, rather than all possible mechanisms of failure. This confirms the value in reviewing and identifying the most important mechanisms of deterioration prior to carrying out any quantitative risk analysis.

Integrated system

The terms of reference for the ongoing research contract require consideration of whether "an Integrated System of assessing possible threats to dam integrity could be devised in which standards for reservoir spillway capacity are determined in association with other features of dam design, condition and operation". On one hand this could be viewed as a system to rank the various threats to dam safety by estimating the overall probability of failure, equivalent to Step 1 in Figure 1. However, this is of little value unless the tolerability, or otherwise, of this probability can be assessed in some way. It is therefore considered that any form of Integrated System is in effect a system for carrying out Risk Assessment, comprising steps 1 to 4 on Figure 1. Such a system would not in itself cover any subsequent steps of assessing how safety could be improved, although it could be used to evaluate the safety of the dam for "what if" scenarios of upgrading works.

Drivers for an Integrated System in the UK

In the UK over the years we have developed a repeatable assessment of the design flood (ICE, 1933 through to ICE, 1996). However, the assessment of the probability of failure due to internal erosion and other internal threats remains the personal opinion of the Engineer making the assessment (with guidance provided by Engineering Guides such as Johnston et al, 1999). There is therefore no repeatable way of comparing these two threats. This is a system that was set up in 1930, with the passing of the first Reservoirs Act, and which has been unchanged in concept since then. In the light of society's response to railway accidents involving fatalities in the last few years, it is suggested that this is not 'good enough' and that it does not compare well with the management of other high hazard installations (HSE, 2000).

A current uncertainty in the UK relating to assessment of the safety of existing dams is the method of flood estimation, in that floods estimated using the new Flood Estimation Handbook (IH, 1999) in some cases have higher magnitude than the previously used method (Flood Studies Report, NERC, 1974). Before embarking on further upgrading of spillways to UK dams it would be of value to compare threats from floods with other threats, to establish where the greatest threats to dam safety at an individual dam lie.

The situation is further complicated by the fact that methods of estimating floods will change further in the next few decades, both because a new technique of "continuous simulation" is likely to produce more reliable estimates and possibly because of better understanding of the impacts of climate change. It could be argued that instead of a prescriptive "design flood" which requires an upgrade in spillway capacity every time the estimated magnitude of the design return period flood increases, we should change to a system where the probability of failure of a dam due to overtopping is estimated. The advantage of the latter is that changes in the estimated magnitude of a given return period flood would not automatically lead to a spillway upgrade, but this change in assessed probability of failure due to floods would be assessed in regard to other threats to that dam.

## SYSTEMS PROPOSED FOR RISK ASSESSMENT OF DAMS

### General

There have been a number of attempts at devising a system for assessing the risk from a dam, selected examples of some of these systems being described in more detail below. As all the systems currently in use are relatively new there is almost no published feedback on any of the systems, the only exception being a paper at this conference (Tarrant et al, 2002) on the CIRIA system. Where comments are made on the strengths and weaknesses of any system these are therefore the opinions of the authors of this paper, rather than published feedback.

### Qualitative

A number of qualitative systems have been devised to allow owners to prioritise maintenance and upgrading work within their portfolio, rather than to identify the risk of failure of individual dams. The approach that has been adopted by these systems to the output required from a quantitative risk assessment is summarised in Table 1, with specific examples given below.

One of the earliest qualitative systems was a system devised by the Bureau of Reclamation (BOR) following the failure of Teton dam in 1976. This has subsequently been replaced by a Risk based profile system (RBPS) (Cyganiewicz, 2000) as a priority ranking tool, to identify and rank concrete and embankment dams with deficiencies, this being available on the internet at www.usbr.gov/dsis/risk/. The overall BOR dams safety programme is discussed in Acterberg (1999).

The main features of the RBPS are:-

a) the RBPS is assessed as requiring about a day or less by an experienced engineer
b) loads are considered under four categories namely, static, hydrologic, seismic, operation and maintenance; 30% of the marks being assigned to each of the first three and 10% to the last category
c) the likelihood of failure is assessed using six predetermined scoring sheets, such that there should be little variability between assessing engineers

Table 1. Methodology commonly adopted by qualitative systems

| Output required for Quantitative Risk assessment | Typical output from Qualitative System of assessing dam safety |
|---|---|
| Annual probability of failure for individual threat(s) | a) based variously on threats, failure mode, indicators and/or intrinsic condition (refer to Appendix A for definitions)<br>b) all are qualitative scores, rather than based on annual probability<br>c) scoring methods used either require the judgement of the assessor or are obtained from predetermined tables of indicators and/ or intrinsic condition |
| Overall probability of failure of the dam | a) Most give some form of 'Condition Index' (qualitative assessment of an overall annual probability of failure)<br>b) Ranking of the annual probability of individual threats varies from an average of the threats considered, through predetermined proportions, to Expert Judgement |
| Hazard class (consequences if dam failed) | Range from assigning to a hazard class based on judgement, to a rapid dam break analysis with associated impact assessment |
| Overall risk | Often multiply Condition Index with the likely loss of life, but as a CI is a qualitative assessment this cannot readily be used with Risk evaluation |

Also in the USA the Army Corps of Engineers have developed a Condition Index system for prioritisation of repair, evaluation, maintenance and rehabilitation (REMR) for many elements of infrastructure, including dams; these being available on the internet at www.cecer.army.mil/fl/remr/remr.html. Development of the system for embankment dams is reported in Anderson et al (1995) with the system described in Technical Report REMT-OM-25 dated Sept 1999. In this system the dam is considered as ten separate elements (defence groups; e.g. spillway capacity, spillway erodibility, crest elevation), each of which is scored in terms of condition using a predetermined scoring system. Expert elicitation is used to determine the importance factor of each of these ten groups for a particular dam, this ranking then being fixed for subsequent annual assessments. An overall condition index is then calculated by multiplying the sum of the condition scores multiplied by importance factor, and a factor for the hazard potential of the dam; comprising low, significant or high. A feature of interest is the use of expert elicitation to rank the relative importance of the ten elements representing the dam, with the assessing engineer only updating the current condition score.

There is no reported feedback on the use of either system. The BOR system allows no room for adjustment of either the relative importance of the four load categories on a dam specific basis, or of the features within each scoring sheet. Although the USACE system has ten pages of commentary on how to complete the form, it nevertheless appears somewhat open to interpretation, such that different assessing engineers may get significantly different results.

A description of Risk management for UK Reservoirs (CIRIA, 2000) is given in Tarrant et al (2002) in this conference and is not repeated here. It also is a qualitative system intended for use as a portfolio ranking tool, with the scoring left to the assessor (with limited guidance on likely score for consequence, depending on age and height of dam).

Other qualitative systems include the Failure Mode Effect and Criticality analysis described in BS 5760: Part 5: 1991 as applied by Sandilands et al (1998), da Silvera et al (1993) developed by the Portuguese regulator, and Keuperman et al (1996) developed by a Brazilian utility company.

Quantitative

The earliest published quantitative systems which have been identified are McCann et al (1985) in the USA and Cullen (1990) in the UK. The former does not appear to have generally been accepted because of the lack of subsequent publications using this approach, whilst Cullen concluded that "Probabilistic Risk Assessment is not yet a suitable tool for inspection work".

Nevertheless in the last decade risk assessment has received more interest, such that currently ANCOLD and ICOLD have draft risk assessment guidelines under peer review, the former recommending a quantitative approach based on both analysis and the use of historical data. It was generally well received at a workshop in Auckland in late 2001, with publication now scheduled for late 2002. The BOR also have a standard methodology for performing a full quantitative risk analysis, as described in Cyganiewicz & Smart (2000). There have also been a variety of published papers on the use of quantitative risk assessment (QRA), including Bowles et al (1998) and Fell et al (2000).

Techniques that may be used in QRA are analysis (e.g. logic diagrams such as fault trees), data from historical performance as well as expert judgement (e.g. Skipp & Woo, 1993).

One of the areas of difficulty with QRA is how to estimate the probability of failure due to internal threats, with both logic diagrams (fault trees) and historical performance being used. Two approaches to the latter are shown on Figure 2, that for McCann being based on adjustment of "a priori frequencies of failure" for an average dam by an adjustment factor derived using expert judgement, depending on the (condition) 'evaluation score'. The approach by Foster et al (1998, ANCOLD conference), gives the average frequency of failure for different types of dam based on the observed performance of Australian and American dams. These averages for each type of dam are then corrected by up to 20 factors relating to intrinsic and current condition. The worst and best possible scores are plotted as condition evaluation scores 1 and 10 for comparison with McCann et al (op cit).

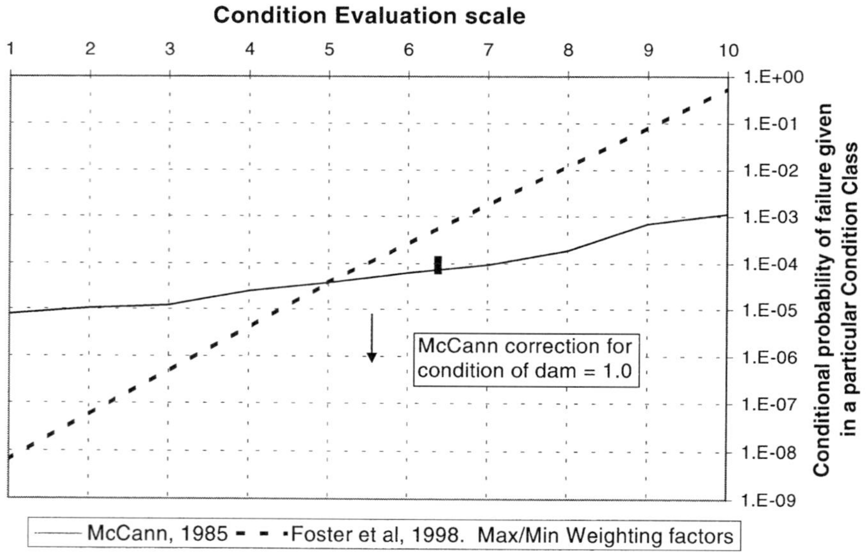


Fig. 2. Example of annual probability of failure vs. Condition Score for internal erosion

Bowles et al (1998) describes the outcome of a portfolio risk assessment for SA Water (previously South Australian Water Corporation) of 17 major dams with a median age of 75 years old, as reproduced in Table 2. The reason for the high life loss risk from internal (static) threats is ascribed to a lack of warning time needed for evacuation.

Table 2. Outcome of portfolio risk assessment for SA Water (after Bowles et al, 1998)

| | Flood | Earthquake | Static |
|---|---|---|---|
| Probability of failure | 91% | 1% | 8% |
| Incremental loss of life | 31% | 1% | 68% |
| Owners Risk cost | 80% | 1% | 19% |
| Total Risk cost | 97% | 0% | 3% |

As well as QRA the system assesses the various aspects of a dam against conventional engineering standards, assigning 'Pass/ No Pass or Apparent Pass/No Pass; the latter where there is insufficient information to make a firm conclusion. The portfolio assessment included ALARP analysis of options to reduce risk presented as reductions in probability of failure and other ranking criteria in Table 2, relative to the existing condition rather than as absolute values.

Overview

The issue of whether it is possible to quantify the annual probability of failure of dams is a subject area that research has been wrestling with for some years now, with some having strong opinions that it is not possible, and others that consider it is.

On a world-wide basis a number of systems have been recently published and are being trialled which are qualitative and focussed at portfolio risk assessment (e.g. USBR, US Army Corps, CIRIA, 2000). None of these is yet well established, or has received feedback from use in a wide variety of situations.

Quantitative methods are being used both in Australia (Bowles et al, 1998) and by the BOR, but neither is documented in the form of an Integrated System. Some may also describe these as a ranking tool used in portfolio analysis, rather than quantitative risk assessment. The closest to what could be described as a quantitative system is currently being finalised, although not yet published (2$^{nd}$ Edition of ANCOLD Guidelines on Risk assessment). This is focussed at preliminary and detailed risk assessment against absolute standards as well as portfolio risk assessment.

## SYSTEMS USED TO ASSESS SAFETY IN OTHER INDUSTRIES

McQuaid (2002) presents a summary of the development of risk assessment and notes that currently there are regulations in place for four industries – nuclear sites, onshore chemical plants (COMAH), offshore and railways. These are described in HSE (2000), which sets out the four principles underlying the approach by HSE to the regulation of higher hazard industries, the first three being reproduced in Table 3 (the fourth relates to the role of the regulator, and is not directly relevant to systems to assess dam safety).

Table 3. Principles of "permissioning" regimes (as HSE, 2000)

1. Through the political process, the regulator and the regulated are subject to society's views about the tolerability of risk:
   - *"Permissioning" regimes are applied to high hazard industries, about which society has particular concerns.*
2. The legal duty to manage risks lies with the organisations that create the risks – "permissioning" regimes require them to describe how, but a description is not sufficient without the active commitment of the duty holder in practice:
   - *Duty holders must identify the hazards, assess the risks, develop effective control measures and keep a current documentary record of all this;*
   - *The control measures must cover design and hardware, systems and procedures and human factors in a coherent whole;*
   - *Duty holders must implement control measures and keep them up to date;*
   - *Duty holders must make and test arrangements for managing emergencies and mitigating their consequences.*
3. A goal-setting framework is preferable to a prescriptive one because it makes duty-holders think for themselves.
   - *The flexibility of goal-setting is more likely to lead to arrangements for controlling risk which are tailored to the particular circumstances, and which through safety case maintenance and re-submission will remain so;*
   - *Within a goal-setting context, "permissioning" regimes define elements of the management arrangements required.*

Some of the general points made in this discussion document by HSE are:

- quantitative risk assessment (QRA) is an important part of the process of analysing, assessing and managing risks
- Risk implies uncertainty; QRA cannot remove uncertainty but it can identify and prioritise uncertainties
- There are no prescriptive standards for whether the level of risk is acceptable; instead emphasis is placed on achieving a balance between the cost of measures to reduce risk and the potential reduction in risk if those measures were implemented (ALARP)

## SPECIFICATION FOR AN INTEGRATED SYSTEM FOR UK

Dam safety in the United Kingdom is implemented through the Reservoirs Act 1975, whereby Panel Engineers, appointed to the appropriate Panel by the Secretary of State, are appointed by dam owners to carry out periodic Inspections of each dam. Any system for use for risk assessment of dams in UK should therefore be suitable for application by a single engineer to a single dam (as well as for prioritising work within a portfolio of dams).

A Section 10 Inspection is only required to assess whether the dam is adequately safe in its current condition and in relation to the hazard it poses. For low hazard dams the answer may well be affirmative, such that it is unreasonable to expect owners to expend major sums on risk assessment. It would therefore be sufficient for any Integrated system to be limited to be a Level 2 analysis as defined by McCann (1998), to allow the dam to be ranked relative to some common standard for UK dams. Ideally this would comprise some form of Rapid Method with standard proformas, similar to those for floods (ICE, 1996, Appendix 1) and earthquakes (ICE, 1998, Table A1), which could be used as the basis of the assessment.

An Integrated system would therefore preferably satisfy the following key requirements

a) provide guidance on the probability of failure from external threats compared to those from internal,
b) indicate an answer to the issue "Is the dam safe enough".
c) at a level 2 analysis not normally take more than a day to carry out
d) identify, for high hazard dams, areas where more detailed analysis is required
e) provide output in a format that could be included as an appendix to the Section 10 report and would be suitable as part of a safety case along the lines of HSE (2000). Such output would provide a benchmark assessment of the dam safety against which the need, if any, for upgrades could be assessed, as well as changes in the condition of the dam with time.

## CONCLUSIONS

Society is now requiring a more transparent approach to issues of public safety, with risk assessment being a commonly used tool. This has focussed attention on the need for a more holistic approach to the overall probability of failure of a dam, rather than considering each threat in isolation, as the traditional approach has tended to. Such a holistic approach needs a method to compare the probabilities of failure due to different threats, with quantitative risk assessment being the preferred approach where practicable.

Although systems have been developed for portfolio risk assessments, these do not provide a quantitative estimate of the probability of failure (and thus any measure of whether the dam is safe enough) and to date no feedback has been published on their value in terms of managing a portfolio of dams.

It is considered that there will be increasing pressure to provide an Integrated System of quantitative risk assessment for assessing the safety of dams in the UK, and that even if it is concluded that this is not feasible at the present time it will be accepted that:

a) this is a medium to long term goal for the dam industry.

b) components of the system can, and should, be developed and put in place now

The difficulties of risk evaluation are noted, and for high hazard dams may require ALARP analysis as a standard approach, as it is in other industries. The challenge for dam professionals is to understand developments in other industries and to consider carefully how risk assessment techniques developed there could be applied to dams.

It is considered that with the stock of British dams, where the original design and construction generally predated modern soil mechanics and hydrology dam risk assessments should continue to be carried out by experienced professionals who are experienced in dealing with such dams. However, it is considered that there is a requirement for some form of integrated system to provide a framework to assist the Panel Engineer in his assessment. The current research contract is examining the feasibility of such a prototype system, and the outcome from this project will be reported later in the year.

ACKNOWLEDGEMENTS
The work described in this paper was carried out as a research contract for DEFRA, who have given permission to publish this paper. However, the opinions expressed are solely those of the authors and do not necessarily reflect those of DEFRA.

REFERENCES

Achterberg D (1999) BOR dam safety programme. 19th USCOLD lecture Series (conference). Pp 41-51

Anderson G R & Torrey VH (1995) Function-based condition indexing for embankment dams. ASCE J Geotech Eng. 121(8) 579-58

Bowles D S, Parsons A M, Anderson L R & Glover T F (1998) Portfolio risk assessment of SA Water's Large dams. Proc ANCOLD Annual meeting, Sydney, NSW, Australia.

BSI (1991) Guide to failure modes, effects and criticality analysis (FEMA and FEMCA) BS 5760 Part 5.

Cullen (1990) Risk assessment of earth dam reservoirs (SAC 9790). Report No DoE 0002-SW/3, DOE ref.: PECD 7/7/191. WRC, Swindon. 69pp

Cyganiewicz J M, Smart J D (2000) BOR use of risk analysis and risk assessment in dam safety decision making. ICOLD Beijing Q76 R23

Cyganiewicz J M, (2000) USBOR risk based profiling system. ICOLD Beijing Selected contribution to discussion on Q76. Vol. V, pp90-93

Da Silveira A F, Redro J O & Gomes A S (1993) Engineering guide to hazard and performance evaluation of dams. Water Power & dam construction. March. Pp35-36

Fell R, Bowles D S Anderson L R & Bell G (2000) The status of methods for estimation of the probability o failure of dams for use in quantitative risk assessment. ICOLD Q76 R 15. Vol. 1 pp 213-236.

Foster M, Fell R and Spannagle (1998) Risk assessment - estimating the probability of failure of embankment dams by piping. ANCOLD 98 Conference on dams. 11pp plus 4page supplement giving example of application of the UNSW method

HSE (2000). Regulating higher hazards: exploring the issues. Draft for discussion. London. HSE. Available at www.hse.gsi.gov.uk.

HSE (2001) Reducing risks, protecting people. Revised version (draft for discussion published 1999)

ICE (1933) Interim Report of the Committee of floods in relation to Reservoir Practice. 42 pages

ICE (1996) Floods and reservoir safety: 3rd edition. Thomas Telford, London. 63pp.

ICE (1998) An application note to 'An engineering guide to seismic risk to dams in the UK'. Thomas Telford, London. 39pp

Institute of Hydrology (1999) Flood Estimation Handbook (in five volumes). Institute of Hydrology, Wallingford, UK

Johnston T A, Millmore J P, Charles J A & Tedd P (1999). An Engineering Guide to the Safety of Embankment Dams in the UK. 2nd Edition. 102pp. BRE 363

Keuperman et al (1996) Dam safety, a new approach. International Water Power and Dam Construction, IWPDC Publishing, March. Pp18-20

Kreuzer H (2000) The use of risk analysis to support dam safety decisions and management. General report. Transactions of 20th International Congress on Large Dams, Beijing, September, vol. 1, pp769-896

McCann M W, Franzinia J B, Kavazanjian E & Shah H C (1985) Preliminary safety evaluation of existing dams. 2 volumes. John A.Blume Earthquake Engineering centre, Dept of Civil Eng., Stanford Univ. under a project supported by US FEMA. Contract EMW-c-0458.

McQuaid J (2002) Risk assessment – its development and some current priorities. BDS Conference. Dublin

NERC (1975) The flood studies report (five volumes). Natural Environmental Research Council, London. 18 supplementary reports issued 1977 to 1988

Sandilands N M, Noble M & Findlay J W (1998) Risk assessment strategies for dam based hydro schemes. Proc 10th BDS Conference, pp13-26

Skipp & Woo (1993) A question of judgement: expert or engineering? Proc Conf Risk and reliability in ground engineering. London. Publ Thomas Telford. Paper 2. Pp 29-39

Tarrant et al (2002) Reservoir risk assessments in the north of Scotland. BDS Conference. Dublin

Tatalovich (1998) Comparison of failure models from risk assessment and historical data for Bureau of Reclamation dam. DSO-98-01. 12 pages and 24 thick appendices. Available on BOR website

Tedd et al (2002) The use of historical data to predict performance of UK dam. Dams & Reservoirs 2002

## APPENDIX A : PROPOSED DEFINITIONS FOR USE IN RISK ASSESSMENT OF DAMS

Table A1: Terminology proposed for use in dam risk assessment

| *Proposed Term* | *Alternative term(s) sometimes used* | *Definition* |
|---|---|---|
| Threat(s) | Cause, Initiating event, Root cause | Random Event (External threat) or Potential Internal Instability (Internal threat) that poses a threat to the integrity of the dam |
| Mechanism(s) of deterioration | System response, Cause, Adverse conditions | Process by which the integrity of the dam is undermined. The mechanism can have a quantitative threshold above which deterioration is likely to occur e.g. slope protection designed to withstand waves due to 100 year wind |
| Intrinsic condition of dam | | Current physical property or dimension of the dam which can be measured and which affects the outcome of the application of a mechanism of deterioration. Although initially determined by construction details; this may change with time due to ageing, neglect, maintenance or upgrading. |
| Indicator(s) | Effect, Damage, Symptom, | Measurable outcome from the application of a mechanism of deterioration e.g. deformation, seepage, instrumentation results. |
| Failure | | Uncontrolled sudden large release of water |
| Failure mode(s) | Effect, Consequences | Means by which a failure (uncontrolled sudden large release of water) may occur; four modes are differentiated namely external erosion (including overtopping), internal erosion, sliding and appurtenant works. |
| Incident | | Detectable change in Indicator causing sufficient concern to lead to some action (three levels are used in the BRE database; emergency drawdown, concern leading to works and concern leading to the involvement of a Panel Engineer) |

Fig A.1 Possible event trains leading to dam failure initiated by extreme rainfall

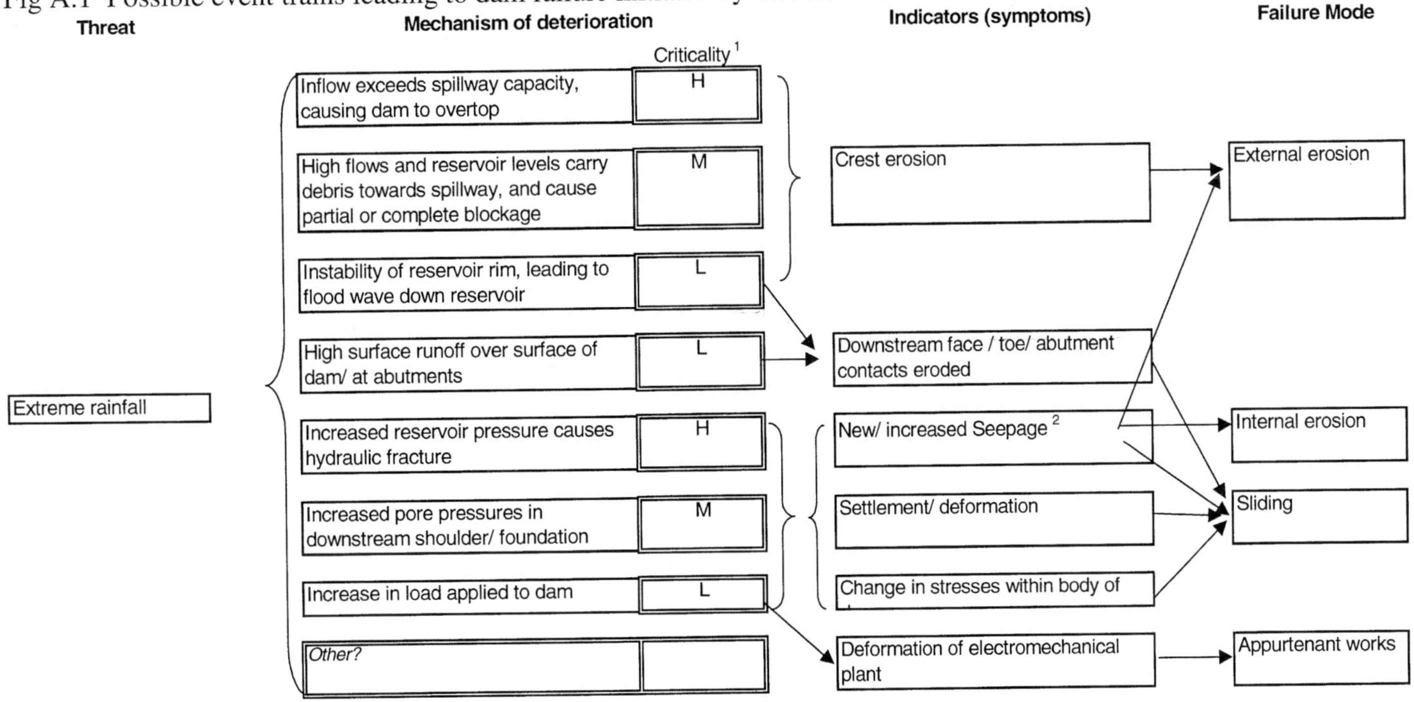


Notes
1. Likelihood that the mechcnasim could lead to the failure of the subject dam? (Factors relevant to the assessment may be recorded on separate sheet)
2. May be temporary, and disappear once threat has passed
3. The Criticality levels assigned here are those which frequently apply to dams in the UK. Individual consideration should be given to the criticality level when a specific dam is assessed.

Fig A.2 Possible event trains leading to dam failure initiated by internal instability (geotechnical)

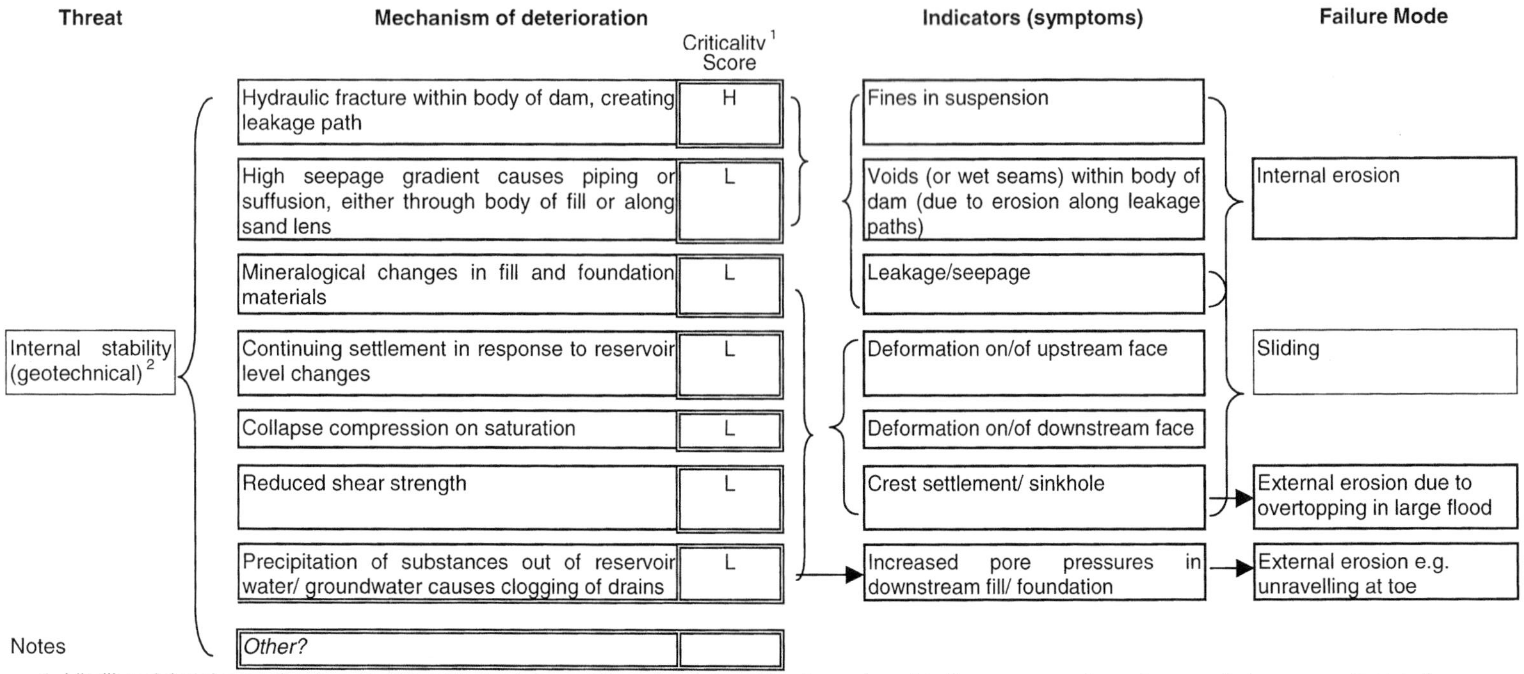


Notes

1. Likelihood that the mechanism could lead to the failure of the subject dam? (Factors relevant to the assessment may be recorded on separate sheet)
2. Includes both time dependent deterioration, and under reservoir load (which may vary seasonally and daily but excludes flood rise)
3. The Criticality levels assigned here are those which frequently apply to dams in the UK. Individual consideration should be given to the criticality level when a specific dam is assessed.

**Author Index**


| | |
|---|---|
| Ackers J C | 61, 551 |
| Airey M | 73 |
| Attewill L J S | 563 |
| Baker E A | 534 |
| Bettzieche V | 155 |
| Bommer J J | 112 |
| Bray C | 444 |
| Bridle R C | 31 |
| Briscoe, A | 236 |
| Brown A J | 602 |
| Bu S | 87 |
| Carter I C | 31, 236, 415 |
| Casey B | 314 |
| Charles J A | 367, 378, 494 |
| Chaudhury A | 61 |
| Claydon J R | 415 |
| Conaty E | 302, 324 |
| Daniell W E | 100 |
| Davis J P | 534 |
| Dedja Y | 289 |
| Dempster K J | 87, 456 |
| Dutton D P M | 394 |
| Enston R P | 218 |
| Evans R | 31 |
| Fawcett S | 31 |
| Fitzgibbon T | 314 |
| Fleming E | 3 |
| Gallocher S C | 87 |
| Gardiner K D | 126 |
| George A A | 236 |
| Gillespie D | 31 |
| Gosden J D | 602 |
| Graham-Smith N | 551 |
| Grundy P | 236 |
| Hacker J N | 247 |
| Hall J W | 534 |
| Hartford D N D | 510 |
| Haugh B | 143, 431 |
| Heitefuss C | 155, 167 |
| Hill M J | 415 |
| Hinks J L | 289 |
| Holohan J | 31 |
| Hopkins J K | 444 |
| Hughes R A N | 224 |
| Jones P E | 61 |
| Kelly P | 224 |
| Kny H J | 167 |
| Kovacevic N | 337, 353 |
| Lannen N | 456 |
| Latham D C F | 218 |
| Lavery S | 193 |
| Le Masurier J W | 534 |
| Lewin J | 193 |
| Long M | 302, 314, 324 |
| Lund K A | 209 |
| Lydon I | 302 |
| MacDonald D E | 274 |
| McCulloch C S | 49 |
| McInerney S | 31 |
| McQuaid J | 520 |
| Moffat A I B | 581 |
| Molyneux J D | 274 |
| Morison A C | 87 |
| Morris M W | 484 |
| Morrison K F | 181 |
| O'Keeffe J D | 15 |
| O'Mahony B | 143 |
| Padgett I E | 181 |
| Parks C D | 403 |
| Penman A D M | 471 |
| Ridley A M | 337, 353 |
| Rigby P | 126 |
| Robertshaw A C | 367 |
| Scholefield I | 262 |
| Scott C W | 112 |
| Smith D A | 31 |
| Spasic-Gril L | 563 |
| Stewart J | 575 |
| Stewart R A | 510 |
| Tarrant F | 551 |
| Taylor C A | 100, 534 |
| Tedd P | 367, 444 |
| Townshend P D | 209 |
| Vaughan P R | 337, 353 |
| Walthall S | 126, 403 |
| Woolf D R S | 247, 262 |

**Subject index**

(The page numbers which are given for a particular subject refer to the first page of each paper in the proceedings which deals with that subject)


Asphaltic concrete 378 415 431 444
Barriers and spillway gates 181 193 209
Catastrophic failures 378 471 494
Concrete 31 73 87 456
Climate change 193 337 353
Culverts through embankments 3 224 236 378
Cut-off trenches 31 581
Dambreak studies 484 551 563
Diaphragm walls 155
Drainage 155
Dynamic behaviour and analysis 87 100
Earthquake damage 471
Emergency plans 563 575
Environment, social, planning 3 15 31 49 61
Flood estimation and hydrology 3 31 262 274 289
Geomembranes 415
Grouting 31 403 415
Guides and surveys 100 112 126
Hazard and risk management 87 100 112 126 289 484 510 520 534 551 562 575 602
Instrumentation 126 155 444
Internal erosion 3 378 394 403 415
Landslide dams 563
Legislation 494 510 520
Movement monitoring 324 367 444
Outlet and intake works 31 100 167 181 218 224 236 456
Overflow works 31 209 247 262 274
Overtopping 471 484
Pore pressure, total stress 126 337 353
Puddle clay 3 224 367 581
Rapid drawdown 302 314 378 403
Rockfill 31 126 415 431
Roller compacted concrete 73
Seepage and leakage 378 394 403 415 431 444 456
Seismic stability 87 100 112 126 143 289
Slope stability 3 87 126 236 289 302 314 337 353 471 563
Tailings dams 471
Uplift in concrete and masonry dams 143 155

**Dam index**

(The page numbers which are given for a particular dam refer to the first page of each paper in the proceedings which deals with that dam)

**(a) United Kingdom and Republic of Ireland**


Abberton 581
Abercribban 581
Aldenham 353 581
Alwen no 2 581
Anglezarke 403
Ardsley 367
Arnfield 126
Balderhead 378
Banbury 581
Bann Reservoirs 3
Barden Upper 581
Barrow Compensation 581
Barrow no 2 581
Barrow no 3 581
Belmont 126
Bilberry 494 581
Blackmoorfoot 581
Blaennant Ddu 581
Blaen-y-cwm 581
Breaclaich 456
Brent 224 581
Brownhill 367
Burnhope 581
Cadney Carrs 218
Cant Clough 581
Carno Lower 581
Carrigadrohid 15
Challacombe 367 581
Cheddar 581
Clattercote 394
Clydach 581
Cod Beck 581
Coedty 494
Colliford 444
Coulter 581
Covenham 218
Cow Green 337
Cullaloe 581
Cwmwernderi 367 581
Cyfartho lake 581
Dale Dyke 494 581
Darwen 494
Digley 367
Draycote 353
Drayton 394
Dungannon Park lake 3
Eigiau 494
Fedderate 551
Fort Henry embankment[#] 302 314
Foxcote 353
Fullerton Pollan 31
Glascarnoch 87 100
Glencorse 581
Golden Falls 15
Gorpley 581
Grassholme 581
Greenbooth 378
Grimwith 581
Hallington West 581
Hanningfield 353
Harperrig 581
Heapey embankment 403
Holmestyes 367
Hurst 126
Hury 581
Inniscarra 15
Kennick 581
Kentmere Head 581
King George V 581
King George VI 581
Knockendon 581
Ladybower 367 581
Lairg 87
Lambieletham 581
Langsett 247 262
Leixlip 15 143
Lluest Wen 378
Loch Mhic Gille-bhride 551
Loch Ordie 551
Loch Smalag 551
Loichel 87
Lockwood 581
Lough Island Reavy 3
Lower Bohernabreena 274
Lower Lliw 581
Lower Vartry 3
Luttrellestown House lake 3
Luxhay 581
March Ghyll 581
March Haigh 367
Monar 87
Monkswood 581
Muirhead 581
Mullardoch 87

Nant 87
Oakdale Lower 581
Ogden 367
Piethorne 236
Pitlochry 87
Pollaphuca 15
Portumna embankments# 324
Queen Mary 581
Ramsden 367 581
Rhodeswood 126 337
Rivelin Upper 581
Roadford 444
Rooden North 126
Rooden South 126
Rotton Park 581
Selset 581
Silent Valley 581
Skelmorlie 494
Sloy 87
Staines South 581
Sutton Bingham 581
Tamar Lake 581
Thames barrier 193
Toddbrook 581
Torside 581
Turlough Hill 15 431
Upper Bohernabreena 274
Usk 581
Walshaw Dean Lower 367 581
Walshaw Dean Upper 581
Warmwithens 378
Widdop 367
William Girling 581
Wilstone 394
Winscar 378 415
Withens Clough 378
Woodburn Middle 3
Woodburn Upper 3
Woodhead 337
Yateholme 367

#flood protection embankments

**(b) Overseas**


Aberfeldie (Canada) 510
Arbon (Spain) 378
Avis (Namibia) 209
Baia Mare+ (Romania) 471
Bigge (Germany) 167
Brucher (Germany) 155
Buget (France) 378
Caspe (Spain) 378
Coquitlam (Canada) 510
Coursier (Canada) 510
Dreilager (Germany) 155
Elbe-Seitenkanal* (Germany) 378
El Cobre+ (Chile) 471
Elsie (Canada) 510
Ennepe (Germany) 155 167
Fishte (Albania) 289
Fonte Longa (Portugal) 378
Fuelbecke (Germany) 155
Fuerwigge (Germany) 155
Ghazi barrage (Pakistan) 61
Gloer (Germany) 155
Goricani (Albania) 289
Gostei (Portugal) 378
Gourdon (France) 378
Grizhe (Albania) 289
Grossee (Austria) 378
Grundsjoarna (Sweden) 378
Hallby (Sweden) 378
Hasper (Germany) 155
Helmes (Albania) 289
Hyttejuvet (Norway) 378
Ibra (Germany( 378
Jubach (Germany) 155
Jukla (Norway) 378
Juktan (Sweden) 378
Kashta (Albania) 289
Kotri barrage (Pakistan) 181
Kurjani (Albania) 289
Lake Sarezx (Tajikistan) 563
La Prade (France) 378
Lavaud-Gelade (France) 378
Liz no 1 (Albania) 289
Lovon (Sweden) 378
Main-Donau-Kanal* (Germany) 378
Malas Grope (Albania) 289
Martin Gonzalo (Spain) 378
Merriespruit+ (South Africa) 471
Mochikoshi+ (Japan) 471
Moehne (Germany) 167
Moravka (Czech Republic) 378
Motru (Romania) 378
Mysevatn (Norway) 378
Nepes (France) 378
Nyrsko (Czech Republic) 378
Peruca (Croatia) 378
Porjus (Sweden) 378
Rengard (Sweden) 378
Sadovice (Albania) 289
Saint Aignan (France) 378
Saint Julien des Landes (France) 378
Saint Pardoux (France) 378
Sapins (France) 378
Selce (Albania) 289
Seitevare (Sweden) 378
Shkalle (Albania) 289
Shtodri (Albania) 289
Songa (Norway) 378
Sorpe (Germany) 378
Stava+ (Italy) 471
Stenkullafors (Sweden) 378
Suorva (Sweden) 378
Sylvenstein (Germany) 378
Taibilla (Spain) 378
Tannur (Jordan) 73
Torcy Vieux (France) 378
Troshan (Albania) 289
Uljua (Finland) 378
Vajont (Italy) 337
Vashaj (Albania) 289
Verse (Germany) 167
Viddalsvatn (Norway) 378
WAC Bennett (Canada) 510


* canal embankment
+ tailings dam
x landslide dam